THE PETROLEUM HANDBOOK

THE PETROLEUM HANDBOOK

SIXTH EDITION

Compiled by staff of the Royal Dutch/Shell Group of Companies

ELSEVIER

1983

Elsevier, Amsterdam — Oxford — New York — Tokyo, 1983

ELSEVIER SCIENCE PUBLISHERS B.V.,
Molenwerf 1
P.O. Box 211, 1000 AE Amsterdam, The Netherlands

Distributors for the United States and Canada:

ELSEVIER SCIENCE PUBLISHING COMPANY INC.
52, Vanderbilt Avenue
New York, N.Y. 10017

ISBN 0-444-42118-1

© Shell International Petroleum Company Limited, 1983

All rights reserved. No part of this publication may be reproduced, stored in a retrieval system or transmitted in any form or by any means, electronic, mechanical, photocopying, recording or otherwise without the prior written permission of the copyright owners, Shell International Petroleum Company Limited, Shell Centre, London SE1 7NA, U.K.

FOREWORD

Some fifty years ago the first edition of The Petroleum Handbook was published, primarily to provide an authoritative book of reference for recruits to the oil industry. Since then the work has been continuously in demand and periodically updated. During recent years the impact of the oil industry on the economies of many countries has been such that an understanding of underlying principles of its activities has become of much wider interest. Governments, the news media and technical personnel outside of the Industry, are all much more concerned to learn of our activities than in the past.

I am pleased then, that after a gap of well over a decade this new sixth edition of the Handbook is now available. The present volume is a product of re-writing rather than revision, for although the basics of geology and of the chemistry of petroleum do not alter (and there are thus some passages included taken from the previous volume), the radical changes that affected the Industry during the Seventies, both in its structure and general environment and also in its technology, have called for an essentially fresh approach.

The concept of the Handbook continues to be that of a technically orientated manual. The aim, however, has been to combine explanations of the processes of today's petroleum industry, from crude oil exploration to product end use, with some historical background and explanation of the economic context in which the oil, gas and petrochemical businesses operate. The authors have had to face up to the hazards of obsolescence and take the risk that views expressed, particularly on the future outlook, may appear all too soon to be lacking in focus.

I welcome this opportunity to thank the large number of Shell contributors who have found time among their other pressing tasks to cooperate with specialist knowledge in the writing and revision of the various chapters. A manual of this nature requires a degree of detailed information and expertise that inevitably calls for widespread team effort, and for this reason the Editors have felt it best to continue the Handbook's tradition of anonymity.

Sir Peter Baxendell
Senior Group Managing Director,
Royal Dutch / Shell Group of Companies

ACKNOWLEDGEMENTS

Thanks are due to various industrial organisations and to the authors, editors and publishers of a number of books and journals for permission to reproduce the figures specified.

Fig. 3.2	P.A. Rona, 1977. *EOS, Trans. American Geophysical Union*, 58(8): 629–639.
Fig. 3.4	A.W. Bally and S. Snelson, 1980. Memoir 6, Can. Soc. Petrol. Geol.
Figs. 3.9	B.D. Evamy, J. Haremboure, P. Kamerling, W.A. Knaap, F.A. Molloy and P.H. Rowlands 1978. *Am. Assoc. Pet. Geol. Bull.*, 62: 1,
Fig. 3.11	D. Roeder, 1970. Summer School Course Notes, Am. Assoc. Pet. Geol.
Fig. 3.18	K.J. Weber, G. Mandl, W.F. Pilaar, F. Lehner and R.G. Precious, 1978. *Offshore Tech. Conf. Proc. 10*, Vol. 4 (10th Annual O.T.C. Conf., Houston, May 8–10, 1978).
Fig. 3.23	United States National Aeronautics and Space Administration.
Figs. 3.61, 3.66, 3.86	Nederlandse Aardolie Maatschappij.
page 236	Aerocamera—Bart Hofmeester.
Fig. 6.4	Trianco Redfyre Limited.
Fig. 6.5	Robey Lincoln Limited.
Figs. 6.6, 6.7	Central Electricity Generating Board.
Figs. 6.12, 6.15	Rolls-Royce Limited.
Fig. 6.13	AB Optimus Limited.
Fig. 6.16	Brown, Boveri and Company Limited.
Figs. 8.4, 8.5	GASCO
Table 10.5	Table 13 from *Plastics: The Energy Saver*, Franklin Associates Limited, Kansas.

PREFACE

Although the last edition of the Petroleum Handbook was published as long ago as 1966, when the petroleum industry was very different from today, the steady demand for copies of a new up-dated edition has encouraged the production of this volume. Parts of previous editions have been retained, but much of the material is completely new, since the technology, structure and political environment in which the industry operates have all radically changed.

It is not intended that this book should be "read at one sitting" but rather provide a source of reference to different aspects of the industry. We hope that the contents will be helpful to those, both within the industry and outside, who seek general information in a field which is not their own speciality.

Although consistency is a virtue, we have not sought to impose a rigid discipline on the authors of the various chapters since, for example, the differences between short tons, long tons and metric tonnes are small in the context of a general work. A more comprehensive description of units used is provided in the "Note on Units of Measurement in the World Energy Industry".

The editors gratefully acknowledge their indebtedness to all those members of the staffs of Shell companies who have so readily assisted as authors or in the submission of material for illustrations or by their critical review of the contents.

Shell companies have their own separate identities, but in the book the collective expressions "Shell" and "Group" and "Royal/Dutch Shell Group of Companies" are sometimes used for convenience in contexts where reference is made to the companies of the Royal Dutch/Shell Group in general. Those expressions are also used where no useful purpose is served by identifying the particular company or companies.

The Editors

CONTENTS

Foreword, v
Acknowledgements, vi
Preface, vii

Chapter 1. The world petroleum industry, 1
   Petroleum, 1
   The significance of oil and gas, 2
      Historical outline, 3
      Early developments, 3
      The period of rapid post-war expansion, 5
      The rise of OPEC and the period of producer country dominance, 5
      The situation at the beginning of the 1980s: precarious balance and uncertain future, 8
   The oil industry today, 10
      Basic characteristics, 10
      Structure of the industry, 11
      Crude oil production, 13
      Oil products sales, 15
      Changes in oil trading, 15
   Future prospects, 16
   Projection of future energy requirements, 19
   The financial implications of the energy prospect, 20
      Oil, 20
      Other energy sources, 21
      Financial impact of other changes, 23
   The petroleum industry and the future, 23

Chapter 2. Oil and gas in the centrally planned economies, 25
   Their significance, 25
   Oil development in the USSR, 26
   Soviet natural gas resources, 28

Future prospects for oil and gas in the USSR, 30
Soviet energy exports, 33
China, 34

Chapter 3. Exploration and production, 35
  Introduction, 35
  Exploration, 36
    Historical background, 36
    Some basic geological facts and principles, 38
    Sedimentary basins, 42
    Hydrocarbon geology, 52
    Exploration methods, 61
    Exploration drilling, 67
    Exploration results, 67
    Successive stages in exploring a sedimentary basin, 69
  Production, 72
    Production development, 72
    Well-site operations engineering, 78
    Petrophysics, 83
    Production geology, 87
    Reservoir engineering — Primary and secondary recovery, 91
    Reservoir engineering — Enhanced oil recovery, 97
    Planning of oil recovery projects, 110
    Production technology — Engineering and chemistry, 112
  Engineering, drilling and production operations, 122
    Drilling, 122
    Marine drilling, 139
    Deep-water drilling methods, 144
    Production operations, 149
  Economic, financial and other aspects of exploration and production activities, 182
    Risks of the business, 182
    Arrangements with governments, 183
    Economics, 187
    Financing of exploration and production activities, 191
    Project management, 193
    Offshore logistics, 197
    Safety and environmental conservation, 204
    Information and computing, 209
  World oil and gas reserves, 214

Chapter 4. The chemistry of petroleum, 221
  Introduction, 221
  Hydrocarbons, 222

Non-hydrocarbons, 225
   Sulphur compounds, 226
   Nitrogen compounds, 226
   Oxygen compounds, 226
   Other compounds, 228
Hydrocarbon reactions, 229
Types of crude oil, 233
   Paraffin-base crude oils, 234
   Asphaltic-base crude oils, 234
   Mixed-base crude oils, 234

Chapter 5. Oil products — Manufacture, 235
   Manufacturing activities, 235
      Physical separation processes, 236
      Chemical conversion processes, 237
      Treating and subsidiary processes, 238
      Control and supervision of refinery processes, 238
      Utilities, 238
   Distillation, 240
      Simple distillation, 241
      Fractional distillation, 242
      Column internals, 242
      Distillation of crude oil, 243
      Vacuum distillation, 248
      Fractionators for conversion units, 251
      LPG recovery/production, 253
   Solvent extraction, 257
      Principle of solvent extraction, 258
      Extraction equipment, 258
      Solvent extraction processes, 260
   Crystallisation and adsorption, 263
      Crystallisation, 263
      Adsorption, 266
   Reforming, 268
      Introduction, 268
      Catalytic reforming, 269
   Isomerisation, 276
   Thermal cracking, 279
      Visbreaking, 280
      Thermal gas oil production, 282
      Delayed coking, 283
      Production quality, 283
      Plant operation/decoking, 284
   Catalytic cracking, 284

Introduction, 284
The Houdry and the Thermofor catalytic cracking processes, 285
The fluidised catalytic cracking process, 288
The modern fluidised catalytic cracking process, 288
Feedstocks and catalysts, 291
Hydrocracking, 294
Basis for the choice of conversion route, 295
Process description, 296
Configurations, 296
Alkylation, 300
Polymerisation, 303
Hydrotreating, 306
Hydrodesulphurisation/hydrotreating of distillates, 307
Hydrotreating of pyrolysis gasoline, 309
Smoke point improvement of kerosine, 311
Hydrodesulphurisation of residual fractions, 311
Hydrofinishing of lube base oils, 313
Wax hydrofinishing, 314
Gasoline treating, 314
Kerosine treating, 316
Gas treating and sulphur recovery, 318
Types of gases and their contaminants, 318
Gas-treating processes, 319
LPG treating, 322
Sulphur recovery and tail gas treating, 322
Treating of base oils, 323
Sulphuric acid refining, 323
Clay treating, 324
Comparison of acid/clay refining with hydrogen treatment, 325
Bitumen blowing, 325
Energy management in refineries, 327
Introduction, 327
Principles of energy saving, 328
Energy and temperature levels, 329
City district heating, 329
Combined heat and power generation, 330
Choice of fuels, 330
Organisation of energy management, 330
Process control and systems technology, 331
Process control — New concepts, 331
Supervision systems, 332
Scheduling programming business operations, 334
Protecting the environment, 335
Gaseous effluents, 335

Aqueous effluents, 337
Oil spills, 340
Noise, 341
Safety, 342
Conception, 343
Design, 343
Procurement, 344
Construction, 344
Commissioning, 344
Operation and maintenance, 344
Static electricity in petroleum liquids, 345

Chapter 6. Marketing of oil products, 349
Marketing organisation, 349
The automotive retail market, 353
Aviation, 356
Domestic heating, 359
Marine, 362
Manufacturing and process industries, commercial road and rail transporters and civil engineering industry, 365
Agriculture, 371
Special product businesses, 372
Distribution and storage of oil products, 374
Planning a distribution system, 375
Transport, 375
Installations and depots, 378
Storage and handling of special products, 385
Safe operating practices, 385
Oil products application, specification and testing, 386
Motor gasoline, 387
Aviation gasoline and aviation turbine fuel, 400
Domestic (illuminating) kerosine, 408
Gas oils and distillate diesel fuels, 413
Residual fuel oils, 420
Liquefied petroleum gas, 423
Energy efficiency, 429
General characteristics of lubricants, 431
Engine lubricants, 440
Lubrication of marine diesel propulsion engines, 444
Gas turbine lubricants, 445
Other lubricants for industry, 446
Petroleum waxes, 458
Bitumen, 464

Chapter 7. Transportation–Marine and pipelines, 479
  Marine, 479
    History and development, 479
    Organisation of the World's tanker fleets, 485
    Class of tanker, 487
    Luboil carriers, 490
    Tankers and the environment, 495
  Pipelines, 496
    Main crude oil pipelines in continental Western Europe, 497
    Oil products pipelines, 500
    Main oil products pipelines in Western Europe, 500
    Natural gas pipelines, 501
    The economics of oil pipelines, 502
    Pipeline legislation, 503
    Planning and preparation, 503
    Materials and equipment, 503
    Construction, 505
    Operation and maintenance, 506
    Safety measures, 507
    The future of pipelines, 508

Chapter 8. Natural gas and gas liquids, 509
  What is natural gas?, 509
    Its composition, 509
    Its origin, 510
  Exploration and production, 510
    Exploration, 510
    Production of associated and non-associated natural gas, 511
  World reserves, 512
  Consumption of natural gas, 514
    World perspective, 514
    The United States, 515
    The USSR, 518
    Western Europe, 519
    Japan, 527
    Other markets for natural gas, 528
  Transport of natural gas, 530
    Economics of gas transport, 531
    Transport by pipeline, 533
    Shipment of liquefied natural gas (LNG), 534
    LNG plant, 535
  LNG shipping and terminalling, 539
    The closed-loop system, 540
  Distribution and marketing, 541

CONTENTS   xv

    Local distribution of natural gas, 541
        Markets for gas, 546
    Development of the international gas trade, 551
        The economics, 551
        Integration, 552
        The growth of international gas trade, 553
    Natural gas liquids and gas-derived liquid fuels, 555
        Natural gas liquids, 555
        Methanol, gasoline and ammonia, 558
    The future, 559
        Cost, price and value, 560
        The energy picture, 560
        The energy picture, 560
        Possible trends, 561

Chapter 9. Oil supply and trading, 563
    Introduction, 563
    Factors and constraints in oil supply, 564
        The geographical factor, 564
        Differences in types of crude oil, 566
        Diversity of product demand in consuming countries, 566
        The price of crude oil, 568
        Transportation costs, 569
        Abrupt changes in production and demand, 570
        Effects of consumer government taxes, 572
        Non-technical constraints, 572
    The oil supply scene post-1973: fragmented with diminishing flexibility, 572
    The supply system in the early 1980s, 574

Chapter 10. Petrochemicals, 577
    The origin of petrochemicals, 577
    The importance of petrochemicals, 579
    Manufacture, 585
        Base chemicals, 585
        Polyethylene, 589
        Thermosetting resins, 590
        Synthetic fibres, 591
        Solvents, 593
        Detergents or surfactants, 595
    The future, 596

Chapter 11. Unconventional raw materials and synfuels, 599
    Introduction, 599
    Characteristics of unconventional raw materials, 600

The availability factor, 601
The hydrogen factor, 602
The mineral factor, 602
Characteristics of synfuels, 604
   Liquid synfuels, 604
   Gaseous synfuels, 606
The status of the technology, 607
   Hydrogen-addition technologies, 607
   Carbon removal technologies, 610
   Biomass technologies, 610
Ranking the options, 611
Future outlook, 615

Chapter 12. Research and development, 617
Introduction, 617
Exploration and production, 618
   Natural phenoma, 619
   Subsurface evaluation techniques, 619
   Supplementary recovery, 620
   Design of offshore equipment and installations, 621
Manufacturing, 621
Oil products, 625
Chemical processes and products, 628
Natural gas, 630
Transport, storage and handling, 632
Basic research and new technologies, 633
Patents, 633

Chapter 13. Environmental conservation, 637
Introduction, 637
Exploration and production, 639
   Seismic exploration, 639
   Drilling, 639
   Production, 641
   Accidental oil spills, 643
   Supporting services for offshore operations, 644
Transportation and storage of crude oil and gas, 644
   Pipelines, 645
   Terminals, 646
   Oil tankers, 649
   Oil-spill clean-up, 653
   Gas carriers, 653
Oil refineries and petrochemical plants, 653
   Gaseous emissions, 654

CONTENTS

      Effluent water, 656
      Waste disposal, 658
      Noise, 659
      Accident hazards, 659
   Distribution and marketing, 659
      Atmospheric pollution, 660
      Spent products, 661
   Industry associations, 662

Note on units of measurement in the world energy industry, 663

Glossary, 669

Subject Index, 699

Chapter 1

THE WORLD PETROLEUM INDUSTRY

PETROLEUM

During this century the petroleum industry has risen from being relatively small, through the stage of being one of many large industries, to a position where whole economies are profoundly influenced by the need for and price of petroleum products. The origins of the industry lie in the product itself.

All over the world, at various depths beneath land and sea, there are accumulations of hydrocarbons formed long ago by decomposition of animal and vegetable remains. Hydrocarbons are compounds of hydrogen and carbon which, at normal temperatures and pressures, may be gaseous, liquid or solid according to the complexity of their molecules. The natural deposits are correspondingly gaseous, liquid or solid, depending on the relative proportion of the various hydrocarbons present in the mixture.

In its widest sense, petroleum embraces all hydrocarbons occurring in the earth. In its narrower, commercial sense, petroleum is usually restricted to the liquid deposits known as crude oil, the gaseous ones being known as natural gas and the solid ones as bitumen or asphalt.

Most crude oils, although liquid as such, contain gaseous and solid hydrocarbons in solution. The gases come out of solution, either on the release of pressure as the crude oil is produced or during the first stages of refining, and contribute to the total natural gas production. Some of the solids are recovered during refining as bitumen and wax, some stay in solution in the liquid oil products. Natural gas may be found associated with crude oil as a gas-cap above the oil or on its own, unassociated with oil.

Crude oil and natural gas are the raw materials of the petroleum industry. It is the business of the industry to find them, to retrieve them from the earth on-shore and off-shore, to manufacture useful products from them and to sell the products in the markets of the world.

THE SIGNIFICANCE OF OIL AND GAS

The twentieth century might be described as pre-eminently the age of petroleum. Although oil was first commercially exploited on any scale in the late nineteenth century, the twentieth century has seen the development of oil into "the biggest business", the growth of a large-scale international petrochemical industry, and the rise of natural gas as a prime source of energy.

The twentieth century dominance of oil and gas in the total pattern of energy consumption is shown in Figure 1.1. From this it can be seen that whereas coal, which in the early decades made by far the major contribution, has simply maintained its world production level, oil and natural gas have been the fuels that have met the vastly increased demand for energy as industrialisation and world economic development have proceeded.

It is true that the share of oil and gas in the total energy spectrum today shows signs of diminishing. Even so it seems certain that until the year 2000, and

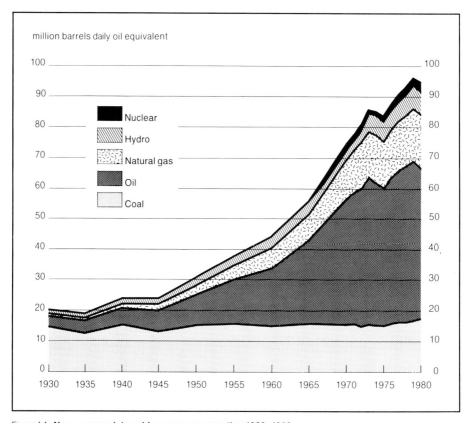

Figure 1.1 **Non-communist world energy consumption 1930–1980**

probably far beyond, they will continue to play the major role in meeting world requirements.

Customer demand has propelled this growth. The exceptional versatility of crude oil as a base material for the manufacture of a very wide range of products, the convenience and cleanliness in use of oil and gas, their ease of transportation and storage, their relative cheapness since the 1940s, their particular efficiency for such special purposes as providing energy for transportation, raw material for lubricants, and feedstock for the petrochemical industry ... these factors have powerfully stimulated growth and given petroleum major importance in the economies of producer and consumer countries alike.

Oil production figures speak for themselves:

	Million barrels per day
1900	0.4
1940	6.0
1950	11.0
1960	22.0
1970	48.0
1980	62.9

This enormous expansion has meant that producer countries have become heavily reliant on oil for national revenue and foreign exchange. Venezuela, for example, has for decades relied on oil exports for more than 90 per cent of its foreign exchange. In most consumer countries, oil has also dominated national economies, as a major component of imports and thus substantially affecting balances of payments. Crude oil price increases have significantly contributed to the growth of inflation, and with it recession and mass unemployment. Oil and gas have also provided finance ministers worldwide with a convenient vehicle for tax-collecting. Today, except in China and some other areas, virtually everyone in society is affected by the availability and price of oil and gas: directly in terms of domestic use and family transportation; indirectly in relation to jobs and to many other aspects of national economies, whether they are buoyant or in difficulty.

HISTORICAL OUTLINE

Early Developments

Petroleum was used for many centuries in Mesopotamia, Egypt, Persia, China and elsewhere for heating, lighting, road-making and building.

In Europe, the northern Italian town of Salsomaggiore, near which an issue of natural gas was known, adopted the crest of a burning salamander in 1226. A small oil accumulation was discovered in 1498 at Pechelbronn in Alsace, and "earth balsam" was mentioned in Poland in 1506. Marco Polo noted "oil springs" at Baku on the Caspian Sea towards the end of the thirteenth century.

In the Americas, Raleigh reported on the Trinidad Pitch Lake in 1595 and there are accounts of visits by a Franciscan to "oil springs" in New York in 1632 and by a Russian traveller to those in Pennsylvania in 1748.

In Burma, oil has long been used and was being produced from hand-dug wells in substantial quantities by the end of the eighteenth century.

Nevertheless, until the middle of the nineteenth century, almost all lighting oil used in the world came from animal or vegetable sources, and early machines were lubricated with castor oil or whale oil. In 1850, James Young of Glasgow introduced a process for the production of lamp oil by the distillation of coal or shale, and this was taken up in the USA where by 1855 several factories were making "coal oil" for use in lamps.

In 1859, Drake drilled the first well to be sunk specifically for oil and struck it at a depth of $69\frac{1}{2}$ feet in Pennsylvania. This is generally taken as the start of the modern petroleum industry, although small quantities of oil were being produced in Russia by 1856 and in Romania by 1857. Developments followed in other countries, including the East Indies, Poland, India and Burma, Japan and Canada, and by 1900 commercial production was averaging just over 400,000 barrels per day.

In those early days, kerosine, as lamp oil, was the important product and the main object of refining was to extract as much of it as possible from the crude. Lubricants and some fuel oil were also sold, but gasoline was burnt off as unwanted and bitumen was also largely useless.

After 1900 expansion was more rapid: Mexico became a producing country in 1901, followed by Argentina in 1907 and Trinidad in 1908. An international trade developed, undertaken by US, UK and Dutch companies, and the names of Rockefeller of Standard Oil, Deterding of Royal Dutch and Samuel of Shell became well known.

By 1910, world production had grown to some 900,000 barrels per day, of which the USA produced 560,000 and Russia 200,000 barrels per day. The Middle East became prominent when oil was found in Iran (then known as Persia) in 1908 and exports began in 1911. Production started in British Borneo in 1911 and in Venezuela in 1914. The internal combustion engine provided a use for gasoline, now becoming a major product, and World War I caused a greatly increased demand for all types of oil, including fuel for shipping. During World War I, the chemical side of the industry was also begun and developed in the USA.

Throughout the 1920s and 1930s demand continued to grow, especially for gasoline, which would have been difficult to supply in the quantities required but for improvements in refining methods and the introduction of cracking, which increased the proportion of gasoline obtainable from a given crude. A demand for bitumen developed for the construction of roads to cope with the increasing needs of motorists. The United States stayed far in the lead and was responsible for most of the expansion in production, supplying its own needs and exporting large quantities. Russia was largely self-supporting, but the rest of the world became increasingly dependent on the international trade in oil in which the Caribbean (mainly Venezuela) was the chief supplier, followed by the United States and the Middle East. Iraq became a producer in 1927 and Saudi Arabia in 1938, and these countries began to export oil in 1934 and 1939 respectively.

The Period of Rapid Post-War Expansion

It was in the late 1940s that the great period of oil industry expansion in absolute terms began, with world consumption more than quintupling between 1950 and 1980, as shown in Figure 1.2. The major international companies greatly developed their operations to meet this demand, including widespread exploration and production activity and the building of new refineries. There was also a spectacular increase in the size of ocean tankers. In the course of these three decades major changes took place in the patterns of the international oil business. Middle East production grew dramatically: the Middle East became the chief supplier of Western Europe and much of the Eastern Hemisphere, and at the same time the area containing the major part of the world's identified crude oil reserves. The USSR meanwhile successfully developed its own production, most of which it consumed itself, with the remainder being exported to the East European countries and to certain others such as Cuba.

The Rise of OPEC and the Period of Producer Country Dominance

Volume growth was one outstanding feature of the late 1940s onwards. Another key development was the formation in 1960, prompted by Venezuela, of the Organisation of Petroleum Exporting Countries (OPEC), whose principal aim was to secure for member countries higher export earnings for what they recognised was a non-renewable natural resource. This desire for increased revenues was enhanced by the observation that governments of consuming countries were collecting far more in taxes from the sale of oil products than did the producers of the original crudes. During the 1960s, the price of oil remained stable or even slightly declined in real terms, and this was due in large part to the fact that the international oil business was mainly in the hands of competing private compa-

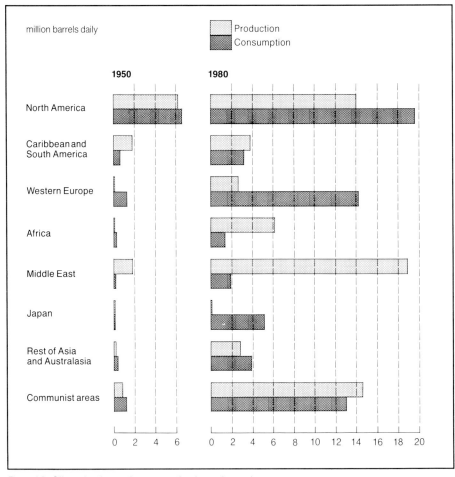

Figure 1.2 **Oil production and consumption by major region**

nies. Moreover, additional supplies were readily forthcoming to meet increased demand. However, by the early 1970s, continuing rapid growth in world oil demand tightened the relationship between demand and the supplies which host governments were prepared to make available. The producing countries, with Libya in the forefront, appreciated that by taking even a relatively small volume of oil out of international trade they would be in a position to upset the traditional pattern of price negotiations with the major international oil companies, and instead to determine themselves the export price of crude oil. During the years 1969 to 1973, the commercial relationships (i.e. over price, government "take" and even ownership of reserves) between the host countries and the oil companies changed dramatically and very much in favour of the host govern-

HISTORICAL OUTLINE

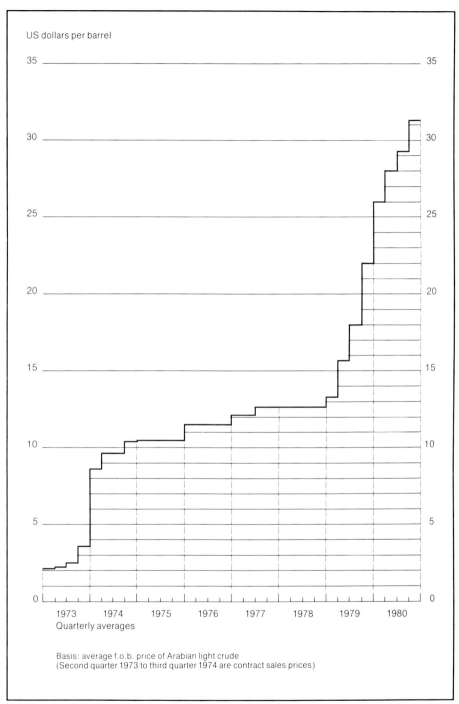

Figure 1.3 **Crude oil prices 1973–80**

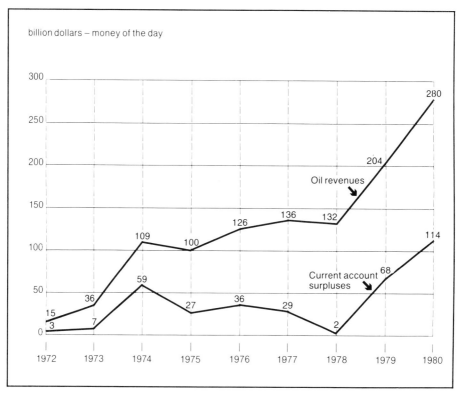

Figure 1.4 **OPEC revenues and surpluses**

ments. Saudi Arabia's special situation with an annual oil production and oil revenues far in excess of immediate needs provided the essential "valve", while at the same time the Arab-Israeli war stimulated collective action. The results were dramatic. In October 1973 and January 1974 the price of oil rose from two to eleven dollars per barrel. The producing countries showed the world that for the time being at least their influence was predominant. These points are illustrated in Figures 1.3 and 1.4.

The Situation at the Beginning of the 1980s: Precarious Balance and Uncertain Future

The dominance of producer countries in the world oil supply scene brought with it a range of problems related to the massive transfer of economic resources that resulted from the rise in the price of crude oil. For many of the producing countries themselves there were the problems of the wise use and investment of their new wealth, and the revolution in Iran illustrated the dangers of popular backlash when rising expectations are not quickly fulfilled and programmes of Western style industrialisation appear to threaten traditional religious and cultural patterns. For the consuming countries, huge increases in oil import bills deepened a recession that severely slowed down economic growth, and the question of security of energy supplies became an important issue, causing a reappraisal of the contribution of coal and calling for other new strategies and initiatives. With the incidence of political risk increasingly apparent, it became more than ever difficult to forecast and plan with confidence.

The further very sharp increases in 1979 and 1980 of the price of oil showed signs of being at last counter-balanced and controlled by market forces, at least to a limited extent. World recession and high inflation rates affecting Western industrialised countries and the hard-hit developing countries brought a halt to the hitherto rising graph of consumption and a return (at least temporarily) to a situation of high stock levels and abundant supply. In understanding the fluctuating price of oil it is a matter of observation that whenever supply in the market has exceeded demand by two million barrels per day prices have tended to fall. However, when the gap between demand and the willingness to supply narrows, spot prices rise dramatically and are usually reflected in overall prices. Almost invariably, a perception of scarcity leads to stock-building, which itself exacerbates the situation. Other factors affecting the position included the drive by consumer countries to find and develop wherever possible new indigenous sources of supply (in practice, mainly very expensive offshore oil) and the search for alternatives to conventional oil. Amidst much uncertainty about the future of the energy balance, one (healthy) outcome was sharper worldwide awareness of the need for improved energy conservation and efficiency, and for realistic pricing that would encourage appropriate end-uses for the various available forms of energy.

One major factor was Saudi Arabia's perception of where the balance should be struck between the needs of the OPEC countries and of consumers generally. Being possessed, simultaneously, of the largest and cheapest reserves and of a small population that had definite limits to its need for funds, the Saudis have had a controlling hand on the world's oil supply "valve", particularly since 1974. As Saudi Arabia could make marginal supplies available more quickly, more cheaply and in greater volume than any other major producer country, its influence within OPEC on oil price decisions has been paramount. Most commentators would agree that it has acted with a due sense of worldwide responsibility. At what level, at what price and in what general political circumstances, the Saudis will be prepared to make supplies available in the future, continues to be the dominating issue in international oil affairs. This is likely to remain so, at least for the 1980s.

THE OIL INDUSTRY TODAY

Basic Characteristics

If customer demand has been the factor that has stimulated the enormous development of the industry and essentially financed its expansion, the properties of crude oil itself have determined the main characteristics of the business. Crude oil is almost useless in the form in which it is found. It requires complex processing in expensive plants to produce technically useful and marketable products. Historically, it has often been found thousands of kilometres away from main centres of consumption. Sometimes it has been very difficult to find at all. Even the United States and the USSR (the two major areas where over the years there has been some degree of balance between production and consumption) conform to the general pattern in that their present main production areas are geographically distant from industrial centres of principal consumption.

As a logical consequence of these key characteristics the industry has been international from its early days. Whereas many of the world's large multinational companies outside the oil industry have developed from a large home base in the United States and have expanded into foreign markets in the search for customers for marginal production, companies within the oil industry have had to search worldwide for the basic raw material and then have been essentially concerned with moving crude and products across national boundaries.

These circumstances have also favoured bigness because high risk exploration projects, expensive production operations, huge-volume sea transportation and costly refining have inevitably called for large scale resources. Companies have had to be strong enough to withstand the exploration disappointments inseparable from risk ventures. They have needed the funds for large capital expenditures on refineries, pipelines, ocean tankers, drilling rigs and distribution networks. Their operations have called for a very wide range of management and specialist skills. The highly technical basis of the whole business of finding crude oil and distributing finished products has also required the backup of sophisticated and expensive research establishments.

Internationalism, bigness, capital-intensiveness are characteristics that have given the industry both advantages and disadvantages. Interdependence between producers and consumers has in the last analysis helped to ensure flexibility and continuity of supply: however, the need to move huge volumes across national frontiers has made for difficulties in a world in which individual countries have tended to become more rather than less nationalistic. Bigness has brought with it the possibility of undertaking large-scale risk projects; it has also been a source of unpopularity in a world in which most large institutions are suspect and become easy targets for public criticism. Capital-intensiveness has meant low unit costs, and hence low prices to consumers, and high employee productivity and thus good wages and salaries. On the other hand, it has meant low direct employment in the industry, a significant difficulty in those areas (and these days these include the so-called developed countries as well as the developing countries) where the creation of employment is often given as high a priority as other economic targets.

Structure of the Industry

The oil industry falls naturally into divisions responsible for exploration, production, manufacture, transport, marketing and research, each of which will be dealt with separately in the pages that follow.

The need for marketers to secure supplies of products, for refiners to secure both supplies of crudes and markets for refined products, for producers to secure outlets for their crudes and for each to ensure adequate transport and storage facilities, with all the inter-related problems of coordination and timing, led to a considerable degree of integration of these various activities by the major oil companies. During the 1950s and the 1960s, the international oil business was largely conducted by seven large companies (known as the "Majors"):

> Standard Oil of New Jersey (Exxon)
> Royal Dutch/Shell
> Mobil
> Texaco
> Standard Oil of California (Chevron/Socal)
> Gulf
> British Petroleum (BP)

With worldwide operations these companies were able to carry out a very efficient exercise in logistics, matching different crude oils to the needs of particular markets, coping with substantial seasonal fluctuations in product demand and even surmounting without significant supply disruption such major crises as the sudden closure of the Suez Canal.

By no means the whole of the industry was thus vertically integrated. The 1960s saw the emergence and development of more than one hundred state oil companies. In the United States, in addition to some thirty large vertically integrated companies, several thousand smaller producers and many independent refinery, transport, and marketing organisations carried on operations.

The gradual erosion of the dominating position of the "Majors" is illustrated in the changes that took place during the 1970s in the ownership of crude oil and in product sales. By 1980 in no phase of the industry had the seven Majors a cumulative share of as much as 50 per cent.

12   THE WORLD PETROLEUM INDUSTRY

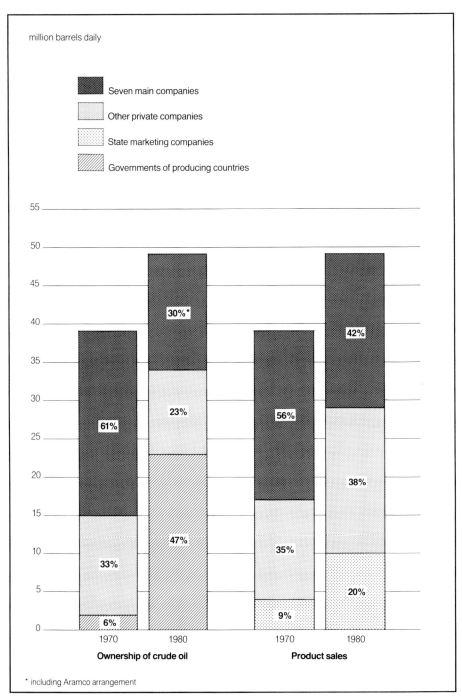

Figure 1.5 **Structure of the oil industry 1970 and 1980**

Crude Oil Production

Figure 1.5 shows the dramatic changes in the ownership of crude oil that took place during the 1970s. This in itself did not mean drastically less private oil company involvement in actual production operations; in many parts of the world the companies moved from a position of ownership to being contractors to Government, and in other countries such as Venezuela they continue to do business under the terms of service contracts to national oil companies.

The relationship of proven oil reserves to current consumption has changed little over the years. Potential oil and gas resources are large enough to sustain present consumption until well into the twenty-first century. The Middle East countries not only produce most of the oil consumed by the world outside the Communist areas; they also possess the largest reserves of oil that have yet been discovered in any part of the world, as shown in Figure 1.6.

	Proven reserves*		Production		Ratio of proven reserves to production
	billion barrels	%	million barrels daily	%	
USA	26	4.5	**8.6**	14.6	8:1
Canada	6	1.0	**1.3**	2.2	13:1
Latin America	71	12.0	**5.6**	9.4	35:1
of which: Mexico	44		1.9		62:1
Venezuela	20		2.2		25:1
Western Europe	17	2.8	**2.4**	4.0	19:1
of which: UK	8		1.6		14:1
Norway	7		0.5		37:1
Africa	58	9.9	**6.0**	10.2	26:1
of which: Libya	26		1.8		39:1
Algeria	12		1.0		32:1
Nigeria	11		2.1		15:1
Middle East	307	52.0	**18.4**	31.1	46:1
of which: Saudi Arabia	113		9.6		32:1
Kuwait	68		1.4		134:1
Iran	40		1.5		74:1
Abu Dhabi	35		1.3		73:1
Iraq	34		2.6		35:1
Asia and Australasia	20	3.3	**2.6**	4.4	20:1
of which: Indonesia	11		1.5		19:1
Communist areas	86	14.5	**14.3**	24.1	16:1
of which: USSR	65		11.8		15:1
China	19		2.1		25:1
World total	591	100.0	59.2	100.0	27:1
of which: OPEC	*383*	*64.8*	*26.7*	*45.2*	*39:1*

* at year end (source *World Oil* 15.8.81)

Figure 1.6 **Crude oil reserves and production 1980** (excluding natural gas liquids)

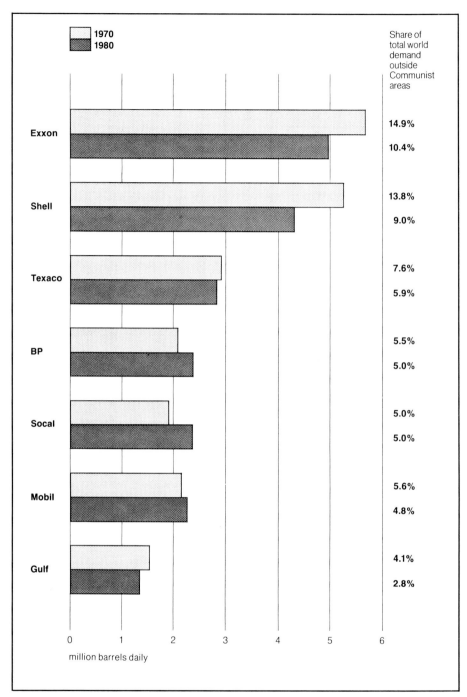

Figure 1.7 **Sales of oil products by seven major oil companies**

Oil Products Sales

Changes in the industry position in relation to the ownership of crude oil have been paralleled by alteration in the pattern of oil products sales. Here too the market share of the Majors has substantially declined. This is shown in the cases of individual companies in Figure 1.7.

Although in percentage terms the erosion of the historic position of the Majors is notable, this has to be seen in the perspective of a giant industry. The huge contribution that continues to be made by the larger international companies can be appreciated by tabulating key operating statistics. Thus, for the Royal Dutch/Shell Group of Companies (smaller in size than Exxon) the extent of operations in 1980 is shown in Figure 1.8.

Changes in Oil Trading

The trend towards fragmentation in international oil trading has recently gathered pace. Governments have increasingly involved themselves; among producers, OPEC as an organisation has become more a forum for crude oil price discussion than a controlling force; consuming countries have attempted to tackle the problem of supply security both by engaging in bilateral deals and by participat-

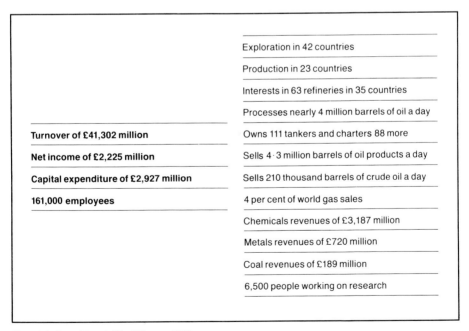

Figure 1.8 **Royal Dutch/Shell Group, 1980**

ing in the IEA (International Energy Agency), which was specifically set up to deal on an international basis with the problem of possible world energy supply emergencies.

Not only in organisational terms, but in many other respects the relatively orderly system of international oil supply and trading has tended to move towards greater uncertainty and instability, made all the more complex by the factor of political risk. The impact of these changes on supply planning is dealt with separately in Chapter 9 (Oil Supply and Trading).

FUTURE PROSPECTS

The future of the petroleum industry will be determined by its ability to adapt to an environment which is radically different from that in which it attained its present size and structure. A major change has been the transition to low or even negative growth in demand for oil products in OECD countries. For an industry accustomed to steady growth, this is proving a traumatic experience. To add to these difficulties the cost of oil is of major international concern. That there is likely to be a continuing demand for liquid fuels is not really open to question. After all, hydrocarbons are a very convenient form of energy, well suited to storage for the sporadic use characteristic of transport and many other applications. To what extent existing major oil companies will be involved in the supply of such fuels will depend on their capability of performing a useful function.

Contrary to the belief of some, the world is not running out of energy, nor is it rapidly running out of oil. The problem is of a different dimension: it is to find and gain access to new reserves of energy before existing sources decline too far, or are denied to consumers, and while possible alternative sources and techniques are being developed.

The fossil energy resource base is in fact ample. Figure 1.9 shows the position for coal, oil, natural gas, tar sands and shale.

The figures indicate reserves already identified and for how many years at current rates of consumption they could be expected to fill world needs. The great reserves of coal (almost 700 billion * tonnes) are sufficient for over two centuries' consumption at current rates, and suggest that coal could in the future return to its pre-eminent position as an energy source at least in a number of countries. Its role, particularly in international trade, is likely to grow steadily over the coming decades. Although coal lacks the advantages of oil as a liquid fuel, a great deal of

---

* 1 billion $= 10^9$.

FUTURE PROSPECTS

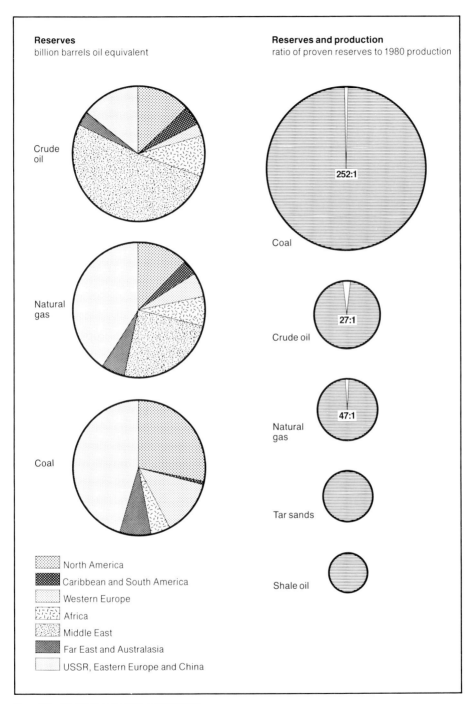

Figure 1.9 **World oil, gas and coal reserves**

research has been carried out, and technically-proven (though still uneconomic) processes exist both for its gasification and liquefaction.

The reserves of oil and natural gas, though smaller, are certainly adequate for the medium-term future, and during the past twenty years the ratio of oil reserves to production has altered very little. There is no ratio shown for shale oil and tar sands: that is because production has so far been largely experimental and in terms of contribution insignificant. With shale oil, the problem of disposal of the rock from which the oil is produced has not yet been solved satisfactorily; with tar sands, although plants have been in operation for some years there are significant impediments to large-scale production. The amount of energy needed to extract oil from both shale and tar sands is itself a major factor. In addition, the availability of technical and managerial manpower could limit the pace at which shale oil and tar sands plants could be brought into production, especially in the early stages of a vigorous development effort by the industry. There are also problems of other kinds (infrastructure, environment and politics) that raise question marks as to whether these huge potential sources of energy will ever be developed on a large scale. The future here will clearly depend mainly on the cost of alternative sources and in the case of the Athabasca Tar Sands, on the wishes of Canadians concerning the speed and manner with which their natural resources are exploited.

The diagram does not include other major sources of energy, actual and potential, of which nuclear power and hydroelectricity are the most important. Nuclear power has the potential to fulfil a very large proportion of the world's energy needs by the year 2000. However, its commercial development is being delayed by political and environmental factors in some countries. Hydroelectricity has a substantial role in some countries and currently provides about 8 per cent of world electricity supplies. With the other alternatives (solar, wind, wave power, tidal power, and geothermal energy for example) considerations such as the high capital costs, the geographic availability and the problem of convenient storage seem likely to prevent them making a large contribution before the next century. They are of use in small, particularly suitable, projects, but cannot in the present state of technical progress make more than a minor addition to world supplies.

Non-commercial energy (firewood, dung and vegetation) is currently of great importance in many developing countries, particularly in Africa and Asia. These resources are not in limitless supply, and are already becoming scarce in some parts of Asia and Africa, such as Nepal, Sudan and the Sahel. Increasingly, they are being replaced by commercial energy sources, particularly kerosine.

Taking all possibilities into account there is rather general agreement that at least for the remainder of the present century, oil, gas and coal will have by far the major roles in filling the world's energy needs. For a variety of reasons, including lead-times, technological problems, and political and environmental considerations, it appears that the spectrum of alternative forms of energy available within the next few decades is narrowing. This being so, it is fortunate that the impression conveyed in Figure 1.9 is pessimistic in the sense that no measure is included of the quantities of oil and gas still to be found. For example, there is undoubtedly much more oil that could become available, though the industry understandably tends to be cautious in its estimates. As a rough indication, it seems likely that the amount of oil that will eventually be discovered and produced is not less than three times the proven reserves shown in Figure 1.9. Some might take an even more optimistic view on the basis that new techniques of recovery may enhance still further the prospects for output from existing fields and those still to be found.

The problem of meeting future energy demand is thus not one of the resource base, but rather one of matching present and future demand for energy with supply on a continuing and viable basis. It is a problem of access and price, not resources.

PROJECTION OF FUTURE ENERGY REQUIREMENTS

The fall in demand for energy between 1979 and 1981 needs to be treated with caution. It is still difficult to evaluate the relative contributions of increased energy efficiency, long-term price elasticity, substitution for oil, industrial restructuring ... and how much the decline reflects reduced levels of economic activity in a time of world recession. In these circumstances, forecasting is even more hazardous than usual. For the industrialised countries, a return to the growth patterns of the past twenty years seems unlikely in the 1980s. In Western Europe, the United States, Canada, and Japan, energy consumption is likely to stabilise or may even decline over the long term. This would represent a shift away from the energy-intensive industries, such as steel-making, towards more capital-intensive, high technology industries (computers, micro-chip electronics, and biotechnology for example) and would also reflect the impact of improved energy efficiency. The position in the developing world (where the volumes of energy involved are much smaller) is different. The developing countries have their individual patterns and prospects, but particularly in the countries of the West Pacific Basin and Latin America higher than average growth rates are expected. With or without rapid economic growth, rates of energy consumption are forecast to increase as a consequence of rising populations, of urbanisation, and of the substitution of commercial fuels for existing traditional, non-commercial sources. In addition, energy consumption may be expected to expand as investments are made in such major industries as iron and steel and cement manufacture.

Energy conservation and improved energy efficiency can play a beneficial role in any future scenario. There is the hope that oil will increasingly be reserved for those applications for which it is best suited. That is to say, it should tend to be withdrawn from under-boiler use and reserved for premium markets (those uses that other fuels cannot serve as efficiently) such as transportation, the manufacture of lubricants and the provision of feedstock for petrochemicals.

Whatever the truth turns out to be about future energy demand, given the incidence of political disruptions and the lack of incentive for some Middle East countries to produce oil at the level of their technical potential, prudence points to the need for the rapid development of the technology needed for all potential sources of energy capable of making an economic contribution. This may be difficult to achieve if relatively low oil prices persist for a period due to an economic downturn. Moreover, it is worth noting that in the event of even low annual percentage growth in world energy requirements, this would mean large additional volumes by the year 2000 (perhaps as much as 50 million barrels per day oil equivalent of total energy supply).

THE FINANCIAL IMPLICATIONS OF THE ENERGY PROSPECT

New supplies of energy (no matter what the source) can only be produced at much higher costs than those to which the world has become accustomed.

Oil

It is estimated that until 1990, 80 to 90 per cent of world oil could come from existing fields, provided that their production is not restricted. This percentage is then likely to drop rapidly and the balance will have to be made up from additional supplies at much higher cost, calling for capital investments up to ten times greater than before.

Low-Cost Oil. An additional 10 to 15 million barrels per day of low-cost oil could be expected, for example from the Middle East, Mexico and some other parts of Latin America. Most of this production is under the control of the producing governments and its rate of introduction is uncertain. The average capital cost of such oil is around $2,000 to $3,000 per daily barrel of production capacity (i.e., for each barrel per day of production), but this is expected to continue to increase in real terms.

Medium-Cost Oil. Medium-cost oil should account for up to 10 million barrels per day by the late 1990s. Capital costs in this category are estimated to increase from $8,000 to $20,000 per daily barrel between now and the year 2000. It will include production from smaller fields offshore, deeper drilling of complex structures and many new projects in Western Europe, the United States and the USSR.

High-Cost Oil. Around 5 million barrels of oil a day up to the year 2000 and a rapidly growing share of world production thereafter are expected to come from high-cost sources, such as enhanced recovery, tar sands and from such hostile and remote areas as Siberia and the North American Arctic. Capital costs of this oil, currently averaging around $18,000 per daily barrel capacity could rise to $35,000 by the turn of the century. On this basis, although high-cost oil would represent only about one-quarter or less of new production, it would begin to take the lion's share of investment. For this reason, even if total volumes of oil were to stabilise or decline within the next twenty years, projections for exploration and production expenditure would still soar.

The costs quoted above (for low-, medium- and high-cost oil) are all expressed in 1980 money (in other words, with no allowance for inflation).

Other Energy Sources

Figure 1.10 shows how the estimated costs of other energy sources compare with those of oil.

Today coal is cheap, particularly in the United States, where vigorous development is expected. Coal imported into Western Europe is still considerably cheaper than fuel oil. Indigenous coal in Western Europe is more expensive, partly for geological reasons. The prospect is one of significant development of international trade between the main producing countries and Western Europe. Export projects vary widely, but one of, say, 5 million tonnes a year, equivalent to some 65,000 barrels of oil per day, involves capital costs (including mines, trains, port facilities and ships) of more than $700 million in 1980 money. To this sum can be added a further $2.4 billion for the power plants needed to convert this coal into electricity.

In the United States, where domestic gas supplies may be running short by the end of the century, the conversion of coal to gas is likely to grow in importance. In the longer term, coal liquefaction may also become important, but this will be even more expensive.

Current volumes of internationally traded liquefied natural gas are comparatively small, but growing fast. Trade in 1980 was some 550,000 b/doe (22.5 million tonnes) but this could rise by 1990 to nearly 1.75 million b/doe (70 million tonnes). This will be very expensive: a typical LNG export project of some 150,000 b/doe (6 million tonnes p.a.) would require capital expenditure of around $4 billion in 1980 money.

Nuclear energy is a fully commercial proposition, but the cost of building the number of reactors needed will be large. Even under a low growth forecast, the United States alone will need by the year 2000 the additional 90 to 100 nuclear plants planned or under construction as well as the 70 already operating in 1980. Completion of the former without substantial delays must be open to question. A

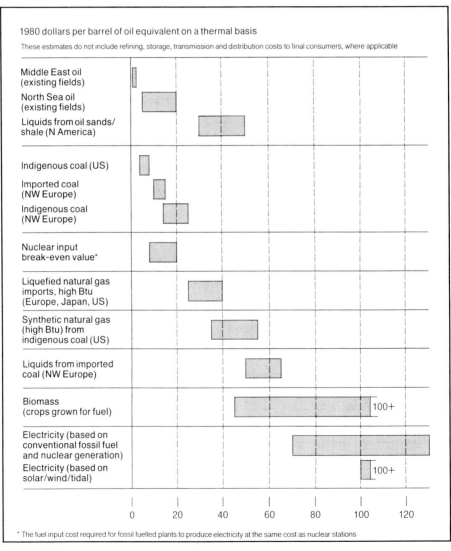

Figure 1.10 **Comparative energy costs**

1,000 MW nuclear plant giving an electricity output of some 10,000 b/doe might call for an investment of around one billion dollars in 1980 money. The three-to-one ratio of fuel input to electricity output does, however, mean that this would actually save some 30,000 barrels per day of fuel oil that might otherwise have been used in a conventional oil-fired power station.

Of the other alternative energy sources, hydroelectricity projects have already been introduced in most sites in the developed countries where conditions are suitable, but there is still considerable potential in the developing countries especially in Africa and South America. The costs of other alternatives (solar, wind, tidal and so on) are high, except under particularly favourable circumstances. Large-scale contributions can only come about if the problem of the cost of storage of electricity is solved.

The indications are that countries will increasingly try to find localised solutions to their energy requirements, seeking to maximise supply security and to minimise balance of payments problems. The building of nuclear plants in France is one example of this; the use of biomass in Brazil another.

Financial Impact of Other Changes

Apart from the very large projected increased capital costs of providing new energy supplies, the financial outlook is complicated by other significant changes. Whereas the rapid expansion of the oil industry in the late 1940s onwards was essentially financed by the Oil Majors' recycling of funds, with capital expenditures largely catered for out of retained earnings, this pattern is likely to be less pronounced in the future. There is a move towards more emphasis on project financing involving greater recourse to borrowing from financial institutions. The provision of finance in the massive amounts necessary has now to be arranged in a less propitious investment climate of worldwide political uncertainty against a background of long lead-times, changing technology, changing markets and the exponential scale of risk involved in large projects.

As long as society accords the necessary priority to the provision of energy the massive financial resources required will doubtless be forthcoming, but in the context outlined above this must result ultimately in customers paying more in real terms for the energy they use.

THE PETROLEUM INDUSTRY AND THE FUTURE

In the past, the petroleum industry has succeeded by creating markets and supplying them with suitable products. It thus worked hand in hand with vehicle manufacturers to provide the right fuels; in conjunction with the aviation industry it developed supplies and facilities around the globe. To meet ever-increasing demand, new petroleum resources were sought and technologies of exploration and production correspondingly developed. For the future, although the outlook for the industry appears much more complex and the lines of development less certain, market influences are likely to continue to be of key importance. The industry will surely prosper if it remains attuned to changes in customer requirements and retains its ability to exploit competitively the basic convenience and effectiveness of hydrocarbon fuels.

Looking ahead, there is a wide range of possible outcomes of the energy scene of almost equal probability, and wise planning must take into account many different scenarios. The petroleum industry today operates in a market that has become highly politicised, a market that is at once of international diplomatic significance and of national economic and social significance. Efficient supplies in the years ahead will more than ever depend on the degree to which the objectives and priorities of authorities, national and international, are reconciled with the needs of the commercial operators (particularly for secure and stable financial ground rules), whose aim is to use resources of men, money and equipment productively. The active intervention of governments in energy matters and the large increase in the number of companies engaged in oil and gas ventures, in one sense spreads the weight of responsibility, but in another and perhaps truer sense makes the planning and coordination of future supply arrangements a greater challenge to which the industry must respond.

As for the large international oil companies themselves, although their role has undoubtedly been eroded, they still see their future as bright with opportunity. With their technological and managerial skills, their solid financial resources, and their capacity to plan and coordinate the implementation of large international supply projects, they are very well placed to make a continuing, substantial contribution to the development of the world economy.

Chapter 2

OIL AND GAS IN THE CENTRALLY PLANNED ECONOMIES

THEIR SIGNIFICANCE

The Centrally Planned Economies are here defined as the USSR, the six other European members of COMECON (Bulgaria, Czechoslovakia, the German Democratic Republic, Hungary, Poland and Romania), Vietnam and Mongolia, China, Albania and the Korean People's Republic. The energy industries in these countries have been developed almost independently of those of the rest of the world. Taken together, these countries account for about one-third of the world's energy; they produce and consume approximately 30 per cent of the world's oil, 33 per cent of the world's natural gas and half of the world's coal.

Within this grouping, the USSR (on which this chapter concentrates) is by far the largest entity in terms of oil and gas. It is the world's largest producer of oil and second largest producer of natural gas and coal. In consumption of energy it is exceeded only by the USA, as shown in Table 2.1.

Table 2.1 **1980 World energy production and consumption** (million b/doe)

	Production	Consumption
USSR	27.8	23.5
European COMECON countries	5.9	8.2
China	8.7	8.3
Rest	1.0	1.1
Total	43.4	41.1
World total	134.9	134.9
of which USA	30.8	36.1
West Germany	2.3	5.6
Japan	1.0	7.3

OIL DEVELOPMENT IN THE USSR

Oil development in what is now the USSR has a long history dating back to the period prior to 1860 when commercial production began. In 1870 the first significant discovery was made at Baku in Azerbaidjan, and by the turn of the century Russia was producing around 10 million tonnes per annum (200,000 barrels per day).

At that time Russia was the world's largest oil producer, and in the early years of the twentieth century it exported between 1 and 2 million tonnes of oil products each year. Subsequently, oil production stagnated and by the beginning of World War I oil exports had fallen to insignificant levels, in part because of the effects of high domestic transport tariffs. During this period the oil industry was in the hands of private companies, including the Royal Dutch/Shell Group of Companies, which took over the Rothschilds' interest in 1912. Following the Revolutions of 1917, the privately held companies were nationalised and the industry went into a period of decline. Oil production fell to 3–4 million tonnes by 1920.

Production was given new impetus with the assistance of Western companies and by the early 1930s the USSR again became a significant oil products exporter, reaching a level of over 100,000 barrels per day. Distribution networks were developed in the neighbouring European countries, and by the beginning of World War II production had expanded to about 600,000 barrels per day.

The ravages of World War II obliged the USSR to re-develop its oil industry for the second time. Prior to the war, production had been centred primarily on Baku and nearby areas, particularly Groznyy. The early post-war period saw the emphasis on development change to the Volga/Urals area, the "second Baku", where oil had first been discovered in 1929. In the twenty years between 1945 and 1965 Soviet production grew twelvefold, from 0.4 to 4.9 million barrels per day, almost entirely as the result of the intensive development of this latter area. By 1975 production in the Volga/Urals area was some 3.5 million barrels per day, and one major field, Romashkino, was producing at a rate of 1.3 million barrels per day, i.e. over 25 per cent of the total daily production in the country. This re-development is shown in Figure 2.1.

Between 1965 and 1980 total Soviet production more than doubled to reach 12.1 million barrels per day. Volga/Urals production, and that of Romashkino, expanded until the mid-1970s before going into slow decline. Emphasis was then switched to the "third Baku", namely Tyumen in Western Siberia.

Oil was first discovered in this area in 1960 and production was developed much more rapidly than had been the case elsewhere, growing from about 20,000 barrels per day in 1965 to 6.3 million barrels per day in 1980. By 1980 the Tyumen area accounted for half the total Soviet production and included the

OIL DEVELOPMENT IN THE USSR

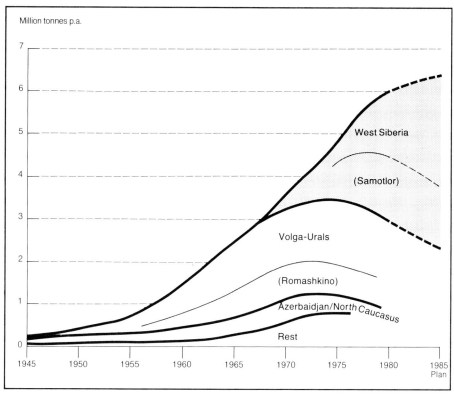

Figure 2.1 **Soviet crude oil production development**

largest field, Samotlor, which was producing at a "plateau" of 3 million barrels per day. Exports had grown to a level of 3.2 million barrels per day, placing the USSR second only to Saudi Arabia as an oil-exporting country.

The future prospects of the Soviet oil industry are uncertain. With about one-third of the world's sedimentary area available, ultimately recoverable resources of oil are potentially huge. However, calculated levels of proven oil reserves are not published, and so there is continuing controversy as to the adequacy of those reserves to meet the needs of future economic growth.

The USSR has been slow in developing its oil resources. Despite making its first significant oil discovery at about the same date as the USA, large-scale production was only started some forty years later, as shown in Figure 2.2. Major oil-producing regions have been discovered and developed sequentially, with each major region being developed more rapidly than its predecessor. The USSR is still in its "third Baku" phase, but production is now showing signs of reaching a plateau after the major expansion period of the 1970s. In 1981 production was only 100,000 barrels per day higher than in 1980. The Eleventh Five Year Plan,

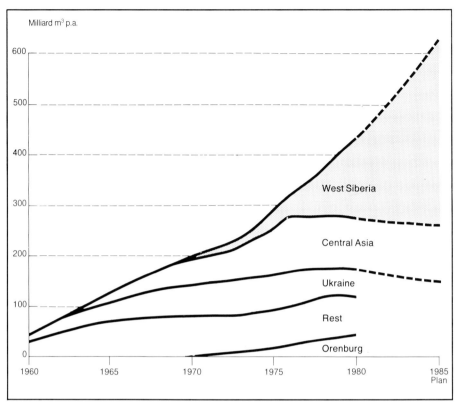

Figure 2.2 **Development of Soviet natural gas production**

even if achieved, allows for an increase only to 12.6 million barrels per day by 1985. Development is still concentrated on newer smaller fields in Tyumen, and production from most of the older areas is now in decline. There are as yet no indications of a "fourth Baku".

SOVIET NATURAL GAS RESOURCES

Whereas, in the past, natural gas had been oil's poor relation, by the early 1980s it has become the USSR's greatest strength in the energy field.

The Soviet system of classification of reserves differs from that used by the Western oil industry, and the category defined by the USSR as $(A + B + C_1)$ used in this chapter gives appreciably higher results than the Western "proven plus probable" category. According to the Soviet method indicated, natural gas reserves were approximately 34,000 milliard cubic metres at the beginning of 1982, which is about seventy times the level of 1981 production. As estimated by the journal *World Oil*, proven reserves represent approximately 40 per cent of the world total, but by Western methods this percentage would be nearer 35.

Serious development by the USSR of its gas resources started much later than of its oil resources. Early production was primarily of gas associated with oil production, and most of this gas was wasted by "flaring". As late as 1955 marketable production was less than 10 milliard cubic metres, equivalent to about 10 per cent of oil production. In the late 1950s production grew rapidly, particularly with the development of the gas fields in the North Caucasus and Ukraine. The early 1960s saw the major development of what is today the USSR's second most important production area, the Central Asian fields of Turkmen and Uzbek. By 1975 when these fields were approaching their peak, the USSR's total annual gas production had reached 289 milliard cubic metres. Thereafter, the growth in gas production has been dominated by the development of the Northern Tyumen fields of Western Siberia, although the development of the Orenburg field in the Volga/Urals has also been a significant achievement. By 1981 Soviet annual production of natural gas had reached 465 milliard cubic metres (equivalent to 8.1 million barrels per day of oil), of which 176 milliard came from Western Siberia.

Western Siberia is in the early 1980s the dominant Soviet gas province. Reserves there are estimated by the Soviet method at 27,000 milliard cubic metres

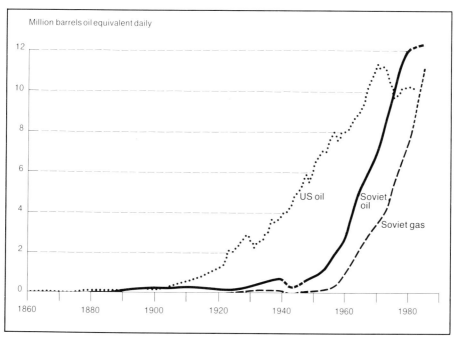

Figure 2.3 **Comparative historical development of Soviet and US petroleum production**

located in several "super giant" gas fields, only three of which (Urengoy, Medvezhe and Vyngapur) have yet been put into production. Between 1965 and 1980 Soviet gas production grew at an annual average of over 8 per cent, or about 50 per cent every five years. The Eleventh Five Year Plan aims for a further 50 per cent expansion in total annual production to 630 milliard cubic metres by 1985. This would be more than covered by growth in Western Siberia, and all effectively from one gas field (Urengoy). The intention is to expand annual production of this field to over 250 milliard cubic metres by 1985.

By 1980 the USSR had also developed into a major exporter of natural gas. Pipeline gas had been sold internationally on a small scale for many years, although in the early 1970s the USSR imported more gas than it exported as a result of two deals made with Iran and Afghanistan to supply gas-deficit areas in the south. With the development of Ukrainian gas production and the use of the Brotherhood (Bratsvo) pipeline, completed in 1967, exports to the six European COMECON partners sharply expanded from 1973. The following year saw the beginning of large-scale exports to Western Europe.

Gas exports to the West have largely taken the form of "compensation deals", enabling the USSR to expand production with the help of Western equipment, mainly high-quality pipe and compressors. Very long-term sales agreements have been made, and because of steeply rising gas prices the value of the gas sold has been many times the cost of the equipment purchases, with consequent benefit to the Soviet economy.

With the development of the pipeline network, gas exports have risen substantially. The part of the network known as the Northern Lights system has made gas available to export markets from Vuktyl (Pechora) and subsequently from Western Siberia. The construction of the Soyuz pipeline from the Orenburg field in the Urals to the Czech border was notable for being a joint effort by the USSR and its six European COMECON partners. Gas was first transported along the line in 1979 and enabled exports to these countries to be doubled (thus paying for the work on the line).

Gas exports have been split approximately equally between the USSR's COMECON partners and Western customers. By 1980 they had increased to an annual level approaching 60 milliard cubic metres (nearly one-third of the equivalent level of those of oil). The USSR had become the world's largest exporter of natural gas.

FUTURE PROSPECTS FOR OIL AND GAS IN THE USSR

By 1980 oil and natural gas had each overtaken coal in primary energy supply to the domestic market, as shown in Table 2.2. The share of natural gas rose from

Table 2.2 **Energy consumption in the USSR**

	Million b/doe			Per cent		
	1960	1970	1980	1960	1970	1980
Oil	2.4	5.2	8.9	26	33	38
Coal	4.9	5.6	6.3	54	36	27
Natural gas	0.8	3.5	6.6	9	22	28
Primary electricity	0.3	0.6	1.1	3	4	5
Other	0.7	0.7	0.6	8	5	2
Total	9.1	15.6	23.5	100	100	100

only 9 per cent in 1960 to 28 per cent by 1980. By 1990 the share of gas should rise to between 30 and 35 per cent, when it should be supplying a significantly greater share of primary energy than oil, despite oil's captive uses in an expanding transportation market.

Gas (including LPG) today reaches almost two hundred million domestic consumers. It has become the principal fuel used in the production of pig iron and steel (some 93 per cent of total fuel consumption in this sector), mineral fertiliser (95 per cent), and in cement (60 per cent). The gas network continues to expand to reach new areas. Gas will, of necessity, supply a greater share of industrial needs and of centralised heat, steam and electricity, substituting for oil in sectors where oil was the major growth fuel in the past decade.

Of this consumption, 41 per cent in 1970 was for under-boiler uses, with the remainder for transport, commercial and technical uses; in 1980 the pattern was much the same, with 43 per cent for under-boiler uses.

Certain features of the present Soviet energy scene will inevitably have a significant influence on the future. These include the huge scale of present development; the enormous resource base; the location of resources at present undeveloped, predominantly in Siberia; the location of major consumption (70 to 80 per cent in the Europe/Urals region); the necessary dependence on natural gas for expansion in the 1980s. The combination of these factors means that while the USSR should not run short of supplies of energy in the future, there are great practical and logistical problems inherent in the situation. Virtually all new supplies of fossil fuels have to be developed in areas with an extremely inhospitable climate and often with permafrost or swamp, sited between 2,500 and 5,000 kilometres from the principal areas of consumption.

In addition, new supplies will have to be considerably in excess of incremental needs to cater for the decline in production of the established areas. For example, at present eight tonnes of oil have to be developed for every one tonne net increase in production. Problems include not only the actual drilling or mining of the resources in Siberia, but also the logistics of equipment supply and the

32 OIL AND GAS IN THE CENTRALLY PLANNED ECONOMIES

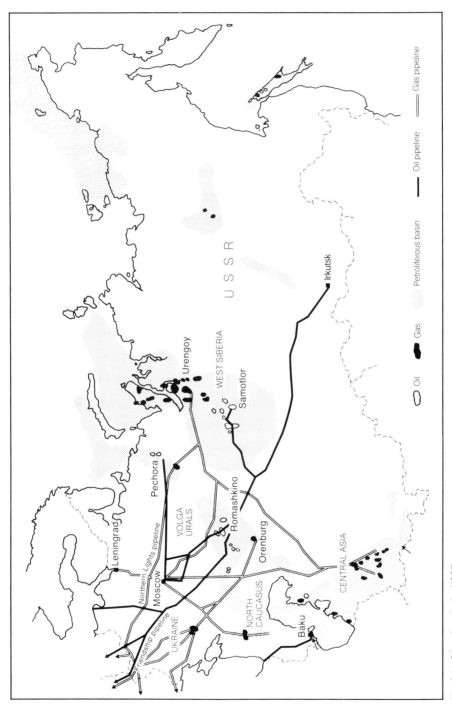

Map 2.1 **Oil and gas in the USSR**

provision of skilled manpower. These problems apply equally to coal as to oil and gas, and so have led to a major programme of investment in nuclear power to meet the growing demand for electricity and centralised heat of European USSR.

The huge land distances between areas of production and consumption have led (as in the USA) to the construction of an enormous network of oil and gas pipelines. By the end of 1981 the USSR had a trunk pipeline system of around 220,000 kilometres of which over 140,000 kilometres were for gas lines. The Eleventh Five Year Plan includes the construction of approximately 10,000 kilometres of trunk pipelines each year. The changing pattern of location of resources means that oil pipeline construction has to continue whilst total production at best increases only marginally. To achieve the targets for gas production in the Five Year Plan will require the completion by 1985 of six major trunk pipelines, each of a length of at least 2,500 kilometres and in some cases much longer.

Coal development in the Eastern USSR has been restricted by bottlenecks in rail transportation, and the development of the large-scale, shallow-depth deposits of Kazakhstan and Southern Siberia (Kansk-Achinsk in particular) is also dependent on the construction of pithead power stations and of long-distance, extra-high voltage, electricity transmission lines.

SOVIET ENERGY EXPORTS

The USSR in 1980 exported energy equivalent to approximately 4.5 million barrels per day of oil; three-quarters of this energy was in the form of crude oil and oil products and the remainder was predominantly natural gas. This export effort has become a major feature in the economies of both the USSR and of its COMECON partners. The development of these exports from virtually nothing in 1950 is shown in Table 2.3.

In 1980 rather more than half of the energy exports of the USSR went to other

Table 2.3 **Soviet energy exports**

	Oil (million tonnes)	Gas (milliard cubic metres)	Coal/coke (million tonnes)	Electricity (billion kWh)	Total	
					(million b/doe)	(% of energy production)
1960	33	neg.	15	neg.	0.8	9
1970	96	3	29	5	2.4	13
1980	163	57	28	20	4.7	17
1980 split:						
COMECON	90	30	18	15		
Rest	73	27	10	5		

COMECON countries, and the oil in particular has been sold at preferential prices. For several years it has been exported at the average of Western prices of the previous five years, resulting in some years (those following the 1973/74 and 1979 OPEC price increases for example) in prices as low as half those typically paid by Western customers. These lower prices have covered about 80 per cent of the energy imports of the other COMECON countries (equivalent to 40 per cent of their consumption). An integrated energy supply network has been developed and this is still being extended. The major Soyuz and Brotherhood gas pipelines are part of a wider grid which includes links to Romania and Bulgaria; the twin Friendship crude oil lines form a major trunk route from the Urals to all the European COMECON countries except Romania and Bulgaria; electricity networks are linked through the MIR grid, and are being expanded with major power lines from nuclear power stations being constructed in the Ukraine, jointly with other partners in COMECON.

Energy, and particularly oil, is the source of approximately half the USSR's hard currency earnings from all sources. Moreover, it has been the only major source of earnings that has increased in purchasing power in recent years. The oil and gas industries thus play a vital role in the economic growth of the USSR. As oil production is reaching a plateau, exports of gas will become relatively more important. The most significant development of the early 1980s has been the controversial plan for a 5,400-kilometre natural gas pipeline from Western Siberia to many West European countries. If the complete project goes ahead, then together with other linked developments it could raise Soviet gas exports to Western Europe by 40 milliard cubic metres to 70 milliard, and thus total Soviet gas exports to all areas would be substantially in excess of 100 milliard cubic metres.

CHINA

In contrast to the COMECON countries, oil and gas in China are of less importance than coal, which accounts for over 70 per cent of energy production and consumption. However, around 2 million barrels per day of oil are produced, and exports at about 300,000 barrels per day are significant. 90 per cent of the crude oil production comes from onshore fields in North Eastern China and more than half from one field, Daqing.

China is a country of large potential energy supplies, but the potential for oil appears to be much more limited than in the USSR. In the early 1980s China's main problem as regards oil will be to develop new fields, following a period during which the policy has been simply to obtain maximum supply from the existing fields without foreign assistance. The major hope lies in exploiting the offshore continental shelf, for which China looks for assistance from Western technology.

Chapter 3

EXPLORATION AND PRODUCTION

INTRODUCTION

Oil and gas, as found in nature, are trapped underground within the myriad microscopic pores of reservoir rocks into which they migrated from source rocks over a period of millions of years. These source rocks were themselves deposited in ancient seas, rivers or lakes. Impervious sediments which were deposited on top of the porous reservoir formations sealed the reservoir underground, preventing the hydrocarbons from seeping away to the surface. Not always, however; all too frequently, a formation which looks like a potential reservoir on the basis of geological and seismic data, once drilled, can turn out to contain nothing but water or perhaps just traces of hydrocarbons.

In exploring for and producing hydrocarbons, the oil industry, in only just over a century, has developed its own special equipment and skills for remotely probing the earth's crust. As the need for energy in easily transported forms has grown in step with the expansion of industrial and transportation activity, so the search for hydrocarbons has intensified. With that intensification has come greater knowledge and understanding of the conditions under which oil and gas were formed and are found, and of the methods by which optimum recovery can be made. In the last few decades, the search has moved into offshore waters, and into ever greater depths. The industry has called for and contributed to advances in knowledge in other fields in order to conduct its business — in diving, medicine, meteorology, engineering design and construction, helicopter operations, subsea pipeline design and construction, and many other activities.

The "spin-off" benefits from exploration and production activities can be immense, providing employment, direct and indirect, both offshore and onshore, in those countries where oil and gas have been found and in others with the necessary industrial infrastructure to meet the needs of the oil industry.

This chapter discusses the geology of the earth with particular reference to the formation and occurrence of hydrocarbons, the methods of surveying the substructure in order to identify hydrocarbon-bearing rock formations, and the process of drilling exploratory wells. The means of developing hydrocarbon discoveries and of producing and treating oil and gas are examined, as are the technologies of enhancing recovery from the reservoirs and of developing discoveries in ever-greater depths of water.

The aim of exploration is to locate new oil and gas in the subsurface in order to exploit these on a commercial basis. This can be achieved only if the host government wishes exploration to take place within its territory and either actively conducts the search on its own behalf (e.g. by way of a national oil company) or grants exploration rights to private companies. These aspects and the economic and financial considerations are also discussed, as is the modern-day concern over how much oil and gas is "left".

EXPLORATION

Historical Background

In some parts of the world, oil and gas have been known to mankind for thousands of years. In ancient times, surface occurrences or seepages of oil, bitumen and asphalt were used for a variety of purposes (medicinal, heating and lighting, caulking the seams of boats) and gas emanations were, and locally still are, venerated as "eternal fires".

The Chinese discovered oil beneath the surface over 1700 years ago while drilling for salt, and in that context, around 600 BC, Confucius mentions wells that were probably a few hundred feet deep, an achievement that was not to be equalled in the west until the 19th century. By about 1100 AD, the Chinese were capable of drilling to a depth approaching 1,000 metres.

During the late 18th century, hundreds of wells were dug to exploit shallow oil at Yenangyaun, in Burma, where annual output was estimated to exceed 250,000 barrels. At about the same time, oil was also actively produced from hand-dug wells in the Caucasus, Romania, Poland and Germany, although production was not as great as in Burma. Geological advice on the location of a series of successful wells was probably first given in 1859, in Germany.

Oil seeps were also widespread in the United States, and it was near to one of these, at Oil Creek, Titusville in Pennsylvania, that the modern oil industry is considered to have begun with the successful drilling of a $69\frac{1}{2}$ feet (21 metres) deep well by "Colonel" Drake, also in 1859. It was not the discovery of oil that heralded the beginning of the modern industry, but rather the establishment of a supply of oil in sufficient quantity to support a business enterprise of some magnitude. By 1860, there were 19 producing wells at Oil Creek, with eight others

EXPLORATION 37

nearby, and production had jumped from a former 2,000 barrels to 500,000 barrels per year. (World oil production in 1980 was about 22,000 million barrels.) By 1865, annual production in the United States, still largely from Pennsylvania, had increased another five-fold and the country became an important exporter of oil.

From the drilling of Drake's well until far into the 20th century, wells throughout the world were still sunk close to seepages (what is now known as "seepage drilling") and there was no exploration in the modern sense until geology was applied to the finding of oil in the late 19th century.

Geologists were not employed in the search for oil in any number until the discovery of the Cushing Field, Oklahoma, in 1912, as the direct result of a geological survey.

The period from 1912 to 1925, during which most of the principal surface anticlines (Fig. 3.1) in the USA were drilled, is known as the "anticlinal period", the end of which marks the beginning of modern scientific exploration.

From the early exploitation of oil from surface or near-surface accumulations, modern exploration methods, including the application of the increasingly im-

Figure 3.1 San Migueleto anticline, California, USA.

portant geophysical techniques (discussed later), have led to the finding of substantial recoverable oil and gas reserves at ever-greater depths down to 7 kilometres or so.

Some Basic Geological Facts and Principles

Exploration for oil and gas is today strongly dependent on the recognition and understanding of some basic geological facts and principles. For convenience, they are listed here in their simplest form; some of the more important will be treated in greater detail later.

Hydrocarbons

Oil and gas are derived from organic-rich source rocks comprising mainly the remains of marine algae and bacteria, and plant matter of continental origin. Oil and gas occur underground in the pore spaces of sedimentary rocks and are trapped there if prevented from migrating further.

Rocks

Rocks are divided into three main groups: *igneous rocks*, which include granites and volcanic rocks consolidated from hot, liquid material; *sedimentary rocks*, which are either fragments of other rocks deposited on land or under the sea by wind and water, chemically precipitated from evaporating waters, or of organic origin; *metamorphic rocks*, which comprise rocks originally of igneous or sedimentary origin whose composition and structure have been profoundly changed by heat and pressure.

Most hydrocarbon accumulations are limited to sedimentary rocks although some significant oil and gas accumulations are contained in fractured igneous and metamorphic rocks. The occurrence of many metals, on the other hand, is largely confined to igneous and metamorphic rocks, with the exception of some iron, and sulphide ores such as those of copper, zinc and lead, and "placer" deposits like those of gold, tin and uranium.

Global Geology

The earth's crust forms a relatively thin and brittle layer of rock, some 10 to 50 kilometres thick, which can be considered as floating on a hotter and more plastic mantle. The crust basically consists of two types, namely oceanic and continental. The transition of the crust to the underlying mantle occurs at the "Moho" (Mohorovicic Discontinuity) (Figs. 3.2 and 3.3).

EXPLORATION

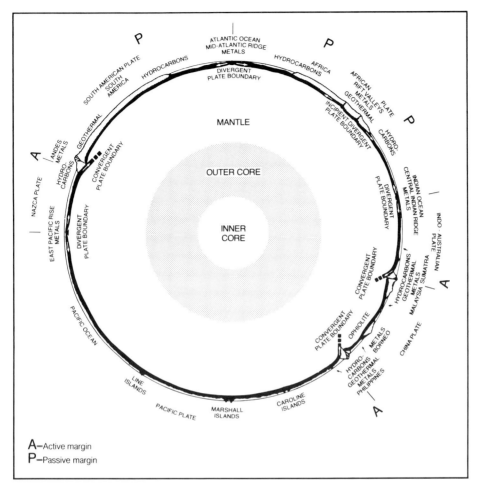

Figure 3.2 **Schematic great-circle section through the equator of the earth, showing lithospheric plate boundaries, and occurrences of mineral resources.** For clarity, the thickness of the lithosphere is expanded by a factor of about 3. From P.A. Rona, 1977. EOS, Trans. Am. Geophys. Union 58(8): 629-639.

Oceanic crust is relatively dense and thin (10 to 20 kilometres), is entirely below sea level, and amounts to almost 60 per cent of the total world crust (Figs. 3.2, 3.3 and 3.4). Continental crust, on the other hand, is relatively light and is thick (25 to 50 kilometres).

Although not proven, it is likely that slow-moving thermally-induced convection currents within the mantle result in the extrusion of new oceanic crust along the axes of mid-ocean ridges and the lateral displacement of earlier-formed crust (sea-floor spreading). The destruction of crust occurs elsewhere in down-going "subduction" trenches at converging plate boundaries (Fig. 3.2).

In many cases, the moving oceanic crust is attached to continental crust, which

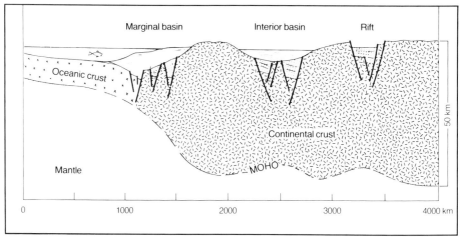

Figure 3.3 **Typical locations of sedimentary basins within continental areas and overlying the transition from oceanic to continental crust.** The sedimentary fill of these basins becomes increasingly metamorphosed with depth.

moves passively with it (Continental Drift). The area of such contact between oceanic and continental crust is known as a passive margin (Fig. 3.2). An active margin occurs where the denser oceanic crust is carried down ("subducted") beneath continental crust.

Because oceanic crust is being created and destroyed continuously, none is known that is older than about 180 million years (Jurassic, Table 3.1 and Fig. 3.4). Continental crust, on the other hand, is continually being generated and/or rejuvenated in fold belts such as the Andes and in island arcs as, for instance, Japan, and also when the low-density sedimentary cover to the subducting oceanic crust is scraped off and accreted to the overlying active continental margin. Continental crust is less dense than oceanic crust, so the former is rarely subducted. Thus in the ancient cores of the continents (shields), rocks occur with an age possibly as great as 4,500 million years (Table 3.1), to which the younger sequences have been added (Fig. 3.4).

The interaction of moving crustal plates, with the associated transfer of heat both across and through the mantle and crust, results in the generation of enormous stresses. These give rise to horizontal and vertical compressional deformations (folding, thrusting, uplift) and to tension (crustal stretching, faulting, rifting, subsidence). The areas of horizontal compression may give rise to mountain ranges, whereas those of crustal subsidence develop into sedimentary basins.

EXPLORATION 41

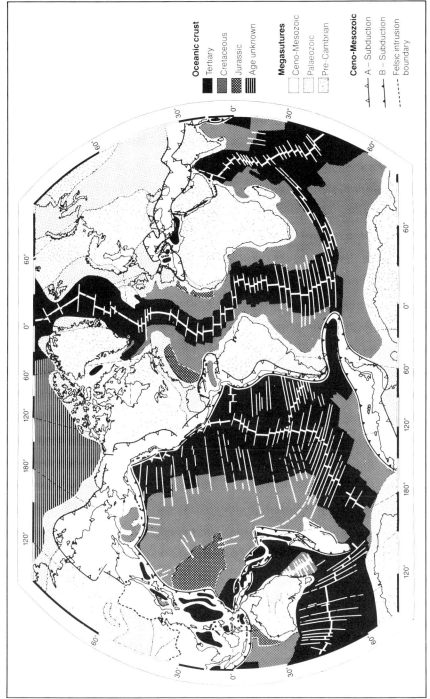

Figure 3.4 **Generalised ages of the world's continental and oceanic crust.** The axes of oceanic spreading in the Atlantic, Pacific and Indian oceans are offset by transform faults (Figure 3.10). The oldest known oceanic crust (Jurassic) occurs in the central Atlantic and western Pacific oceans and in the Indian Ocean west of Australia; the cores of the continents (shields) extend back into the Pre-Cambrian era. From A.W. Bally & S. Snelson, Memoir 6, Can. Soc. Petrol. Geol. 1980.

Table 3.1 **The geological time scale**

Eras	Periods and epochs	Derivation of names	Approximate age in millions of years	Major events in evolution of life
CENOZOIC (Cenos = recent) (Zoe = life)	Quaternary		2.8	Man
	Tertiary			Rapid development of mammals, birds and flowering plants
			65*	
MESOZOIC (Mesos = middle)	Cretaceous	Creta = chalk		Flowering plants become dominant
			143	
	Jurassic	Jura Mountains		Origin of birds
			200	
	Triassic	Threefold division in Germany		Earliest mammals
			245*	
PALAEOZOIC (Palaios = old)	Permian	Permia, ancient kingdom between the Urals and the Volga		
			289	
	Carboniferous	Coal (carbon)-bearing		Earliest reptiles
			367	
	Devonian	Devon		Origin of amphibians
			416	
	Silurian	Silures, Celtic tribe of Welsh Borders	446	First plants and animals adapted to life on land
	Ordovician	Ordovices, Celtic tribe of North Wales	509	Oldest known fishes
	Cambrian	Cambria, Roman name for Wales		Diverse marine life-forms with exoskeleton
			575	
PROTEROZOIC (Proteros = earlier)	Upper			Development of multi-cellular organisms without exoskeleton
			1650	
	Lower			Unicellular organisms; development of photosynthesis
			2600	
ARCHAEAN (Archaeos = primaeval)			> 3000	Origin of life?
			± 4500	Oldest rocks?

* Major changes in life due to extinction of many species

Sedimentary Basins

Hydrocarbons are found in sedimentary basins. It is important, therefore, to understand something of the origin of sediments and of the basins in which they accumulate.

EXPLORATION

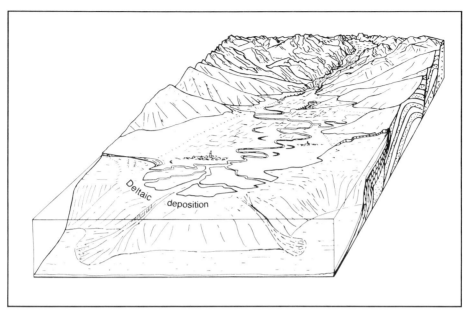

Figure 3.5 **Erosion, transport and deposition of sediment**

Erosion, Sediment Transport and Deposition

Wherever rocks are elevated and exposed to the elements, they become subject to weathering and erosion. Assisted by the force of gravity, the products of erosion are carried away by water, ice and wind and are deposited as sediment in the valleys and plains and in the seas beyond (Fig. 3.5).

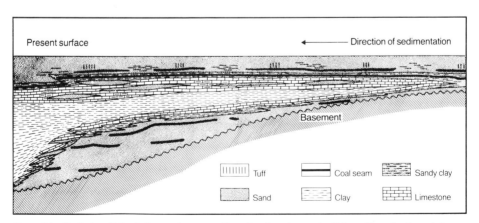

Figure 3.6 **Vertical and lateral changes in composition of sedimentary rocks.** Schematic section through the edge of the southern Sumatra basin. Note: Tuff is another name for volcanic ash.

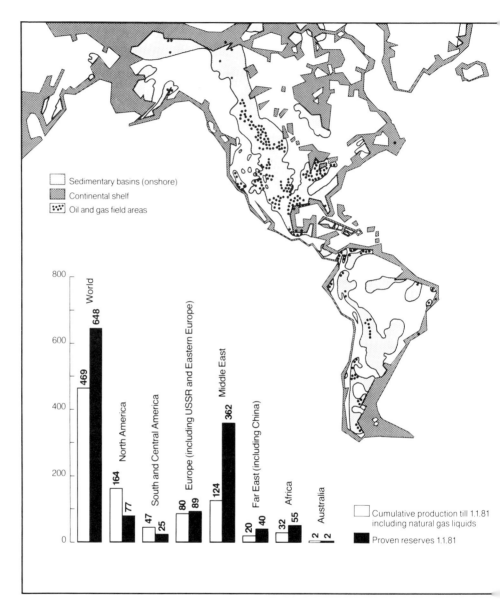

Figure 3.7 **Sedimentary basins and continental shelf areas.** Histograms of the cumulative oil production per area, and the remaining proven reserves at 1.1.1981. (Barrels × 10^9).

Under arid climatic conditions, salt and gypsum deposits, referred to generally as evaporites, may form by the evaporation of sea water in shallow lagoons, for instance, or in desert basins of inland drainage. The shallow waters of warm, clear tropical seas favour the growth of corals and algae, which are important contribu-

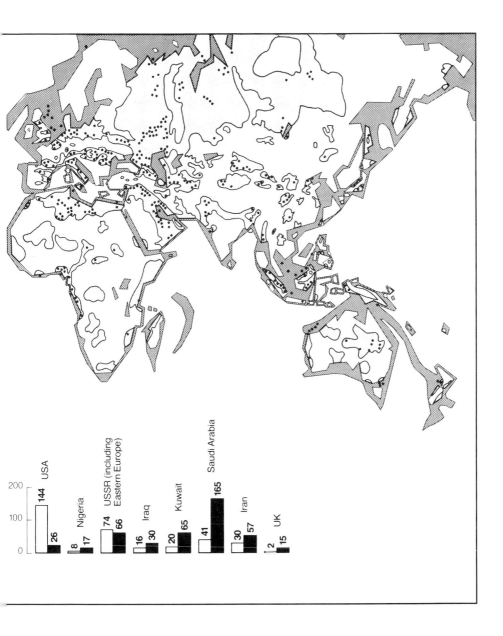

tors to the formation of carbonate rocks such as limestone and its related alteration product, dolomite. Where vegetable matter accumulates, peat is formed which, after burial, converts into lignite and eventually coal. The beds of sedimentary rocks deposited in this way are seldom uniform in thickness, com-

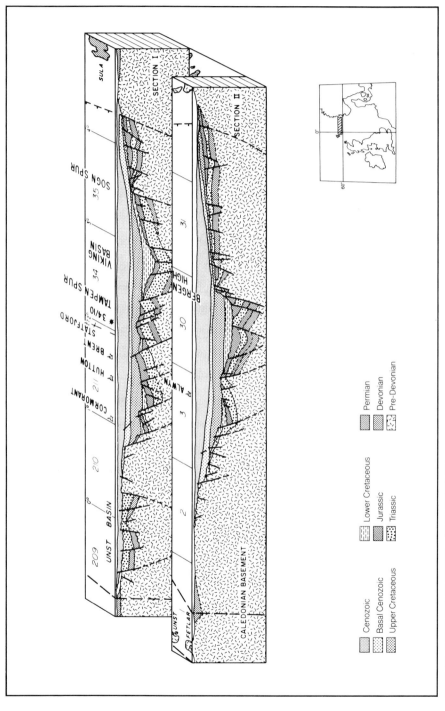

Figure 3.8 **Isometric block diagram** — Northern North Sea

EXPLORATION   47

position or texture. The variations may be small or large depending on many factors, and a section through a series of sedimentary rocks usually shows lateral and vertical changes in both lithology and thickness (Fig. 3.6).

Geological Age

Traces of animal and plant life are often preserved in sediments. These fossils record the evolution of life on earth and at the same time provide the geologist with a means of assessing a relative age for the sediments in which they occur. The absolute age of the rocks can be determined from the degree of decay of contained radio-active minerals, and a geological time scale can be established as in Table 3.1.

Basin Development

Sediments are deposited preferentially in topographic depressions, which may then be referred to as sedimentary basins. The close relationship between sedimentary basins and the distribution of oil and gas fields is clearly seen in Figure 3.7.

Sedimentary basins are of several types. Rift basins (e.g. Fig. 3.3) are relatively

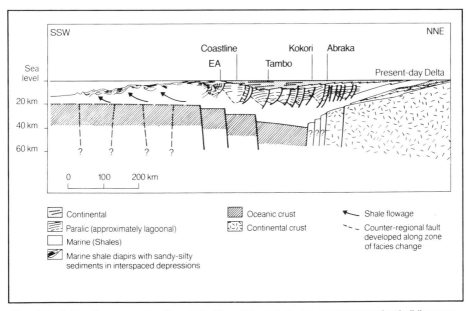

Figure 3.9 **Schematic cross-section through the Niger Delta, showing the present stage of outbuilding over the oceanic crust of the south Atlantic Ocean.** The sedimentary sequence includes the development of growth faults in the right-central area, and shale diapirs to the left. (From Evamy et al, AAPG Bull. 62, 1978).

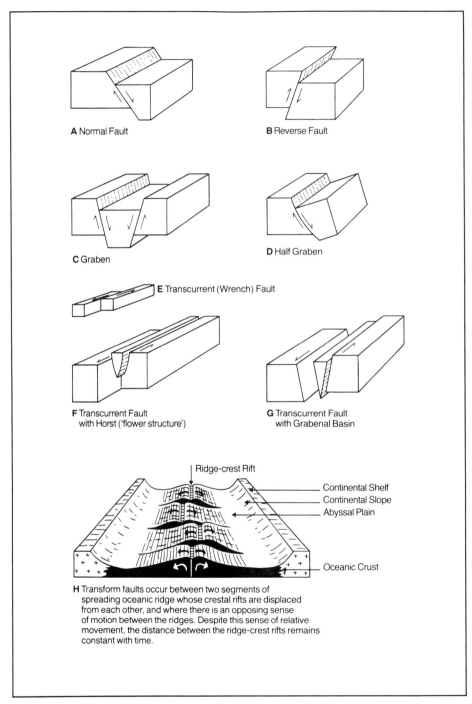

Figure 3.10 (part 1) **Fault and fold types**

EXPLORATION

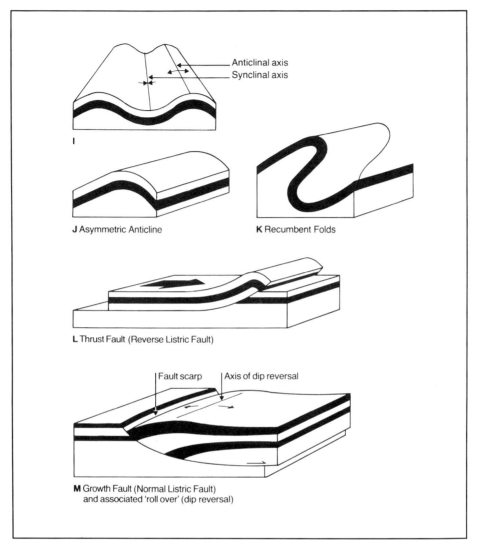

Figure 3.10 (part 2) **Fault and fold types**

long and narrow because they occupy the depressions formed when the crust fractures and pulls apart under tension. The unsupported sides of the rift collapse towards its axis and, because these fault blocks rotate slightly along faults (Fig. 3.12), they fill the floor of the rift (e.g. North Sea rifts, Fig. 3.8, Section I).

Because of later burial, many intracratonic basins have no obvious faulted margins at the surface (e.g. Fig. 3.8, Section I). As sediment accumulates in the basin, its weight causes the crust to subside and thereby to leave space for more sediment to be deposited. By a combination of crustal sagging and sediment-load-induced subsidence, basinal areas increase with time and some basins can acquire a sedimentary fill in the order of 10 kilometres or more (e.g. Amazon Basin).

Another area of sediment accumulation occurs over the transition between continental and oceanic crust. Given time and a sufficient supply of sediment, the surface of the sedimentary wedge will extend out from a passive continental margin as a large delta (Fig. 3.5) such as the Niger (Fig. 3.9). Below sea level, however, high-density sediment flows (turbidity currents) may transport sediment hundreds of kilometres across the ocean floor (e.g. Indus Cone).

Diagenesis of Sediments

After burial, sediments react to increasing temperature and pressure, and commonly also to the corrosive action of water saturated with other chemicals. The result is a change in the composition of the rock (diagenesis) which, in general, has the effect of reducing its porosity. Limestones, however, may alter to dolomite with the creation of extra porosity. If subjected to very high temperatures and pressures, the buried sedimentary rock may be completely recrystallised; it is then said to have been metamorphosed from, say, sandstone to quartzite, or limestone to marble.

Folds and Faults

Basins develop under essentially tensional conditions, and the unsupported sides of the fracture tend to subside along "normal" fault planes so as to fill the gap (Fig. 3.10A). Most tensional situations in the crust are relieved by faults of this type. Depending on other factors, a group of faults either have the same sense of movement (e.g. from Cormorant to Tampen Spur in Fig. 3.8, Section I) and result in overall subsidence by the creation of a series of "half grabens" (Fig. 3.10D), or the sense of movement alternates to form a series of "horsts" (highs) and "grabens" (lows) (Fig. 3.8, Section I, west of Cormorant, and Fig. 3.10C).

Horizontal stresses sometimes result in major linear "wrench" faults (Fig. 3.10E). If accompanied by local lateral tension, then fault-bounded subsidence of the crust may occur (Fig. 3.10G), over which a basin will develop. If, on the other hand, shearing is accompanied by local lateral compression, then linear slivers of the more rigid rocks will be squeezed upwards as narrow horsts bounded by outward-curving faults ("flower structures", Fig. 3.10F) and the overlying sediments will be uplifted ("inverted") into a series of fault-bounded anticlines. Provided other criteria for the accumulation of hydrocarbons are met, these anticlines can form traps for oil and gas.

With mild horizontal compression, sediments deform into a series of anticlines

EXPLORATION

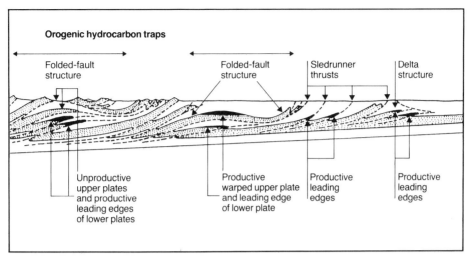

Figure 3.11 **Sequence of thrust sheets in the style of the southern Canadian Rockies.** Traps, mainly for gas, occur in folded thrust sheets. From Roeder, AAPG Structural Geol. School. 1977.

and synclines (Fig. 3.10I). If compression continues, a plane of detachment may develop between the relatively plastic sedimentary sequence and the underlying more rigid crystalline basement. Under these circumstances, folds may go through the stages of being asymmetric (Fig. 3.10J) and recumbent (Fig. 3.10K), until finally the upper limb of a recumbent fold shears at the point of maximum curvature and a thrust develops (Fig. 3.10L). Repeatedly folded and thrust sequences of this type develop over the subduction zones of active continental margins. The thrust sheets along the eastern side of the Canadian Rockies (Fig. 3.11) developed as a result of strong compressional deformation of the western margin of the North American continent following its collision with an island-arc

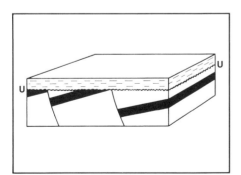

Figure 3.12 **An unconformity (U) coincides with the plane of erosion between the underlying sequence of deformed rocks and its sedimentary cover**

Figure 3.13 **Unconformity between rocks of Eocene and Miocene age, State of Lara, Western Venezuela**

52 EXPLORATION AND PRODUCTION

system marking the subduction of the Pacific plate (Fig. 3.2). The anticlinal crests of some of these thrust sheets form traps for oil and gas. As can be seen from Figure 3.8, rift basins are commonly subjected to more than one phase of sedimentation, with intervening periods of uplift and erosion. The interface between the truncated sequence and the overlying horizontally-bedded sediments is known as an unconformity (Figs. 3.12 and 3.13).

Hydrocarbon Geology

Source Rocks, Oil Generation and Migration

Hydrocarbons are formed by the thermal conversion of organic matter trapped in sedimentary rocks (source rocks). With increasing burial, the temperature of the source rock increases and at a given temperature the organic matter (kerogen) transforms into oil and gas. After expulsion from the source rock, oil and gas migrate to the reservoir formations (Fig. 3.14).

In aquatic environments, abundant life is seen only in the upper water layers (down to 100 metres depth), where algae and bacteria create organic matter by photosynthesis from carbon dioxide and water. Under normal circumstances, the

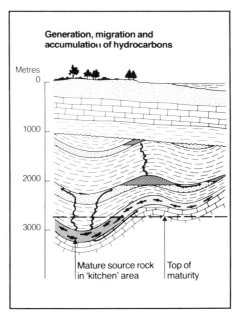

Figure 3.14 **The burial/temperature-related maturation of a source rock and the migration of the generated hydrocarbons into both structural and stratigraphic traps**

organic matter is recycled into carbon dioxide and water by oxidation and by aerobic bacteria.

Only under certain conditions, for example when photosynthetic activity is abnormally high, will organic matter accumulate on the sea floor in large quantities. This is related to an unusually large supply of nutrients, supplied, for instance, by up-welling water. In such cases bacteria and algae consume so much oxygen that it causes a drastic reduction of the aerobic zone of the water.

The algae and aerobic bacteria die and are deposited in large amounts on the bottom. Here, anaerobic bacteria convert the degradable parts into carbon dioxide and water, simultaneously synthesising their own bodies. In this way, the original organic matter is reworked and converted into a (sapropelic) biomass. This biomass consists of bacterial bodies (the "reworkers") plus the microbially resistant part of the planktonic bodies (lipids) together with the resistant organic matter originally present or supplied from elsewhere, such as pollen, spores, plant waxes and plant resins. As the anaerobic degradation is relatively slow and the biodegradation of bacteria and algae in these circumstances very incomplete, organic-rich sediments (3 to 20 per cent carbon) can be deposited. The type of organic material contained in such rocks, which are termed "source rocks", determines whether they are capable of generating predominantly oil or gas. Thus hydrogen-rich, amorphous, sapropelic organic material is an excellent source for both crude oil and gas, while hydrogen-poor, coaly organic matter is mainly a source for gas. The quality and composition of crude oil is also to a large extent determined by the type of organic material. Structureless organic matter of marine origin will result in an oil different from that formed from organic matter of lacustrine (i.e. from lakes) or terrestrial origin containing plant waxes and resins.

Rich oil-source rock, generally known as oil shale, is exposed at the surface in a number of places in the world (e.g. Colorado, Morocco and Australia). If heated in a retort, it will give off oil. Oil shales formed the basis of local synthetic oil production in several areas in the past (e.g. Scotland and Germany) and, in a few areas, production continues today largely on a pilot-plant scale (e.g. Colorado, China and Australia).

Coal is also an important source of hydrocarbons, especially gas. It originates largely by burial of forest and swamp types of vegetation. As the result of progressive depth/temperature-related carbonisation of the original plant material, the volatile gaseous and liquid hydrocarbons are driven off to leave a sequence that ranges from peat, through lignite and the bituminous coals, to anthracite, which is almost pure carbon. In Europe, the bulk of the coal is of Carboniferous age; coal is not found in the older Palaeozoic sediments as land plants did not develop in quantity until the late Silurian or early Devonian (Table 3.1).

54   EXPLORATION AND PRODUCTION

Hydrocarbon production from the North Sea consists almost entirely of gas in the south and largely of oil in the north. This contrast can be ascribed to a difference in source rock type; the source for the southern gas fields is in the

Figure 3.15 **Trinidad Pitch Lake.**

underlying Carboniferous (Table 3.1) coal measures, whereas the oil of the northern fields is derived primarily from Late Jurassic marine source rocks (Fig. 3.8). The few gas fields in the northern North Sea seem to have been charged from marine source rocks that were buried so deeply that they entered the "gas window" (see below).

Oil and/or gas is said to migrate when it leaves the source rock in which it was generated and moves either to a reservoir rock where it is trapped, or to the earth's surface where it escapes as a seepage (Fig. 3.15). Migration is not fully understood, but is thought to take place along faults and minor fractures within the rock sequence. As hydrocarbons are lighter than water, they generally migrate in an upwards or sideways direction from areas of higher to lower pressure.

As with the synthetic production of oil from oil shales, the natural generation of oil starts when the respective source rock is buried sufficiently for the temperature to reach the critical value at which the organic content is converted to oil and gas. In nature, crude oil is generated by thermal decomposition of organic matter in source rocks. As the process is dependent on both time and temperature, the critical temperature at which crude oil is formed depends on the length of time during which the rocks have been buried (e.g. for Tertiary rocks some 130°C is needed, while for Cambrian rocks 65°C is already sufficient). Initially oil is generated, but at greater burial temperatures, gas will be formed. Ultimately, a source rock will be burnt out (carbonised) if it is exposed to too great a temperature. A source rock is referred to as "mature" when the hydrocarbon-generation processes have started, and as "post mature" when it is burnt out. The area in which a source rock is mature is referred to as a "hydrocarbon kitchen".

By contour-mapping the distribution of hydrocarbon kitchens in relation to potential traps (see below), an idea can be gained of which traps are most likely to have received a charge of oil or gas. A source rock of given area and thickness can produce only a finite volume of hydrocarbons. If the outlined kitchen area is small, then only the nearest traps are likely to be oil-bearing. If, however, the kitchen area is both large and deep, gas, the bulk of which presumably was generated after the oil, may have replaced the oil in the traps closest to the kitchen, and the oil will have been forced to migrate to more distant traps.

If source rocks mature before traps have been formed, the migrating hydrocarbons will escape upwards to the earth's surface. In the search for hydrocarbons, therefore, it is vitally important to ascertain the history of trap creation relative to that of kitchen development.

Accumulation of Oil and Gas

Migrating oil and gas can accumulate if the following essential geological conditions are satisfied:

- The presence of reservoir rock, i.e. formations containing interconnected pores (e.g. sands and sandstones) or cracks and voids (e.g. some limestones).
- The presence, at the top of the reservoir rock, of a formation that is impervious to the passage of hydrocarbons (e.g. anhydrite, salt or shale). When this lies directly over the reservoir rock, as in most oil and gas accumulations, it is called a "cap rock", "roof rock" or "seal".
- The presence of a trap, i.e. a geometrical configuration of the reservoir rocks and seal that prevents the lateral escape of fluids, such as when the cap rock is concave when viewed from below. Geometrical shapes of this type are depicted on maps as being enclosed by a series of depth contours: thus these potential hydrocarbon traps are commonly referred to as "closures".

These conditions define a potential trap in which oil and gas migrating from the source rock may accumulate. Because of a difference in density, oil will displace downwards the water previously filling the reservoir, and free gas (gas not dissolved in oil as a result of high pressure), if present, will collect in the highest part of the reservoir to form a "gas cap" with the oil below it. Below the oil, the pores in the reservoir rock will remain full of formation water, usually saline.

Oil and Gas Traps

Oil and gas accumulate in many types of traps, which can be divided broadly into structural and stratigraphic traps.

Structural Traps. Structural traps result from local deformation, such as folding, faulting or both, of the reservoir rock and cap rock. Figure 3.16 shows an asymmetric anticlinal trap in which a reservoir sandstone and a reservoir limestone are capped by impervious beds, which also cover the flanks of both reservoirs, thereby providing closure and preventing the horizontal escape of oil and gas. The upper part of each reservoir contains gas underlain by oil; the upper sandstone reservoir is filled to spill point, with the pore space of the lower part filled with salt water.

Figure 3.17 shows a trap in which the fault provides the closure for the tilted sandstone reservoir by bringing an impervious layer alongside it on the up-dip side; this is not the case for the limestone reservoir, however, in which oil and gas could not accumulate because they would escape across the fault plane and move updip through the juxtaposed sand. In some cases, the fault plane itself can act as a seal.

Another type of anticlinal trap, colloquially called a "roll-over", is associated with "growth faults" (Fig. 3.10M). If part of a sheet of near-surface sediment slips slightly on a gently inclined gliding plane, a concave-shaped (listric) fault plane

EXPLORATION

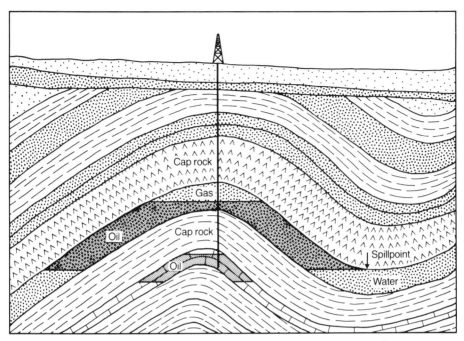

Figure 3.16 **Anticlinal traps and spillpoint.** Note that in this example, the structure of the hydrocarbon-bearing strata is not apparent at the surface because of the intervening unconformity.

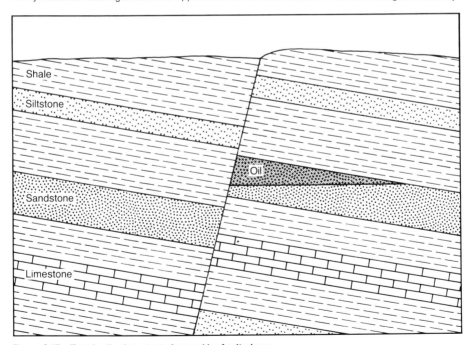

Figure 3.17 **Trap in dipping strata formed by fault plane**

develops at the upper end of the sheet and flattens at depth into the glide plane. To fill the void so created, the bedding overlying the listric fault plane sags slightly (to create the roll-over), and the overlying depression becomes the site of further sedimentation. The weight of added sediment is responsible for further slippage along the glide plane with a resulting growth of the fault. Many of the hydrocarbon traps in the Niger Delta are of this type (Fig. 3.18).

Traps also form because of the ability of some rocks to flow plastically at relatively low temperatures. Salt has this ability, and at depth is commonly also less dense than the immediately overlying sediments. Because of this gravitational instability, salt tends to rise vertically. Lateral imbalance in the overburden results in the flow of salt from areas of heavier to lighter load. Localised subsidence attracts further sediment, which may be derived by erosion of the adjacent salt-induced uplift. Eventually, these movements can result in the salt punching its way up through the overlying sedimentary column. If oil is generated in the subsiding "rim syncline", it may accumulate in the upturned beds against the salt plug (diapir), or above the plug in reservoir formations that were folded as the

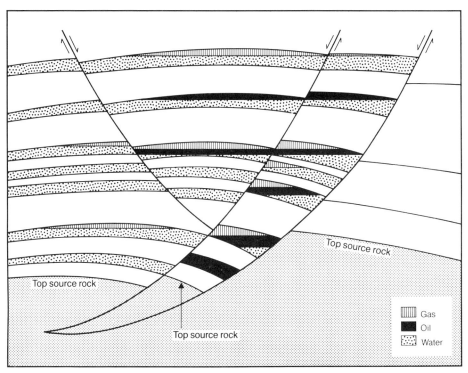

Figure 3.18 **Schematic section of Nigerian oil and gas accumulations in 'roll-overs' associated with growth faults.** In some cases, the faults act as seals to oil or gas; in other cases, they provide migration paths from underlying source rock. (Modified from R. G. Precious, et al., Offshore Tech. Conf., Houston, 1978).

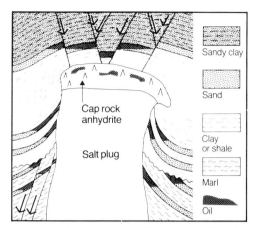

Figure 3.19 **Idealised section through a Gulf Coast salt-dome field, showing traps associated with diapiric salt intrusion.**

plug rose diapirically. Oil fields of these types occur over and around cylindrical salt plugs in Texas (Fig. 3.19). In the North Sea area, salt intrusions commonly take the form of linear salt walls that seem to be aligned with underlying faults. Small gas accumulations occur over some of the large elongate salt "pillows" (an early stage in salt-dome formation) in the southern North Sea, and oil is trapped in chalk reservoirs of dome-shaped fields over deep-seated diapirs in the central North Sea.

Shale diapirs are known to complicate some structures associated with growth faults in deltaic provinces such as Nigeria (Fig. 3.9), and diapiric shales are involved in some of the producing anticlines in Brunei and Sabah (N.W. Borneo).

Stratigraphic Traps. Reservoir rocks can pinch-out laterally or change to non-porous rocks. Such lateral changes in composition can result in stratigraphic traps, which are often associated with the wedge-out of a sand layer in an up-dip direction (A in Fig. 3.20) and its replacement by impermeable clay or shale (C in Fig. 3.20). Oil accumulations also occur in traps formed by lenticular sand masses completely enclosed in tight (impervious) sediments.

Limestone is often impervious but may contain fissures and cavities that can form stratigraphic traps. The remains of an ancient coral reef buried by impervious sediments can also form a stratigraphic trap (B in Fig. 3.20).

A different kind of stratigraphic trap is formed when a succession of layers, including a potential oil reservoir, has been deformed and eroded, and finally overlain by impervious sediments that act as a cap rock (Fig. 3.21). Many of the oil fields of the central and northern North Sea are essentially of this combined structural and stratigraphic type (Fig. 3.8).

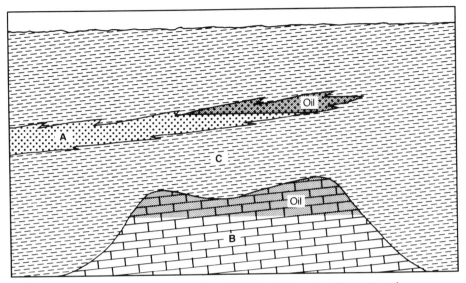

Figure 3.20 **Stratigraphic traps.** Organic reef embedded in shale, and wedging-out sand.

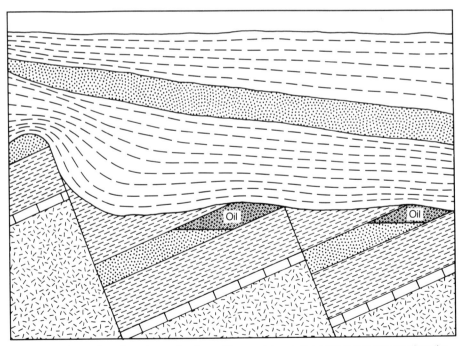

Figure 3.21 **Unconformity or truncation traps.** The overlying shales provide the upper seals to the tectonically tilted and partly eroded sandstone reservoirs.

Exploration Methods

The successful exploration for hydrocarbons depends on the ability to predict their presence with accuracy. An accumulation of oil or gas can be considered as the product of a whole series of chance events that resulted in the presence of a source rock, a kitchen, a reservoir and a trap in the correct geometrical and historical relationships to each other. The structural and historical variations are so numerous, however, and the chance so great that one of the critical factors will be missing, that prediction is far from simple.

Field Geology

The traditional method of finding oil in the first half of the century was by mapping the surface geology and studying the relationships to each other of the various geological units. Field geology is now used largely to understand and predict the types of rock one may expect in the subsurface, in structures outlined by geophysical means.

Airborne Imagery

Before the commencement of exploration of new land areas it is customary to photograph the whole area from the air. An aircraft fitted with a high-resolution wide-angle camera flies stripwise over the area and takes photographs, each of which overlaps those adjoining. By a stereoscopic study of these photos, a fairly

Figure 3.22a **Asymmetrical anticline.** Aerial photograph of Timimoun area, Algerian Sahara.

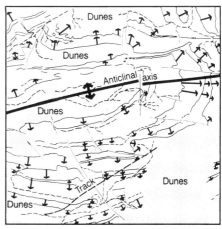

Figure 3.22b **Photogeological interpretation of Figure 3.22a.**

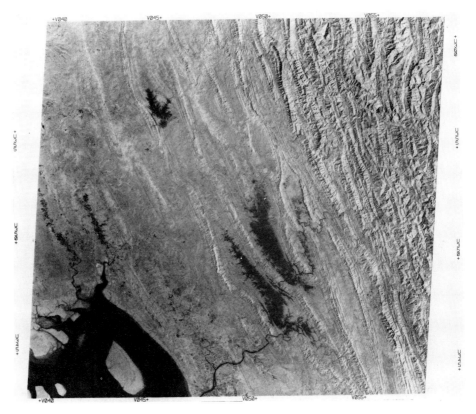

Figure 3.23 **Landsat scene, Chittagong Hill Tracts, Bangladesh. Note that this photo-like image is not a rectangle, due to the combined effects of earth's rotation and near-polar orbit of the satellite which carries the scanning receptor.** (Source US National Aeronautics and Space Administration).

accurate topographic map, and a map showing the geological surface features observable from the air, are constructed (Fig. 3.22).

Since the mid-1970s, orbiting space craft have been digitally recording data from the surface of the earth. These are transmitted to receiving stations, which convert them to a photograph-like image (Fig. 3.23) with a resolution of about 100 metres. By the mid-1980s, a new series of satellite images will have a resolution of about 15 metres and will be capable of producing stereoscopic images worldwide.

Geophysical Exploration

In many areas, the deeper structure of the earth bears no resemblance to that seen on the surface. The truncation of upturned beds such as those seen in Figure 3.13 cannot be suspected if the beds exposed at the surface are all horizontal.

EXPLORATION 63

The objective of geophysics is to determine the properties and structure of the rocks within the earth by the quantitative measurement of physical fields at the surface. Measurement of the passage of seismic (sound) waves through the earth is the most widely used geophysical technique in exploration, although gravimetric, magnetic and electric methods are also employed. Geological information from outcrops and wells is used to calibrate the geophysical data so that a prediction of the rock properties can be made in locations distant from the geological control.

The Gravimetric Method. The gravimetric method depends on measurements of very slight variations in the force of gravity at the surface of the earth. Such measurements can be made with great precision on land and at sea by means of highly sensitive instruments known as gravimeters. Recent developments are making it possible in favourable circumstances to record gravity variations from

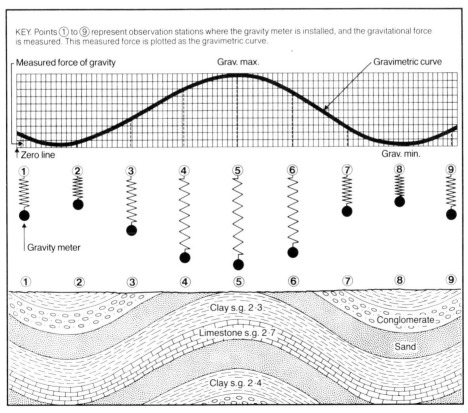

Figure 3.24 **Gravimetric survey.** Schematic relationship between rock structure and vertical gravitational pull, based on the assumption that the deeper rock horizons are more dense than the shallower ones.

gravimeters in helicopters. The force of gravity at any point on the earth's surface is influenced in magnitude and direction by the distribution of rocks of different densities underlying the area (Fig. 3.24). For example, the older rocks that underlie and flank a sedimentary basin generally have a stronger gravitational pull than the sediments of the basin itself. Thus mathematical analysis of the variations can yield evidence of the presence of concealed geological structures and/or larger regional elements such as basins and grabens, which can be further explored by the seismic method.

The Magnetic Method. The magnetic method depends on measuring variations in the intensity of the earth's magnetic field. This is done by sensitive instruments known as magnetometers, which have been designed for use on land, at sea (towed behind survey ships) or in aircraft. In the latter mode, large areas can be surveyed and evaluated quickly. Localised magnetic variations come from the varying magnetic properties of the underlying rocks. For instance, ocean-floor lavas have a stronger magnetic susceptibility than marine limestones. From analysis of the observed variations, the dispositions of the more magnetic rock elements can be determined. Aero-magnetic surveys are particularly useful in outlining the regional framework of sedimentary basins before expensive seismic surveys are undertaken.

The Seismic Method. The seismic method involves the measurement at the surface of acoustic waves which have been generated by a source and have travelled through the earth. There are three stages in the seismic technique: acquiring the data, processing and interpretation.

Seismic surveys are performed in a great variety of operating environments, both on- and offshore. However, there are always three components in the data acquisition system: the source, the detector and the recording equipment.

The seismic source transmits a pulse of acoustic energy into the ground. Before 1960, dynamite was used almost exclusively as a source. Due to the adverse environmental, safety and cost aspects of explosives, there has been a trend towards more efficient lower-energy sources. The air gun, which is basically a valve releasing compressed air, has proved very effective offshore. The vibroseis technique is growing in popularity on land and can even be used in urban areas. It sends a long "sweep" of vibrations into the ground, which after processing can be shortened into a pulse comparable with other sources.

The pulse of acoustic energy travels as a wave into the earth. At each interface between rock formations with different acoustic properties, part of the energy is transmitted down to deeper layers within the earth, and the remainder is reflected back to the surface where it is picked up by a series of detectors (Fig. 3.25). The arrival times of each "reflection" give an indication of the depth of the reflector.

EXPLORATION

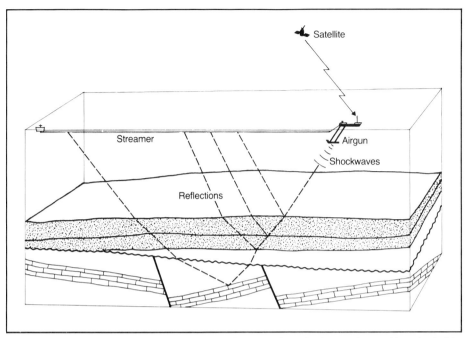

Figure 3.25 **Seismic survey.** Seismic boat, airgun and streamer, the main elements for acquiring seismic data offshore; the ray path of a sound wave; navigation using satellites or fixed locations on the surface (land and/or production platforms).

The difference in rock properties above and below the reflector controls the amplitude of the reflected wave, but it is also affected by other processes, e.g. spreading and absorption losses during transmission. The detectors transform the reflected seismic waves into electric signals. On land they are called geophones and operate on a moving coil and magnet principle similar to a microphone. At sea small piezo-electric crystals are contained within an oil-filled cable called a "streamer". Groups of detectors are connected together to form "patterns" so that the reflected wave is enhanced compared with other disturbing arrivals. Each pattern or station transmits its signal down a cable, normally one to three kilometres long, to the recording instrument. The recording instrument is basically a high-fidelity tape recorder. The signals from up to 1000 stations are stored on magnetic tape in a digital form similar to digital recording on audio discs. These magnetic tapes are sent to the computer centre where the signals are processed to improve their quality. Major processing steps include stacking, which reduces noise by adding corrected traces together, deconvolution, which improves vertical resolution, and migration, which focuses the seismic wave into its correct subsurface position. The resulting seismic section, such as that in Figure 3.26, clearly shows the subsurface structure, variations in the thickness of individual rock units, and major unconformities.

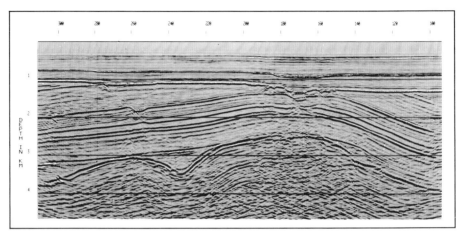

Figure 3.26 **A seismic section after migration and depth conversion, showing clear structural details**

The objective of interpretation is to translate the pattern of recorded seismic signals into a geological cross-section along the trace of the respective seismic line. From well results, it is possible to find out which reflection corresponds to which rock formation. Reflections are followed around a grid of seismic lines and the geological structure can be mapped out. Care must be taken to transform the reflection times to depths, which requires a knowledge of the velocity of seismic waves through the rocks. This was already carried on for Figure 3.26 during the final stage of processing.

Recently, much attention has been paid to acquiring a close grid of seismic data locally, so that a full three-dimensional picture of the subsurface can be made. In oil or gas fields such a technique will give a better structural image and thus allow for the planning of the optimal development of a field. By using numerical techniques it is sometimes possible to make lateral predictions of rock properties or the fluid content of reservoirs away from a control well.

As new oil and gas fields become increasingly more difficult to find, the further development of the seismic technique towards providing a clearer picture of the subsurface will certainly continue.

Electrical Methods. Electrical methods depend on differences in the resistance to electric currents by rocks of various types. They can disclose useful information about rocks buried beneath a relatively thin cover of soil. The methods are only rarely used as surface exploration techniques. In the subsurface, however, these same principles are applied extensively in various well-logging tools, which help the identification and correlation of formations drilled through, and permit an assessment of their fluid content (see Logging and Formation Evaluation section in Petrophysics, p. 83).

Exploration Drilling

It has been mentioned that, in many cases, surface structures bear no relationship to those at greater depth, some of which are likely to be the target of exploration drilling. Because of the high cost of deep exploration wells, especially in offshore areas, it is essential that all other suitable exploration surveys be undertaken before drilling begins; in this respect, seismic surveys will provide the depth and three-dimensional shape of the potentially oil-bearing traps. Ultimately, however, the selection of an exploration well location will depend on a geological analysis of all the available data (e.g. kitchen maps, lithofacies (rock type) maps, porosity/permeability trends, burial history, any pertinent outcrop or well data, as well as the structural configuration of the area).

When all data have been assembled and a prognosis has been made of the layers to be expected at depth, a location is selected which, it is hoped, will prove conclusively the presence or absence of hydrocarbons in the target formation. The first well in a new exploration area may reveal the presence of a column of hydrocarbons that could prove to be commercial. In many cases, however, the first well is "dry" but gives information of direct geological significance, or perhaps has just sufficient shows of oil or gas to justify further drilling in the area. All wells drilled to discover accumulations of hydrocarbons are "exploration wells", commonly known, especially by drillers, as "wildcats" (a designation possibly dating back to the first decade of the century) which emphasises the hazardous and speculative nature of drilling in a new area. A successful wildcat is a "discovery well"; an unsuccessful one is a "dry hole". After oil has been discovered, new wells must be drilled to establish the limits of the field; these are known as "outstep" or "appraisal" wells.

Once the size of a field has been established, and a decision has been made to produce its hydrocarbons, all subsequent wells will be known as "exploitation", "development" or "production" wells.

If the first wells prove the presence of a commercially exploitable oil or gas field, plans are then made for development. With fields located on land, the initial "wildcat" and some or all of the appraisal wells could be used as production wells. In offshore locations, however, especially in deeper water, it may be more economical to abandon these early wells in the interests of the efficient development of the field from later production platforms.

Exploration Results

"Oil is where you find it". In the early days of wildcatting, the success ratio (by which "success" implies that oil or gas was found, but not necessarily in commercial quantities) of "finds" to "dry holes" was low.

Because of intense competition between oil companies, more exploration wells are drilled annually in the United States (about 16,000 in 1981 in contrast to over 60,000 development wells) than in the rest of the world put together (in 1981 around 1,300 exploration wells in the world outside the Communist areas and North America), and this has probably been the case ever since "Colonel" Drake's first. From 1946, when one discovery well was drilled in the United States out of every 9.4 wildcats, the ratio improved to 1 in 5.2 in 1980. Over the same time interval, the average depth to which wildcats were drilled had increased from 1,200 metres (4,000 feet) to over 1,900 metres (6,000 feet).

Of greater relevance to the world's future supplies of hydrocarbons, however, is the ratio of significant oil and gas discoveries to wildcats drilled. In the United States, most of the easy-to-find oil in large, relatively shallow, structural traps has already been discovered. This is also apparent from the annual decline in the volume of new oil and gas discovered for each foot of wildcat drilled. The exploration effort is moving increasingly towards the discovery of deeper and smaller traps, many of which are of a more subtle nature (e.g. stratigraphic traps) and are therefore much harder to find. There is thus a declining ratio of significant oil discoveries to wildcats drilled. Although the statistics that support these statements are most readily obtained for the United States, the same trend is apparent in the rest of the world. Oil is currently being consumed at a greater rate than new reserves are being found. For the years 1978 to 1980, the ratio of consumption to discovery of new reserves in the United States was about 7 to 2.

Because of their high cost of development, offshore fields must be much larger and more prolific than those on land if they are to be commercially viable. In the whole of the North Sea area, about 1,500 exploration wells were drilled between the beginning of this offshore venture in 1964 and the end of 1981. These have resulted in the discovery of around 90 named oil and gas fields (including some additions to earlier discoveries) that are currently in production, are being developed or are likely to be so in the not-too-distant future. There are, in addition, another 40 or 50 unnamed discoveries. In technical terms, this offshore exploration venture must be considered highly successful but its overall economic success will not be known until its history of production is much older.

The rapid increase in our understanding of the earth's geological processes since World War II is matched by, and is partly the result of, our relative success in finding hydrocarbons. With the advent of more sophisticated exploration equipment and techniques, and our computer-aided ability to handle enormous volumes of factual geological data, the search for oil and gas is becoming much more labour-intensive, with each specialist member of an exploration team adding his or her own quota of skill, knowledge and ideas to the search. This trend of increasing labour intensity is likely to continue during the next few decades as the exploration for hydrocarbons extends to some of the less predictable traps.

Successive Stages in Exploring a Sedimentary Basin

Industry experience shows that the historic development of oil and gas exploration, like oil production, often passes through three successive stages which have occurred in a similar way in many basins and in different countries and environments. These stages can be described as the *pre-mature, mature* and *post-mature* stages, and can be illustrated by a typical example (Fig. 3.27) where the rates of discovery of gas are shown for a particular sedimentary basin which has passed through all three stages.

In the pre-mature stage, the industry is learning how to explore a new sedimentary basin, discovering whether petroleum is present at all and, if so,

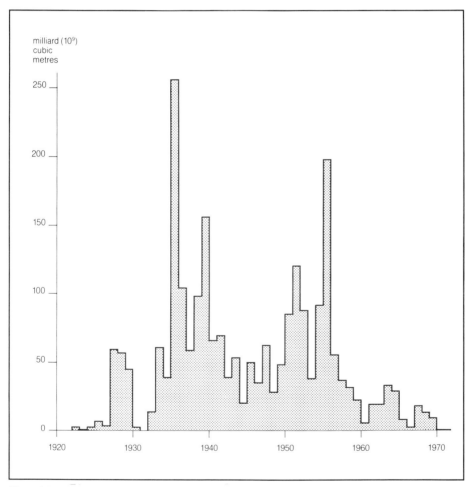

Figure 3.27 **Total gas discovered each year in South Louisiana, USA**

which are the most successful techniques for finding it under the geological conditions of that basin. This may be described as finding the "exploration key" for a particular basin, where the relevant geological concept is either stumbled upon by good luck or, more often, is arrived at by the application of imaginative geological thinking to the results of the latest exploration techniques. A number of promising ideas are then tested by drilling expensive exploration wells, which may not validate any of these concepts within the first few trials, if at all.

An example of such an "exploration key" is shown (Figs. 3.8 and 3.28) for the northern North Sea area. Here it was recognised that the sub-cropping of the Middle Jurassic reservoir sands against an unconformity closure provides the ideal trapping environment for hydrocarbons. Once this concept had been vali-

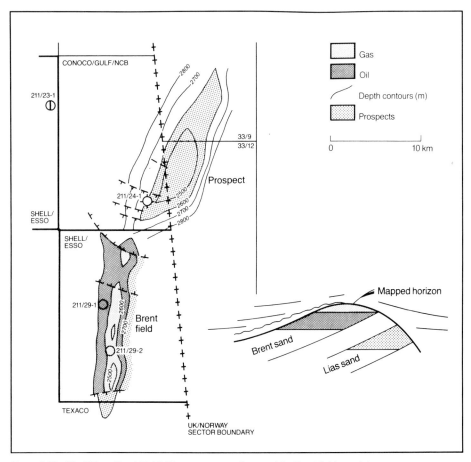

Figure 3.28 **Shell/Esso Brent Field.** The prospect, north-east of the Brent field, was later brought into production as the Statfjord field.

dated for this area, then there followed the quick succession of discoveries which have so largely contributed to the self-sufficiency of the United Kingdom.

The pre-mature stage often shows a gradual increase in the proportion of successful wells, and it is frequently at this stage that the very large or "giant" fields are discovered. It may, of course, end in failure to discover any petroleum at all.

During the mature stage of exploration, the industry makes the fullest use of its now successful petroleum-finding techniques in a particular basin, refining them as time goes on. This stage is often one in which the average amount of hydrocarbons discovered per exploration well (or "discovery ratio") remains roughly constant over a number of years. In other words, the exploration key is being used.

By contrast, the post-mature stage of a basin is characterised by a marked decline in the amount of new hydrocarbons discovered per exploration well, reflecting the fact that the easier, larger and more accessible fields have already been discovered and only the smaller, technically complex and more marginal fields remain as objectives.

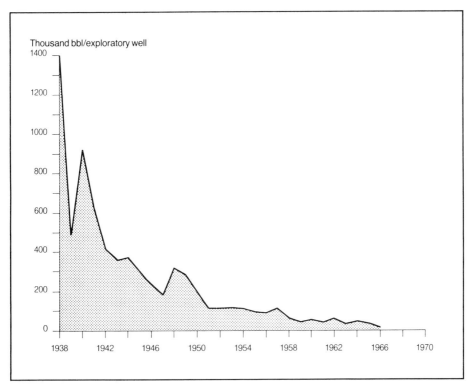

Figure 3.29 **Proved crude oil reserves rolled back to year of discovery per exploratory well drilled.**

An example of the post-mature stage, in the United States, is given in Figure 3.29. This, of course, is the aggregate result of a large number of individual basins which may each be at different stages of maturity. The trend of the total picture, however, is obvious. The exhaustion of physical possibilities now becomes the limiting factor on exploration discoveries. In practice, the post-mature stages of many basins have proved to have remarkably long lives, exploration being periodically revived by increasing prices, and also by new or more precise exploration techniques, together making the search for ever smaller fields potentially profitable.

PRODUCTION

While they remain in the subsurface, the hydrocarbons are essentially the responsibility of petroleum engineers, whose main task is the technical evaluation and planning process leading to the efficient delineation and development of the oil and gas accumulations and, once the field is on-stream, ensuring that the recovery of oil and gas is maximised at an acceptable economic return.

Specialists within the engineering, drilling and production operations departments are responsible for the drilling of the wells and the design, construction, monitoring and maintenance of the process and transportation facilities for the fluids coming from the wells.

Production Development

After a discovery of oil or gas has been made in an exploration well and is judged sufficiently promising to justify further expenditure, the information available at that time forms the basis for an evaluation study and planning phase.

The first task is to determine whether development is likely to be economically viable and, invariably, additional information is required to answer this question with sufficient confidence. Accordingly, appraisal drilling and additional data gathering is carried out in order to assess the development potential more accurately.

The size of the accumulation is estimated, the expected production rates of wells predicted and assessments made of the number of development wells required, and their type. A broad outline can then be given of the related size and nature of surface production and treatment systems.

Feasibility and cost estimate studies of the development drilling and surface installations are contributed by the field engineering, drilling and production staff. Economic yardsticks are then extracted from the foregoing studies to provide a basis for development decisions.

PRODUCTION 73

There is, in practice, considerable overlapping and repetition of these exercises; for instance, the early tentative plans for the development well drainage pattern may be changed by the results of later appraisal drilling. Again, a favourable production system may emerge, from several feasible alternatives, through economic considerations of the projected operating costs balanced against the initial capital expenditure. This evaluation and planning phase therefore requires a fully integrated approach by the individuals of the different disciplines.

Pre-development Studies

A typical study organisation chart is shown in Figure 3.30 which broadly illustrates how the different engineering specialists integrate their specific types of

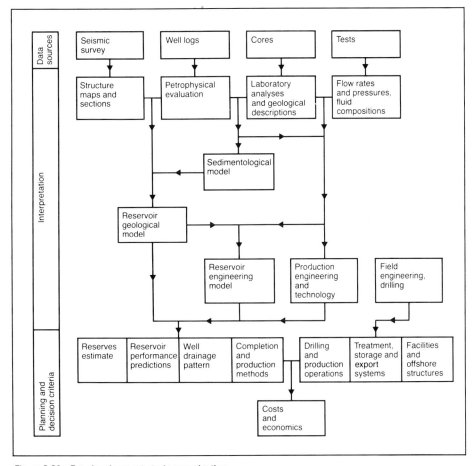

Figure 3.30 **Pre-development study organisation**

information to arrive at the answers which will determine an optimum development scheme. Each field or reservoir has its own characteristics which extend from subsurface features, such as oil density and rock properties, to surface conditions, such as terrain, environment and, in the case of offshore discoveries, water depth and distance from shore. All these aspects must be taken into account before development can be planned and implemented.

The investigation starts with the petrophysical assessment of reservoir rock properties and fluid content obtained from measurements made in the discovery well. Potential reservoir intervals are selected for testing the production of oil or gas; flow rates and reservoir pressures are measured and samples collected for laboratory analysis. These types of data are the particular concern of the reservoir engineer and the production engineer/technologist.

Meanwhile, the reservoir geologist collaborates with the exploration geologists and seismic interpreters to assemble a geological picture which will incorporate the results of the discovery well. Almost certainly the density of the seismic coverage will have to be increased and appraisal wells drilled to refine the structural interpretation and investigate accumulation conditions over a wide area of the potential field. The first one or two wells, sited by the study team, normally will be dedicated to establishing reasonably confident estimates of the oil or gas reserves of the field. Subsequent wells may be required to investigate such features as variations in rock properties or to follow up conditions found in the earlier wells that were unexpected.

To evaluate homogeneities in the rock fabric and to establish the physical properties of the rock, coring may be required during the appraisal phase. Further production testing will also be undertaken in order to provide more representative information from the reservoir as a whole.

Onshore appraisal wells can be used later for production, but offshore appraisal wells are usually abandoned after testing. Current research developments of underwater completion methods may, in the future, provide the means for using such wells for production.

The geologist constructs the reservoir geological model consisting of a series of maps, cross-sections and other illustrations which define the accumulation in a quantitative manner. That is to say, the heterogeneous rock types of the reservoir are not merely classified and their distribution indicated; their dimensions are also calculated and numerical values ascribed to their properties of porosity and permeability, together with volumetric estimates of their contained fluids.

The geological model provides the reservoir engineer with a framework for evaluating reservoir pressure data, fluid densities and viscosities, and flow rate behaviour obtained from production tests. Guided by experienced assumptions about the natural production mechanisms of the reservoir and the appropriate production methods required, the reservoir engineer predicts the performance of the reservoir in terms of flow rates over long periods of time. He also studies the theoretical effects of applying artificial methods of improving recovery, such as water injection. In the past, the flow of fluids within the reservoir during production was calculated on the basis of averaged rock properties, but this did not satisfactorily account for the effects of actual rock heterogeneities which occur. Within the last two decades computer programs have been gradually developed which will numerically simulate complex reservoir conditions and allow more realistic predictions of performance. Collectively, these manual and computer-assisted calculations constitute the reservoir engineering model.

Meanwhile, the production engineer extracts the information he requires to select the appropriate downhole production equipment and to define optimum casing designs and methods for completing the wells. He is concerned with stimulating inflow, if necessary, by acid treatment or hydraulic fracturing of the reservoir rock in the vicinity of the well bore. Forethought will also have to be given to problems such as minimising damage to the reservoir caused by the drilling process and to later production phase problems such as preventing formation sands from being washed into the well bore and excluding excessive water entry.

Once the subsurface aspects of the study are well advanced, the reserves have been estimated, the number of wells and their pattern established and a suitable production profile agreed, the provisional designs for the surface production facilities can be made. These cover gas and water separation equipment, gathering and storage systems with pumps and pipelines to a major terminal. At this stage, the studies will be aimed at feasibility and cost estimates for materials, equipment, design, construction and installation. The sequences and timing for drilling and production build-up will emerge from such practical considerations as drilling rig availability, materials ordering times, manpower requirements and local climatic conditions.

The profile of expected annual production and the development costs of the project form the basis for the economist's forecasts of profitability. In addition, he will have to estimate operating costs over the life of the project and make assumptions about future oil and gas prices. The effects of the current fiscal regime will have to be accounted for and possibly even forecasts of changes that may be imposed at later stages in the life of the project.

It will be appreciated that the several parts of such a complex study cannot be pursued in isolation. The geological model may show up anomalies in the initial petrophysical evaluation which will disappear through adjustment of certain factors in the calculations. The geological model may not at first fully account for reservoir pressure behaviour and require modification. The theoretical production profile could clash with field engineering feasibility and logistics. Finally, the economics may indicate marginal profitability, and alternative development schemes, less optimal than the first but still technically sound, may have to be worked up.

It also must be realised that the models and assumptions are fallible and that there is a strong element of subjective judgement being applied throughout the exercise. Risk can be reduced by organised planning but it cannot be eliminated.

Once a development project has been approved, all or parts of the pre-development study phase will be repeated in greater depth and detail to provide an actual development plan which will then be implemented. The project then moves into the design, construction and installation phase and the bulk of the effort will devolve on the field engineers.

Evaluation of Gas Fields

The development of a large natural gas accumulation is, in several respects, a special case. Oil can be produced, transported, refined and sold as products in the normal course of integrated oil industry activity. This is what the industry has been accustomed to do, virtually from its inception, and considerable infrastructure has been developed over the years to facilitate these activities. Until relatively recently, on the other hand, a significant gas discovery could only be developed if there was an adequate potential local or nearby market of industrial and perhaps also domestic users. For example, gas discoveries in the industrialised USA could be utilised in this manner through a huge network of pipelines which exist to transport gas from the wellhead to the consumer. In the Netherlands the Groningen gas discovery in 1959 and in the UK the North Sea discoveries in the mid-1960s could be linked to the existing town gas distribution systems, although not without considerable effort. However, large gas discoveries in areas remote from major markets could not be developed until the problem of long-distance transportation of natural gas was solved. This required (1) improvements in pipeline/compression technologies to move natural gas via pipelines over great distances or through deep waters, and (2) the development of gas liquefaction processes and transport over the high seas in purpose-built thermally insulated tankers. The transport of natural gas from Siberia to Western Europe and the movement of gas under the North Sea are examples of improved pipeline technologies being deployed. The transport of liquefied natural gas (LNG) from Algeria to the European continent and from Brunei to Japan are examples of the latter approach to moving remote gas supplies to major markets. The vast majority of internationally traded gas moves to market via pipelines.

Such projects can be extremely costly. A critical factor in such schemes, particularly in the case of LNG projects, is the size of the gas accumulation(s), the gas reserves, to support the huge investments required. For example, a typical

LNG scheme would require recoverable reserves of above 200×10^9 cubic metres (7×10^{12} cubic feet). A market has to be established before development can be contemplated. Potential industrial users have to be identified and committed to the project, and they, in turn, seek a firm assurance that supplies will be uninterrupted and adequate over a long period of time, say 15 to 20 years. Furthermore, the capital cost of the pipeline or of a liquefaction plant and the building of a dedicated fleet of tankers (five or six vessels) has to be covered before any gas is sold, so outside financing may have to be sought. The financial institutions also require very firm assurances of reserves and long-term gas deliverability.

These requirements have an effect on the pre-development study phase. The confidence level on the gas reserves estimate has to be very high and provision made for maintaining a steady level of deliverability by phasing development and by introducing compression facilities at appropriate times in the project life. Also, the specialists responsible for the siting, design and construction of the pipeline or liquefaction plant become involved at an early planning stage; marketing and, in the case of LNG, marine transport advice will also be required at the inception of the project study. As these large schemes have an important impact on both the producer and the consumer countries, governments also become heavily involved, particularly on matters of safety and the environment.

The economic development of smaller gas accumulations may be possible if the reservoir fluid has a reasonably high content of condensate. In this situation, the hydrocarbon compositions are such that, under certain conditions of temperature and pressure, the condensate will exist in the gaseous phase (and therefore mixed with the gas) in the reservoir. Below a certain pressure two phases, gas and liquid (condensate), exist. Once gas is produced to the surface, the condensate can be separated as a liquid fraction that can be handled like a very light oil. Gas re-cycling is a procedure to produce condensate and re-inject the processed lean gas into the reservoir. This has the advantage of storing temporarily unsaleable gas and maintaining reservoir pressure. The latter is desirable to prevent liquids forming and separating from the gas in the reservoir, an occurrence known as retrograde condensation, where subsequent recovery of the liquids under most reservoir conditions will not be possible.

Underground gas storage is another method of temporarily retaining gas that cannot be immediately utilised. Near-surface underground caverns close to the point of sale offer one solution; another is re-injection into a reservoir rock formation near the point of production.

All these systems require specially tailored pre-development studies along the general lines discussed above.

Well-Site Operations Engineering

Drilling Logs

During the drilling of a well, records of information against depths are obtained. The procedure is termed logging. The presence of oil, gas or salt water in the formations being drilled may be revealed as traces in the drilling mud which is continuously being pumped down the drill pipe, through the bit and back through the annular space to the surface. Traces of oil will fluoresce under ultra-violet light, and gas can be detected by blowing air through a mud sample and passing it over an electrically heated platinum filament. Any inflammable gas present will cause a measurable rise in temperature. A rough analysis of gas constituents (methane to pentane) can be made, and scanning equipment to give early warning of poisonous hydrogen sulphide (H_2S) is an important safety feature.

Drill-cuttings are returned to the surface by the mud and are separated, washed, dried and examined under a microscope by the well-site engineer or geologist. Traces of hydrocarbons can be detected under ultra-violet light, either directly or after extraction with a solvent such as chloroform. Geological descriptions are made of the cuttings and these provide the basis for a lithological log in which standard symbols are used to represent different rock types, such as sandstone, shale, coal and limestone.

Both the mud log and drill-cuttings record may require a depth correction as there is a time lag while the mud returns to the surface from the bottom of the hole. During this time lag additional footage will have been drilled, and this can be appreciable in fast-drilling formations at considerable depth.

Logging by these methods does not require elaborate equipment and provides on-the-spot information for making immediate drilling decisions and for later interpretation. For example, a suitable formation in which to set casing can be recognised and drilling halted in time, or oil indications may prompt the decision to take a core.

Drill-cuttings can be examined later in the laboratory for their fossil content. Pollen and micro-fauna, extracted from the samples, can be classified to indicate the geological age of the formations. Work of this nature is particularly important in exploration and appraisal wells where the geological uncertainty is still high and re-interpretations are most likely to be required. In development wells, this information is of less value but it can be argued that the records should be kept in case the hole should be lost for mechanical reasons before further records and measurements can be taken (see Logging and Formation Evaluation under Petrophysics, p. 83).

Wireline Logs

At suitable intervals during drilling a suite of wireline logs is taken. With the drill string out of the hole a recording device known as a "sonde", is lowered to the bottom on an electrical multi-conductor cable (Fig. 3.31). While the sonde is pulled up steadily, measurements of particular rock properties against the corresponding depths are recorded. The properties measured are electrical, acoustic or radio-active and different sondes are used, some combining several different sets of readings.

The particular property recorded is not of direct significance but can be related to reservoir properties such as porosity or the fraction of the pore space which is saturated with hydrocarbons. The type of rock can also be identified fairly accurately from the logs.

Further detail on wireline logs and on the techniques used to interpret them is given in the section under Petrophysics (p. 83).

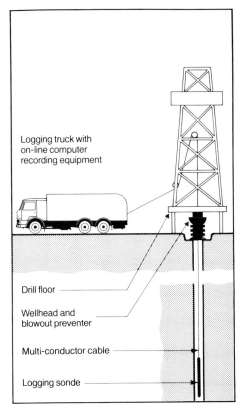

Figure 3.31 **Wireline logging operations**

Coring

When rock properties have been derived from logs, it is desirable to check the results against values accurately measured on actual formation samples from corresponding depths. Samples big enough for laboratory examination are obtained from known depths by "coring". The drilling bit is replaced by a "core barrel" and a "core head" by means of which a cylinder of the formation being drilled passes through the core head into the core barrel where it is retained and brought to the surface (Fig. 3.32). In some formations the cores are badly broken and in soft sandy formations they may be washed away by the drilling fluid and a representative sample cannot therefore always be obtained. Moreover, progress is usually slower than when drilling normally, and the drilling string has to be pulled more frequently because of the limited length of core barrel. Coring is thus an expensive and uncertain process. It is used in the early stages of field appraisal, more so than in exploration wells, where the location of formations of interest is not known until they have been passed, by which time it is too late to core them.

Figure 3.32 **Core samples.**

These disadvantages can be overcome to some extent by "side-wall" coring whereby small cores can be taken at any chosen depth in an uncased hole during periods when the drill pipe is out of the hole and in conjunction with a wireline log. Several hollow cylinders are shot into the wall by means of explosive charges from recesses in a steel cylinder, the "carrier". The charges are detonated by an electric current transmitted by a conductor cable on which the carrier is lowered. The cylinders are connected to the carrier by steel wires and are pulled out of the rock and to the surface with the carrier. Side-wall samples are, however, small and often badly contaminated by mud. Rock properties such as porosity and permeability will have changed due to the impact of the sampling bullet. Their use therefore is limited mainly to determining the lithology and age of formations and the presence or absence of oil, which assists in evaluating logs.

Cores and side-wall samples in conjunction with wireline logs are an invaluable source of information for the reservoir geologist in determining type of lithology, quality of reservoir rock and environment of deposition; and finally for the reservoir model that will be used as input in computer reservoir simulation work. In this respect coring is increasingly being used in development wells to provide the detailed information that is now required over wide areas of the reservoir.

Formation Tests

Although the above methods may give evidence of the presence of hydrocarbons, or even indicate the amount, they may not always be able to distinguish between oil and gas or yield direct information about the rate at which a reservoir formation can produce fluids. In addition, pressure information is required to determine the possible vertical extension of hydrocarbon columns and differentials between reservoirs with differing pressure regimes.

Prospective producing zones, particularly in exploration and appraisal wells, may therefore be tested soon after being encountered. There are three different techniques available, namely: the wireline repeat formation test, the drill stem test and the production test.

Wireline testing is by far the fastest and least costly method of gathering data on the fluid content of a prospective interval, its pressure, the fluid properties and productivity. The tool is run normally in open hole on a regular logging cable and hydraulically set at the required depth.

By means of a test probe which protrudes into the formation, fluid from the formation can enter a sample chamber which is subsequently sealed by closing a valve (Fig. 3.33). Pressure measurements are made during the test.

The principal advantage of this type of tool is that a number of intervals can be tested in a matter of hours. A disadvantage can be the limited size of the

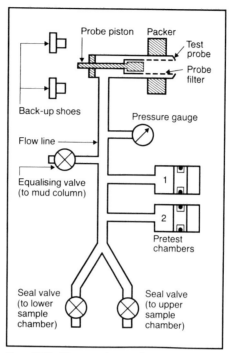

Figure 3.33 **Diagram of repeat formation tester**

sample, which therefore may not always be representative of the formation contents.

The principle of "drill-stem testing" is to isolate a part of the hole containing a possible oil- or gas-bearing formation and then to establish a direct connection between this part of the hole and the surface.

A "formation packer" and "drill-stem tester" are lowered on the drill pipe. The drill-stem tester contains valves that can be controlled from the surface so that no drilling fluid can enter the pipe while it is being lowered into the hole. First the packer is set above the section to be tested, sealing off the space between wall and pipe; then by slightly lowering the drill string the main valve in the tester is opened, thus bringing the isolated section of the hole into communication with the empty drilling pipe. Formation fluid then enters the pipe under the pressure in the formation, depending on which it will partly or completely fill the pipe. In the latter case the surface flow is controlled by valves.

After completing the test, which may last anything from a few minutes to a day or more, the drill pipe is sealed at the bottom by closing the main valve, and drilling fluid from the annular space around the drill pipe is admitted below the packer through the equalising valve. The packer can then be pulled loose and the whole assembly withdrawn from the hole.

Testing is not limited to formations close to the bottom of the hole or to uncased holes. It is possible to set packers both above and below a formation to be tested (straddle test) or to set them in the casing to test gun-perforated intervals.

A production test is a test with tubing and packer in a permanently or temporarily completed cased well. This is the most costly way of testing a well, but also the most complete one, since it may give information from which well and reservoir performances under future operational conditions can be predicted.

Petrophysics

Logging and Formation Evaluation

Wireline logs nowadays are taken in every well at convenient intervals. Since most types of log are only effective in open hole, they are run just prior to setting casing. They might also be run at intermediate stages of drilling a long open hole interval, in order to obtain at least a partial record in case the hole is lost for mechanical reasons. Again, it is often desirable to stop drilling and log over an interval in an exploration or appraisal well when there are indications of the presence of hydrocarbons from the mud log or drill cuttings.

Wireline logs measure electrical, acoustic and radio-active properties of rocks which can be indirectly interpreted in terms of rock type, porosity and fluid content.

Apart from identifying the presence of hydrocarbons, it is important to distinguish between reservoir rock and non-reservoir rock. Shales are composed of argillaceous or very fine grained minerals and are hence impermeable, thus forming the major category of non-reservoir rock.

Shales can usually be identified by the gamma-ray log. This records the natural gamma radiation of the formations, and shales generally have a much higher level than other rock types. Not all of the latter will be reservoir rocks, and other logs will be required to make the differentiation.

At the contact of impervious shales and permeable formations, variations in electrical potential may be observed in the borehole. A record of these potentials, the spontaneous-potential log, will also assist in delineating permeable formations. Moreover, the magnitude of the potential variations will often allow the resistivity of the formation water to be estimated.

The gamma-ray log is an example of a nuclear log. It is often run in combination with a neutron and density log which serves to estimate the porosity of a formation. The neutron sonde contains a radio-active source that emits fast neutrons which penetrate the formation and are there slowed down by collisions with atomic nuclei. At each collision they lose energy and are finally captured by the nuclei of formation atoms which thereupon emit gamma radiation. A detector mounted at a short distance above the source measures the intensity of the signals. The neutron log responds primarily to the amount of hydrogen present in the formation. Thus, in clay-free formations, the pores of which are filled with water or oil, the neutron log reflects the amount of liquid-filled porosity. The density log or gamma log contains a source emitting gamma radiation. A detector measures the gamma radiation scattered back to the sonde by the formations. The signal recorded reflects the density of the formations, from which the porosity may be deduced.

Such nuclear logs can also be run in cased holes. The cement and casing steel will attenuate the recorded signals and for quantitative evaluation the logs may still be useful, but less accurate. They will serve mostly for correlation, however.

An "acoustic" or "sonic" log is another type of porosity log which is frequently run. It records the time that it takes an acoustic wave to travel over a certain distance through the formations. The speed of propagation of acoustic waves is much higher in solid rock than in fluids. Consequently, the recorded travel time will reflect the porosity of the formations.

Although the above logs will assist in determining the quality of possible reservoir rock they do not reveal whether the porous formations contain any hydrocarbons. To obtain this information, it is necessary to run resistivity logs which record the resistivity of the formations. The ability of a formation to conduct electric current varies inversely with the resistivity of the formation water and depends on the amount of water present in the pores, which is determined by the porosity and the fraction of the pore space occupied by non-conductive hydrocarbons. Provided that porosity and formation water resistivity are known, a resistivity log will enable the hydrocarbon content of a formation to be calculated.

Determination of a rock's electrical resistivity for water saturation determination requires measurement in open hole. In a cased hole, water saturation can often be determined (though somewhat less accurately) with a pulsed neutron log. High-energy neutrons are generated by physical means in bursts of short duration. These neutrons are rapidly "thermalised" due to collisions with elements. A cloud of thermal neutrons decays due to capture reactions with nuclei with corresponding emission of gamma rays. The decay rate is measured, which depends on the rock's capture cross-section, which latter is primarily dependent on the chlorine content. This chlorine is present as sodium chloride in the formation water. Hence, if the salinity of the formation water is known, the rock's water saturation can be calculated. The log is frequently used for monitoring the water-front movement in producing reservoirs. Differences in capture cross-section between oil and gas may also allow monitoring expansion of gas caps in producing reservoirs.

PRODUCTION

In addition to the logs described above, a dipmeter log is frequently run to investigate structural dip and sedimentary features. The tool uses four pads, 90° apart, mounted on hydraulically actuated arms. On each pad a micro-resistivity device records a curve. The four curves are correlated and from depth shifts between the curves, angle and direction of formation dip can be determined.

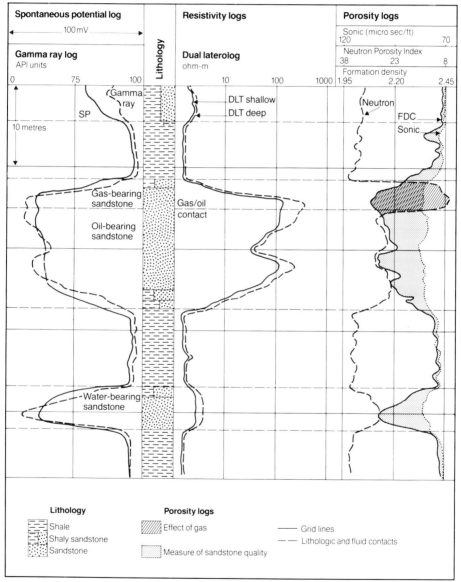

Figure 3.34 **Response of different logs.** Overplotting the various types of logs reveals lithologic changes and gas/oil contacts.

Continuing improvement of logging tools and evaluation techniques allows reliable determination of several different rock properties. For this purpose combinations of several logs are run (Fig. 3.34). Rock lithology, i.e. the constituent minerals and their volumetric proportion, can often be reliably determined, together with their porosity, hydrocarbon saturation and kind of hydrocarbon (oil or gas).

From the density log and the sonic log a rock's acoustic impedance (the product of density and specific sound velocity) can be calculated along the hole. In a seismic survey, seismic reflections occur when adjacent earth layers differ enough in acoustic impedance. Hence, logging can be used for identifying the reflection points in a seismogram, but also for quantitative calibration. By studying the influence of the presence of either gas, oil or water on the logged acoustic impedance, predictions can be made laterally on the seismic record on the extension of hydrocarbon-bearing intervals.

In the past, all wireline logs were recorded as curves, on film. To meet requirements such as high-density storage of large quantities of data and easy input into modern computer and data handling systems, and to facilitate fast transmission of data via telecommunication networks, well logs are now recorded on digital tapes.

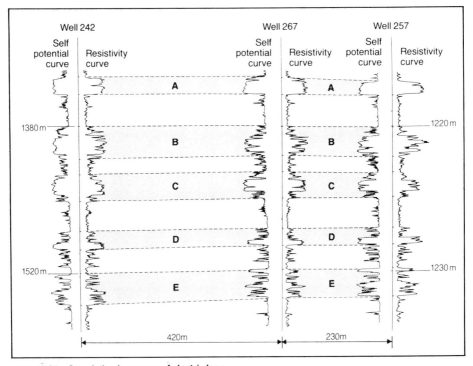

Figure 3.35 **Correlation by means of electric logs**

Production Geology

This is a branch of geology that is wholly concerned with the appraisal and development of oil and gas fields after the discovery has been made. The geologists work in an engineering environment and their basic aim is to assist in solving engineering problems. The methods used are the application of geological principles to explain well data in terms of both the external geometry of the reservoirs and their internal rock properties.

Operations

At the operational level, updating of the geological interpretation as drilling proceeds is mostly accomplished from wireline logs (electric, acoustic and radio-active). The log curves are "correlated", or matched, from well to well and a number of marker levels identified in each log (Fig. 3.35). From the correlations,

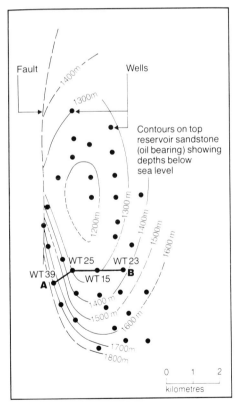

Figure 3.36
Structural contour map of an oilfield (simplified)

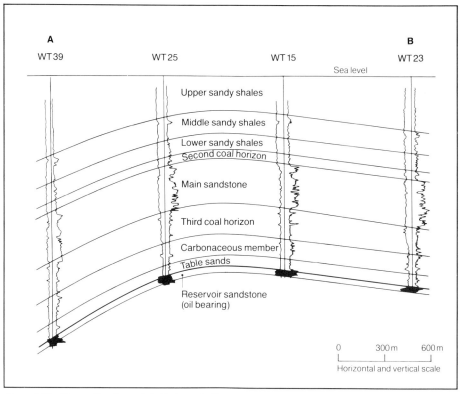

Figure 3.37   **The section through A-B shown in Figure 3.36**

the geologist constructs the structural contour maps (Fig. 3.36) and cross-sections (Fig. 3.37) to illustrate the three-dimensional distribution of the oil and gas in the accumulation. In addition, more detailed studies are carried out to describe the distribution and variation in lithology and reservoir rock properties (permeability, porosity) throughout the field. Using these data, the volumetric estimates of oil and gas can be calculated and further wells planned to delineate efficiently the size of the field and provide sustained production of oil and gas.

Fields under development require continuous attention and the geological interpretation at any one time must be regarded as a working hypothesis, subject to revision as new wells are drilled. If a new well, on logging, is found to have missed its objective, it may be possible to plug back the hole and re-drill it as a sidetrack (Fig. 3.38). In this case the operations geologist must be on hand for a rapid revision to set a new target.

The question of reserves is most important and estimates are kept up to date, not only for the field as a whole, but for individual reservoirs and fault blocks.

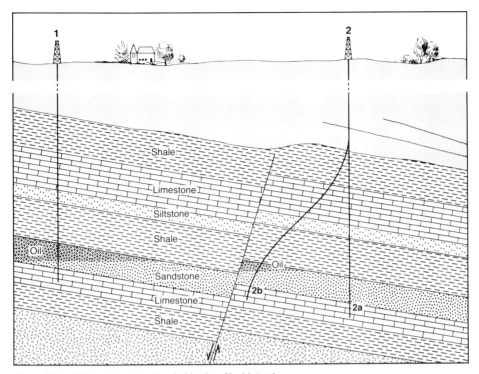

Figure 3.38 **Sidetracking: 2a missed objective; 2b sidetrack**

This controls, with some other considerations, the number of wells required for proper drainage.

Production Seismology

Within the past decade, data acquisition and processing techniques have so improved that it is now feasible to use seismology to assist the production geologist. In the past, for example, areas with complex faulting showed only blurred results and the seismic method was essentially an exploration tool for providing regional structural pictures, adequate for identifying prospects but not for revealing their details.

In the main, seismology still only provides structural interpretations. However, improvements to signal characterisation and resolution can, in favourable circumstances, give an indication of lithological variations within a formation. Fluid contacts can sometimes be recognised as horizontal anomalies, particularly in gas accumulations since the density and velocity contrast between gas and water is large and causes significant differences in the acoustic impedance, which is the product of sonic velocity and formation density.

Production seismology is now an established discipline and is particularly valuable in the appraisal phase of development when well control is limited. As mentioned in the section on Pre-Development Studies (p. 73), additional seismic lines are required to augment the original exploration coverage; lines as close as 500 metres are commonly used. Recently, techniques have been developed to interpret a three-dimensional grid system with spacings in the order of 50 to 100 metres. Under suitable conditions, remarkable detail and accuracy can be achieved.

Reservoir Geology

In the past, reservoirs in the subsurface were frequently depicted as single layers of homogeneous rock. In reality this is seldom the case. Their depositional and diagenetic history generates a wide range of heterogeneities which control the distribution of porosity and permeability and thereby affect reservoir performance and fluid flow behaviour. It is the task of the reservoir geologist to subdivide the reservoir and characterise its constituent components in terms of geometry, reservoir properties, and impermeable layers which cause barriers to the flow. Of prime importance to primary and secondary recovery is a proper understanding of the lateral and vertical distribution of permeability.

Carbonate reservoirs are particularly sensitive to diagenesis with consequent effects on the reservoir properties. The factors controlling their formation are different in significant aspects from those controlling the deposition of clastics (sands and shales) and this has given rise to carbonate geology as a separate discipline from clastics sedimentology. In terms of reservoir geology, the problems are similar and the following outline of reservoir geological modelling applies to both carbonates and clastics.

Sedimentology forms an important step in the development of reservoir models because of the fundamental control exerted by the depositional environment on the size, geometry, orientation, continuity and porosity/permeability characteristics of reservoirs. It is for this reason that the first step in any reservoir modelling exercise is a detailed analysis of the sedimentary facies (sediment associations) in the cored reservoir. The results are then integrated with well log data so that an overall picture of the areal and vertical variation in reservoir quality can be established on a field-wide scale. The degree of confidence in the validity of the model is dependent upon the available data base, particularly well density, well log quality, core control, pressure and production data. In a recently discovered field, such as found in the North Sea in the 1970s, the initial data base was limited to a few appraisal and even fewer production wells. Reservoir models at this stage of field development are of necessity highly conceptual and draw heavily on modern analogues of similar reservoirs and experience from fields in other parts of the world.

PRODUCTION 91

As the well density increases in later stages of field development, so the data base expands to allow more detailed subdivision of the reservoir and characterisation of its individual components. These are quantified and their precise extent painstakingly mapped.

The final step in reservoir modelling is taken by the reservoir geologist and engineer working together as a team. They have to reach a realistic compromise between the natural complexity of the reservoir and the capabilities of the computer in order to arrive at a representative and usable computer reservoir model.

Reservoir Engineering — Primary and Secondary Recovery

Reservoir Fluids and Characteristics

As was stated earlier, an oil reservoir is a porous sedimentary rock formation, capped with a layer of impermeable rock through which liquids and gas cannot pass (Fig. 3.39). The shape of the reservoir must allow oil (or gas) to accumulate, and the cap rock is essential to prevent them from migrating further upwards. Because of capillary forces, some of the water originally in the pores could not be displaced by the accumulating hydrocarbons. This immovable water is called connate or interstitial water (Fig. 3.40). The volume of all the pores and openings in a reservoir rock (porosity) is normally expressed as a percentage of the total volume of rock. The larger the porosity, the more oil can be stored; it ranges from 10 to 30 per cent of total rock volume.

If oil is to flow through the reservoir, there must be a free connection between

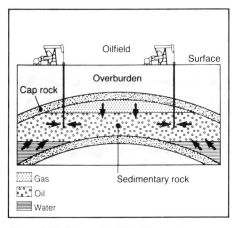

Figure 3.39 **Reservoir with bottom water and a gas cap**

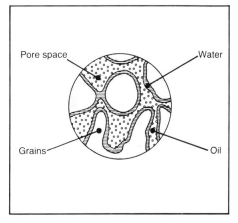

Figure 3.40 **Reservoir on a microscopic scale with connate or interstitial water**

the pores, and the ability of a rock to allow the passage of fluids (called permeability) depends upon the size of the inter-connecting channels between the pores.

Both porosity and permeability vary throughout a rock formation; consequently, wells in different parts of a reservoir may have widely different production rates. Oil reservoirs are encountered from quite near the surface to as deep as 6,000 metres, and pressure can vary from near atmospheric for reservoirs close to the surface to more than 15,000 psi (1,000 bars) for deep reservoirs.

The oil contains dissolved gas, the maximum amount depending on reservoir pressure and temperature. If the oil cannot dissolve any more gas under the prevailing reservoir pressure and temperature conditions, it is said to be saturated; excess gas will then move to the top of the reservoir, where it will form a gas cap. If the oil can dissolve more gas under these conditions, it is described as undersaturated and no gas cap is initially present.

Oils vary in nature from the very heavy, viscous type (with a specific gravity close to that of water and a viscosity of 10 times to 100,000 times as high), usually found in shallow reservoirs containing little or no dissolved gas, to the extremely light, low-viscosity type found in deep reservoirs, containing a large volume of dissolved gas. The less viscous the oil, the more easily will it flow through the interstices of the reservoir rock to a well.

Natural Production Mechanisms

If oil is to move through the reservoir rock to a well, the pressure under which the oil exists in the reservoir must be greater than that at the well bottom. The rate at which the oil moves towards the well depends on the pressure differential between the reservoir and the well, permeability, layer thickness, and the viscosity of the oil.

The initial reservoir pressure is usually high enough to lift the oil from the producing wells to the surface, but as the oil and gas are produced the pressure decreases and the production rate starts to decline. Production, although declining, can be maintained for a time by naturally occurring processes such as expansion of the highly compressible gas and influx of water.

The major natural production mechanisms are water drive, solution gas drive and gas cap drive.

Water Drive. Most oilfields are underlain by water (aquifers) and as the pressure in the oil reservoir drops, the water starts flowing and enters the reservoir, as a result of the expansion of the water and the reduction of pore volume (compaction of the rocks) (Fig. 3.41). This water encroachment maintains the reservoir pressure to a greater or lesser extent, depending on the size of the

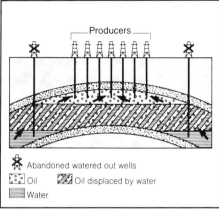

Figure 3.41 **Field after production for many years, with strong waterdrive**

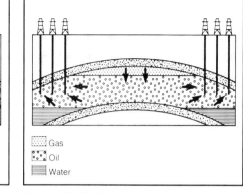

Figure 3.42 **Field after production with gas cap expansion and weak waterdrive**

aquifer. As the volume of the aquifer is often many times greater than that of the oil reservoir, it delivers a substantial amount of energy to an oilfield.

Solution Gas and Gas Cap Drive. As oil is produced, and reservoir pressure drops, gas is liberated from the oil. The gas forms small bubbles, which gradually displace the oil (solution gas drive) (Fig. 3.42). The size of individual gas bubbles increases until they join together to form a continuous phase and the gas begins to flow. Part of the liberated gas moves to the producing wells, but some of it segregates to the gas cap at the top of the reservoir or, if no gas cap was present initially, forms a secondary gas cap. Advancement of the primary or secondary gas cap results in displacement of the oil by gas (gas cap drive). When a large gas cap is present (or formed), its high compressibility makes it a useful source of energy for the production of oil.

Reservoirs rarely fit neatly into any one of these categories. In most of them some or all drive mechanisms play a part.

Natural production mechanisms contribute to what is known as primary recovery. Depending on the type of oil, the nature of the reservoir and the location of the wells, the recovery factor (the percentage of oil initially contained in a reservoir that can be produced by these mechanisms) can vary from a few per cent for a solution gas drive to as high as 30 to 35 per cent for a water or gas cap drive. Worldwide, primary recovery is estimated to produce on average some 25 per cent of the oil initially in place.

Secondary Recovery

Over the years, petroleum engineers have learnt that the application of techniques for maintaining reservoir pressure can yield more oil than is obtained by primary recovery alone. By such techniques (known as secondary recovery) the reservoir's

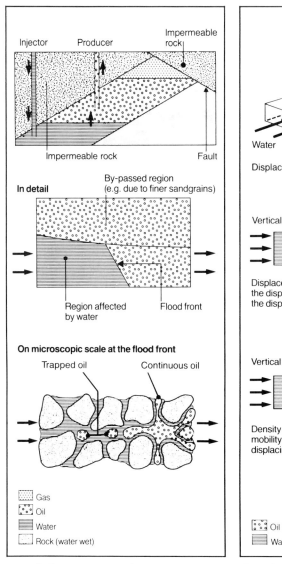

Figure 3.43 **Displacement of oil by water**

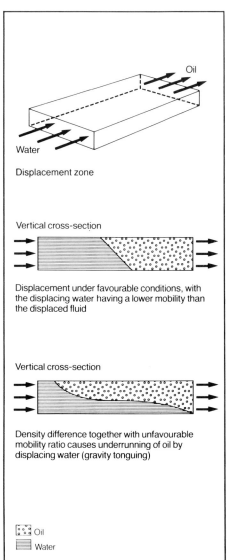

Figure 3.44 **Theoretical example to show the effect on displacement efficiency of viscosity and density differences between oil and water**

natural energy and displacing mechanism, which is responsible for primary production, is supplemented by the injection of water or gas. The injected fluid does not, however, displace all the oil. An appreciable amount is trapped by capillary forces in the pores of the reservoir rock and is bypassed. This is known as residual oil, and it can occupy from 20 to 50 per cent of the pore volume. Moreover, because of permeability variations, the injected water may bypass certain oil-bearing regions (Fig. 3.43).

The total efficiency of a displacement process depends not only on the number and location of injectors and producers and the reservoir characteristics (permeabilities and residual oil), but also on the relative mobilities of the displacing fluid and the displaced oil. If the mobility ratio is less than a factor of one (that is when the displacing fluid has a lower mobility than the displaced one), sweep or displacement efficiency will be high and a large amount of oil will be moved. An example is the displacement of a light, low-viscosity oil by water. When the mobility ratio exceeds a factor of one (that is when the displacing fluid has a higher mobility than the displaced fluid), the sweep will be less efficient. Because of the difference in density of the two fluids, gravity segregation generally occurs, and the higher mobility of the displacing fluid will cause it to flow faster than the oil. These effects (Fig. 3.44) cause earlier breakthrough of the displacing fluid and so reduce the efficiency of the process.

The viscosity of the displacing fluid used in secondary recovery operations is most important. Ideally, it should not be significantly lower than that of the fluid being displaced. This is in fact the case with water in a light-oil reservoir, as water

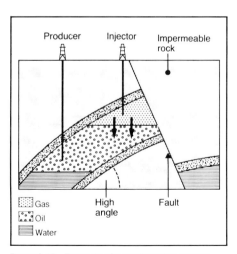

Figure 3.45 **Example of a gas injection project in a steeply dipping reservoir with gas/oil gravity segregation**

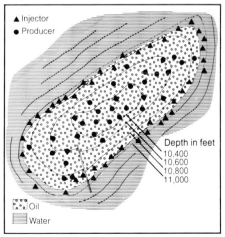

Figure 3.46 **Areal view of a reservoir with water injection wells placed at the periphery**

and light oil have approximately the same viscosities. Natural gas, however, has a viscosity much lower, and thus a mobility higher than that of any oil; consequently, when used as a displacing fluid, it tends to bypass the oil. Gas is usually considered only if the reservoir is steeply dipping (Fig. 3.45) and/or has a high permeability. Under these conditions, displacement of oil by gas is dominated by gravity. Even where conditions are less favourable, gas injection is sometimes required to conserve the gas temporarily. Since gas is valuable, and water is often more efficient in displacing oil, water injection has become the more conventional recovery process and the one most widely used.

A fundamental consideration in designing an effective and efficient secondary recovery project is how to displace oil from as large a volume of the reservoir as possible. Injection wells may be located on the periphery of the reservoir or in a pattern of various configurations, dependent on reservoir fluid and fluid characteristics. Figure 3.46 shows an example of a peripheral water-injection project.

The major practical problems experienced in carrying out a secondary recovery operation result from the stratification of reservoirs and variations in permeability, which make it difficult to control the injected water so as to avoid trapping of oil. In modern projects wells are carefully monitored, and even completed in such a way that selective injection into, and selective production from, sublayers in the reservoir is possible. In this way, optimum use is made of a displacing fluid.

Recovery Factors

The proportion of oil in place that can be produced by different recovery processes varies widely. This is due to a number of factors, including the viscosity, gas solubility and gravity of the oil (Table 3.2); the presence or absence of a gas cap; the presence and strength of an aquifer; the depth, pressure and degree of complexity of the reservoir; the permeability and porosity of the rocks.

In Table 3.2, the low values for each oil type are mostly associated with low-gas-solubility oils in unfavourable reservoirs. The high values relate to high-gas-solubility oils in favourable reservoirs. Primary recovery can be even higher than indicated when a strong aquifer is present; in such circumstances a secondary recovery operation is not attractive.

Table 3.2 **Range of recovery factors for various types of oil**

Oil type	Primary recovery (% of oil in place)	Secondary recovery (% extra of oil in place)
Extra heavy	1– 5	–
Heavy	1–10	5–10
Medium	5–30	5–15
Light	10–40	10–25

Reservoir Engineering — Enhanced Oil Recovery

Enhanced oil recovery is the description applied by the oil industry to non-conventional techniques for getting more oil out of subsurface reservoirs than is possible by natural production mechanisms or by the injection of water and gas. The oil not producible, or left behind, by these conventional recovery methods may be too viscous or too difficult to displace. It may also be trapped by capillary forces in the flooded parts of the reservoir or bypassed by the injected water or gas. In general, the aim of enhanced oil recovery techniques is to recover more oil by improving the displacement efficiency.

The terms primary, secondary and tertiary (commonly used in the past) indicate the order in which these recovery processes were originally applied. Today, secondary and tertiary recovery processes are sometimes applied from the start of production. Consequently, the term "conventional" is nowadays preferred for primary and secondary processes, and "enhanced" for tertiary processes. Enhanced oil recovery techniques can be conveniently subdivided into three categories: thermal, miscible and chemical.

Thermal processes aim to recover more oil by reducing viscosities by injecting or generating heat in a reservoir.

In *miscible processes*, the displacing fluids mix with the oil in such a way that there is no sharp interface between the injected and displaced fluids. Consequently, the capillary trapping capacity is reduced since there is no interfacial tension between the fluids, and more oil can be mobilised.

In *chemical processes*, fluid components are chemically altered, or chemicals are added to the displacing fluid to change its physico-chemical properties and those of the oil. The main aim is to reduce capillary forces and/or to increase the viscosity of the displacing fluid, thus improving displacement efficiency and recovering more oil from the reservoir.

Thermal Recovery

Principles. All current applications of thermal recovery are aimed at producing heavy oil with viscosities and flow resistance from 100 to 100,000 times greater than water. The pronounced effect of an increase in temperature on oil flow properties such as viscosity is illustrated in Figure 3.47. This shows that oil with an API gravity of 12° (density close to that of water) will have a viscosity of 1,000 centipoises at 50°C, but that this viscosity will be reduced by a factor of 500 to 2 centipoises if its temperature is raised to 250°C. It is also apparent from Figure 3.47 that, in general, this effect is more marked with heavier oils than with light or medium oils or water.

In addition to reducing oil viscosity and improving displacement efficiency, the

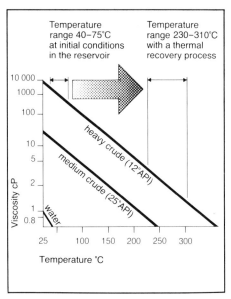

Figure 3.47 **Effect of temperature on crude oil viscosity**

application of heat may have other beneficial effects on the recovery processes. It may, for example, cause lighter components to evaporate, or promote swelling of residual oil. These effects do not necessarily all occur at the same time, nor are they equally effective in all reservoirs.

Heat Generation. Two principal methods are used to generate and transport heat for thermal recovery operations, namely surface generation and underground *in situ* combustion. By far the more common is surface heat generation, in all cases using water in one form or another, to transfer heat to the subsurface reservoir.

Water is not only cheap and readily available, it is also an excellent heat carrier. It has a specific heat among the highest of all available fluids. The additional heat required to effect the change from water to steam is called heat of vaporisation or latent heat. This latent heat can be high, especially at lower pressures, and this is one of the advantages of using steam instead of water.

Heat Transfer. Heat flows from hot to colder parts of a reservoir, and it does this in two ways:

Conduction — in which heat passes through the rock or the oil without movement of fluids, and

Convection — in which heat is transferred by relative movement of the fluids flowing through the reservoir.

Energy Balance. The ratio between the energy or fuel required for a thermal recovery project and the additional oil produced is a direct measure of a project's attractiveness. For steam injection projects, this can be expressed as the ratio of additional oil produced to the amount of steam injected (e.g. in barrels of oil per tonne of steam). For all practical purposes, 0.6 barrels of oil is required to convert one tonne of water into steam, and so more than 0.6 barrels of oil must be produced per tonne of steam injected.

The energy balance depends primarily on reservoir characteristics, heat losses and on the prevailing displacement mechanism. Only a small fraction of the energy injected is actually used to heat the oil in the reservoir. For example, one cubic metre (1,000 litres) of reservoir rock has to be heated to the same temperature as the 100 to 200 litres of oil it may contain. Moreover, heat losses to the underlying and overlying formations play an important role. The reservoir must therefore be thick enough for the process to be efficient.

The energy efficiency of thermal processes is reduced not only by heat losses to adjacent formations, surface facilities and injection wells but also by losses due to production of hot drive fluids. And, since one of every two or three barrels of oil produced may be needed as fuel for a thermal recovery project, these losses are a critical factor. For screening purposes, thermal projects with yields lower than 1.2 barrels of oil per tonne of steam are not normally considered to be attractive.

At present, depth and pressure are still major constraints in thermal projects. Excessive heat losses and mechanical problems are liable to occur in injection wells at depths greater than 1,000 to 1,500 metres and temperatures above 320°C. However, downhole steam generators, for generating steam at the bottom of an injection well, are being developed and field tested.

Methods. Three main thermal recovery techniques can be distinguished, and are discussed in the following pages. They are hot water drive, steam drive and steam soak.

Hot Water Drive. Hot water drive used as a follow-up to conventional cold water injection in a heavy-oil project has probably been tried out for almost as long as cold water flooding, but its earlier applications are not documented. As an injection fluid, hot water is operationally simpler than steam, but the latter is more effective as a heat carrier due to its latent heat. In the reservoir, steam maintains a constant temperature (its boiling point at reservoir pressure) until all of it has condensed into water, all the while giving up its latent heat to the cooler reservoir rock and fluids. In similar circumstances, the temperature of hot water drops steadily from the very start. As a result, hot water has a lower displacement efficiency than steam.

Where it is impracticable to apply steam (for example where the fresh water required for steam is incompatible with the reservoir fluids/formation and causes plugging, or where fresh water is not available, or in reservoirs with pressure so high that steam temperatures become excessive), hot, and sometimes also saline, water is occasionally injected.

Steam Drive. Because of its relatively low density and viscosity, steam tends to bypass oil along the top of the reservoir. This tendency is greatly reduced, however, by the fact that steam condenses as it releases heat to the colder parts of the reservoir rock and fluid. A schematic two-dimensional illustration of a steam drive is shown in Figure. 3.48. The steam not only reduces the viscosity of the oil, by increasing its temperature, but may also cause it to vaporise, and when the steam condenses a hot, low-viscosity oil is formed in front of the condensing zone.

In laboratory process experiments, recoveries of between 60 and 100 per cent of oil in place have been obtained with steam drive, though in actual practice (with economic constraints) reservoir heterogeneity and heat losses normally prevent the attainment of such high recovery levels. Even so, in suitable reservoirs (for example where gravity segregation in dipping reservoirs or compaction play a significant role), recoveries of up to 60 per cent of oil in place may be expected,

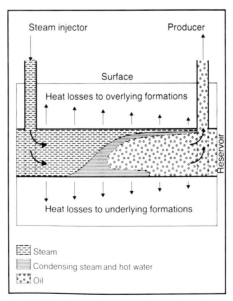

Figure 3.48 **Reservoir cross-section: schematic representation of steam drive**

with oil to steam ratios of 1.2 to 4.0 barrels per tonne of steam. This compares very favourably with the primary recoveries from shallow heavy-oil reservoirs of 1 to 10 per cent of oil originally in place.

A potential application of steam drive is in the enhanced recovery of residual light oil. The mechanism of this process is that oil evaporates when exposed to steam, and the resulting vapour of light components is transported (along with the steam) to colder parts of the reservoir. The trapped oil is stripped by steam until a non-volatile residue is left. An oil bank forms ahead of the steam-condensation front and is driven to the producing wells. Although this method seems technically feasible as a means for recovering light oil left behind by conventional recovery methods, its energy balance is critical and in most cases unfavourable. Unless a cheap and otherwise unattractive energy source is available, therefore, steam drive does not seem a suitable method for enhanced recovery of residual light oil.

Steam Soak. The steam soak process, often referred to as cyclic steam injection (or, more colloquially, "huff and puff") was developed in the late 1950s by Compañía Shell de Venezuela. There are three stages in the application of this process, as illustrated in Figure 3.49. First, a quantity of steam, usually from 500 to 5,000 tonnes, is injected. This is followed by a waiting ("soak") period of one to two weeks, during which the heat is transferred by conduction and convection and the oil viscosity is reduced, and then by a production period of one to two

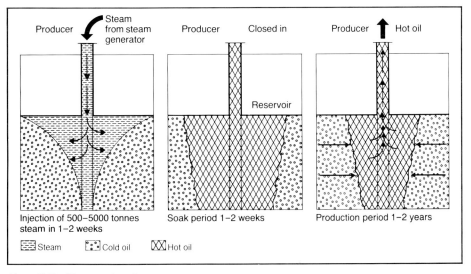

Figure 3.49 **Steam soak cycle**

years. The heated rock serves as heat exchanger for the cold oil flowing into the hot zone around the well.

As the producing well cools down, the oil viscosity increases, and the production rate drops until a point is reached when another steam soak cycle is necessary. These cycles can be repeated until the production rate is no longer economically viable. Steam soak is essentially a method of improving well productivity, and thus always requires an additional displacement mechanism. In fact, steam soak will tend to activate "natural" drive mechanisms (e.g. compaction of the rock, solution gas drive or gravity segregation) which may be dormant under "cold" conditions. Also steam soak is often applied in combination with steam drive. Oil yields of 5 to 50 barrels per tonne of injected steam can be obtained by the steam soak method, for which only a relatively small part of the reservoir has to be heated.

Underground Generation of Heat. The principle of what is usually called *in situ* combustion (or fire flooding) is to generate heat by injecting air, and burning part of the oil in the reservoir. The main advantage is that oil is burnt that would otherwise be left behind. The remainder evaporates and moves with the combustion gases towards the producing well. The vaporised hydrocarbons condense in the colder parts of the reservoir and form an oil bank, as shown in Figure 3.50. As air is injected, the combustion zone (with a temperature of 400 to 800°C) moves through the reservoir.

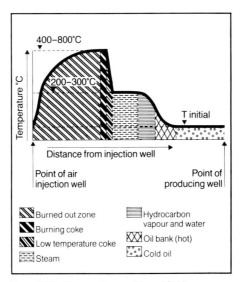

Figure 3.50 **Temperature profile and fluid distribution in a reservoir during laboratory test of an *in situ* combustion recovery process**

Two principal techniques have been tested, both in the laboratory and in the field, namely forward and reversed combustion. In forward combustion, the combustion front moves in the same direction as the injected air, combustion gases and fluids. In the reversed combustion process, the front moves against the flow of injected air, combustion gases and fluids. Forward combustion is now generally preferred to reversed combustion. A further development in *in situ* combustion was the simultaneous injection of air and water (wet *in situ* combustion). The water evaporates in the combustion zone, and this serves to scavenge the heat and bring about a more uniform heat distribution. Most projects now employ some form of wet *in situ* combustion.

Although *in situ* combustion works well in laboratory experiments, mechanical and operational difficulties have been encountered in field tests. It is difficult to control the movement and direction of the combustion front, and mechanical failures have occurred in production and injection wells as a result of extremely high temperatures and corrosive combustion gases.

Miscible Recovery Methods

While the thermal recovery methods described above are mainly used to increase the recovery of heavy oils, the main objective of the miscible and chemical methods is to increase recovery of medium and light oils.

Principles. Fluids are said to be "miscible" (or "miscibility exists") when they are able to mix totally with each other in all proportions.

In miscible recovery processes, a drive fluid is used which mixes with the oil and forms a mixing zone, in which a gradual change in composition and properties from oil to drive fluid takes place in such a way that no sharp interface exists between them. Since there is no interfacial tension between the fluids, capillary trapping capacity is absent (Fig. 3.51). Hence miscible recovery processes can recover the oil left behind by conventional processes.

Two types of miscibility can be distinguished: direct miscibility in which the fluids mix in all proportions, and developed miscibility where the fluids are not directly miscible but develop miscibility as a result of component exchange between two fluids.

Processes of this kind, based on the injection into the oil reservoir of a suitable solvent or gas, have been studied since the early 1920s, and in theory they can recover all the hydrocarbons left behind by conventional recovery methods. But, since the miscible drive fluid is often more mobile and less dense than the oil it has to displace, it tends to bypass the oil by over-running it or fingering through it, thus leading to low displacement efficiency. Miscible processes are therefore best applied in dipping, highly permeable reservoirs, where the displacement

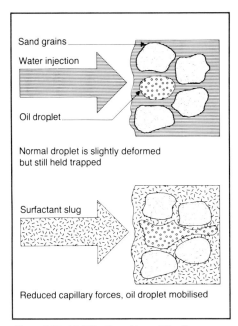

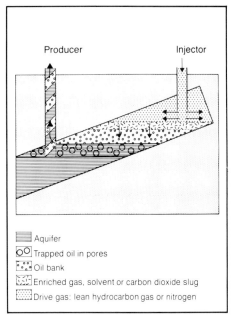

Figure 3.51 **Mobilisation of immobile oil**

Figure 3.52 **Gravity stable displacement by a miscible slug drive in a watered out reservoir**

process is assisted by gravity segregation. If there is a likelihood of over-running or fingering, as in most present projects, water can be injected and a repeated solvent/gas slug injection scheme applied. This technique may be used to redistribute the oil and to improve the displacement efficiencies.

There are five main miscible drive fluids available: hydrocarbon solvents, enriched hydrocarbon gas, high-pressure hydrocarbon gas, carbon dioxide, or nitrogen. The applicability of these fluids depends largely on reservoir pressure, availability and cost.

Hydrocarbon Solvents. Any solvent miscible with oil (e.g. LPG, kerosine or gasoline) can be injected into a reservoir. But, since the cost of such refined products is high, the miscible fluid is not injected continuously but in the form of a slug (typically 10 to 20 per cent of the reservoir pore volume) followed by a gas or water drive. Figure 3.52 illustrates gravity stable displacement by a miscible hydrocarbon solvent. This technique is effective at relatively low pressures and temperatures.

Enriched Hydrocarbon Gas. The hydrocarbon gas produced in association with oil is not usually miscible with the oil to any significant degree at prevailing reservoir pressures. If, however, this gas is enriched with intermediate hydrocarbon components such as propane, butane and pentane, it may become miscible. When the enriched gas contacts the oil, the intermediate components are condensed in the oil; a transition zone is formed between the gas and the oil, which at the downstream end of the transition zone may become miscible with the oil.

Because of the high cost of the hydrocarbons required, enriched-gas injection is also conducted as a slug process, the slug being driven through the reservoir by a lean hydrocarbon gas, usually methane (Fig. 3.52).

High-pressure Hydrocarbon Gas. Although a lean gas is not normally miscible at low pressures, it may be miscible (or become so) if reservoir pressure is high enough. In the case of developed miscibility, the injected gas contacts the oil and the intermediate hydrocarbon components evaporate from the oil into the gas. At the displacing front a rich gas mixture develops, and this may become miscible with the oil (Fig. 3.53).

The difference between this method and the one using enriched gas is the direction of the intermediate hydrocarbon component transfer. In the enriched-gas process, these components move from the gas to the oil; with the high-pressure lean gas, they move from the oil to the gas.

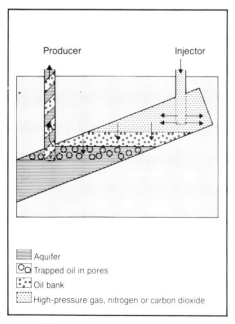

Figure 3.53 **Gravity stable displacement by a high-pressure gas, nitrogen or carbon dioxide drive**

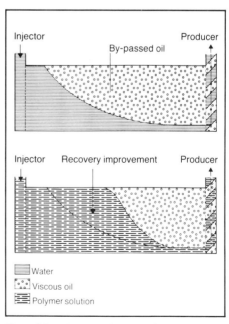

Figure 3.54 **Principle of polymer flooding**

Carbon Dioxide. At low pressures, carbon dioxide is not immediately miscible with oil. But, in the same way as lean gas, it can develop miscibility with suitable light oils at higher pressures. Even if carbon dioxide is immiscible and does not develop miscibility, injection of carbon dioxide could still have a favourable effect, since some of it dissolves readily in the oil, causing it to swell and decreasing its viscosity, both effects which can improve recovery.

Two methods of carbon dioxide injection have been tested. The first involves injection of a carbon dioxide slug displaced by a gas drive in a steeply dipping reservoir under gravity stable conditions (Fig. 3.52). The second method involves continuous injection (or large slugs) of carbon dioxide in relatively low-dip, low-permeability reservoirs. Apart from the technical problems encountered with carbon dioxide flooding, such as its corrosive effect on injection and producing wells, the availability of carbon dioxide is also a limiting factor. It is estimated that from 0.3 to 1 tonne is required for every barrel of oil recovered by this method. Large-scale projects depend on the availability of a natural carbon dioxide reservoir within a reasonable distance. For small-scale projects, an alternative possibility may be the extraction of carbon dioxide from industrial waste-gas or from hydrocarbon gas streams.

Nitrogen. Increased interest in the use of nitrogen developed when it became economically more attractive to manufacture and inject this inert gas than a hydrocarbon gas. At first waste gases such as stack gas, flue gas and exhaust gas were considered and field tested. But the problem with these is that they contain waste products such as nitrogen oxides and sulphur oxides which give rise to corrosion and pollution problems. Later, attention was given to producing nitrogen cryogenically.

Although nitrogen is not miscible with the reservoir oil at low pressures, it can develop miscibility at sufficiently high pressures. Its application is not, however, limited to miscible displacement processes; in view of its unlimited supply it can be used to replace non-miscible hydrocarbon gas injection in secondary recovery projects, or as a drive fluid for more expensive miscible slug systems.

Chemical Processes

Principles. Chemicals can be added to change the physico-chemical properties of the displacing fluid and those of the oil. The primary objective is to reduce capillary forces and/or to increase the viscosity of the displacing fluid and so improve displacement efficiency and recover more oil.

Chemical recovery methods employ polymers, surfactants or caustic soda. A new technique that has shown promise is the use of foam as a mobility-reducing agent in steam and miscible drive projects. This is very much in the experimental stage.

Polymer Flooding. If the oil in a reservoir is less mobile than the displacing water, then the water will tend to bypass the oil. This can lead to early production of water, poor sweep efficiency and low oil recovery. It is for such reservoirs that polymer flooding can be beneficial. A polymer (a long-chain molecule) dissolved in the injection water will thicken the water, reduce its mobility and prevent bypassing of the oil; consequently, oil recovery will be improved. The principle is illustrated in Figure. 3.54.

The ability of a polymer to thicken the injection water depends on the type of polymer and the reservoir conditions. A number of chemical structures have been suggested as suitable polymers for enhanced oil recovery, the principal candidates being polyacrylamides and polysaccharides.

Polyacrylamides. Polyacrylamides can be very effective where the salinity of the reservoir brine is below about 1 per cent (compared with 3.5 per cent in sea water). In reservoir water of much higher salinity (e.g. in the North Sea, where some reservoir brines contain up to 25 per cent sodium and calcium chlorides) polyacrylamides are unsuitable because they lose their thickening power. Another problem associated with these long-chain molecules is that they are prone to shear degradation in less permeable reservoirs. In a number of field trials they were broken down by shear forces, and the viscosity of the displacing fluid irreversibly reduced.

Polysaccharides. Polysaccharides are produced as an extracellular coating by bacteria in a fermentation process. One such product of the bacterium *Xanthomonas campestris*, usually referred to as Xanthan gum, has been found to have useful properties for oil recovery. It is much less sensitive than the polyacrylamides to shear degradation and salinity and can therefore be used in most reservoirs with moderately saline reservoir waters.

But polysaccharides need protection against biological degradation and a biocide therefore has to be injected with them. Their long-term stability at reservoir temperatures is under appraisal. The increased viscosity of the displacing fluid resulting from the addition of polymer leads to more efficient displacement of the oil but reduces the fluid's injectivity (ease of injection). Depending on oil viscosity, a polymer flood project could double the recovery obtainable with a conventional water drive. With the high cost of polymers and reduced injectivity, however, there is a limit to the maximum concentration that can be used. For all practical purposes the application of this technique is restricted to reservoirs containing oil with viscosities in the range 10 to 100 centipoises at temperatures below 80°C.

Surfactant Flooding. Surfactant flooding aims at producing the residual oil that is left behind by water drives. This oil, in the form of immobile, capillary-trapped

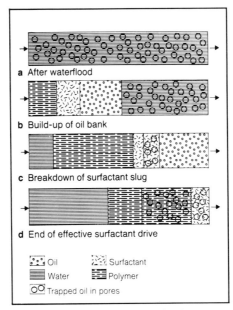

Figure 3.55 **Principle of surfactant flood**

droplets, can be mobilised by injection of suitable surfactant solutions; these interact with the oil to form a micro-emulsion that reduces the capillary trapping forces to a very low level (Fig. 3.51).

Once mobilised, the oil forms a growing bank that theoretically leaves almost no oil behind in the flooded part of the reservoir. The different stages in surfactant flooding are shown in Figure 3.55. Since the oil bank precedes the surfactant, which serves only to lower the interfacial tension behind the bank, it is not necessary to inject expensive surfactant continuously throughout the flood. So, when a certain volume of surfactant solution has been injected, it may be followed by a cheaper fluid of the same viscosity, such as water thickened with a polymer. For economic reasons, the concentration of polymer is often reduced gradually so as to achieve a gradual transition from the high viscosity of the oil/water emulsion to that of the plain water following the polymer.

Surfactants are soaps, or soap-like chemicals. Their molecules consist of a hydrophilic part, attracted to water, and a lipophilic (or hydrophobic) part, attracted to oil. Because of this amphiphilic nature, even at small concentrations, they can greatly reduce the interfacial tension between oil and water and form micro-emulsions.

Factors that influence the formation of oil-in-water or water-in-oil emulsions are the composition of the oil, reservoir temperature, reservoir brine salinity and the type and concentration of surfactant.

At present, systems containing specifically tailored surfactants can be designed for application in sandstone reservoirs at temperatures up to 80°C. Oil viscosity preferably should be low, and the reservoir brine not too saline. Excessive clay, because of its cation exchange capacity, can be harmful to the surfactant slug.

The overall recovery efficiency of a surfactant flood could be of the order of 30 to 60 per cent of the oil left behind by conventional recovery methods. The main problem in surfactant flooding is still to maintain the integrity of the surfactant slug while displacing it through the reservoir.

Caustic Flooding. Caustic flooding is an enhanced oil recovery method based on the principle that the petroleum acids naturally present in some oils can react with the alkali in a caustic solution. This reaction leads to the *in situ* formation of surfactants and emulsification at the oil/water interface. The result is a decrease in interfacial tension between the oil and the water, comparable to that effected by surfactant flooding.

Depending on the salinity of the caustic solution, the addition of surfactants and the temperature, either an oil-in-water or a water-in-oil emulsion can be formed. A pre-condition for *in situ* emulsification is the presence of sufficient petroleum or organic acids in the oil. This is almost exclusively the case with medium and heavy oils.

The caustic solution reacts not only with the petroleum acids in the oil but also with the reservoir rock and brine. Consequently, it is rapidly depleted, and it is this effect that complicates the design and control of caustic flooding projects. More laboratory and field testing will have to be done before this method, which is promising in principle, can be implemented on a large scale.

Recovery Factors and Costs

The maximum additional oil that can be recovered by chemical, miscible and thermal methods varies widely with the type of oil, the depth and characteristics

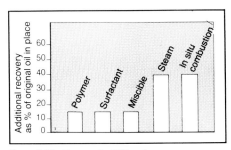

Figure 3.56 **Range of additional recovery by enhanced methods**

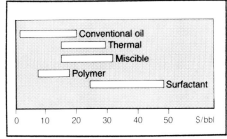

Figure 3.57 **Range of total production costs, 1982**

of the reservoir and the original production mechanism. Figure 3.56 shows the range of recovery factors for each process, in addition to what could be obtained by conventional methods.

For chemical and miscible methods, which are normally applied in light or medium oil reservoirs with high primary and secondary recovery potential, the additional oil recovered can be up to 15 per cent of the original oil in place. For thermal methods, applied in heavy oil reservoirs with low primary and secondary recovery potential, the additional recovery can be up to 40 per cent of the original oil in place.

The ranges of production costs for the various enhanced recovery methods are indicated in Figure 3.57. It can be seen that, although, under favourable conditions, these production costs approach those for conventional oil, they are in general much higher.

Planning of Oil Recovery Projects

The capital investment required for primary field development is often high; for the implementation of secondary or enhanced recovery schemes, it is even higher. Recovery projects, therefore, have to be extremely carefully planned. This may involve field tests, laboratory research and computer simulation of reservoir performance. The information obtained from these various approaches enables decisions to be made, for example, on the optimum number of injection and production wells, optimum production rates, and on what field facilities and pipelines are necessary.

The field tests may be pressure tests on a single well or special multiple well tests. The aim is to characterise the reservoir and to identify possible barriers to flow. Other tests may be necessary to determine the injectivity of water or gas injection wells, and sometimes a pilot waterflood project is carried out in part of the field before a field-scale project is started.

One aim of the laboratory research is to determine rock fluid parameters. This involves, for example, measuring the residual oil to be expected under water or gas drive conditions, and the permeabilities of the reservoir rock to oil, water and gas, which depend on how much of the total volume is occupied by each of these fluids.

Sometimes it may be desirable to carry out physical model experiments, scaled to represent the actual reservoir and fluid conditions, and reservoir simulation with computer models provides a powerful additional tool for field development planning.

In a reservoir simulation model, the physics of multiphase fluid flow is applied to yield a mathematical formulation of the recovery process that can be solved numerically. This approach has a number of advantages over the more conventional methods used to design and monitor primary and secondary recovery projects. Whereas the actual reservoir can be produced only once and at high cost, the simulation model can be used many times, and at a relatively low cost, to evaluate probable results for a variety of production policies.

Reservoir Modelling

The construction of a "reservoir model" is a joint effort of petroleum engineers and geologists. It takes into account geological information obtained from the wells as well as interpretations of pressure test data and laboratory measurements of rock and fluid characteristics. Volumetric calculations are made to determine the amounts of oil and gas initially in place. Porosity and permeability maps are prepared for the various strata that can be defined in the reservoir. Sealing shales between these strata, and other barriers to flow, such as faults and field boundaries, are identified. The initial distribution of water, oil and gas throughout the reservoir is calculated.

Reservoir Simulation

The calculations described in the previous section give a considerable insight into the characteristics of the reservoir concerned, and enable the reservoir engineer to assess at least qualitatively the relative merits of various methods for producing oil and gas from it. At this stage he will be able to define clearly the objectives of his study and to choose a modelling approach in line with these objectives.

Firstly, a choice has to be made from a variety of reservoir simulation computer programs. One class of models ("black oil" models) takes account of only two hydrocarbon components in the liquid phase (oil plus its dissolved solution gas) and one in the vapour phase (the free gas). This type of model is quite adequate for the development and planning of many reservoirs. For other reservoirs, however (for example those containing very light oil), a more sophisticated compositional model may be required, which takes into account the individual hydrocarbon components in the liquid and vapour phases.

Secondly, a decision has to be taken on the size of the model in relation to the amount of detailed information available and the objectives of the study. This choice also depends on whether the entire field, a single well or a representative symmetry element is being modelled.

Various techniques have been (and are being) developed to facilitate the preparation of data input and the display of output for reservoir simulation studies. For instance, to improve understanding of the reservoir processes Shell companies have developed a dynamic colour display technique, which allows the distribution in the reservoir of oil, water and gas (as well as the pressure and temperature distributions) to be displayed on a video screen. This approach is a valuable aid in optimising oil recovery, in diagnosing and checking simulator performance and in presenting final study results.

A thorough analysis of past reservoir performance is then made. The amount and quality of these data can differ considerably between individual reservoirs. For example, for a recently discovered North Sea reservoir only limited data may be available from a few appraisal and early production wells but fortunately these data are usually of high quality. Production data may cover only a short period (say one year) and relate to only a small fraction of the oil initially in place. On the other hand, there may be an abundance of data (partly contradictory and partly of doubtful quality) for a large reservoir that has been producing for a number of years. These data on past reservoir performance may concern pressures and individual well production rates (including gas/oil and water/oil ratio trends).

Calculations are then made to compare "current" and "initial" fluids in place with cumulative production, with the objective of estimating the possible amount of water that may have entered the reservoir. These calculations also indicate the relative importance of the various reservoir mechanisms (water drive, solution gas drive, gas cap expansion).

In all these calculations, mini-computers are of vital importance. These mini-computers are equipped with peripherals (printers, plotters, visual display screens, hard copy units) and are linked with data base management systems on large computers. In this way adequate analysis, processing and display of reservoir data, both for study and reporting purposes, is assured.

Production Technology — Engineering and Chemistry

This branch of petroleum engineering is concerned with the design and layout of downhole equipment for production wells and the treatment of the formation around the borehole. Usually the production engineer deals with the mechanical aspects of such designs while the production chemist looks after the chemical aspects. The latter will also be involved with the chemical aspects of the drilling of wells (e.g. drilling fluids) and of oil and gas handling (e.g. dehydration).

Completion Methods

Should logging and testing indicate the presence of a potentially productive formation, the well must be completed in a manner which will permit the production of oil or gas. The walls of the hole must be supported against collapse. The entry into the well of fluids from formations other than the producing layer, and the flow of the oil from the producing layer into other formations via the well, must be prevented.

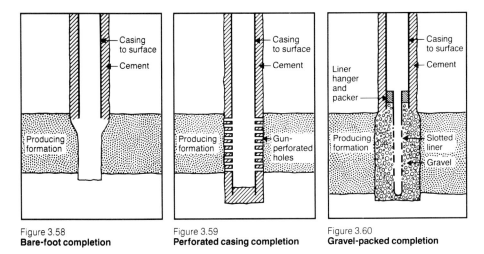

Figure 3.58
Bare-foot completion

Figure 3.59
Perforated casing completion

Figure 3.60
Gravel-packed completion

According to the nature of the producing formation, different completion methods are used, but a string of casing is always run and cemented, at least to the top of the producing layer.

When the producing layer is firm and not liable to cave in, the casing can be cemented immediately above it, leaving it unsupported. This method is sometimes used in wells producing from limestone formations and is called a "bare-foot completion" (Fig. 3.58). If the producing layer is not firm enough, as is usually the case, it must be supported by casing. Casing is necessary anyway when there are several producing layers and any of these is to be excluded from production.

The casing is run through the producing layer to the bottom and cemented, after which holes are shot through the casing opposite the producing layer (Fig. 3.59) by means of a "perforating gun". Most types of gun are lowered on a conductor cable into the hole and carry explosive charges that are electrically detonated. Small explosive charges are used, so shaped that the explosive force is focussed and intensified into a small-diameter jet which penetrates casing and cement.

In poorly consolidated sand formations, sand may be carried into the well with the oil and gas. This should be prevented, or at least kept to a minimum, to avoid plugging the well and to lessen wear of such items as pumps, valves and liners. The erosion of large quantities of sand can also cause caving of the overlying strata and damage to the casing. Various types of liner are used to prevent ingress of sand into the well (e.g. slotted pipe, wire-wrapped pipe and liners in which fine gravel is washed into place between the liner and the formation wall or casing to form a sand-excluding screen) (Fig. 3.60).

The sand can also be consolidated by means of special chemical products such as epoxy resins, to bind the sand grains together without materially affecting the productivity of the well.

Marine Completion and Production

In general, the same methods used for completing and producing wells on land are used for wells in the open sea. This is particularly true for wells which can be completed with their wellheads above the water surface, which is possible where operations are handled from fixed platforms. However, with floating platforms, where wells are completed with their wellheads on the sea floor, the inaccessibility of the well control equipment in deep water beyond the reach of human divers requires the use of remote controls which can be operated either automatically or by surface-actuated underwater robots. (See Offshore Oilfield Development, p. 156).

Stimulation and Removal of Impairment

The formation immediately around the wellbore is a bottleneck in the production process. The flow of oil (or gas) from the producing formation converges towards a small hole (the wellbore) and the situation is much like a large crowd of people trying to go through a door. At this point therefore the formation is extremely sensitive to any form of impairment. Solids left behind by the drilling fluid or loose clay particles from within the formation may partly block the pores and severely reduce the well's production. Such impairment may sometimes be removed by squeezing acids and/or other chemicals into the formation.

Even if not impaired, many formations of low permeability need some form of stimulation to increase their production to an economically acceptable level. Stimulation is usually done by increasing the inflow area for the oil from the formation to the wellbore. Returning to the analogy of the crowd going through a small door, it is simply like increasing the size of the door. In limestone formations, which dissolve easily in acids, the formation is fractured open by pumping acids into it under high pressure. The acid etches away part of the walls of the fracture and conductive channels remain after the fracture has closed upon release of the pumping pressure. Through these channels the oil now flows towards the wellbore more easily.

Sandstone formations (which do not dissolve in acid) are fractured in a similar manner, but to prevent closure of the fracture, sand, sintered bauxite pellets or other material is mixed with the fracturing fluid. Due to the high pressures and large fluid volumes involved, the larger of these fracturing jobs require an equipment set-up which is very impressive indeed (Fig. 3.61).

Figure 3.61 **Close-up view of some of the large amount of equipment required at the well-site for an hydraulic fracturing job.** (Photo: NAM).

Production Modes

After a producing well has been completed (see Completion Methods, p. 112) the means must be provided to bring the oil to the surface. In most fields, the earlier wells will produce by natural flow, i.e. the oil will flow to the surface without assistance. At a later stage, as the reservoir pressure decreases, artificial lift such as gas lifting and, later, pumping may have to be employed. Some fields, especially those producing very viscous crudes, will require artificial lift immediately after completion. The three systems, natural flow, gas lifting and pumping, are described below.

The oil is usually brought to the surface via a string of pipe up to 7 inch (18 centimetres) diameter, called the "tubing", which is of smaller diameter than, and separate from, the casing. This tubing is run into the well and hung from the wellhead with the bottom just above the producing formation. The function of the tubing varies with the type of production method. In general, it forms a replaceable string that enables production methods or equipment to be changed at will, protects the casing from wear or corrosion, and enables the well to be filled with water or drilling fluid should it be necessary to "kill" the well to effect repairs.

Figure 3.62 **Christmas tree for offshore use being prepared onshore.**

Natural Flow. With natural flow, the reservoir pressure forces the oil from the bottom of the well to the surface. The size of the tubing plays an important part in determining the pressure loss as the oil flows upwards through it, and consequently influences the production rate. Gas coming out of solution in the oil helps it to rise up the tubing. Too large a tubing diameter would allow much of this gas to bypass the oil without aiding its upward movement. Too small a diameter would result in too high a friction between fluid and tubing. There is thus an optimum size of tubing for any given conditions.

Offtake from the well is controlled at the surface by varying the size of a choke or "bean" through which the fluid passes. The assembly of valves and fittings at the wellhead, whereby flow can be diverted through alternative chokes or the well can be closed in, is known as the "Christmas tree" (Fig. 3.62).

Many flowing wells, particularly gas wells, have very high pressures at the surface, 10,000 psi (700 bars) or more, and adequate precautions must be taken against the well getting out of control. Equipment is provided which automatically shuts off production in the event of damage to or failure of the wellhead, automatic surface safety valves at the wellhead and, particularly for offshore operations, special valves installed in the well itself (subsurface safety valves).

An oilfield may contain more than one producing horizon, each with marked differences in pressure, specific gravity of oil, and other variables, and thus

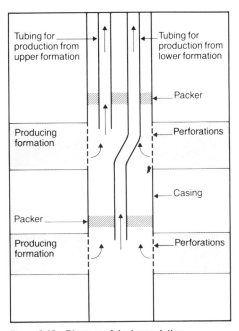

Figure 3.63 **Diagram of dual completion**

needing separate production. This may sometimes be done within the same well by means of a "dual" or "multiple" completion. A simple form of dual completion is shown in Figure 3.63. This type of completion is frequently much cheaper than the alternative of drilling separate wells to each layer or exploiting the layers consecutively within each well. However, flow rates are more limited, artificial lift is more complicated and costly, and repairs to any of the intervals are costlier and result in temporary loss of production from the other intervals. The overall economic picture is therefore not always as favourable as that indicated by the savings during the completion stage.

More oil is produced by natural flow than by all other methods combined; it is a high-capacity method which is both simple and cheap. However, its efficiency diminishes as the reservoir pressure and flow rate decrease, and eventually a stage will usually be reached when production can only be maintained by installing gas lift or a pump, as described below, or by a form of reservoir pressure maintenance such as water injection.

Gas Lift. Production by gas lift is, in effect, an extension of natural flow. The amount of gas produced with the oil is artificially increased by injecting gas into the flowing column, usually by means of special valves set at various depths and controlling the amount of gas entering the flow stream through ports in the tubing. This increase in gas/oil ratio reduces the pressure needed to lift the oil to the surface, delaying the necessity for the installation of pumping equipment.

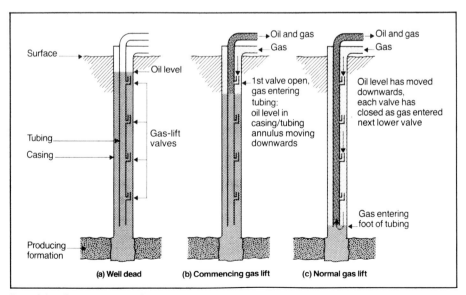

Figure 3.64 **Operation of gas-lift valves**

To commence injection at the foot of the tubing in a "dead" well partially filled with "dead" liquid (containing little or no gas), an initial injection pressure would be required substantially higher than that subsequently needed to maintain production. To avoid the need for a high-pressure gas supply solely for starting, gas-lift valves are often installed in the tubing at predetermined points when the tubing is run into the well (Fig. 3.64). These valves allow the gas to enter the tubing and to blow out the dead liquid at progressively lower points until the well starts to produce. Eventually all the gas enters the flowing column at or near the foot of the tubing, where it is used with the greatest efficiency.

A gas-lift installation often requires a large capital investment to provide a gas compression plant when high-pressure gas is not available from the wells, but it is relatively easy to maintain and operates at low costs per unit of production. Its chief application is where large production rates have to be handled and it may be the final lifting method in a well that produces large quantities of water in the later stages of its life.

Pumping. When the formation pressure diminishes to the point where insufficient liquid flows from the formation to the borehole, gas-lift gas starts to slip continuously through the fluid column to be lifted, and the process of gas lift becomes ineffective. It is then necessary to introduce an alternative form of artificial lift. The method normally adopted is pumping, of which three main types are available: beam or rod pumping, hydraulic pumping, and submersible electric pumping. The system chosen for use in a particular field depends on various operating factors, such as gas/liquid ratio, depth, sand, deviation, space, power and workshop availability.

The simple reciprocating plunger pump with surface power unit connected by "sucker rods" to the foot of the tubing is still the most widely used. The modern beam system comprises: prime mover, pumping unit, sucker-rod string, and subsurface pump. Figure 3.65 shows the layout of a conventional beam pumping installation.

The surface pumping unit (Fig. 3.66) changes the rotary motion of the power unit to an up-and-down motion of the "sucker rods" at the required speed. The size of power unit (usually an internal combustion engine or an electric motor) depends on the power necessary to lift the fluid to the surface. The weight of the sucker rods and the fluid being lifted is counterbalanced by an adjustable weight. The pump lifts the oil up the tubing while the casing forms a passage for the gas which would otherwise seriously interfere with the pumping of the oil.

Despite their superficial simplicity and wide acceptance, rod-operated plunger pumps are subject to certain disadvantages: alternate stretching and contracting of the sucker rods leads to a reduction in the length of the plunger stroke, a reduction which increases with the depth of the pump. Also rods sometimes break

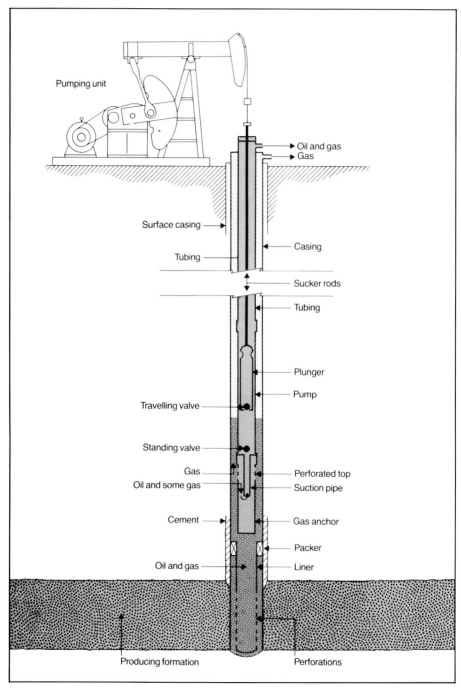

Figure 3.65 **Diagram of pumping installation**

PRODUCTION   121

Figure 3.66 **Surface pumping units in the Schoonebeek oil field, the Netherlands.** (Photo: NAM).

as a result of cyclical stresses and of wear caused by friction between the rods and the tubing.

Hydraulic pumps are similar to rod pumps in that they are also of the plunger type. However, there the similarity ceases. Power is transmitted from the surface pump to the reciprocating-type subsurface hydraulic engine/pump assembly by means of hydraulic fluid under high pressure (power fluid). The hydraulic control valve (also called the distribution or engine valve) directs the flow of power fluid alternately to each side of the working cylinder or engine. This is connected to the single or double acting production cylinder or pump by a rod and thus produced fluid is pumped to the surface. This fluid may be mixed with hydraulic fluid (the open system) or the hydraulic fluid may be returned by separate line (the closed system). The ratio of engine to pump plunger size determines the efficiency of the pump.

The subsurface pump can either be installed at the bottom end of the tubing (fixed type) or may be pumped down through the tubing into a locking device at the end of the tubing (free type). In this latter type, reversing the flow of hydraulic fluid brings the pump back to surface. It is obvious that the free pump is attractive from the point of view of maintenance. However, the size of the fixed pump is not restricted by the size of the tubing and it can therefore handle larger volumes of liquid. A special type of hydraulic pump, the jet pump, operates without moving parts by converting pressure energy into kinetic energy through a nozzle.

One of the main disadvantages of the subsurface hydraulic engine is that it will not withstand sand or other abrasive materials in the power oil.

Submersible pumps are centrifugal pumps submerged in the well fluid and driven by an electric motor installed immediately below the pump. Power is transmitted to the motor via an electric cable clamped to the tubing.

The pump may be installed on the bottom of the tubing (tubing-suspended system) or run on the electric cable into a locking device in the bottom of the tubing (cable-suspended system). This latter method is applied in large capacity wells (up to 5,000 barrels a day) and is well suited for use in crooked holes. However, the electric cable is a disadvantage in running and pulling the pump and is a possible source of failure. Moreover, this type of pump is sensitive to solids in the produced fluid and shutdowns resulting from power cuts or fluctuations. Among the advantages of this type of pump is the fact that the power supply to the pump can be easily varied, and along with it the pump's throughput, over a wider range than is possible with other types of pump.

ENGINEERING, DRILLING AND PRODUCTION OPERATIONS

Drilling

Historical

The technique of well drilling goes far back into history. It is first mentioned in ancient Chinese manuscripts, which describe wells drilled as early as the third century AD to tap underground strata for brine (Fig. 3.67). The wells were drilled with a heavy "bit", which hung from a rope and was jerked up and down by relays of men bouncing on a spring board. The weight of the periodically rising and falling bit drove it deeper and deeper into the ground. This method was the precursor of the "cable tool" method, a percussion system used in oil well drilling in the 19th century and predominantly in the first two decades of the 20th.

ENGINEERING, DRILLING AND PRODUCTION OPERATIONS 123

Figure 3.67 **Early Chinese drilling rig.**

The cable tool system was essentially a method of pounding out a hole by repeated blows with a bit attached to a "drill stem", a heavy length of steel suspended from a wire rope. The drill stem provided the weight to force the bit into the ground, and the hole was kept empty except for a little water at the bottom. After drilling a few feet, the bit was pulled out and the cuttings removed with a "bailer", an open tube with a valve at the bottom. Steel pipes known as casing, of progressively smaller diameter, were run from time to time to prevent the hole from caving in and to keep back any water flow.

Cable tool drilling was cheap, simple and effective for shallow wells, but progress was slow, and no means were provided for stemming the flow of high-pressure oil and gas when encountered. In such cases the wells blew out and spewed quantities of oil and gas over the countryside. The "gushers" of these early days were spectacular but wasted a lot of oil and gas, and were a serious fire hazard.

The present-day method of drilling, known as the "rotary" method was introduced at around the turn of the century. It was first successfully used for the discovery well of the famous Spindletop field in the Gulf Coast region of Texas. With this method, the bit is attached to the bottom of a string of steel pipes and

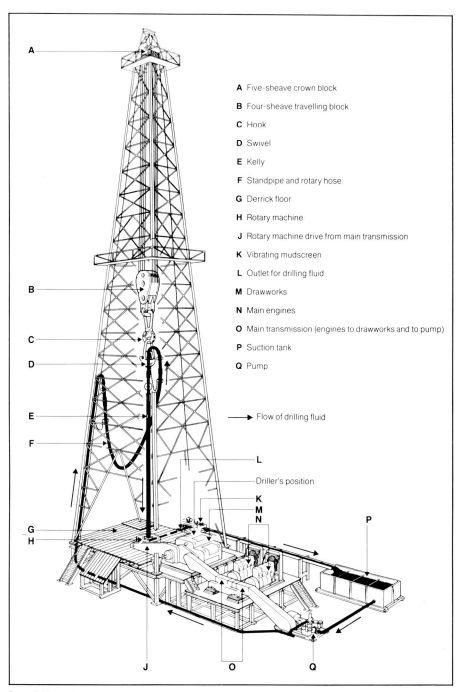

Figure 3.68a **Diagram of rotary drilling rig**

ENGINEERING, DRILLING AND PRODUCTION OPERATIONS

rotated by means of a rotary table which turns the uppermost pipe or "kelly". "Drilling fluid" or "drilling mud" is continuously circulated down through the hollow drilling string, through the bit and back up to the surface through the annular space between drilling string and borehole wall. The drilling mud flushes the cuttings out of the hole and the hydrostatic pressure of the mud column normally slightly exceeds the pressure of the fluids in formations penetrated by the bit, thus greatly reducing the risk of a blowout.

A variant of rotary drilling is "turbo-drilling". In this method, the bit is rotated at the bottom of the well by means of a fluid motor or turbine powered by the mud stream. The drill pipe does not transfer torque from the surface down to the bit, although it is usually kept rotating slowly to prevent it from sticking against the wall of the hole.

The Modern Rotary Drilling Installation

A rotary drilling installation consists essentially of bit, drilling string, rotating equipment, hoisting equipment, mud circulating and treating equipment, prime

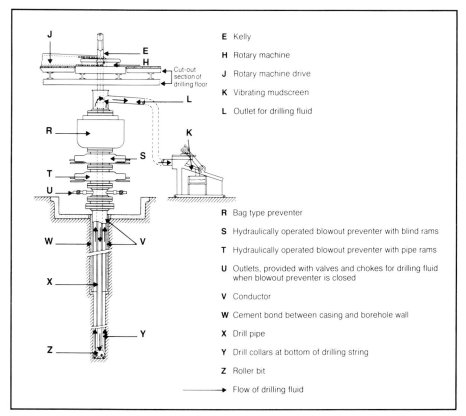

E Kelly
H Rotary machine
J Rotary machine drive
K Vibrating mudscreen
L Outlet for drilling fluid

R Bag type preventer
S Hydraulically operated blowout preventer with blind rams
T Hydraulically operated blowout preventer with pipe rams
U Outlets, provided with valves and chokes for drilling fluid when blowout preventer is closed
V Conductor
W Cement bond between casing and borehole wall
X Drill pipe
Y Drill collars at bottom of drilling string
Z Roller bit
→ Flow of drilling fluid

Figure 3.68b **Diagram showing equipment at and below the derrick floor, the borehole and drilling string**

movers and transmission, and an installation for pressure control (Figs. 3.68a and b show the assembly).

Bit and Drilling String. The bit is screwed to the bottom of the drilling string, made up of lengths of special steel pipe (drill pipe), 32 feet (ca. 10 metres) long and mostly 5 inches (127 millimetres) in outside diameter. (Drill pipe, bit and casing sizes are still given in feet and inches in many countries, as a result of the US origins of the drilling business.) Each length or joint is equipped with special steel couplings (tool joints) having a coarse tapered thread and square shoulder to ensure leak-proof connections that can transmit torque and yet be made up and broken repeatedly, rapidly and safely.

The top joint (length of pipe), or "kelly", passes through the rotary table and is used to transmit the torque from the rotary machinery to the drilling string and thus to the bit. This kelly is square or hexagonal in cross section and sets of horizontal rollers are attached to the rotary table by means of a special housing known as the kelly bushing. The sets of rollers within this bushing form a square or hexagonal aperture through which the kelly passes and is driven by the table, thus giving an almost frictionless drive as the kelly is lowered and drilling proceeds. Heavy, thick-walled tubes (drill collars) are used at the bottom of the drilling string, just above the bit. As the drilling string is lowered, the bit touches bottom and starts to take load, thus throwing the bottom of the string into compression. Ordinary drill pipe, which is designed only for tension loading, would soon fail under the buckling and fatigue stresses which occur at the bottom of the string.

In fact, all components of the drilling string must be of the highest quality and finish to withstand the high stresses imposed while drilling. Ordinarily several hundred horsepower may be transmitted mechanically to the bit by means of the drilling string while rotating. The high stresses are due partly to the extraordinary proportions of this string. If a scale model were made of a string 4,000 metres (13,124 feet) long and 12.7 centimetres (5 inches) diameter, and a knitting needle of normal cross-section (2 millimetres or 5/64 inch) were used to represent the string, the length of the needle would be 63 metres (207 feet).

The drilling string with bit attached is rotated at between 75 and 250 revolutions per minute, with loads as high as 36,300 kilogrammes (80,000 pounds) bearing down on the bit from the weight of the drill collars.

The ability to apply such loads on fast-drilling formations depends on the mud-flushing action at the bit, since drilling will slow down rapidly unless the hole bottom is properly scavenged. To achieve the proper scavenging action, the bit is provided with hardened steel nozzles through which the drilling fluid is ejected downwards at a velocity of 90 to 120 metres per second (300 to 400 feet per second), just ahead of the rotating cones of the bit. Some 300 to 450 kilowatts

Figure 3.69 **Three-cone roller bit attached to drilling string.**

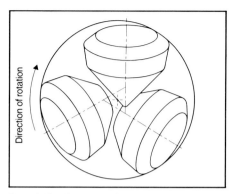

Figure 3.70 **Disposition of cones in bit for soft formations**

(400 to 600 horsepower) may be used in achieving this jetting action on the bottom, in addition to the power required for circulating the mud down the hole and back up to the surface.

The type of bit normally used is known as the three-cone roller bit, and is shown in Figure 3.69. Fewer and longer teeth are used on the cones of bits intended for softer formations, while shorter teeth are used where formations are hard, the shorter teeth allowing for a larger and heavier bearing. For the softer formations it is also customary to set the cones with their axes to the left of centre, as shown in Figure 3.70, which increases the scraping action on the bottom. Of more recent design are the special tri-cone bits with studded tungsten carbide inserts, and bits provided with thin plates consisting of pressed and bonded fine artificial diamonds (Fig. 3.71). Application of these bits is increasing and no longer limited to use in hard, abrasive formations only. These bit types can be run for up to 100 hours and more (Fig. 3.72).

Hoisting Equipment. The hoisting equipment consists of a block and tackle system supported by a "derrick" or "mast". Wire rope, 2.5 to 3.8 centimetres (1 to $1\frac{1}{2}$ inches) in diameter, is wound round a drum and reeved over a fixed sheave assembly (the crown block) at the top of the derrick or mast and a travelling block with a hook (Fig. 3.73). The fixed end of the wire rope is attached to the base of the derrick and to it is clamped a "weight indicator" which measures the tension. This measurement is translated into the weight suspended from the travelling block and, by subtraction, the load on the bit during drilling. A comprehensive record of drilling operations is made by recording the loads on a moving chart.

In a modern rig the hoisting gear (drawworks) is driven, through a system of chains, gears and clutches, by a diesel-electric power system providing a lifting

Figure 3.71 **Diamond drilling bit, unused.**

capacity of up to 500 tonnes. In essentially the same way that a car driver changes gear for different speeds and engine loads, the driller can select the appropriate gear ratio for a particular operation. The winch drum is controlled by a heavy-duty braking mechanism which allows the driller to feed a steady load to the bit during drilling.

Figure 3.72 **Diamond bit, after 160 hours of use in hard formations.**

Rotating Gear. The rotary table is also driven by the prime mover system through a reduction gear which allows different speeds to be selected. The kelly is suspended by a swivel from the travelling block hook.

ENGINEERING, DRILLING AND PRODUCTION OPERATIONS

Figure 3.73 **Downward view from derrick to drill floor shows travelling block supporting drill string.**

Drilling Fluid Circulation System. The drilling fluid is pumped under high pressure from a suction tank outside the derrick, up a standpipe in the corner of the derrick, through the rotary hose and swivel to the hollow kelly and drilling string. After leaving the drilling bit, the fluid, carrying the drill cuttings, ascends to the surface through the annular space between the drill string and the wall of the borehole. At the wellhead, the drilling fluid passes through a vibrating screen (the shale shaker) which removes most of the drill cuttings, and then returns to the suction tank via an intermediate treatment tank.

At the heart of the circulation system are two (sometimes more than two) reciprocating, gear-driven, triplex or duplex mud pumps. As its breakdown during drilling might have serious consequences, including the loss of at least part of the hole, the pump must be of exceptionally sturdy construction, capable of sustained service under heavy loads. Pumps with input rated capacity of 1,200 kilowatts (1,600 horsepower) or more are commonly used on the deeper wells, with working pressures in the range of 3,000 to 3,500 psi (200 to 240 bars).

Prime Movers and Transmissions. Most modern rigs utilise diesel engines as prime movers. These drive either DC generators and motors, or AC generators, AC–DC converters and DC motors.

The power plant of a typical drilling installation using diesel engines may consist of three or more engines, each self-contained on its own skid.

The drawworks, mud pumps and rotary table are normally directly driven by DC motors which can be controlled and regulated from the driller's console on the drill floor.

Pressure Control Equipment. Oil and gas occurring at shallow depth are usually associated with pressures at or about the equivalent of a column of salt water (hydrostatic pressures) reaching from that depth to the surface. Gas, however, will tend to expand when being brought to the surface, and can easily eject some of the annular fluid column if not handled with dexterity. Furthermore, a bit, when rapidly withdrawn from a hole which contains viscous mud, can exert a powerful swabbing action causing gas or other fluid to enter the bore hole. Care therefore is required when pulling the pipe. At greater depths, oil, gas or salt water may be encountered unexpectedly at pressures in excess of that of the hydrostatic column. Rig personnel must therefore be continuously alert for such emergencies.

To handle such an event, a system of control equipment is installed at the wellhead after setting the surface casing. This control equipment, commonly termed the "blowout preventer stack" (BOP stack) can close off the annulus between drill pipe and casing, and can hold pressures up to 10,000 psi (700 bars) or more depending on their size and rating. The preventers are operated by hydraulic pressure which causes horizontally opposed pistons to close rams around the drill stem. Similar rams of different shape are used to close up against one another and thus shut off the entire opening at the wellhead should the pipe be already out of the hole. Furthermore, a bag-type preventer which can close off any shape of pipe or even the entire borehole is installed on top of the ram type preventers. The arrangement is shown diagrammatically in Figure 3.68b.

Drilling the Well

The Technique of Drilling. The drilling of a well is a round-the-clock shift operation and usually continues without interruption from the moment of drilling the first metre (spudding in) until completion. Under the supervision of a drilling supervisor (toolpusher), the crews, each comprising a driller and four or five men, normally work three eight-hour or two twelve-hour shifts. A mechanic and an electrician attend to all the service equipment and other specialists are called in as required. While the drillers are responsible for the mechanical operations of drilling, the engineer in charge has to see that the drilling programme is carried out to the best advantage (see Wellsite Operations Engineering in Production section, p. 78).

During the drilling operation drill pipe joints are screwed to the top of the string as required. When pulling the string out of the hole, to change the bit for example, the drill pipe is unscrewed in approximately 96 feet (ca. 29 metres) stands of three joints each. The stands are stacked at one side of the derrick floor, being racked at the top by a member of the crew occupying a small platform high up in the derrick. Before unscrewing each stand the drill pipe is wedged by "rotary slips" inserted in the bushing of the rotary table. The reverse procedure is applied to run the pipe back into the hole; the entire operation is known as a "round trip" and is a lengthy business. On a well-organised rig drilling at 14,750 feet (4,500 metres), it may take $5\frac{1}{2}$ hours to pull out and $3\frac{1}{2}$ hours to run back in again.

Casing the Well. A well is started with a relatively large hole, $17\frac{1}{2}$ inches (44.5 centimetres) or more in diameter, which must be lined as soon as possible with steel pipes. In shallow development wells this conductor casing (usually called the conductor) may be run to a depth of some 50 metres or less, but deeper wells, especially exploration wells, may need a conductor to a depth of 300 metres or more before drilling proceeds in the next section of the hole. This casing prevents the upper hole from caving in and water from entering or mud from leaving the hole. It also provides a firm base and anchor for the blowout preventers and for the long strings of casings which may be run later to "case" the lower part of the hole. This casing is designated by its outside diameter, for example 20 inch, $13\frac{3}{8}$ inch, $9\frac{5}{8}$ inch and 7 inch (50.8 centimetres, 34 centimetres, 24.4 centimetres and 17.8 centimetres respectively).

The conductor is rigidly secured by filling the space between the casing pipe and the borehole wall with cement. Cement slurry is fed into the casing pipe, a rubber plug is placed on top of it, and drilling fluid then pumped in, so forcing the cement down inside the pipe and up again between pipe and wall. When the plug reaches bottom, pumping is stopped and the well is left standing long enough to allow the cement to set. Drilling is then resumed using a smaller bit, e.g. $17\frac{1}{2}$ inch (44.5 centimetres) through 20 inch casing, $12\frac{1}{4}$ inch (31.1 centimetres) through $13\frac{3}{8}$ inch casing, and $8\frac{1}{2}$ inch (21.6 centimetres) through $9\frac{5}{8}$ inch casing. If the well is successful and oil or gas is met, a further string of casing may be cemented at or near the bottom.

Casing is expensive and its cost may be an appreciable proportion of the total cost of the well. For development wells, especially where conditions are already rather well known and where production rates do not call for large-diameter completion strings, smaller clearances and reduced diameters may be warranted to reduce capital costs.

Drilling Fluids (Muds). The progress and efficiency of drilling depend also on the use of the right drilling fluid for the rock being drilled. Normal drilling fluids usually consist of colloidal suspensions of clays in water, with chemical additives to control viscosity and other properties. Under some conditions the use of an oil emulsion, or an oil-base mud, is advantageous. Air, gas or aerated liquids can also be used in certain circumstances and result in very rapid drilling. In extremely deep wells, the mud must remain fluid at temperatures of up to 400°F (205°C).

The properties of the drilling fluid will vary with its ingredients, but in the main the fluid serves to:

- Assist in maintaining maximum drilling rates compatible with safety. To achieve this, the drilling fluid must be of such a "weight" (density) that it will only just prevent uncontrolled influx of gas, oil or water from the formations into the borehole. In addition, the solid content and viscosity of the fluid must be kept to a minimum.
- Remove drill cuttings from the bottom of the hole and the face of the drilling bit and carry them out of the borehole.
- Support and protect the wall of the hole against caving or collapse by the pressure of the fluid column. In addition, a protective sheath (or mud cake) is deposited on porous formations.
- Keep the drill cuttings in suspension when circulation is stopped or when replacing a worn bit. To do this effectively, the drilling fluid should stiffen or gel when at rest and become fluid again when put in motion.
- Cool the bit. Considerable heat is generated by a bit drilling under heavy load.
- Enable satisfactory electric logs to be obtained. This demands that the fluid has certain properties of electrical conductivity or resistivity.

When highly permeable formations are penetrated, drilling fluid may escape into them and either part or all of the fluid stream may be lost. To combat this, fibrous, flaky or granular "lost circulation" materials (e.g. mica, cellophane flakes or walnut shells) may be added to the drilling fluid. If losses cannot be stopped by this means, a slurry of cement, bentonite and water or diesel oil can be forced into the rock to plug the pores and fissures. Alternatively in certain circumstances, drilling can be continued without any return of the drilling fluid to the surface.

Deviated Drilling. Whenever possible, wells are drilled vertically, but from offshore platforms particularly it is necessary to drill wells deviated from the vertical towards widely spread targets at reservoir level. As a rough guide, the cost and time of drilling a deviated well is approximately 30 per cent greater than that of drilling a vertical well of the same hole depth. This is because of the slower drilling rate and the time required to make surveys of the course of the well, and to correct this course where necessary.

Normally the well is drilled vertically for a short distance, and cased, before deviation is begun. Before 1960, the most common method to deviate a well utilised a steel wedge, or "removable whipstock", which is orientated in the required direction and set on bottom. The bit follows the wedge and starts the

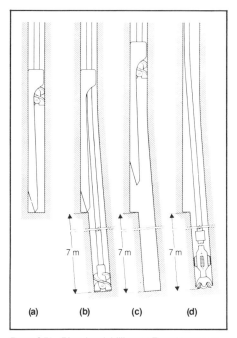

Figure 3.74 **Directional drilling:** (a) The whipstock in position; (b) A short length of deviated hole is drilled; (c) The whipstock is pulled; (d) The deviated hole is reamed, and drilling is eventually continued

deviation which is subsequently built up with flexible assemblies. Once the required angle is reached, a stiffer assembly maintains direction and angle (Fig. 3.74).

The engineer directing these operations is thoroughly familiar with the effect of various assemblies and drilling techniques on the probable course of the hole, and the whipstock, once used for the initial deflection, may only be required occasionally to correct large deviations from the desired course of the well.

The well is surveyed at intervals by running a wireline measuring device incorporating a magnetic compass and plumb bob system or gyroscopic instruments, to record the angle of hole deviation and its direction. A recent development is the use of inertial navigation equipment to give the most accurate survey, but its use is limited by its size to cased holes with diameters down to $13\frac{3}{8}$ inches (34 centimetres). Nowadays, down-hole motors together with a "bent sub" have replaced the whipstock in deviated drilling.

One special use of deviated drilling is to control a blowout or "gusher", by drilling a relief well from a safe distance away (say 400 metres). The relief well is deviated to encounter the flowing formation as close as possible to its position in the out-of-control wellbore. Large quantities of heavy drilling fluid are pumped down the relief well to "kill" the flow in the main well. Clearly, great skill is required to reach the desired target which is sometimes several thousands of metres below the surface.

Drilling Hazards

The drilling of a well is not always simple and one or other of the following hazards may be encountered.

Blowouts. A blowout (Fig. 3.75) can occur when a high-pressure oil or gas accumulation is encountered unexpectedly and the mud column fails to contain the formation fluid which erupts from the wellhead. The fire hazard is great and severe pollution of the surroundings can occur rapidly. Nowadays, improved techniques, training and equipment have made actual blowouts comparatively rare.

The first sign of trouble is often an increase of the drilling rate accompanied by an increase in the mud return flow, indicating that formation fluid is entering the wellbore. The driller must be constantly alert to spot these symptoms and to take emergency action without delay.

The blowout preventers must be closed immediately and the surface pressure reading checked. From the surface pressure a calculation can be made to determine the density of mud required to control the well. A weighting material, usually barytes (barium sulphate), is added to the mud in the storage tanks and

ENGINEERING, DRILLING AND PRODUCTION OPERATIONS

Figure 3.75 **Blowout near Long Beach, California, USA.**

the new mud is pumped into the well while a back-pressure is held on the mud-return line. Once the weighted mud has filled the hole the pressure on the formation should prevent further fluid entry and drilling can be resumed.

Lost Circulation. When a very porous formation, fissured rock, or rock containing cavities (such as limestone) is encountered, the mud seal on the borehole may be ineffective in preventing the escape of drilling fluid into the formation. Circulation will diminish or cease and drilling may then become impossible. Materials that plug the formation are added to the drilling fluid. If this does not have the desired effect, cement or other means may be required. In extreme cases an extra string of casing may have to be set to cover and close off the lost circulation zone.

The simultaneous occurrence of lost circulation and blowout conditions is extremely difficult to handle since the zone of mud loss must be plugged to a degree sufficient to bear the additional weight of the mud column, now weighted to counter the higher pressure.

Stuck Drill Pipe. The drilling string may become stuck in the hole as a result of mechanical obstruction such as a broken bit cone, excessive drill cuttings or collapsing formation. Further, the difference in pressure of the hydrostatic column and the formation pore pressure can cause the pipe to stick against the side of the hole. This phenomenon is called "pressure differential" sticking.

The stuck pipe may sometimes be freed by "spotting" a slug of oil opposite the stuck portion, and allowing this to soak while pulling and jarring on the string. If the string cannot be freed in this way, the stuck point may be established by instruments, and a small explosive charge lowered just above the stuck point inside the drill string. The explosion loosens the tool joint sufficiently to allow it to be unscrewed easily by rotation at the surface. The stuck portion must then be "sidetracked", as described under "deviated drilling".

Stuck pipe is not as common now as in the past, due to improvements in mud treatment and to the use of "stabilisers" in the drilling string which help prevent it from pressing into the mud cake. Special spiral grooved drill collars also help in preventing "pressure differential" sticking.

Fishing. A fish is the term used to describe a part of the drilling string, bit cones or similar junk, left in the hole through mechanical failure.

Efforts may be made to remove the fish, or it may be sidetracked. Small parts may be fished with a magnet, or a "junk basket", while the larger fishes involving pipe require an "overshot" or "tap". The overshot is used to grip the pipe on the outside, while the tap screws into heavy-wall pipe such as drill collars and grips it with its tapered surface (Fig. 3.76). Nowadays less time is spent attempting to recover a fish than formerly, since it is usually cheaper to abandon the fish and drill a sidetrack.

ENGINEERING, DRILLING AND PRODUCTION OPERATIONS

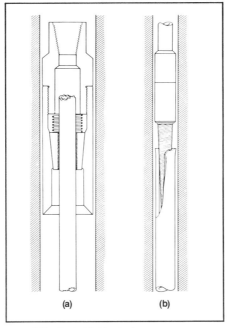

Figure 3.76 **Typical fishing tools:**
(a) Non-releasable overshot; (b) Fishing tap

Marine Drilling

Before World War II, little effort was made to drill for oil in the underwater regions of oil basins. In the 1880s, shallow wells were drilled from piers off the coast of southern California, near Rincon. In the late 1920s, immense developments started in Lake Maracaibo in Venezuela, and the seabed extension of the Huntington Beach field in southern California was tapped by means of deviated wells drilled from beach locations.

After World War II, the search for oil was intensified, and the prospects revealed by geophysical surveys in the Gulf of Mexico soon led to the drilling of the first offshore well in those waters. Since then, the technique of offshore drilling has developed rapidly, and many types of drilling rigs have been constructed which are suitable for operations far from shore in water as deep as 150 metres.

Sit-on-bottom Barges or Submersibles. These mobile structures are provided with ballasting facilities so that they can be floated to the location, and then ballasted down on to the seabed to provide a stable base for drilling operations.

Figure 3.77 **Submersible barge with drilling barge coupled to it, operating in an artificial bay in the River Niger, Nigeria.**

When these operations are completed, they are deballasted for relocation. Buoyancy and stability of the submersible structure derives from widely spaced columns which support the superstructure. Footings on the columns support the unit when ballasted on to the seabed (Fig. 3.77). Sit-on-bottom barges or submersibles are only suitable for water depths of up to 20 metres and for a mild weather environment.

Jack-ups. The jack-up unit is a mobile floating structure with retractable legs connected by a jacking mechanism to the hull, which also serves as the working

Figure 3.78 **The jack-up rig *Charles Rowan* under tow.**

deck. On location, the legs are lowered through the hull to the seabed to provide a stable base for drilling. The hull is then jacked up on the legs to an elevation clear of wave action. Jack-ups can operate in water to a depth of 90 metres and are suitable for a severe weather environment (Fig. 3.78).

Ships or Barges. Ship-shape or barge-type vessels can be used with anchors and cables to maintain station in water depths of up to 360 metres but not in severe environments (Fig. 3.79). These vessels can also be equipped with computer-controlled propulsion systems to hold station at a particular location, in which case they are then suitable for operation in water depths of up to 1,200 metres. However, with their conventional ship-shape hulls, they are more susceptible than semi-submersibles to bad weather.

Semi-submersibles. Semi-submersibles are mobile structures with a superstructure supported by widely spaced columns which sit on lower hulls below the depth of wave action. Together the columns and hulls provide buoyancy and stability for the unit in all operating conditions. Semi-submersibles are suitable for operation in a severe weather environment. Anchors and cables can be used to keep the vessel on location in moderate water depths of up to 360 metres in a

Figure 3.79 **Drillship *Petrel* operating in deep water.**

ENGINEERING, DRILLING AND PRODUCTION OPERATIONS

Figure 3.80 Semi-submersible drilling rig *Stadrill*, riding on deballasted pontoons. Anchor winches, chains and guides can be seen on main columns.

severe weather environment (Fig. 3.80). The dynamically positioned semi-submersible can hold station over a particular location without the use of anchors. This is achieved by a computerised positioning control system, which regulates thrusters or propellers, in response to signals from a position reference

Figure 3.81 **The dynamically positioned semi-submersible drilling rig *Sedco 709*.**

indicator, such as an acoustic beacon on the seabed or a microwave system. This facility makes the unit suitable for operating in water depths up to 1,200 metres.

A recent example of these vessels is the dynamically positioned semi-submersible Sedco 709 (owned jointly by Nautilus B.V. and Sedco Inc.). The power plant is capable of generating 3,300 kilowatts (25,000 horsepower) of which 2,700 kilowatts (20,000 horsepower) is available for positioning. This vessel, which is 91 metres long, 76 metres wide and 34 metres high, is extremely stable and can work in very rough environmental conditions (Fig. 3.81).

Deep-Water Drilling Methods

Successful drilling operations were carried out offshore California and in the North Sea in the late 1960s, in up to 185 metres of water using traditional systems where re-entry was made with guidelines between the drilling unit and the wellhead on the seabed, a task requiring diver assistance. Beyond this depth, however, divers could not safely operate and the use of anchoring systems and steel guide wires became impractical, so new systems had to be developed to permit operations in deeper waters.

It was clear that operating in great water depths would increase exploration and development costs considerably, but the drilling industry felt that the steady price increases of crude oil during the early 1970s, together with the trend of increasing world demand for energy, could make deep-water operations economically viable. Before a comprehensive exploration programme in deep water could be undertaken, a number of systems needed further development. These were guidelineless re-entry, blowout preventer control, risers, and dynamic positioning.

Re-entry Systems

The first guidelineless re-entry system used for exploration was developed for the Sedco 445, a dynamically positioned (DP) drillship completed end-1971 (Fig. 3.82). This system uses a sonar scanner to detect the distance from the bottom of the riser assembly to the wellhead. Figure 3.83 shows a typical re-entry operation for a BOP stack. The operation is monitored on a screen on board ship, and by manoeuvring the ship, the re-entry is accomplished.

Later, underwater television cameras were introduced to monitor re-entry

Figure 3.82 **The dynamically positioned drillship *Sedco 445*, shown at anchor.**

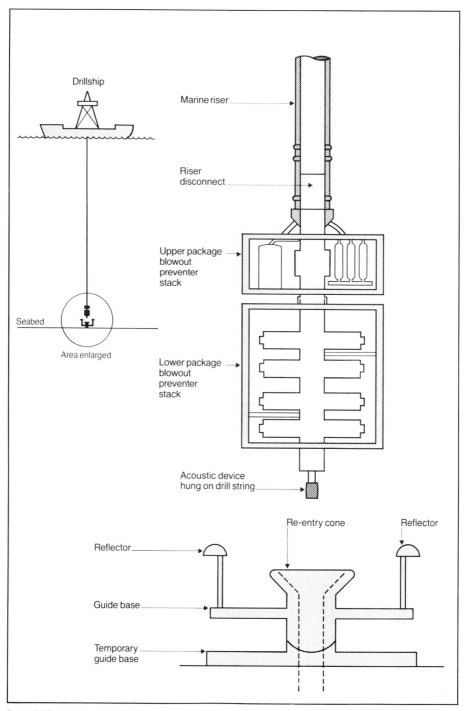

Figure 3.83 **Typical guidelineless re-entry operation**

Figure 3.84 **Underwater control stack, being prepared onshore, has two control pods on top, one on either side of the riser connector (centre).**

operations and experience showed that it was possible to make re-entry without using a scanner. Modern units such as the Sedco 709 and some French drillships use almost exclusively television for re-entry.

Blowout Preventer (BOP) Controls

In water depths of from 450 to 600 metres, conventional hydraulic control systems provide acceptable response time to operate BOP rams and valves. The increased length of control hoses required in deep water, however, caused response times to become unacceptably long. To overcome this, an electro-hydraulic system was developed. A hydraulic accumulator power source is mounted on the BOP stack installed on the seabed wellhead, and control commands are given from the surface via an electric cable with up to 70 separate wires, one for each function. Improved designs were developed around 1975 which introduced multicomplex electronic controllers that transmitted the control signals in coded form on a few single wires, thus providing a simpler and lighter system. Such systems are used on the Sedco 709 and other modern drillships.

Figure 3.84 illustrates a typical underwater control pod attached to a BOP stack. A duplicate unit is always provided for safety.

Marine Risers

A marine riser is a large-diameter tubular connection between the drilling vessel and blowout preventer stack, tensioned at the top through a heave compensator to allow for the vessel's movement on the surface. The riser is normally made up in 50 feet (15.2 metres) joints and fitted with special couplings. Risers used with dynamically positioned vessels have a quick disconnect system, enabling the riser to be disconnected from the seabed wellhead in case of an emergency, such as a failure of the DP system.

In very deep water, buoyancy materials are required to support part of the submerged riser's weight in order to keep top tensioning within practical limits.

The marine riser has to be strong enough to withstand the forces of waves, currents, and induced forces from ship motions, together with the contained weight of drilling mud. Complex computer programs are used to determine operational limitations for any given water depth and environmental condition.

Dynamic Positioning Systems

Dynamic positioning (DP) is the technology of maintaining a vessel's position by means of thrust, generated by a number of propellers. Conventional drilling units are anchored by a number of heavy chains attached to anchors, but for deep water this becomes uneconomic and impractical. There is no exact water depth limitation for anchoring systems, but in most cases beyond 600 metres the DP system will be more attractive.

Prior to 1970, several coring vessels were using an elementary form of DP, but positioning accuracy was not very high. Developments in the Sedco 445 and later in the Sedco 709 and other DP ships, provided a much more reliable and accurate station-keeping system, allowing exploration wells to be safely drilled to great depths.

The Sedco 445 is capable of remaining on station within 5 per cent of the water depth in winds of up to 50 knots, a significant wave height of 3.7 metres and a 3 knot current. A computer receiving information from an acoustic beacon system placed on the seabed in conjunction with a taut wire system controls several propellers or thrusters to provide the correct amounts of thrust to counteract the effects of wind, currents and waves on the vessel.

Operational Experience with Dynamically Positioned Drilling Units

For the Shell Group, experience with DP drilling units dates back to the early activities of Sedco 445 which was used from December 1971 for a four-year worldwide exploration programme. Fourteen wells were drilled in seven deep-water concession areas in this period. World records were set in 1974 and 1975 when wells were drilled off the west coast of Africa in water depths of 640 metres and 701 metres, respectively. Further records were set with the drillship Seven Seas operating at a depth of 1,980 metres off Surinam in 1979.

Production Operations

Production operations involve the management of hydrocarbons from the reservoir to the initial customer. This in effect means producing, processing and delivering the correct quantity and quality of product and ensuring that all the production systems are optimised. The total production system can be divided into three distinct subsystems: subsurface (wells), surface handling (process facilities) and storage and sales metering (terminal).

Wells

There are various types of wells: producers, from which the hydrocarbons are obtained, and injectors, through which reservoir maintenance is achieved (secondary and enhanced oil recovery).

Unfortunately, wells do not continue to produce without attention until the hydrocarbons are exhausted. The amount of attention required can vary enormously and is influenced by such factors as reservoir characteristics (type of formation), the nature of the produced or injected fluids and the production method. Safety valves, gas lift valves and other flow control devices have to be installed and replaced; downhole pressure, temperature and flow surveys have to be conducted; sand, scale and wax can cause considerable problems and may have to be removed; corrosion has to be combated and monitored through downhole inhibitor injection and inspection techniques; subsurface pumps have to be maintained and broken sucker rods, which operate the plunger pumps, replaced.

In addition to the more routine well servicing or maintenance already referred to, more radical repairs may be needed. These vary considerably in nature and extent, from sealing a leaking wellhead or replacing a corroded or leaking tubing conduit to setting a new liner or replacing a failed gravel pack. A typical repair could involve shutting off water that is entering the well from a section of the producing formation and this is often effected by squeezing cement into the formation at the offending point. This may require very high pressures, for which special pumps have to be provided, and packers set in the hole around the tubing to contain these pressures and ensure correct cement placement.

Much routine well servicing work is carried out through the tubing by means of small-diameter wireline. This is used to run, set and manipulate various tools, flow devices and measuring instruments. These operations are carried out under pressure, at depths of up to 4,000 metres or greater, where along-hole depths can exceed 5,000 metres due to well-bore deviation. A more recent development, replacing wireline for specific applications such as underwater completed wells, where the wellhead is not directly accessible, is "through flowline" (TFL) or "pump-down" (PD) well servicing. With this method, the tools are pumped along the flowline and down the well and reverse circulated out again, utilising some suitable fluid medium. Essentially, the same routine well servicing work as conducted by wireline can be achieved by the TFL method.

In wells operated by pumps another routine operation is the pulling of sucker rods in order to replace broken rods or to change the subsurface pump. In the case of fields on land this operation is carried out with a hoist and telescopic mast mounted on a truck (Fig. 3.85). Such a hoist can also be used to pull shallow tubing strings, and its simplicity makes it less expensive to operate than a full-size drilling rig.

Sand may still be produced into the well-bore in spite of various types of liners or sand-consolidation techniques. In addition to eroding pump parts and even wellhead fittings and surface equipment, sand can fill the well-bore to a considerable height. Small quantities of sand can be removed by means of an open-ended tube with a check valve at the bottom, lowered on a wireline (bailing). This is a laborious operation and the well may be out of production for several days. For

Figure 3.85 **Truck-mounted telescopic mast.**

large volumes it may be more economical to circulate the sand out by means of small-diameter continuous tubing (usually ca. 2.5 centimetres), which is unreeled from a large-diameter storage drum and lowered inside the production tubing under pressure. Continuous tubing is also used during stimulation work on wells for placing acids or other chemicals into the formation and during sand-consolidation operations.

Wax from some types of oil is deposited on the wall of the tubing in the upper, cooler part of the well. This wax can often be removed by mechanical methods, but sometimes the tubing must be pulled out and steam-cleaned. To avoid wax formation, the temperature of the oil can be raised by electrical heating of the upper part of the tubing or by injecting hot oil into the well. Alternatively, chemicals may be injected to slow down the rate of wax deposition.

Non-routine well servicing work generally involves more radical repairs to the well, such as redrilling sections of the hole (side-tracking). Such work often necessitates use of a full-size drilling rig, which in the case of offshore operations is particularly expensive to operate.

Process Facilities

Oil. The oil produced at the wellhead is associated with a certain amount of gas and, possibly, water. Facilities have to be provided to separate the gas and water from the oil and to gauge the production of all three streams. The oil is then transferred to some intermediate storage or direct to the main storage tanks or oil terminal, where it awaits delivery to the initial customer by pipeline, tanker or, less frequently, road or railcar.

In disposing of the formation water produced, which is usually saline and contains minor amounts of emulsified hydrocarbons, full consideration is given to avoiding any adverse effect on the environment, on- or offshore (see Environmental Impact, p. 208).

If the water associated with the oil is fresh, it will be properly treated to remove entrained oil and drained. If it is saline, it will be treated for both oil and solid entrainments before being reinjected into a suitable reservoir.

The associated gas in recent years has become a valuable commodity. Prior to sale of the gas, some will be used for power generation, heating and gas lifting. In cases where sales are not possible, the gas can be reinjected into the reservoir for pressure maintenance.

Commonly, the production of each well is led through a "flowline" from the wellhead to a gathering station. The gathering station collects oil from a number of wells into separators, in which gas, and often water, is separated from the oil by stepwise pressure reduction. The oil is collected in tanks, where water can further settle out and the oil quantity can be gauged. In order to monitor the behaviour of the individual wells and of the underlying reservoir, a separate set of test separators and tanks is provided, into which the production of each well can be switched and gauged for a limited period.

A typical field would combine the production of a number of gathering stations and direct it towards the main production station, where the oil is treated to meet the specifications required for onward shipment by pipeline, tanker or other means of transport.

Oil and Gas Separation. Basically, oil and gas separators consist of vertical or horizontal cylindrical vessels containing baffles. The detailed construction of separators varies widely according to the capacity required and the operating pressure, which may be from a few pounds to several thousand pounds per square inch. The separated gas may be transported by pipeline and sold outside the field. However, this gas can also be a valuable source of energy on the oilfield (e.g. for use in heaters, gas engines, gas lift installations and for injection into the oil-bearing formation). The gas may also contain valuable liquid components which can be extracted in a gas treatment (fractionation) plant.

Dehydration. In most cases salt water is produced with the crude oil. The presence of this water is economically undesirable as it would occupy space in ships, pipelines and storage facilities, which could otherwise be occupied by crude oil. Furthermore, to assist efficient processing at the refinery, crude oil should not contain more than 1 per cent of water and 50 milligrammes of salt per kilogramme. Consequently, most of the water produced with the oil is removed before shipment from the oilfield. Water which separates freely from the crude oil is initially drained off at the gathering stations and subsequently at the main storage tanks. Unfortunately, water often occurs as microscopic droplets in a water-in-oil emulsion. These water droplets will not readily settle out and must be induced to coalesce into larger drops which will freely settle out. Special treatments are usually required to achieve this, and so the "wet production" is normally pumped from the gathering station to a dehydration plant. In order to promote coalescence of the water droplets, the emulsion may be heated and chemicals added, or it may be passed between electrodes maintained at a high alternating potential of 15,000 volts or more. Disposal of the large quantities of water that are often produced is sometimes difficult. The water often has a high salt content and would contaminate surface drainage systems. However, this water, after being cleaned, may be injected into the formation under a "water flooding" secondary recovery project.

Gas. The gas separated from the oil during oil processing as well as the gas produced directly from gas and gas-condensate wells may be saturated with water. Not only can this water cause severe corrosion under specific conditions, but in conjunction with certain components of the gas it can form "hydrates", crystalline compounds rather like snow, which can plug valves, chokes and gas transport lines.

Gas processing facilities are primarily designed to remove the water vapour and associated hydrocarbon liquids as well as "heavy" hydrocarbon components from the gas. This is done to prevent the formation of hydrates and the corrosion of downstream facilities and at the same time to maximise revenue by providing sales quality gas. In some cases, additional processing is necessary to remove undesirable components such as hydrogen sulphide.

The most widely utilised processes in treating gas involve either low-temperature separation, by which most of the water and also certain hydrocarbons are separated from the gas as liquid, or separation to remove any free liquids followed by glycol contacting to remove water vapour from the gas.

Low-temperature separation involves cooling the gas below some specifically required dewpoint by expanding the gas across a choke or control valve. The resultant liquids (water and hydrocarbons) are then heated by means of a heating coil through which the incoming stream of high-pressure gas from the well is passed; this will melt any hydrates which have formed in the liquid phase. Sometimes glycol is utilised in conjunction with this process to prevent premature hydrate formation and thereby to assist in the dehydration process. Figure 3.86 shows part of the processing facilities of such a plant required for the production of gas from the large gas reserves in the northern part of the Netherlands (Groningen).

Glycol gas processing, which follows primary separation of free liquids, necessitates the gas being bubbled through a series of trays filled with glycol.

Figure 3.86 **Gas processing plant at a gas production location in the Groningen field, the Netherlands.**
(Photo: NAM).

ENGINEERING, DRILLING AND PRODUCTION OPERATIONS

Glycol, being hygroscopic, absorbs the water vapour from the gas, thereby drying it. Dry glycol is continually circulated into the trays inside the contactor tower from a stripper unit, where the water is stripped from the wet glycol by the application of heat.

Following gas processing, the sales quality gas may well have to be recompressed to a higher pressure to meet the contractual sales specification, to optimise pipeline capacity or to permit reinjection into an oil-producing reservoir.

Before delivery of the gas to the initial customer, which may be a gas distribution company or a liquefaction plant, the gas is delivered to a metering station or sales point where quality and quantity are carefully measured.

General Facilities. In addition to the equipment and installations directly connected with the production or handling of oil and gas, general facilities are also required. In remote areas their provision can constitute a substantial part of the total development cost of an oilfield. For example, a large stores organisation with well-stocked warehouses may be required to ensure that essential projects are not held up by the lack of materials or equipment. Water supplies and all forms of communication have to be provided or developed. A power station, sometimes quite large in size (Fig. 3.87), may have to be erected to provide electricity. Fully equipped machine shops and vehicle repair depots are essential.

Housing, social, educational, hospital and recreational facilities have to be provided for the staff. In fact the development of an oilfield in remote areas requires not merely the installation of oil production facilities, but the construction of a small town with all the necessary services.

The difficulties of remote area development has perhaps reached its present-day zenith in the northern North Sea oil and gas fields. Within the limited space of the offshore structure it has been necessary to incorporate all gas, oil and water

Figure 3.87 **Power station, Sullom Voe oil terminal, Shetland islands, UK.**

processing and handling facilities as well as the related ancillary services to support these facilities. Added complexity arises from the necessity to drill, produce and sometimes construct concurrently.

Operations of this nature may involve a resident complement of up to 400 persons, for whom life support, safety and logistics systems have to be provided (see Offshore Logistics, p. 196). Because of the complexity and integrated nature of some northern North Sea platforms, it has been necessary to introduce centralised control of processes and systems involving increasing application of computer-assisted technology.

Terminals

A terminal, as the name implies, is the ownership transfer point. In the case of gas it is the metering point at which final quality and quantity are established. Oil terminals are designed for the preparation and storage of crude oil to meet shipment patterns. Facilities are provided to ensure that the crude oil quality and quantity are sufficient for export and that the total storage capacity within a terminal allows for final dehydration, if required.

Provision is also made within an oil terminal for the receipt, treatment and disposal of ballast water transferred from tankers before loading. (Bunkering facilities may also be provided.) The water treatment systems installed within a terminal ensure that drain and ballast water are oil-free before final disposal.

Offshore Oilfield Development

As exploration activities advanced from the land areas to offshore and on into deeper and rougher waters, so the structures, equipment and techniques required to exploit the prospects discovered had to be developed to meet the new conditions. These developments can be grouped as follows: fixed, compliant and floating installations; underwater equipment; marine production risers; underwater support (diving and submersibles).

Fixed Installations

The first fixed offshore installation in the Gulf of Mexico was placed in 1945 in six metres water depth to drill an exploration well. The structure was made of timber and supported a converted land rig. In 1947, the first producing well in the area was drilled from a steel structure in 10 metres water depth. From then on, there was a gradual but steady progress into deeper waters.

In 1960 the first permanent structure was installed in over 50 metres water depth, and in 1967 the 100 metres depth mark was passed. These structures were

ENGINEERING, DRILLING AND PRODUCTION OPERATIONS

piled into the seabed to provide the foundation to carry the load of the structure and its support equipment. They were designed to resist the forces exerted by waves, wind and currents, and in some cases ice or even earthquakes.

The following types of fixed installations have been used.

Steel-piled Structures. This type of installation has proved the most commonly used substructure for permanent structures. The technology associated with this

Figure 3.88 **Simple platform structure supporting a four-well cluster in a shallow river-mouth, off Nigeria.**

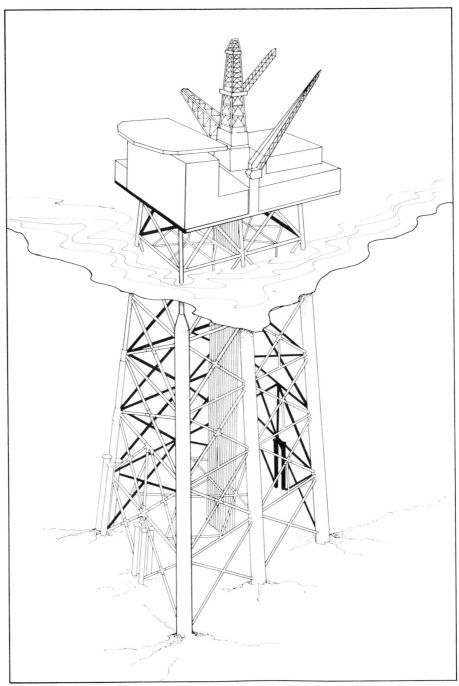

Figure 3.89 **Example of a complex steel structure.** The Shell/Esso drilling and production platform Brent A, UK sector North Sea.

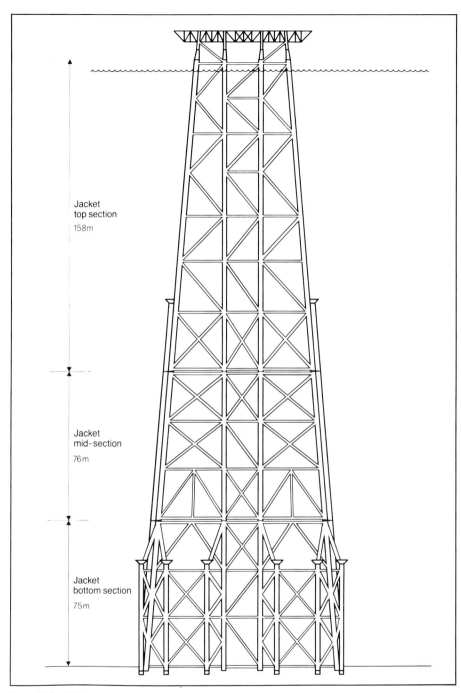

Figure 3.90 **Multi-part steel jacket structure.**

type has been developed very successfully from simple structures of 50 tonnes weight in very shallow water to complex structures of up to 60,000 tonnes in 300 metres water depth; while there is no indication that 300 metres represents an upper limit technically, this depth does represent an approximate limit beyond which other types of structure may be more economic.

All steel platforms consist of three basic parts: superstructure, "jacket" or tower, and foundation (mainly piling). The type of superstructure depends to a large extent on the function of the platform. Consequently, the superstructure can range from a simple deck (Fig. 3.88) to a multi-level fully integrated drilling and production deck structure with a large number of deck modules such as drilling packages, process facilities, living quarters, platform cranes, helicopter landing deck and a flare tower to burn off gas whenever it cannot be used otherwise (Fig. 3.89).

Initially, jackets on the seabed were used to support the deck above the sea surface. A jacket is a welded space frame designed as a template for pile-driving, and to provide the lateral stability for the platform. The piles carry the top loads into the subsoil. Later, towers were also developed. These carry the vertical loads through a relatively small number of large-diameter legs. Hybrid structures combining jacket and tower characteristics have also been developed.

Jackets are usually transported to location on a barge and either lifted off or launched into the sea. Once floating in the horizontal position, they are upended by selective flooding of legs and bracings. Some shallow-water jackets may be placed directly in the vertical position. Tower and hybrid structures are usually floated to location on their large-diameter buoyancy legs and then upended in a similar manner to the larger jackets.

Generally, hollow steel piles are used to pin the structure to the seabed, and are driven in with a pile-driving hammer. The load-bearing capacity comes from outside wall friction in combination with end bearing. The pile-to-jacket connection is made above water by welding shim plates to the pile and the jacket leg. In certain designs, the annulus between pile and leg is fully grouted, and the bond between steel and cement forms the connection.

The latest generation of water depth record-breaking platforms are made of multi-part jackets which are joined together offshore. One of these is the 300 metres water depth jacket for the Shell Oil Cognac field installed in the Gulf of Mexico in 1978. This structure consists of three separate parts that were launched and installed sequentially, one on top of the other, and then rigidly connected (Fig. 3.90). Sheltered deep-water locations for joining the three parts prior to installation could not be found in the Gulf of Mexico, so the parts had to be joined vertically *in situ*. It is expected that steel-piled structures will continue to be used for deep-water applications, although the method of installation may vary depending on available fabrication facilities, installation equipment and the environmental conditions at the platform location.

ENGINEERING, DRILLING AND PRODUCTION OPERATIONS   161

Concrete-piled Structures. This type of platform was most popular in the early days of the offshore industry in Lake Maracaibo and in the Caspian Sea (Fig. 3.91). In both these areas, a calm weather environment and shallow water

Figure 3.91 Concrete piled structure supporting a drilling derrick. The concrete piles are driven through and cemented into tubular guides. A drilling support barge is anchored alongside.

provided the opportunity to venture "offshore", while using simple onshore technology. Nowadays, there are hardly any applications for this type of structure.

Concrete Gravity Structures. A gravity platform "sits" on the seabottom by virtue of its own weight, and all vertical and horizontal loads are transmitted to the top soil layers which have to be well consolidated if such a structure is to be used.

Concrete gravity platforms were originally developed for the northern North Sea and came to maturity there in the 1970s. The first concrete gravity platform was placed in 70 metres water depth in the Ekofisk field in 1973. By 1982, some 14 concrete gravity structures had been installed in the North Sea in water depths varying from 100 to 150 metres (Fig. 3.92).

The development was triggered by a combination of harsh environmental

Figure 3.92 Shell/Esso's *Brent B* production platform, a concrete gravity structure, on tow to the Brent field in August 1975, towers 153 metres out of the waters of the fjord near Stavanger, Norway.

conditions and hard soils. This combination made the installation of conventional steel-piled structures extremely costly due to prolonged pile installation, uncontrollable weather downtime and the limited "weather window" for offshore installation. The presence of deep-water construction sites and deep tow-out channels, together with the fact that the North Sea is surrounded by countries with great technological abilities, favoured the development of gravity structures.

A further advantage of these structures is that they can be towed out and installed with a large portion of the topside facilities already in place, thus reducing the time and cost of offshore hook-up and commissioning of the platform. They also offer the possibility of oil storage in the large cells which form the lower portion of a gravity platform. The combination of all these factors appears unique for the North Sea which is probably why concrete gravity structures so far have not found application in other areas.

Since 1975, concrete platforms have become less attractive because of escalating fabrication costs, whereas the development of large-capacity semi-submersible installation vessels and more powerful pile-driving hammers have offset the earlier installation drawbacks of the conventional steel-piled structures. Also a number of pipelines have now been laid in the North Sea thus reducing the need for offshore oil storage.

Steel Gravity Structures. By 1982, four steel gravity platforms had been installed in 85 metres water depth offshore Congo, where the hard dolomite seabottom would have made the installation of conventional steel-piled platforms extremely expensive. Another application of this design is the Maureen field structure in 100 metres water depth in the North Sea (Fig. 3.93). The relatively higher costs of such a platform, however, are a limiting factor in future applications.

Compliant Installations

The size and weight, and therefore the cost, of fixed structures grows with increasing water depth. With a view to reducing the required structural strength, and hence the weight of the structural steel required, more recent offshore field developments have used compliant structures. Rather than being able to withstand rigidly the forces of wave, wind and current, such structures are allowed to move in a restrained manner to comply with these forces, so reducing the strength required.

Buoyant Towers. In 1968, came the first test of the buoyant tower concept, with the installation of the ELF-Ocean tower in 100 metres water depth in the Gulf of Biscay. This type of structure, also called an articulated tower, consists of a slender truss with a universal joint at the seabottom and a large buoyancy

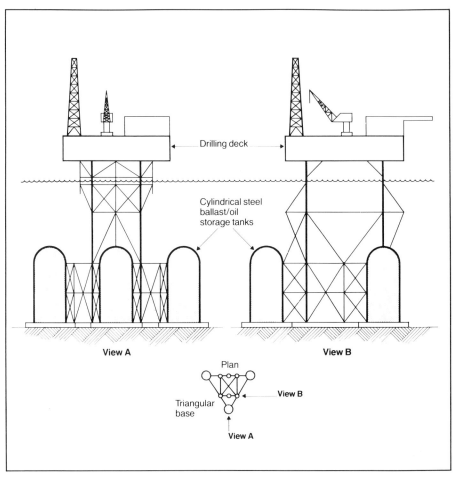

Figure 3.93 **Steel gravity platform**

chamber near sea level to keep it upright. The structure oscillates around the universal joint, and the restoring force is generated by the buoyancy of the chamber when the structure is tilted.

Two buoyant flare platforms in the Frigg and Brent fields, and two offshore loading platforms in the Beryl and Statfjord fields, have been installed in water depths of 130 to 145 metres (Fig. 3.94). Articulated towers have not been used yet as drilling or production platforms but operating experience with the offshore loading towers could be a valuable boost for further development of this type of structure. However, for deeper water the slenderness of such a structure poses problems, particularly during transport to the location. Also, for larger structures, problems can be foreseen in the scaling up of the seabottom universal joint which

ENGINEERING, DRILLING AND PRODUCTION OPERATIONS

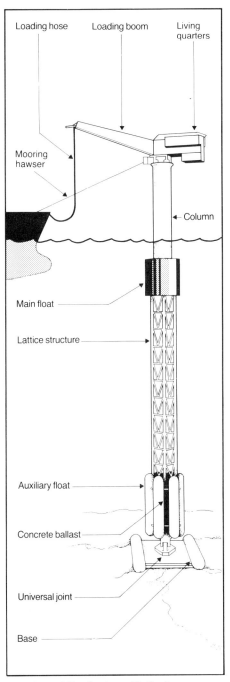

Figure 3.94 **Oil loading tower** (Statfjord)

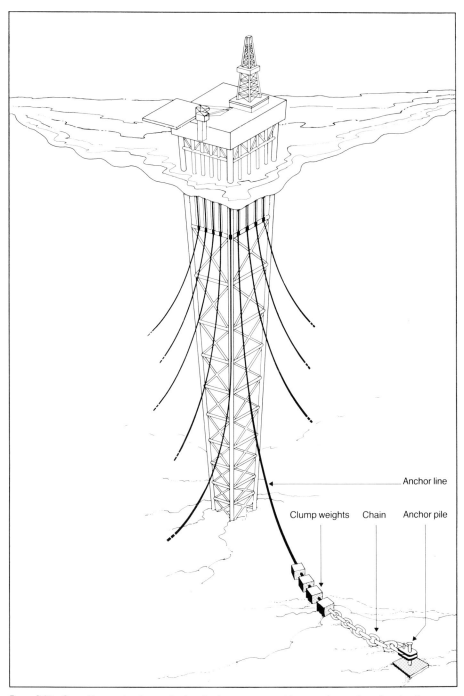

Figure 3.95 **Guyed tower, showing seabed anchoring arrangement for one of a number of anchor lines.**

ENGINEERING, DRILLING AND PRODUCTION OPERATIONS

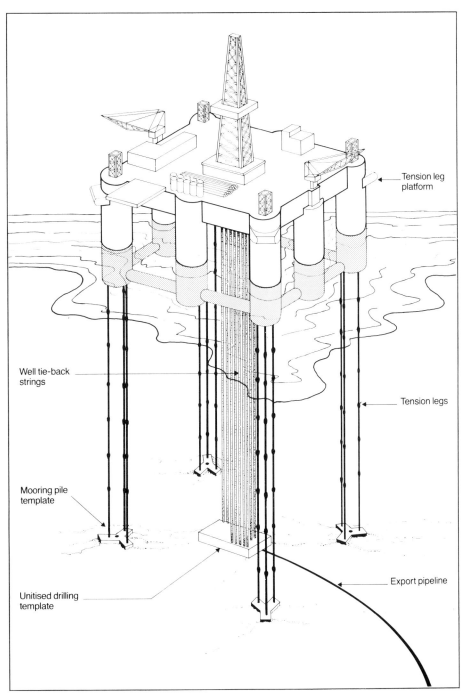

Figure 3.96 **Tension leg platform**

forms the vital feature of this structure. The uncertainty about long-term reliability of the universal joint which has little "redundancy", will probably lead to a cautious approach by the oil industry to adopting this concept on a wider scale.

Guyed Towers. By 1982, there were a number of designs for guyed towers. One was for 300 metres water depth in the Gulf of Mexico and was under construction. Another design was for 450 metres, also in the Gulf of Mexico, and a scale model of this design had been installed in 100 metres of water. This structure consists of a four-legged, slender steel space frame. The vertical forces on the structure are taken by a piled foundation base or, alternatively, a gravity foundation base. The horizontal forces are taken by a number of guy wires (Fig. 3.95). This type of structure is designed for drilling and production with the wellheads on the platform deck level. One of its drawbacks is possible fouling of the guy wires by anchored service vessels, derrick barges or pipelay barges.

Tension Leg Platforms. Tension leg platforms differ from conventionally moored floating structures in that they have excess buoyancy which is restrained by taut vertical cables or steel tubes (Fig. 3.96). With such a structure, heave, pitch or roll motions are virtually eliminated. It is therefore possible to bring the wellheads up to deck level.

Several designs for a tension leg type platform have been proposed and a one-third scale model of a prototype has been tested in 60 metres water depth offshore California. By 1982, the concept had found only one application, with a unit being constructed for installation in Conoco's North West Hutton field in the North Sea. Although a tension leg platform does not provide for storage and offloading, this concept is being considered for developments in extremely deep water, where the reduced efficiency of catenary anchoring systems may dictate a fundamental change in the approach to the design of floating systems.

Floating Installations

With the increased demand for crude oil and the sharp rise in prices during the 1970s, there were incentives to develop small offshore fields which had previously been uneconomic. New concepts in floating production platforms are being developed for fields in deeper and more hostile waters where fixed or compliant platforms could be uneconomic or technically impractical. These can also allow for early production while more permanent installations are designed and constructed.

In particular, conversion of semi-submersible drilling vessels and of crude oil tankers have provided floating installations for production, storage and/or offloading facilities. These have been used with single point moorings (SPM) and shuttle tankers to provide a complete production system.

ENGINEERING, DRILLING AND PRODUCTION OPERATIONS

Figure 3.97 **Artist's impression of the Tazerka Field development, offshore Tunisia.**

A tanker moored to a catenary anchor leg mooring (CALM) system was used as a stationary storage facility for the first time in 1964 by Shell in the Halul field (Qatar). By 1977, 14 similar units were in operation worldwide, some incorporating such variations as a hinged rigid yoke instead of a bow hawser mooring.

A rigid or single anchor leg mooring (SALM) is a feature of a number of floating systems for production and/or storage. In this, the rigid leg is connected by a universal joint to a base fixed or ballasted on to the seabed. The tanker yoke is connected by a swivel arrangement which allows produced oil from a subsea well to flow through a riser inside or attached to the rigid leg and into the tanker's treatment facilities continuously, even as the vessel rotates around the anchor under the influence of wind and current (weathervaning). For a single well field such a swivel was relatively simple. For more than one well a "multi-bore" swivel was required, and such a system for up to eight wells was developed for the Tazerka field offshore Tunisia. Installed and brought into production by a Shell company in 1982, it consists of a 200,000 dwt. tanker connected by a yoke to the top of a single anchor leg riser in 140 metres of water (Fig. 3.97).

Converted semi-submersible drilling rigs have been used in the North Sea in Hamilton Brothers' Argyll field and BP's Buchan field. However, a semi-submersible unit has limitations with respect to storage and loading. Moreover, the time and cost required for converting an existing unit indicate the need for a

Figure 3.98 **Artist's impression of the *Semi Spar* floating production, storage and offloading unit.**

"custom designed" new floater. A modified version being studied by Shell is a semi-submersible unit called the Semi-Spar (Fig. 3.98).

ENGINEERING, DRILLING AND PRODUCTION OPERATIONS

Underwater Equipment

Unless an offshore field can be produced by means of a fixed platform from which all production and reinjection wells can be drilled, it is necessary to make use of wells completed on the seafloor. These are then connected by flowlines and controls to the surface facility. The connection of the flowline or flowlines from seabed to the surface, usually vertically, is called the riser. When several flowlines are used in combination, a subsea manifold may be required.

One of the first underwater completions (UWC) was made in 1943 in 11 metres of water in Lake Erie. Since then, more than 300 UWCs have been made there, representing the largest concentration of UWCs in the world. These wells were equipped with simple land-type Christmas trees and required divers to install them, connect flowlines and operate the valves.

Development of deep-water subsea wellhead equipment and completion technology for the open sea did not start until the mid-1950s. The early seafloor Christmas trees were installed by divers and operated by hydraulic remote controls. The evolution of seafloor well technology was a slow process, accelerating in 1979 with 21 wells being completed by oil operators. The technology has now been developed sufficiently to be applied in deep water or hostile environments.

A single well completed on the seafloor is termed a "satellite" well. Most of the subsea completions to date have been satellite wells connected by flowlines to a platform in shallow water. These wells have been used in outlying areas of fields which could not be reached from the central platform.

A schematic subsea completion is shown in Figure 3.99. The downhole completion below the mud line is similar to an ordinary land completion, and consists of casings of the various sizes required to maintain structural integrity of the well and to allow well control during drilling (as described in Production Technology—Production Modes, p. 115).

As in a land completion, the string of tubing is suspended from the wellhead to the producing formations, the wellhead supporting the casing and tubing at the mud line. Valves start with the downhole safety valve which is used to shut off the flow in an emergency. The Christmas tree valves are mounted on top of the wellhead. The master valves are used to secure the well in normal operations after flow is stopped by the wing valves. A crossover valve is provided in this case to allow connection of the annulus between the tubing and casing with the flowline. Swab valves allow vertical entry into the well from a drilling rig or service vessel located overhead. Most valves are hydraulically operated by controls from the surface. They are so designed that if they fail, they do so in the safe position (i.e. shut off) if hydraulic control pressure is lost.

In some cases two flowlines are used. One flowline is connected directly to the

172 EXPLORATION AND PRODUCTION

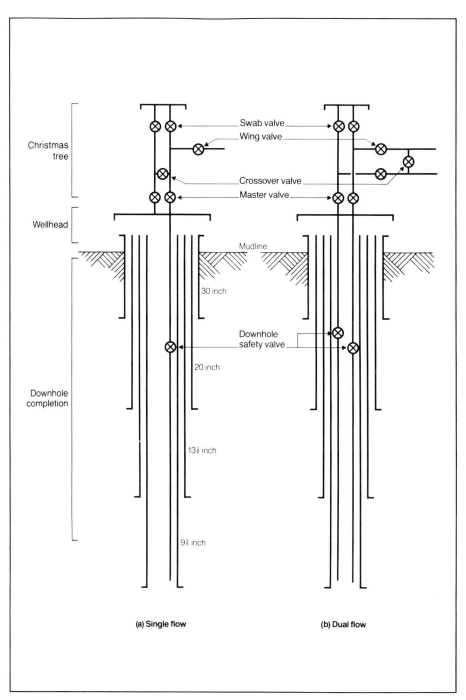

Figure 3.99 **Flow diagram of seafloor and downhole completion**

ENGINEERING, DRILLING AND PRODUCTION OPERATIONS

production tubing and the other to the annulus with a crossover between the flowlines. This allows monitoring of the annulus pressure without stopping production, and provides a circulation path; it also allows production to continue if one flowline is damaged.

An example of a satellite well system with dual flowlines is shown in Figure 3.100. The Christmas tree is designed to be installed using guidelines from a surface rig, and is attached to the wellhead by a hydraulically actuated connector. The flowline connections are at the edge of the guide base near the mud line. Installation of the controls and connection of the flowline require divers. Downhole equipment is maintained by re-entry through the top of the tree. Units like this were installed by Mobil oil in the Beryl field and by Shell in the Cormorant field in the North Sea.

More recently, remotely controlled equipment and tools and flowline connectors have been designed to allow installation of subsea trees without use of divers.

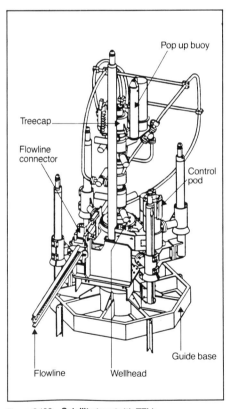

Figure 3.100   **Satellite tree (with TFL)**

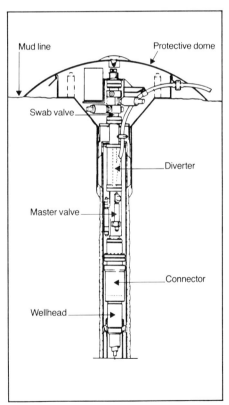

Figure 3.101   **Insert tree (with TFL)**

In addition, tools have been developed which can be pumped down the flowline and into the wells to perform operation and maintenance duties on downhole equipment (as described in Production Operations — Wells, p. 149). The application of this TFL or PD method always requires dual flowlines and dual tubing, while all bends in the flowlines must be five foot (1.5 metres) radius or larger.

A recent development is a subsea tree which is installed below the mud line to give more protection against damage by fishing equipment, anchors, icebergs and the like. As shown in Fig. 3.101, the tree connects to a wellhead located about 20 metres below the mud line.

Satellite Well Flowlines. These are sized and pressure rated for compatibility with the flow. Conventional flowlines are constructed of small-bore steel pipe welded offshore on location, or onshore and then transported to the offshore site. Pipe joints of 40 to 80 feet (12 to 25 metres) in length can be welded together on a floating work deck, and then lowered to the seabed in a controlled configuration to prevent overstressing. This is generally termed the conventional lay method.

Flowlines welded onshore can be transported to the offshore site in lengths of up to several kilometres by reel barge, surface tow or bottom tow. The reel barge transports a continuous string of flowlines coiled on to a reel. At location, the flowline is uncoiled on to the seabed. The wall of the pipe has to be relatively thick to prevent excessive pipe flattening during this process. Recently, flexible flowlines consisting of a nylon sheath in a steel carcass have been developed and proven for use. These are particularly suitable for reel barge installation. The surface tow and bottom tow methods involve pulling the flowline to location in a positively buoyant or negatively buoyant condition respectively.

Trenching and/or burial of flowlines is sometimes required for mechanical protection from trawl boards, or for improved thermal insulation. An unburied steel line can result in the well's production being cooled to seabed ambient temperature within approximately 300 to 500 metres from the well, which could lead to hydrate or wax formation. The soil cover resulting from burial does improve insulation but only to a limited extent. Flexible flowlines with improved insulation layers have been developed which may be buried by ploughing simultaneously with laying. Such a line was installed by Shell to connect a satellite well to the Cormorant South production platform in the North Sea. Foam insulation of steel flowline bundles within carrier pipes, which could provide mechanical protection from trawl board impact, are being installed, using the mid-depth tow method, in the Central Cormorant field.

Present methods for trenching consist primarily of using high-pressure jets to scour soil out from under the flowline which has already been laid, permitting the flowline to settle into the trench. More recently, ploughs have been developed which run along the flowline, ploughing a furrow into which the flowline settles,

ENGINEERING, DRILLING AND PRODUCTION OPERATIONS

and then infilling over the flowline to complete the burial process. Laying and burial of steel flowlines can be done simultaneously.

The Subsea Manifold. The manifold is the central subsea point at which flowlines and export pipelines come together. The design of a manifold must be developed from the outset in conjunction with a maintenance system.

Manifolds come in a variety of shapes and sizes, from a relatively simple arrangement connecting a few subsea flowlines with a basic multibore riser which relies on diver assistance for installation and maintenance, to the giant Underwater Manifold Centre (UMC), installed by Shell in the Shell/Esso Cormorant field in 1982, which connects a remote subsea development to an existing platform. The UMC (Fig. 3.102) weighs 2,200 tonnes, covers an area equal to half a football pitch and has a height equivalent to a four-storey building. It is

Figure 3.102 **Diagram (not to scale) of the Underwater Manifold Centre positioned on the seabed some four miles from the *Cormorant A* production platform (right). The diagram also shows (top left) the maintenance vessel with the Remote Maintenance Vehicle deployed above the UMC.**

designed for remote maintenance techniques. It is anticipated that the UMC concept will have considerable application, both in deeper water tied to floating production systems where conventional platforms would be uneconomic, and for developing areas beyond the reach of existing platforms.

When used as a component of a floating production system, the manifold functions as the interface between the production riser and the flowlines or export lines (lines to the terminal). For this application, design and testing work is directed at establishing the suitability for service of these assemblies which, in addition to allowing remote connection and disconnection of a number of lines, must also provide long-term sealing against all fluids transferred to and from the floating unit. Furthermore, they may also have to provide a solid anchor base for a riser to the surface facilities.

Valves on subsea manifolds are controlled from the surface by direct hydraulic or electro-hydraulic systems. Development and testing of specific items of subsea equipment such as TFL selectors, subsea chokes, pipeline pig/sphere launchers and manifold control systems, is being carried out in a continuing quest for improved reliability. The availability of such components, proven for this service, will lead to the extension of floating production capability for larger and more complex field developments.

Marine Production Risers

As previously mentioned, the term riser is used for the vertical section of flowlines or export lines connecting the seabed equipment to the surface production or offloading facilities. Risers take their simplest form in fixed installations as rigid pipes supported by the main structure.

With floating installations, tension has to be applied at the top or bottom of the riser to prevent buckling and to keep stresses due to wave and current action within acceptable limits. This may be achieved by a constant tensioning system consisting of hydraulic pistons, which compensates for the heave motion of the floater. For large heave motions, such a system can become too bulky and heavy for the floater. To overcome this, alternative methods are being sought, for example, by rigidly attaching the riser top to the floater, with a hinged connection at the lower end, connected to a boom. A ballast weight at the lower hinge of the riser provides the required tensioning force. It is also possible to have a freestanding riser, connected at the top to a large tensioning buoy. Flexible fluid transfer lines are required with all these risers to overcome the relative motions involved with floating units. These last two types of risers are as yet only in the development stages.

Oil or gas conduits on a compliant structure such as a buoyant tower can be supported along their length by the structure itself. However, a flexible joint has

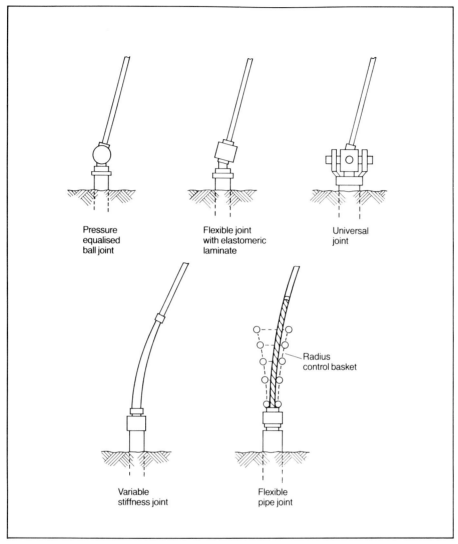

Figure 3.103 **Types of bottom joints for single-tube production risers**

to be introduced at the bottom to accommodate the movement of the structure when responding to environmental forces. Several types of articulated joints were developed for this purpose (Fig. 3.103).

The evolution of riser technology has been a limiting factor influencing the rate of growth of subsea technology, particularly in the area of floating production systems. Consequently, riser technology is the subject of much development work.

Diving and Submersibles

Historically, experience and advances in diving techniques and equipment came mainly from naval institutions and primarily for salvage and rescue operations. These techniques were adopted and the equipment adapted to support offshore oil exploration and drilling operations. Divers are still used extensively for many underwater tasks, although much remotely controlled or operated equipment has been developed because of the limitations that are inherent in diving.

As a diver descends, the hydrostatic pressure of the water surrounding him increases, so to prevent damage to the body cavities such as the lungs and the inner ear, the gases that he breathes must be pressurised. This leads to difficulties.

The first is associated with breathing compressed air. The nitrogen in the air produces a narcotic effect on the body similar to the initial stages of anaesthesia. This effect limits practical compressed air diving to 50 metres. Deeper diving is possible if another gas, such as helium, is substituted for the nitrogen in the breathing mixture. However, with increasing depth the oxygen content of the mixture needs to be reduced to prevent oxygen poisoning.

A further difficulty arises because gases breathed under pressure dissolve throughout the body. If the diver ascends too quickly, the dissolved gases can form bubbles causing the decompression sickness commonly known as the "bends". To avoid this, normal diving has to be followed by regulated decompression to allow these gases to dissipate harmlessly from the body. This was traditionally done by bringing the diver to the surface with a series of stops. This process is time-consuming and the deeper the dive, the longer the decompression period. This factor limits the depth at which safe effective work can be done by this diving technique to a little under 100 metres.

In the mid-1960s, the US Navy developed saturation diving techniques, in which the diver is pressurised to the equivalent of a particular depth and becomes saturated with the inspired gases. With special equipment he can then be kept saturated for up to several weeks at a time, returning to a pressurised chamber on the surface for rest and food after each working dive. Transport of the diver between the underwater work site and the pressurised surface facilities takes place in a pressurised diving bell (Fig. 3.104). Bells can also be used as observation chambers. This technique extended the range of effective diving to around 230 metres and had a significant impact on offshore oil developments in water depths of this magnitude, particularly in the North Sea.

Gases for the dive are supplied through "umbilicals", which can also incorporate heating, power and communications. The protection given to the diver varies from the heavy rubberised twill and woollen garments of a helmet diver, to the foamed neoprene skin of a "wet" suit. A "dry" suit contains gas as a heat insulant. A "wet" suit fits closely to the body and a small amount of water is

ENGINEERING, DRILLING AND PRODUCTION OPERATIONS 179

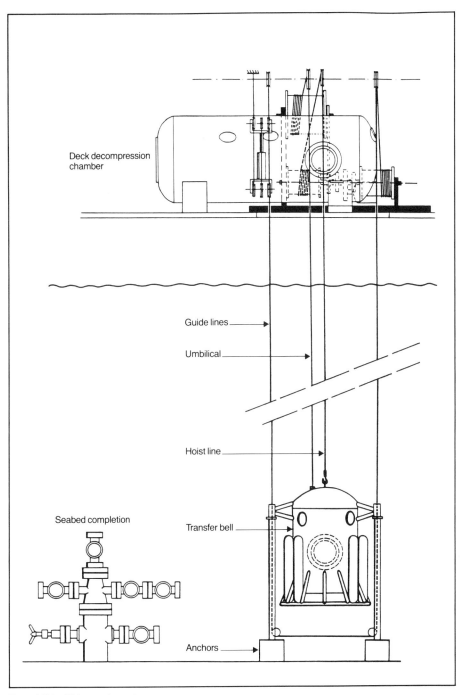

Figure 3.104 **Diving equipment**

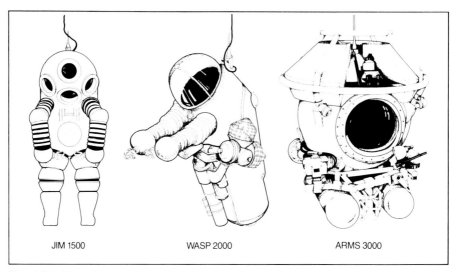

Figure 3.105 **Manned one-atmosphere units**

Figure 3.106 The PC 1805 submarine carried by the multi-functional service vessel *Stadive* can operate with pilot and co-pilot to a depth of 200 metres.

trapped between the suit and the diver's skin; this does not circulate and is soon warmed up. Under cold conditions or when helium is added to the breathing mixture causing a high heat loss, it may be necessary to keep the diver warm (electric heating or circulation of hot water) and to preheat the inhaled gas.

Communications have advanced from "pulls" on a lifeline to telephone systems. However, breathing apparatus limits the rate of speech and gas pressure distorts the voice, particularly when helium is used (the so-called "Donald Duck" effect). "Helium unscramblers" are available, which improve communication from the diver.

Figure 3.107 **Multi-functional service vessel (MSV)** *Stadive* **can perform a broad range of tasks including diving support for underwater construction and maintenance, and comprehensive emergency support.**

Difficulties associated with pressurised diving can be overcome by allowing the diver to breathe atmospheric air supplied by an umbilical from the surface. This requires a capsule built to resist the pressure of the surrounding water. Such systems are described as one-atmosphere units. Figure 3.105 shows JIM 1500, WASP 2000 and ARMS 3000, which can perform limited tasks using manipulators in up to 600 metres.

There has been significant development of tools and systems aimed at minimising the use of divers. Unmanned submersibles, which are controlled from surface vessels, are now available. These vehicles can be directly tethered to the vessel, or to a protective launching device which is suspended from the vessel. Equipped with underwater TV, these vehicles can be used for inspections. They can also be fitted with manipulators to perform light tasks.

Manned submersibles or mini-submarines (Fig. 3.106) are also available for underwater inspection, some including lock-out facilities for divers. In view of their limited range and life support system, these submersibles operate with a surface vessel standing by.

Operating companies with extensive offshore production usually employ full-time diving vessels for underwater inspection and maintenance, like the sophisticated DP semi-submersible diving vessel, Stadive, which will operate in the northern North Sea for Shell (Fig. 3.107).

ECONOMIC, FINANCIAL AND OTHER ASPECTS OF EXPLORATION AND PRODUCTION ACTIVITIES

Risks of the Business

All businesses are risky but some are more so than others and by most standards exploring for oil and gas must be one of the riskiest. To the businessman the idea of "risk" reflects his uncertainty or lack of foreknowledge of the results of a particular decision or action. A low-risk business is one in which the financial results can be forecast with a fair degree of confidence one or more years in advance, and are rarely influenced by external factors over which the businessman has no control. The situation in exploration and production is very different, with a multitude of unknown and uncontrollable factors which can affect the fate of a project, such as:

- will exploration be successful in discovering a new energy resource?
- if successful, will it be oil or gas?
- what quantities will be found?

- can they be developed and brought to market economically with existing technology?
- if new technology is required, will it work and how much will it cost?
- what will be the future market price of any oil or gas found?

In addition to these risks specifically related to exploration and production, such ventures are also subjected to the general economic and political risks affecting any business, with the additional uncertainty that in many countries oil and gas production are of national economic importance and are therefore particularly exposed and vulnerable to government policy changes such as changes in taxation.

Arrangements with Governments

In the main, petroleum operations can only be conducted by virtue of licences, permits and leases granted by governments, or through contracts concluded with state corporations. Over the years those legal relationships have been subject to significant developments. Where countries are fortunate to have a major or potentially major hydrocarbons resource base, petroleum operations are invariably a vital part of their economies, and governments therefore have increasingly extended their involvement in and control of such operations. This greater involvement has resulted in increasing state participation, establishment of state corporations and increased government sharing in the financial benefits arising from these activities.

Petroleum exploration and production arrangements between a state or its national oil company (NOC) and oil companies can be split, very broadly, into either risk-bearing investment agreements or non-risk-bearing services or technical assistance agreements. Most of these agreements are the former, and can be further sub-divided into concessions or leases, production-sharing agreements and risk-bearing service contracts.

However, from country to country (and even within a country) contracts within each of these groups may vary considerably to suit particular circumstances. There may also be many overlapping features, and many principles are common to two or more particular types of contracts.

The application by a government of any particular type of arrangement, and ultimately its financial format, may be the result of detailed legislation or of free negotiation. It is normally reached after weighing a great number of parameters, for example: historic reasons such as a well-established extractive industry; entrepreneurial freedom or dirigisme; mature local oil industry or lack of expertise; system of taxation; petroleum imports or exports; cost of exploration and production; and last but not least, geological prospectivity.

Fundamental Aspects

Four important general aspects may be discerned with varying emphasis in each individual agreement, namely the management of the operations, the investment at risk, the access to petroleum and the economic return.

The management or the control over operations may range from virtual freedom under a concessionary or lease arrangement, without State participation, to little or no control and operational responsibility under certain forms of technical service agreements. Very generally, the degree of control left to the oil company will be commensurate with the degree of investment risk undertaken. Furthermore, the venture operator will be subject in any event to the (petroleum) legislation in force, which normally provides for considerable government supervision and control over working methods and the execution of work programmes.

Under a concession or lease the producer will obtain the totality of production, but may have to offer a proportion for sale to the NOC. In the case of State participation such a right will be proportionate to its equity share in the joint venture. Production-sharing contracts allow the oil company a certain percentage of production in repayment of its expenditures, and the remainder (the "profit share") is split with the State or NOC in a pre-arranged manner. An absolute right to petroleum is not a logical feature of a technical service agreement, but often access to at least part of the production is assured by way of a concurrent purchase-and-sale agreement.

Clearly, while the economic return aspect is the decisive factor in reaching an exploration and production agreement, that same aspect is also not greatly dependent on the actual form of the agreement. For competitive reasons the minimum requirements of oil companies in this respect differ little, and governments make use of this. On the other hand, the economic results of the venture as a whole will vary considerably from case to case; as a consequence, the "government take", that is the share of the economic results accruing to the host state by way of such items as taxes, royalties, profit sharing and production sharing, will also vary. Viewed from the cost side, the economic results will be determined largely by physical factors, such as geological conditions (field size, reservoir performance), geographical situation (onshore/offshore), accessibility and environmental constraints. These factors determine the cost of finding and, in the event of success, the cost of producing the hydrocarbons and so dictate the limits for the government take.

However, before any decision to embark on a venture can be taken the probability of finding commercially exploitable deposits will have to be assessed. The risks that can be taken depend very much on the expectations regarding the economic return. When a government lays down the financial, fiscal or production-sharing terms (terms which determine the government take and thus the economic return), it must take these factors into account. If its demands are too high, it must expect little interest from the oil companies or, at best, a minimum exploration programme aimed only at a better assessment of the area's hydrocarbons potential. In such cases, a government would have to moderate its (standard) financial conditions/desiderata in order to attract oil companies to mount a sustained exploration effort, or to develop modest discoveries.

Other Factors

Of course there are many other aspects customarily provided for in petroleum agreements, often in addition to the provisions of the general mining or petroleum law, and they are the same in essence, irrespective of the type of agreement. These are, on the one hand, the terms and conditions which are aimed at keeping a tight governmental rein on the operations, namely: duration of exploration, development and production periods; area size; relinquishment obligations; minimum work/expenditure commitments; approval of work programmes; reporting obligations; preference for local goods and services; tender procedures; training of the company's national employees and those of the NOC; and domestic supply obligation. On the other hand, to afford protection to the investor, there will be freedom from certain duties and taxes; freedom to remit and dispose of profits; security of tenure; and provision for applicable law and international arbitration.

The provisions may not be fundamental for the economic viability of the agreement, but they are essential for the proper conduct of operations under the contract.

Taxation

Taxation of income is dependent on the type of agreement. Where a large share of the economic results of the venture is already surrendered to the government by agreement, there is not much room left for taxes higher than those applying generally in the country; the government take would become unreasonably large and the interest of oil companies in such agreements would diminish correspondingly.

As a rule, the most extensive taxation of income under concessions or leases is found where the only other provision for government take consists of the payment of royalties. Oil companies working under a production-sharing or risk-bearing service agreement usually will be subject to the generally applicable income tax and often will be exempted from the payment of other taxes. Only very seldom is there a total exemption from taxes.

In many older production-sharing contracts, the oil company, while liable for income tax, will not actually pay the tax; it is instead included in the government's profit share. Where the government take consists mainly of taxes and royalties (as under a concession or lease) the following conditions may apply:

- General corporation tax (in most countries around 50 per cent).
- Special taxes in addition to corporation tax. These may be a deductible item for the calculation of corporation tax or be levied independently.
- Special profit share. Corporation tax remains payable and forms part of the total profit share accruing to the government.
- Windfall profits tax. All or part of the difference between the sales price and a certain (indexed) base price to be surrendered to the government. As it is not related to profit, this type of tax is more a royalty than a tax on income. A variant of this method of taxation, which makes the levy more profit-dependent, is to allow the producer a minimum return on his investment before the levy is triggered. This modified system finds ever wider applications since it avoids too heavy taxation on smaller, more costly and marginally economic fields, the development of which needs to be encouraged. This tax system can exist in combination with production sharing.

The taxes under the first three forms are alleviated by the introduction of measures reducing taxable income, for example, by excluding part of the proceeds from the calculation, or by allowing additional deductions expressed as a percentage of deductible expenditures, resulting in a deduction of such expenditure more than once. Royalties form an essential feature of a concession or lease. They are determined by the rate, assigned royalty value, timing and place of delivery. They are payable in cash or in kind, usually at the option of the government. If payable in cash, the determination of the royalty oil value is of great importance. The rates may vary from zero per cent to 40 per cent, and may be fixed or be dependent on quantity produced. Usually royalty is a deductible item for the calculation of income tax. Governments may waive or reduce the royalty as an incentive for the development of marginal fields.

The evolution of participation and fiscal terms can be attributed mainly to the basic principle of "copying". Governments have become increasingly aware of the value of the potential maximum economic rent to be derived from their hydrocarbons properties, particularly during the oil "crises" of the 1970s, and in "leasing" to the oil companies have sought to maximise their own return by examining the methods used by other countries, particularly those countries which are members of OPEC. This has led to a steady increase in the control of the resources and in "take" by governments in general. The issue this raises is the question of what is a fair level of return on investment and of reward for risk by oil companies which invest heavily in searching for such resources, and which continually improve the technology required for the development of resources from increasingly remote and hostile regions.

Economics

Benefits *versus* Costs

For any business activity to be justified the expected benefits must exceed the expected costs. In the case of an exploration and production venture the principal benefit is the profit arising from the sale of any oil or gas discovered. The costs are those of the initial exploration programme plus, in the case of success, the usually much higher investment in developing the resource, including such items as wells, pipelines, oil or gas treatment facilities and ocean terminals, as well as platforms and other marine structures for an offshore field. Before an investment is made in a new project, both the benefits and the costs are unknown and can only be estimated with great uncertainty in view of the risk factors mentioned above. Nevertheless, because of the immense sums of money involved in such projects and the long periods for which the cash is "locked in", such estimates have to be made, and this is the task of the exploration and production economist.

Exploration Economics

In most countries, the petroleum laws provide for governments and exploration companies to negotiate contracts which, typically, oblige a company to carry out an exploration programme of agreed scope and cost, and entitle it to participate in the development and production of any oil or gas discovered, in return for an agreed share of the rewards (see Arrangements with Governments). Negotiations on acquisition of new exploration acreage often involve careful consideration of the size of the exploration programme to which a company is prepared to be committed, and of alternative profit-sharing mechanisms. In addition, in some highly prospective areas arrangements may include the payment of a signature bonus to the government, which counts as an additional cost to the venture.

The size of the reward in the case of success depends on the geology, which, before exploration starts, is usually only known in broad outline, and is evaluated as accurately as possible using the quantitative and qualitative methods described in the Exploration section.

Economic analysis of an exploration prospect is then based on a computer model simulating the expenditures and revenues of a successful oil or gas production project in the area concerned, including exploration, development and a realistic forecast of production levels. In view of the many technical and other uncertainties, a wide range of financial results is possible, and it is usual to calculate the results of various different sets of assumptions before settling on the range which seems most realistic.

The calculations used are of the "discounted cash flow" type and are standardised in a manner permitting comparison and ranking of alternative investment opportunities in different countries. This involves, for example, using single forecasts of crude oil prices and inflation levels for all comparable projects. One important objective of these calculations is to estimate a range of values of the "real rate of return" and "net present value" of the project (calculated in constant money, eliminating the effects of inflation).

Exploration Costs. With regard to the costs of future exploration, the best guide is probably the actual experience of ventures in progress, or recently concluded. Some unit cost figures for various exploration projects are given in Table 3.3. They have been extracted from the financial returns of a number of Shell exploration companies and are expressed in 1982 US dollars.

Data Acquisition Costs. Actual costs for exploration data acquisition vary widely, depending on such things as the geographical nature of the terrain (offshore or onshore, low-lying or mountainous, swamp, jungle, desert or cultivated), on the available infrastructure, and on the complexity of the geology being surveyed or drilled. Variability is generally greater onshore than offshore.

Offshore seismic costs in 1982 were in the range of $700,000 to $1,000,000 per crew-month, or $600 to $1,200 per kilometre surveyed (averaging about $800 per kilometre).

For onshore seismic the following costs are typical:

	US $ per crew-month	US $ per kilometre
Desert areas	450,000	3,000
Western Europe, populated	400,000	4,000
Tropical jungle	1,200,000	up to 25,000

Offshore drilling costs in 1982 were estimated in the range of $4 million to $6 million per month, including ancillary services but excluding overheads. These

Table 3.3 **Venture costs for some new exploration areas**

Venture	Type	Number of wells	Cost (1982 US $ million)
A	Offshore	1	25
B	Onshore, desert	2	25
C	Offshore	9	90
D	Onshore, desert	5	110
E	Onshore/offshore Africa	4	40
F	Onshore Latin America	1	10

costs would be much higher if a rig had to be mobilised for a small programme of, say, one or two wells in a remote area without infrastructure.

For onshore drilling the corresponding figures for a heavy land rig, such as might be used for a deep exploration campaign in a new area, were in the range of $1 million to $2 million per month.

Other Costs. The costs of all special studies and technical analyses, as well as the share of back-up research necessary to maintain them, vary greatly according to the particular technical problems involved. On average, however, these costs, together with purely administrative overheads, amount to about 20 per cent of the data acquisition costs in any given venture.

Venture Costs. When considering entirely new exploration programmes in areas or countries without ongoing exploration activity, it is useful to have some typical costs for an entire venture, including data acquisition studies, prospect appraisal and overheads. Table 3.3 shows the total costs incurred by Shell companies in the late 1970s and early 1980s in six such ventures. The number of exploration wells drilled is given in each case as an indication of the size of the venture.

These high venture costs were, in effect, incurred while testing the validity of a set of geological concepts. As it turned out, all six of these particular ventures happened to be unsuccessful and had to be abandoned without any return on the investment involved. Had they been successful, the exploration expenditure would merely have been the first stage of a much larger investment in the appraisal and subsequent development of the oil or gas discovered.

Production Economics

Once oil or gas has been discovered and confirmed by appraisal drilling, a new round of cost estimates and economic calculations has to be made in order to help to decide whether further investments are justified. After successful drilling the technical information available is much more detailed than in the exploration stage. Drilling results will have confirmed the depth, thickness, porosity, productivity and pressure regime of the oil- or gas-bearing reservoir and the appraisal wells should have outlined the lateral extent of the deposit and set a minimum value for the size of the reserve. Engineering studies are now carried out (as described in Production Development section) to determine the most efficient way to develop the reserve. Since this is both a technical and an economic problem, engineering and economic studies are carried out in close cooperation.

Economic evaluations at this stage are usually based on computer models simulating a range of technical and financial conditions. Although similar in concept to the economic models used for exploration projects, the production economics models tend to be more complex, reflecting the much greater amount of detailed information now available and incorporating revised assumptions on development costs and project performance.

Development Costs. Development costs of production projects are estimated as far as possible by reference to known costs of existing projects carried out under similar conditions. This is not as helpful as it sounds, since many projects involve some element of novel technology. Moreover, standards are changing in matters relating to the "environment" in the broadest sense, including pollution and noise abatement, visibility of facilities and quality of crew accommodation, all of which influence the capital costs. Having said this, there are some simple "rules of thumb" about development costs:

- The least expensive developments are fields located onshore close to a coast accessible to tankers of appropriate size. Costs increase in an inland direction (because a pipeline is required), and seawards, with rapid increases according to water depth.
- Large fields generally have lower unit costs than small fields, reflecting scale economics resulting from greater throughputs in such items as pipelines and terminals.
- Reservoirs capable of high production rates for each well require relatively few wells, with a favourable impact on costs.
- Technically simple production projects using primary reservoir energy to produce the reserves are always cheaper than more complex multistage developments with a longer lifetime and higher recovery factor, although the latter are becoming more common as increasing attention is paid to maximising oil recovery.

The development costs themselves are often quoted in thousands of US dollars per barrel per day (MDBD) of installed capacity, a measurement which can also be applied to gas fields if gas volumes are restated as barrels of oil equivalent (see Table 3.4).

Table 3.4 **Typical MDBD costs (in constant 1982 US dollars)**

Case	Development costs ($1000s per barrel per day)
Small onshore oil field, Western Europe	5–8
Small offshore oil and gas field, North Sea	7–35
Small onshore gas field, Western Europe	7–10
Large deep-water field (300 metres water depth) in non-hostile climate	20–30
Onshore heavy oil field, Middle East, including steam injection facilities	10–15

Real development projects may of course be subject to inflation of the "constant money" costs. In the late 1970s this relative inflation was particularly high, for reasons related to the two main components of development costs:
- "international" construction costs (e.g. platform construction costs, pipe-laying barge costs).
- local costs incurred in the country where work is being done (e.g. local labour costs).

Of these the "international" costs show high inflation rates at times of rapid industry growth and competition for oilfield construction services, while the local costs come under particular strain in periods of accelerated local economic activity in oilfield areas (e.g. in the cities of Aberdeen and Stavanger). In the late 1970s, oil and gas project costs' inflation, for these combined reasons, has tended to greatly outpace general inflation, as measured, for example, by retail price indices in the producing countries. This is distinct from a second major cause of project budget over-runs, namely the underestimation of technological complexity during the early stages of project development. This is also a common feature of the introduction of advanced technology.

Financing of Exploration and Production Activities

Where then does the oil industry obtain the large (multi-billion dollar) sums necessary to finance the search for, and development of, hydrocarbons resources? There is no one simple answer. The sources of finance may vary from company to company and the financing methods also. The age and size of the company will also influence its ability to finance its activities.

As a start to providing an answer, we should first look at the way the large international companies in general find the money to finance their activities. The methods they use are by and large the same for all established oil companies and for that matter for any well-established company active in another field but entering the oil business. It is only with new companies specially formed to enter the oil business that significant variations from the general pattern are found.

To launch into business, large companies raise funds by issuing shares and use this money to finance their activities. Once a company is established and actively doing business, funds coming directly from shareholders or raised by issuing new shares become of less significance in financing expenditures. To find the money it needs, a company looks first to two main sources of internally generated funds. These are net income (profits) and depreciation. The first needs little explanation. It is the money remaining from the income received from the sales of products after the company has paid taxes and royalties and met all its operating expenses including depreciation. From its net income a company will normally pay a dividend to its shareholders and whatever remains is available to reinvest in the business.

Depreciation is the amount of money a company provides to replace assets at the end of their useful working lives, and/or to spread the costs over the lives of the assets. On the production side of the oil industry one most important source of depreciation is that arising from the actual production of crude oil or natural gas. This depreciation (called "depletion" by the oil industry) is the result of taking a factor resulting from the ratio of production in any one year to the reserves of hydrocarbons remaining to be produced and applying it to the costs which have been necessary to develop the reserves. These amounts of money are available for any increase in working capital necessary and then either to spend in exploring for or in developing sources of hydrocarbons.

So, in order to finance its future activities, an oil company will have available two main sources of internal funds: the remaining net income after it has perhaps paid a dividend to its shareholders and depreciation/depletion. However, a company may still not have enough money from these two internal sources to undertake the activity it wishes to carry out. In that case, it simply borrows.

Until the late 1960s and early 1970s, when the oil industry moved into large-scale and expensive offshore exploration and development, its exploration and production activities, broadly speaking, had been self-financing. The industry had been able to use the funds generated by its existing activities to finance new ones. However, it takes time (perhaps five to six years) to develop, build and commission an offshore platform and enormous sums of money are required. In addition, during that time money is going only one way, namely into the project. None will come out until production actually starts, which will be after completion of the platform and the drilling of the first production wells. The industry was unable to generate all the money it required to finance its programmes, and so it borrowed to finance the shortfall between its own funds and the funds it actually required. Borrowing, of course, can take many forms, and much imagination and ingenuity was evidenced in the manner in which the oil industry and the financial community sought and found methods to make available the funds required.

In many cases the largest companies were able to raise loans guaranteed by the inherent financial strength of the company as a whole. The loan monies were made available to the company and not tied to any particular aspect of that company's activities. In other cases, loans were tied to particular developments (project financing) with the revenues, or a portion of them, from the production expected from the development being pledged to repay the loan. In other cases, companies would sell in advance part of the production expected from the development in return for funds. Today, a loan agreement for a major project usually will involve not just one bank but a consortium of banks or financial institutions and may include aspects of all the various methods outlined above. However, even with the increasing resort to outside financing, which has become necessary to cover the ever-increasing costs of new oilfield developments, especially as the industry moves into development of smaller accumulations of hydrocarbons in more difficult areas, the industry itself still provides the greater part of the money. The industry ploughs back much of the funds it has available. In recycling these monies, it is backing its own abilities and expertise in minimising the risk of failure, both in finding new resources and in developing new and increasingly complex reservoirs, which may be in deep water or in a hostile environment.

Many of the larger and more costly projects have been financed in this way because major companies which compete with each other to market the oil products, cooperate in joint ventures in order to share the burden of providing the necessary funding for large developments. Such developments may each take several billions of dollars before oil or gas starts flowing and several billions more to complete.

For the smaller companies entering the oil business the necessity for borrowing has led to some extremely intricate arrangements. In turn, banking and financial institutions need to understand the risks and the geological and technical complexities of the exploration and production business in order to link repayment schedules to production of a resource which, in its early stage of discovery and appraisal, is to some extent an unknown quantity.

Project Management

As the technology of the industry has become more sophisticated, so the management techniques needed to monitor and control activities have become more complex. Nowhere is this more pronounced than in the management of offshore construction projects, particularly in the North Sea. So vast are these developments that they have earned the title "mega-projects".

Forty years ago, few management techniques were available to the construction engineer. There was a dramatic improvement in the late 1950s when critical path analysis was introduced in the nuclear submarine construction programme in the USA. This powerful technique flourished with the advent of computers, particularly mini-computers. Critical path analysis and other management techniques have been employed and further developed in the construction programmes of North Sea mega-projects. This section reviews some of the management techniques in use in the industry in the early 1980s.

Planning

Modelling is perhaps a more accurate description of the present generation of planning techniques. Immense computer models of projects are created and maintained in network form. The project is analysed into literally thousands of inter-dependent activities, and the network describes the logical relationship between these activities in time series. To each activity is assigned a duration, a cost and a resource level. The model can determine the schedule, highlighting the critical path, i.e. the schedule of activities which may delay the overall completion of the project. It can also provide cash flow forecasts and expenditure estimates.

Resource levels are also plotted across activities with time in order to pinpoint any deficiencies or excesses. The great advantage of this technique is that the modelling can be iterated, altering the logic or sequence and shifting resources, until the optimum result is obtained. This then forms the "Reference Model", against which progress and performance are measured.

This model can also be linked to an economic model which simulates the entire life of the project. For an offshore oil-development project, this will include such variables as production and oil price forecasts, operating cost projections and expected tax structures. By combining this model with the schedule and cash flow forecasts the profitability of the project can be assessed. It follows that changes, both proposed and actual, during the design and construction phases of the project can be evaluated in terms of an increase or decrease in the project's profitability.

These models can also handle probabilistic data, which are a range of possible durations/costs/resources and their respective assessed probabilities. Then by means of computer-generated simulation using random numbers, probability distributions of schedule/cost/profitability can be produced.

These techniques are still in their infancy, but the mega-projects of the North Sea have contributed significantly to their development.

Procurement

Procurement is another area where new computer systems have been developed and implemented for North Sea projects. The earlier systems tracked materials from the purchase order stage through stock to the issue stage. The new systems pick up a requirement from the moment it is identified by the designer and follow it through to the point of installation.

Such systems are on-line and can be accessed from fabrication yards across the globe. Material identification at all stages is vital for certification purposes. It is necessary to be able to identify every piece of steel in a structure and to evidence this by means of test certificates. This presented a filing/retrieval problem of monumental proportions, which has again been overcome by the use of computers and microfilming.

Contracting

Partners to contracts have agreed on new sets of contract forms for mega-projects, because the old "standard" forms in use in the civil and mechanical engineering industries proved inadequate for the type and magnitude of contract works and services being undertaken. New procedures for the tendering, evaluation and award of contracts have also been developed to cope with the considerable internal control problems of contracting on such a scale.

Accounting

Accounting for mega-projects nowadays demands the services of a new type of accountant. The complexity of modern-day taxation regimes makes accurate accounting of expenditure essential. Information demands from fiscal authorities and government agencies are becoming more and more detailed, making even greater demands on today's accountants.

Control of commitments has replaced control of expenditure as the first-line budget control tool. Commitments are the monetary expression of contracts entered into and purchase orders placed. Collecting, coding and recording all this information has exceeded the capacity of conventional accounting systems.

Most mega-projects are undertaken by joint ventures in order to share the risks or meet local legislation or enjoy more favourable fiscal terms. This adds a further dimension to the accounting problems in that cash has to be called from partners to meet their shares of forecast expenditures, and partners' shares of actual expenditures must be reported to them. Typically, the operator of a joint venture must retain two sets of books, namely a joint venture set in 100 per cent terms and a corporate set in operator's share terms.

Quality Assurance

Cost, time and quality are perceived as three competing forces in any project. Cost can often be reduced, but only at the expense of time and quality. Quality costs money and in some cases time as well, but it is often forgotten that lowering quality can result in the loss of time and consequent loss of money.

The aim of quality assurance (QA) is to monitor and control quality within pre-determined parameters, a difficult task since there is no convenient measure, whereas cost and time are fairly easily measured. The QA specialist monitors design to ensure compliance with standards and specifications, undertakes hazard and risk analyses to determine "what happens if ...", looks after material and site inspection, and performs technical audits. Weight control, often critical to an offshore installation, is also the responsibility of the QA specialist in many instances.

Organisation

The management of mega-projects generally calls for the formation of a multi-discipline project team. Disciplines represented can include design, construction, drilling, petroleum engineering, and materials personnel, economists, accountants, contracts engineers, quantity surveyors, and many others.

There is typically a wide geographical spread of activities. For example, the team may be located in the United Kingdom, design work be undertaken in both the UK and the USA, construction work be carried out in continental Europe, and materials may come from Japan, the USA and Western Europe. The project finally comes together 200 kilometres or more off the coast of Scotland. The organisation and procedures must cope with these distances, otherwise insurmountable communication problems may result.

Insurance

Mega-projects require "all risks" construction insurances of a magnitude hitherto unequalled. Specific insurance policies have had to be developed to serve this purpose. Such insurances are underwritten on a worldwide basis.

Offshore Logistics

Any exploration venture or production project requires support to some degree. The degree depends on the remoteness of the location and the hostility of the environment.

One of the most hostile environments in which large-scale exploration and production has taken and is still taking place is the North Sea, and that area provides many examples of the logistical problems that occur in any cold, stormy and deep stretch of water.

On land, the field geologist may have to leave his wheeled transport and go on foot through rough or roadless terrain. The seismic crew may have to cut a path through undergrowth, and drilling crews often have to work for weeks or even months far away from civilisation. Sometimes, it may be necessary to build roads and even airstrips in order to bring in water, food and equipment.

Offshore, such problems are magnified and new problems arise, especially where good weather is relatively infrequent. The exploration drilling rig may be many kilometres from the nearest harbour and supply vessels have to make regular deliveries of all that is required to keep the rig operating: drill pipe, casing, mud, cement, fuel, water, chemicals and foodstuffs. The duty crew spend one or two weeks on the rig and have to be ferried in and out by helicopter (Fig. 3.108).

Figure 3.108 A Sikorsky S61N helicopter delivers a relief crew to a semi-submersible drilling rig.

Although the weather may not be severe enough to prevent a rig from drilling, it may be bad enough to prevent the supply vessels (Fig. 3.109) from maintaining station alongside the rig or stop operation of the rig's cranes. So both rig and supply vessel may simply have to wait out the storm. In the North Sea, particularly in winter, "waiting on weather" is a common phrase in the daily reports of rigs and vessels.

The logistics problem becomes even greater in the development phase of offshore oil and gas fields. A major advantage for operating companies is that the countries surrounding the North Sea have suitable harbours, transport and industrial facilities (Fig. 3.110 shows a typical supply base). The major coastal

Figure 3.109 **A North Sea supply boat holds station alongside a production platform.**

Figure 3.110 **The Shell UK Exploration and Production supply base in Aberdeen harbour.**

towns and cities selected by these companies for the setting up of the initial forward bases have the required infrastructure (such as housing, shops and schools) for staff to live with their families. As exploration led to discovery and subsequent development, so these bases have become full operations headquarters.

At first, the logistics support required was light, with only a few drilling rigs operating, and these mainly in the "weather window" period of April to September, when the least downtime due to bad weather is experienced. Discoveries of gas in the southern North Sea in the 1960s were followed by oil discoveries further north in the early 1970s.

The discovery of a number of oil and gas fields within a few years sharply increased the momentum of activity, as several operators set about the task of designing and developing the offshore structures required to drill for and produce hydrocarbons from reservoirs lying up to 6,000 metres below the seabed in up to 200 metres of water and perhaps 250 kilometres from the nearest support base. Huge steel and concrete structures were built at the coast, towed out and installed on the seabed. This was an entirely new venture and an expensive one, and many lessons were learned. Delays were the rule rather than the exception. There was a

Figure 3.111   **A steel jacket structure. The Shell/Esso North Cormorant platform jacket on tow to its location.**

lack of suitable offshore construction vessels, because the existing monohull derrick barges required fairly steady sea states and low winds to lift the massive equipment packages (modules) and place them accurately on the decks of the structures.

Concrete structures have the advantage that much of the superstructure can be built and assembled inshore and the entire platform towed out and ballasted down on site (Fig. 3.92). Steel jacket structures, however, need good weather for towing out, launching, tilting to the vertical and emplacement (Fig. 3.111). The good weather has to last long enough for the structure to be piled into the seabed to secure it, before the equipment modules can be lifted into place. Delays in floating out these huge structures were often compounded by missing the weather window, which usually meant waiting until the following spring. Entire projects slipped by a year and even longer.

The basic aim of constructing the equipment as packages or modules was to minimise the number of men required to be offshore during construction, hook-up (linking together of equipment packages) and commissioning of the drilling and production installations. However, the workload was extended by the limitations imposed by the weather and, again, delays would ensue. Limited accommodation on site was a major problem. Offshore accommodation modules are generally designed for a drilling crew and production team of up to say 100 men at any one time, and are too small to cope with the number of men required (say up to 400) to hook up and commission the many complex power, process and life-support systems. Expensive construction barges could not be retained on location simply to house construction workers and the transfer of men from barge to platform by gangway is only possible in reasonably good weather.

Gradually, thanks to a temporary surfeit of semi-submersible drilling rigs, a new type of vessel came into service. With their derricks and even their drilling equipment removed and accommodation modules installed, several of these units became temporary accommodation vessels or "flotels", housing up to 500 men.

Anchoring them alongside fixed installations with a gangway in between, allowed the workforce to be housed conveniently close to the work site. In bad weather the gangway is lifted and the flotel pulls back along its anchor chains a safe distance to avoid the risk of collision. This could have interrupted the work schedule, but for the helicopters which often can fly even when conditions prevent the gangway from being connected. Thus the "shuttle" between flotel and installation came into being, with small "field-based" helicopters capable of carrying a dozen men at a time.

The increasing use of in-field helicopters demanded suitable offshore bases to avoid the need to return to shore for maintenance, which resulted in unproductive flying time. In some cases, therefore, hangars were built on the decks of flotels and helicopter engineering teams were added to the variety of skilled support

ECONOMIC, FINANCIAL AND OTHER ASPECTS

personnel who came to work offshore. These are too numerous to discuss in detail, but range from crane operators, radio operators, medical staff, cooks and stewards to divers, electricians, mechanics and painters.

East Shetland Basin

In 1978, the construction of the Shell/Esso Brent, Cormorant and Dunlin fields in the East Shetland Basin of the North Sea was at its peak, with as many as 4,000 men offshore at any one time and living on the (at that time) six production platforms and in four or five flotels. In addition to the large S61 helicopters flying men to and from the mainland of Shetland at the end or start of their offshore stint, a fleet of Bell 212 and Bolkow 105 helicopters was employed to ferry personnel and equipment from installation to installation.

Flotels became increasingly sophisticated. One, Treasure Finder, a converted drilling rig, has twin helicopter landing decks between which is a massive hangar capable of holding five of the Bell 212s (Fig. 3.112).

Figure 3.112 **Aerial view of the accommodation vessel *Treasure Finder* anchored alongside the concrete gravity platform *Brent B*.**

Several phases of offshore activity began to merge, creating the new problem of deciding on work priorities. Construction may have to be carried out on a platform which is still drilling new wells and at the same time producing and processing oil and gas and delivering them to shore. Added to all this is the need to carry out maintenance and modify equipment or to add new equipment packages for unforeseen tasks and processes.

Helicopter traffic, including that for other operators' fields under development in the area, grew to such an extent that it called for a full air traffic control system to ensure safety in the air. This system in one peak summer month handled 22,000 air movements, only 4,000 fewer than London's Heathrow airport in the same month.

The surface of the sea around the platforms also became increasingly congested with flotels, safety vessels, derrick barges, diving support vessels, tugs, anchor handling and supply vessels and the regular shuttle tanker taking Brent oil from the Spar loading facility. A marine coordinator was therefore brought in to decide priorities and calculate anchor patterns to avoid moorings becoming entangled or, more seriously, an anchor being dragged across one of the interfield oil and gas pipelines.

New Developments

Rapid development in the North Sea and elsewhere has led to tremendous innovation in many areas of activity. The semi-submersible principle of drilling rigs was applied to derrick and pipelay barges, giving them greater stability, and this enabled construction engineers to challenge, and to a great extent overcome, the limitations imposed by the April to September weather window.

The need to keep men moving to and from the mainland regularly, despite bad weather (including fog at airports or in the field), led to the development of a large helicopter which can fly 44 men direct from Aberdeen across the 480 kilometres of sea to the Brent, Cormorant and Dunlin fields in less than $2\frac{1}{2}$ hours. The alternative is a staged journey by fixed-wing aircraft from Aberdeen to Shetland and by S61 helicopter (19 passengers) to the fields which, in good weather and counting waiting time, can take three hours, but in bad weather stretches to several more hours or may even leave men stranded in Shetland. In all these North Sea endeavours, the "learning curve" has been one of the steepest encountered around the world. The challenges have been enormous and have called for ingenuity and innovation at almost every stage.

On land, the wellheads, pipework, power and process equipment of an oil field may be spread out over 25 square kilometres of ground. Cramming that hardware on to a single offshore platform with a deck area of only half the size of a football pitch, large though that may seem, has called for new thinking.

ECONOMIC, FINANCIAL AND OTHER ASPECTS

One such platform may have to cope simultaneously with the drilling of one of up to 30 or even 40 wells, almost all of which are deviated to reach various reservoir targets; the production of oil from several of these wells; the treatment of the oil to separate out any gas or water; the treatment of sea water for injection; the re-injection of gas at pressures up to 6,000 psi (415 bars); the delivery of oil and gas by pumping to shore; the generation of up to 14 megawatts of electricity (enough to light a small town) to power all the systems. In addition, up to 200 men have to be housed, fed and even entertained in their off-duty periods, supplies have to be lifted on board from supply boats, and helicopters must land and take off with men and equipment.

Developments in communications have been stimulated by the need for contact between the platforms, and from the platforms to the ships and flotels around them, to the helicopters in the air, and to the headquarters, airports and other locations ashore.

The management of the production of oil and gas from an area as complex as the East Shetland Basin, where several operators have elected to share oil and gas pipelines to shore, has called for new thinking and new methods of operation. For example, the Brent System involves a number of platforms sharing an oil pipeline to Shetland and a gas pipeline to the UK mainland.

To produce and deliver the oil and gas in the most economic manner, and

Figure 3.113 **Production Coordination Control Room in Shell UK Exploration and Production Northern Operations Centre, Aberdeen, UK.**

simultaneously to ensure a fair share of pipeline capacity, have called for a sophisticated computer-assisted monitoring system. This involves taking information automatically from up to 2,000 separate instruments on each platform, as frequently as once every 10 seconds, processing the data by on-board computers and transmitting them by microwave links to a central platform. The collated information is then transmitted by a tropospheric scatter radio system to Shetland, from where British Telecom microwave and telephone cables relay it to the Shell operating company's headquarters in Aberdeen. There in the Production Coordination Centre (Fig. 3.113) it is further processed by computer to provide an overview of the entire system and, on demand, a "close-up" view of individual systems on individual platforms.

One of the centre's major functions is to monitor pipeline integrity, a mandatory requirement, which ensures that any damage to a pipeline is detected almost immediately, so that appropriate action can be taken to minimise any effect on the environment should oil escape into the sea.

The Cost

The cost of all these activities, both capital and operating, is enormous. The development of the Brent field alone, with its four platforms and its separate oil and gas pipelines to shore, has cost more than £3,500 million. It may cost £5 million to drill just one exploration well, £350 to fly one man to his offshore work location and back, and £50 to ship one tonne of cargo from shore to platform.

The result of these activities in the UK and Norwegian sectors of the North Sea is that both countries have become more than self-sufficient in oil in a relatively short period, a fact which marks the size of the endeavour. It has also paved the way for future exploration and possible production in equally or even more hostile environments.

Safety and Environmental Conservation

By its very nature, the search for and production of hydrocarbons poses many problems both in terms of safety of operations and in the potential for damaging the environment, whether onshore or offshore. Governmental and public concern over these problems has increased over the years and the surge of offshore activity close to developed, popular shores has heightened that concern. Everyone desires the benefits derived by governments and individuals from the oil industry's effort to develop national resources and bring the product to market, but quite independently, many apply pressure to prevent the most economic method of delivery.

It is against this background that industry in general and the oil industry in particular has developed a strong awareness of matters relating to safety and the environment. It is common practice now to employ specialists whose sole concern is to identify and eliminate the causes of accidents and to prevent damage to the environment. The exploration and production side of the oil industry has to be in the forefront of this type of activity. Heavy machinery is operated, and the hydrocarbons produced and handled are not only inflammable but often under high pressures. With the general move into offshore operations, marine hazards now also require most serious attention.

Although prevention of accidents to the person is obviously the prime objective of safety measures there is also an important contribution to be made towards cost saving, both in the capital cost of replacing an item of equipment that has failed and in the operating expenditure incurred by the consequent down-time. Not all equipment failures will cause accidents but the potential is there, so it is necessary to investigate the initial causes of failure back at the design level.

Design Safety

As equipment becomes increasingly complex and subject to more severe operating conditions the operator relies heavily on the expertise of the designer. Inherent weaknesses in this development are the interfaces, first of all between components of the design, but also between designer and operator. It is essential to create a system of review both as an operating routine and as a means for auditing equipment safety for management purposes.

Traditionally, engineering design is based on sound engineering judgement with recognised standard specifications, engineering codes, checklists and guides. In spite of many decades of successful practice, some 25 to 30 per cent of accidents are still due to "technical failure", and it is there that engineering design can make significant contributions to improved quality. For instance, the early identification of hazards may lead to their elimination, while a quantitative reliability analysis of components can establish likelihoods of failure, and thus enable the design engineer to make a judicious choice between equipment alternatives or to select a different design altogether. Such an analysis requires a large set of basic "failure data" for such items as pressure vessels, pipes, switches and valves. The collection and diagnosis of operational failure data from maintenance and repair records will provide information on the reliability of, for example, subsurface safety valves. Such studies are enhanced by the fact that industry-wide data banks are also accessible.

Hazard assessment of a complete installation can be achieved by several techniques, both qualitative and quantitative. Qualitative approaches aim at identifying hazards in the design. "Hazard and Operability Studies" (HAZOPs) for instance, is a technique to systematically analyse the system components for their behaviour/failure in circumstances widely deviating from the normal, a so-called "what if" analysis. Likewise, the "Failure Mode and Effect" approach analyses the effects of failure of system components on the performance of the entire system.

Hazard identification nowadays is a "must" in new engineering designs. Such studies as HAZOPs are carried out not only in the design stage (e.g. on compressor facilities, flow-stations, production facilities or on offshore gas processing facilities), but also during major modification or renovation of existing facilities (e.g. an oil terminal extension, modification of gas production units).

Quantifying any identified hazards by estimating their likelihoods of occurrence takes assessment a step further. Actual major accidents are (fortunately) rare, so that generally there is insufficient specific data for a reliable statistical estimate of probability. In most cases such quantification must be done "bottom-up", arriving at an aggregate probability from the probabilities of individual components' failure. Obviously, in many cases such probabilities will necessarily have to be expert estimates, with factual data lacking. Moreover, "human failure" (which is a major contributing factor to many serious accidents) is a hazard that is extremely difficult to quantify. Therefore the quantitative techniques which provide an aggregate likelihood of system failure, should be used cautiously and critically. The main value of such exercises is in providing an order of ranking of design alternatives.

Human Safety

The previous section reviews attempts to decrease the likelihood of technical failures and accidents by using inherently safer designs. However, in 70 to 75 per cent of accidents, human failure (inattentiveness, poor judgement or just plain negligence) is the decisive factor. Everyone knows from experience examples of such human weakness. In most instances these will be classified as "operational errors" or "operational accidents". A diagnosis of the chain of events leading to such accidents shows that many of these could have been prevented by judicious design, by adequate procedures and precautions, or by specific training.

"Designing out" or decreasing the likelihood of human errors can be achieved by a detailed analysis of the man/equipment interface at, say, the drilling floor, the control station, or the construction site. The application of such an approach to exploration and production operations is still in its infancy, but it is expected to provide useful leads for the improvement of safety in operations.

Adequate operational procedures and their enforcement are vital to human safety. Manuals, guidelines or checklists need to be available for the entire range of activities, specifying existing standards and codes, actions to be taken, approaches to be followed, and identifying hazards and possible preventive measures.

Specific training for operational jobs is another "must". When hazardous situations develop, the competence of the operator is of decisive importance. His analysis of the hazard and its causes, his assessment of the possible consequences and his overall judgement of the situation will dictate whether the correct action is taken and a potential accident is therefore prevented. To achieve such competence, both an understanding of the process and extensive practical experience are necessary. Only in-depth training and re-training, theoretical and practical, can provide these. Such specific training must be available to operating staff at basic, advanced and specialist levels, with regular exercises and tests on-site of simulated "deviations from normal operation" supplementing the formal training. For instance, drilling crews carry out blowout practices as a routine, to test the men as well as the equipment.

In addition to measures in design, operational procedures and training for safe practices, there is a need to promote "safety consciousness" in all staff. In contrast to the other approaches, this quality cannot be developed entirely by teaching or acquired by experience; it is a matter of personal attitude.

Environmental Conservation

Exploration and production operations inevitably have an impact on the environment. A balance, therefore, must be struck between the need for oil and gas and the costs of protecting the environment, and these costs are reflected in what the consumer ultimately pays.

During the 1970s responsible stewardship over resources and the environment became of increasing general concern. Poor stewardship will provoke an adverse reaction both by ever more perceptive governments and by the general public, making it increasingly difficult for the oil companies to pursue their business.

As governments, often in response to public pressure, continue to evolve standards and legislation for environmental conservation, the oil companies must cooperate in the process so that the standards set are realistically achievable, both technically and economically.

Today, it is part of the oil industry's basic business policy to carry out its operations safely and cleanly. The appropriate technology, company procedures and training are applied to achieve an environmental practice adequate to implement this policy and to comply with prevailing standards. Where no environmental legislation exists, the individual company's standards apply. Most major oil companies support environmental studies, at local universities for instance, and have their own environmental research and development programmes.

Environmental Impact

Almost every aspect of exploration and production has an impact on the environment. Examples are the noise from a vibroseis survey or from an onshore drilling location close to (or even inside) a built-up area, the visual impact of a drilling derrick or a pipeline being laid, or the small quantities of hydrocarbons which may be discharged with production water (even after the water has been treated).

Responsible environmental management demands continuous control of a company's impact on the environment throughout all its activities. Particular concern needs to be given to the fate of operational discharges (production water, test production, disposal of cuttings, sludges and waste materials), and to contingency measures for major pollution accidents.

The standards for an acceptable performance will have been laid down either in government regulations or work permits, or by the individual company's own environmental organisation. Where a government is in the process of formulating its standards, the industry can contribute its know-how and experience to a technically well-considered government decision.

Specific environmental concerns depend on the actual operation. For major ventures, or even small ones in ecologically sensitive areas, an advance environmental impact assessment will provide the data for well-considered decisions and negotiations. The early involvement of interested parties ("open planning") is clearly to be recommended.

During drilling, continuous attention is given, for instance, to the disposal of drill cuttings, to ensuring the mud pit is environmentally safe, to the careful use and disposal of drilling and completion chemicals, and to keeping the drilling floor and site as clean as possible. When the drill site is in a populated area, measures are taken to reduce the noise nuisance. Fluids from a production test are contained and removed. After completion of the job the site has to be cleared completely, and it may be necessary to re-plant vegetation to avoid scarring of the landscape.

In production operations, strict housekeeping procedures lead to clean work sites, flow stations, and flowline tracks. Special attention and preparatory measures are required for certain production activities, such as enhanced oil recovery by chemical methods. The hazards of an extra environmental load on the receiving waters are carefully assessed and contingency measures are taken to contain any accidental discharge of process chemicals.

As in safety management, environmentally acceptable operational practices require specially designed engineering and equipment, procedures and training. However, the actual selection of a specific measure depends entirely on the local circumstances.

To provide a balanced proposal for "clean" operating practices, it is necessary to:
- know the engineering and equipment alternatives;
- assess the related potential environmental impacts;
- develop tailored procedures for operations, maintenance and monitoring;
- provide specific training for operators.

Offshore Oil Spill Contingency Planning

While onshore spills do occur, they can normally be contained within a reasonably small area; however, a separate area of concern is contingency planning for major pollution accidents offshore, as might arise from a large oil well blowout or a pipeline leak. Sophisticated mechanical or chemical techniques are now available for cleaning up spills, but each technique has to be assessed for effectiveness under the particular set of local conditions. Then appropriate strategies for coastal protection have to be selected in conjunction with government agencies.

Essential to such an approach is the estimation of pollution risk for a specific stretch of coast, taking into account such factors as tidal currents and the depth of near-coastal waters. A forecasting model, SLIKTRAK, has been developed by Shell companies to provide an estimate of pollution risk, of the arrival times and volumes of oil, and of the clean-up costs to be expected. It was successfully tested and updated during and after the 1977 Ekofisk blowout in the North Sea and is now widely accepted as a useful tool. In this connection it is desirable that the utmost advantage should be taken of such unfortunate occurrences to collect, document and analyse the data in a scientific manner.

The measures for protecting a particular coastal area have to be adjusted, not only for the existing ecosystem and its resilience to natural or man-inflicted disasters, but also to government decisions on the value of that ecosystem to society and the costs justified to protect it. Ideally, priority should be given to those areas where crude oil is likely to become concentrated, such as in sheltered bays and estuaries commonly used by the public for recreational or commercial purposes.

It is essential therefore that contingency plans are agreed and implemented, and regular practices carried out, well in advance of a mishap.

Information and Computing

In several sections of this chapter mention is made of the use of computer systems in support of different activities carried out in exploration, petroleum engineering, engineering and in production operations. Examples include:
- acquisition, processing and interpretation of seismic data;

- simulating complex reservoir conditions in order to predict reservoir production performance;
- digital recording of well log data and processing these data to establish formation and reservoir properties;
- assisting engineering management in controlling the planning and progress of large construction projects;
- evaluating the economics of exploration prospects and of the development and production of oil and gas reserves;
- monitoring and supervision of production and pipeline facilities;
- dynamic positioning of deep-water drilling vessels.

This dependence on computing arose over the past 10 to 15 years, but has accelerated in recent years. Computerised information systems are now used in support of all exploration and production activities. This change has occurred for a number of reasons:
- the very rapid advances in computer technology;
- the decreasing cost of computing hardware;
- the ever-increasing demand for accurate and up-to-date information for decision taking, which computer systems are ideally suited to supply;
- the development of new techniques from research efforts and other activities involving complex calculations which are only practical to handle with the aid of a computer;
- the need to improve the effectiveness and productivity of scarce and expensive skilled staff.

Computing, in fact, has become a major contributory factor to the competitive position of any company within the oil industry. This applies to its use for technical calculations but even more to its power to help administration, by supplying information to management, to planners and to staff in the field.

The strong reliance on computer systems has emphasised the need for staff who are, so to speak, bilingual, understanding both the business and the various aspects of computing. Furthermore, experience indicates the need for users to take part in systems development, particularly in determining requirements.

Use of Computing for Technical Applications

Since computers were first introduced into the business world, there has been continued growth in their use for technical applications which often involve lengthy and complex calculations.

With the availability of more advanced hardware and software the degree of sophistication has increased recently and this trend is continuing. Many systems can be used "on-line" (while the user waits) or even "interactively". Through the latter method a user is able to interact with the system, for instance by evaluating the effect of changing the value of one or more variables in a particular process. A good example is an interactive well log analysis system used by petrophysicists for log evaluation, to make operational decisions during drilling.

A great variety of technical applications has been developed, serving the whole range of exploration and production activities. Because of their technical nature the development of the programs is sometimes undertaken by the users concerned. However, in view of the complexity of current hardware and software, most development is now carried out by computer professionals.

A great deal of exploration and production data is more easily digested when represented graphically rather than numerically, and so computer graphics are widely used. There are two main types of equipment:
- plotters (which may be mechanical or electrostatic) to produce graphs and maps;
- interactive graphics systems which allow a drawing to be displayed on a VDU (Visual Display Unit) screen, and the drawing to be modified in an interactive manner.

Technical applications use the total range of computer hardware currently available:
- *mainframe computers* for running most of the technical application programs. Standardisation of equipment facilitates the exchange of programs between several locations.
- *mini-computers* as equipment dedicated to certain applications, e.g. seismic data processing, reservoir simulation, interactive graphics.
- *micro-computers* (or desktop computers) to perform fast calculations in connection with activities such as topography, production technology and reservoir engineering. These are particularly useful for taking into the field.
- *super computers* (array processors) which are faster than mainframe computers for applications like reservoir simulation which involve the processing of large arrays of data.

Use of Computers for Handling Data and Information

In the increasingly complex business of finding and producing oil and gas, the handling of data in order to provide information for decision-taking has become a major application of computers. The complexity of a large company is nowadays such that it is simply not possible to handle the required flow of data and information by traditional methods.

Initially the computerised data and information systems were mainly designed for individual applications, and there was much duplication of data between systems, with the inherent risk of inconsistencies. Data transfer between systems was often difficult because of the differences in data terminology, timing of updating and computing technology.

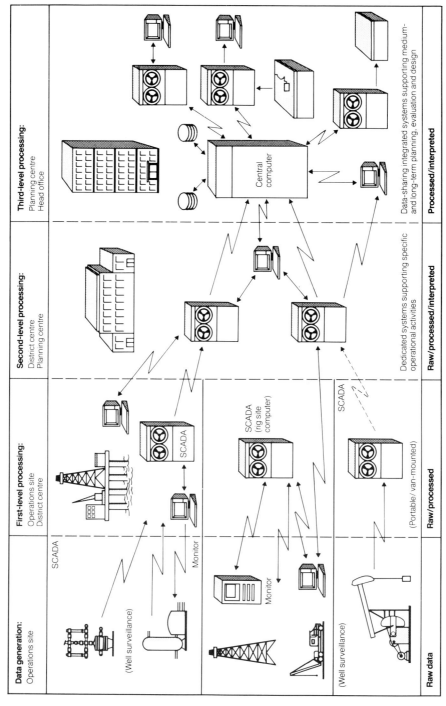

Figure 3.114 Information systems in support of exploration and production operations, planning, evaluation and design

ECONOMIC, FINANCIAL AND OTHER ASPECTS

Nowadays the development of systems takes into account a company's total needs for a particular type of information, rather than those for a single activity. However, current thinking does not favour large integrated systems (because they are usually not practical, either to develop or to use), but rather the use of efficient interfaces between systems in general, and in particular standard data definitions and coding across systems.

The resultant strategy is to plan the integration of all the information systems throughout a company's organisation, thus providing a flow of data and information from operations to tactical and strategic planning. A three-level information systems concept has been developed, which is illustrated in Figure 3.114.

The *first-level* systems provide data and information for monitoring and control, as well as for further processing by higher-level systems. These first-level systems are sometimes referred to as SCADA systems (Supervisory Control and Data Acquisition). They are operated on computer hardware (usually a mini-computer) dedicated to the individual application and use special "real-time" software. The common characteristic of these systems is that the data from the instruments are acquired directly and processed by the dedicated computer system, virtually at the instant of generation at the operations site, rather than the data being collected intermittently by staff and then fed into the computer. Many companies in the industry operate SCADA systems in support of their field production and pipeline operations.

The *second-level* systems also support specific operational activities, by processing and administering operational and technical data and information. They are used for tactical planning (e.g. production programming, maintenance planning, log evaluation, and seismic operations). They are usually on-line systems, operated on dedicated mini-computers, which receive their data from various sources (e.g. SCADA systems, digital recordings, keyboard data entry or batch-type systems). In addition to serving the operational activities, the second-level systems transmit data and information to the third-level systems.

The *third-level* systems are basically concerned with strategic planning, evaluation and design activities (exploration, petroleum engineering, field engineering and engineering design). The majority of these systems run on mainframe computers at computer centres. This equipment is used for the storage and retrieval of both technical and administrative data, with facilities for archiving and integrating important data.

With the increasing number of information systems there is a continuing need for a central general-purpose computer, where data can be combined as required and the resulting information made available in an efficient way, where and when needed. This now requires more sophisticated software techniques such as database management systems.

Need for Coordination

With the rapidly increasing use of computer systems in support of exploration and production activities, particularly where a number of locations and companies within a group may be involved, coordination of the development and use of these systems has become important. This avoids duplication of development and support work, and also improves both the quality of systems and the reliability of the data and information handled by those systems, by ensuring the rapid dissemination of new developments.

Coordination also ensures that information systems serve the overall needs of the whole organisation as well as particular short-term and specific local needs. It has been found that the savings thereby generated well outweigh the additional costs incurred.

Examples of coordination are drawing up coherent plans for all systems development work, establishing responsibilities and procedures for systems development and its control, defining data and codes, and updating information and retrieval systems.

WORLD OIL AND GAS RESERVES

In Chapter 1 the future of the world's energy resources was discussed briefly, and reference made to the concern which frequently centres on the question of how much oil and gas is "left". However, to believe that the supply of oil is just a matter of looking at the remaining reserves in the world is to fail to understand the full technical, political, economic and social dimensions of oil supply.

One frequently held view is that most of the world's reserves of oil and gas are nearly depleted and that we will have to face a physical shortage of oil in the very near future; at the other extreme, it is claimed that there is no physical shortage of oil, and that the remaining reserves are really much bigger than the oil industry is willing to admit. Neither viewpoint stands up to closer inspection; the reality is more complex.

Over the past 40 years, many forecasts have been made about the total amounts of ultimately recoverable conventional oil and gas (excluding oil recoverable from oil shales and tar sands). Invariably, such forecasts have to start from so-called "proven" reserves, proceed to include probable and possible reserves (including secondary and tertiary recovery), and finally allow for such additional amounts as the forecaster believes could still be discovered in new fields. Such estimates are strongly dependent upon the forecaster's assumptions about technological and economic developments, and not mainly upon geological reasoning.

The term "resources" is used to describe the sum total of cumulative produc-

WORLD OIL AND GAS RESERVES

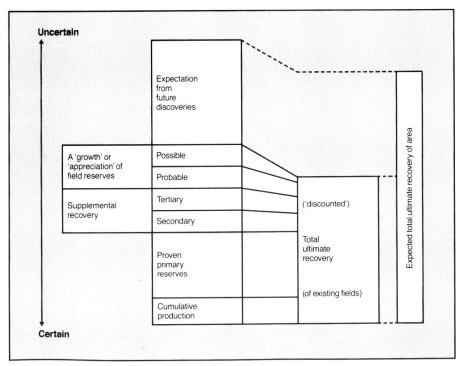

Figure 3.115 **Crude oil reserves terminology**

tion to date, existing (proven, probable, possible) reserves, and expectations from future discoveries (Fig. 3.115). With technological and economic developments, however, resource estimates tend to increase (Fig. 3.116).

In 1977, an enquiry in the form of a so-called Delphi exercise was conducted by the World Energy Conference. A number of leading oil companies and independent experts were asked for their assessment of the remaining crude oil resources in the world. Confronted with the results of a first round of answers, they were asked to reassess their earlier assessments; in the end the result was the range of answers shown in Figure 3.116. In this manner an upper quartile consensus was generated of ultimately recoverable crude oil resources of about 2,250 billion barrels.

A reasonably optimistic forecast of world ultimately recoverable oil resources (Table 3.5) can be derived from the Delphi exercise. The remaining reserves in the fields which have already been discovered (around 700 billion barrels of oil) would, on paper, be sufficient to sustain present production levels for about 30 years, well into the next century. However, the near certainty that such reserves exist is no guarantee at all that it will also be possible to produce them at sufficient daily rates when and where the oil is required.

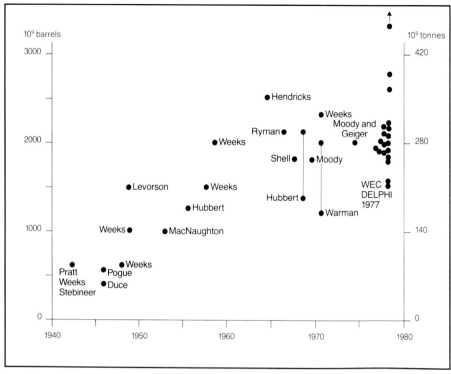

Figure 3.116 **Estimates of world recoverable resources of conventional crude oil** (including production to date)

Similarly, forecasts have been made over the years of ultimately recoverable resources of natural gas (Fig. 3.117). These forecasts resulted, by and large, in ultimate recovery figures of much the same size as ultimately recoverable crude oil resources when expressed in barrels of oil equivalent (BOE).

Table 3.5 **Worldwide crude oil resources (excluding oil shale and tar sands).** Based on crude oil value by the year 2000 of $20 per barrel in 1976 dollars.

1 *Fields already discovered*	
(a) Production to date and remaining reserves (on average 25% recovery of original oil in place)	1150×10^9 barrels
(b) Future improvement of recovery from 25% to 40%	650×10^9 barrels
2 *Estimate of future new discoveries*	
(a) On land	250×10^9 barrels
(b) Offshore on the shelf	200×10^9 barrels
(c) Offshore in deep water and in polar regions	350×10^9 barrels
Total	2600×10^9 barrels

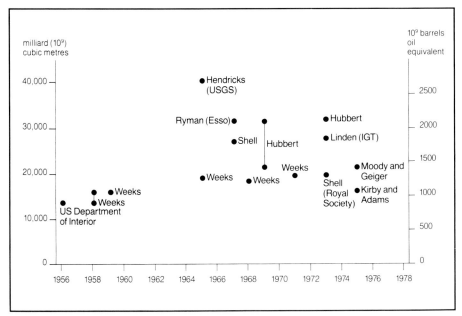

Figure 3.117 **Estimates of world recoverable resources of natural gas** (including production to date)

Nobody can be sure just how much additional oil and gas will eventually be discovered and produced in addition to what has been found so far. The large differences between individual estimates are thus hardly surprising.

Future Application of Modern Recovery Techniques

The application of secondary and tertiary recovery techniques to existing fields is potentially significant, in that such techniques could extend the world's reserves of ultimately recoverable oil by some 650 billion barrels, as indicated in Table 3.5.

However, secondary and tertiary recovery operations require sophisticated techniques and expertise. Moreover, they are often more costly than the initial investment for primary recovery. Frequently, it will be necessary to drill additional wells for injecting water, gas, steam or chemicals. Such processes also require most careful monitoring and control. The injected fluids are often much less viscous than oil, and may channel their way straight to the wells into which they are supposed to push the oil, thereby leaving most of the additional recoverable oil behind them forever. The history of secondary and tertiary recovery is littered with disappointments. It has taken some of the best brains in the oil industry, huge sums of money and the better part of this century to come to grips with the intricacies of these techniques, and to apply them with some confidence today.

If such advanced techniques could be applied in all the oil fields which are known at present, then it should be possible, in theory at least, to push the average recovery factor from these fields up from about 25 per cent to about 40 per cent. If such a feat were to be achieved, it would mean that another 650 billion barrels or so of oil would be added to the reserves which at present are considered as proven, and thus extend the total amount of remaining recoverable oil reserves to more than 1,400 billion barrels. This would be sufficient to sustain the present production level of oil until well into the 21st century.

To make such advances possible, both the political and economic conditions have to be right. Such complex technologies require extensive research, large-scale field trials, massive long-term investments and the deployment of substantial numbers of highly qualified personnel. Without deploying such an effort, it is impossible to realise fully the huge potential of advanced recovery techniques.

About 15 per cent of currently known reserves lie in the Communist areas, particularly the USSR and China. In the foreseeable future it is by no means certain that these countries will be able either to develop for themselves or to "import" the technical and managerial expertise which will be required to produce all the oil which they could technically recover.

A great number of the fields outside the Communist areas lie in the Middle East. These contain about 1,500 billion barrels of "oil in place" of which, so far, only some 25 per cent or 375 billion barrels are deemed recoverable. These fields will remain a major source of the global supply of oil for many years to come. If political and economic circumstances would permit the steady development of these fields to full production capacity using tertiary (enhanced oil recovery) techniques, they could ultimately yield some 600 billion barrels of oil, more than 25 times the oil consumption of the whole world in 1980.

With potential reserves of this magnitude, nearly every one of these countries could maintain its present production level (even without additional discoveries) for 30, 40 or more years to come. For the Middle East as a whole, a reserve/production ratio of well over 50 years could be maintained throughout the 1980s. Whether this amount of oil will indeed become available in the near and medium-term future is, however, not only a matter of technical capability but, even more, one of adequately meeting social and environmental demands.

Continued access to the huge oil reserves of the Middle East is absolutely essential for the oil-importing nations. This is undeniably true for the medium term when no alternative energy source can readily provide more than a tiny fraction of Middle East supplies. It will still remain true in the longer term when hydrocarbons may gradually be relieved of their present role as a source of bulk energy and become increasingly devoted to the production of highly efficient transport fuels and of nitrogenous fertilisers and other chemicals.

To provide these supplies, a political framework has to be established in which the continued production from the Middle East fields is assured. To maintain this production, greatly increased technical efforts are required. Efforts to prepare for secondary recovery have not yet gained their full momentum, and attempts at tertiary recovery have not yet advanced beyond a few isolated instances. Depending on price developments, the producing countries have reasonable expectations that even a gradually declining production of oil will provide them with sufficient foreign exchange to cover their immediate needs. However, for the consuming countries, nothing less than the assurance of continued supplies at about present levels will be sufficient to sustain their economies at acceptable levels or will satisfy the continued growth in energy needs of the LDCs.

Estimate of Future New Discoveries

The story of future oil and gas availability would not be complete without a survey of what may be discovered by new exploration in accumulations not yet known. For these future expectations, a total amount of 800 billion barrels has been included in Table 3.5, split almost equally between onshore, conventional offshore (up to 200 metres water depth) and deep water and polar regions. Actually to find these 800 billion barrels in increasingly hostile conditions will be a task of major dimensions; more than anything else, it will take time.

Earlier in this chapter, the successive stages in investigating a sedimentary basin were described in the section on Exploration. Although it appears that we are probably past the global pre-mature stage of discovering the global "giant" fields, the oil industry's discovery rate of "normal-sized" fields has been roughly constant for some years. This is probably because there are enough new prospective areas entering the pre-mature phase to balance the effect of older basins becoming post-mature, and also because of improvements in technology, particularly in the use of seismic surveys to arrive at a much more detailed picture of the underground geological structures. These technical improvements permit a more precise selection of locations at which to drill exploration wells.

During the 30 years between the end of World War II and the mid-1970s, exploration for petroleum resulted in the discovery of about 15 to 20 billion barrels of oil per year, far more than ever before. With hindsight, many experts would ascribe these exceptional successes to the favourable combination of a generally stable political environment and a number of major breakthroughs in exploration technology.

For the future, expectations are rather more modest, but additional discoveries of some 200 billion barrels of oil in countries outside the Communist areas until the turn of the century (i.e. at an average annual discovery rate of some 10 billion barrels from 1980) are still considered feasible by many exploration specialists. This assumes that continued access to prospective areas remains possible, and that the potential rewards are sufficient to justify the effort.

As far as potential future discoveries are concerned, therefore, it would appear that exploration on a world average basis has now entered the mature stage which could carry on for several decades before reaching the post-mature stage. Many exploration prospects remain to be investigated between now and the turn of the century.

Chapter 4

THE CHEMISTRY OF PETROLEUM

INTRODUCTION

This chapter gives a brief outline of the chemistry of petroleum, which should help to explain how it is possible for various oil products and chemicals to be manufactured from crude oil and natural gas.

Matter is not infinitely divisible: there is a limit beyond which a substance no longer exists as such. The smallest possible unit of a pure substance, like water, still possessing its characteristic properties is a molecule. All water molecules are identical.

On further division, a molecule disintegrates into a characteristic number of atoms, the smallest possible units of a limited group of substances, the elements. All atoms of an element such as oxygen, carbon or hydrogen are identical and unite according to fixed rules. The disintegration of the atom is beyond the scope of this book.

The characteristic properties of a substance depend on the type, number and arrangement of the atoms composing its molecule. Water, for instance, is H_2O, H–O–H, the capitals H and O symbolising atoms of hydrogen and oxygen, respectively, the short lines representing the chemical "forces" or "bonds".

Crude oils are mixtures of many substances, often difficult to separate, from which various oil products (such as liquefied petroleum gases, gasoline, kerosine, gas oil, fuel oil, lubricating oil, wax and bitumen) have to be manufactured.

These substances are mainly compounds of only two elements, carbon (C) and hydrogen (H), and are therefore called hydrocarbons. Other elements may be ignored at this stage since they are present in only small quantities, although some of them, such as sulphur (S), have an important effect on product quality. Two kinds of processes for manufacturing oil products are used. By the physical methods, the hydrocarbons in the raw materials are merely shuffled into technically useful groups without disruption. By the chemical or conversion methods, the more complex hydrocarbons are broken down into simpler ones and rearranged in different patterns of technically useful groups.

Chemical products manufactured from petroleum cover a much more varied range of molecular types than merely hydrocarbons. They do not occur as such in crude oil or natural gas, but they are all compounds of carbon and hydrogen, most but not all of which are combined with other elements, such as oxygen (O), nitrogen (N), sulphur (S) or chlorine (Cl).

The manufacture of oil products is so different from that of chemicals, both in processes and equipment, that it will be described in separate chapters. However, in broad outline the chemistry of petroleum is essentially the chemistry of hydrocarbons as given in this chapter.

HYDROCARBONS

Hydrocarbons may be gaseous, liquid or solid at normal temperature and pressure, depending on the number and arrangement of the carbon atoms in their molecules. Those with up to four carbon atoms are gaseous; those with twenty or more are solid; those in between are liquid. Liquid mixtures, such as most crude oils, may contain either gaseous or solid compounds or both in solution. For example, oil from the Schoonebeek field in the Netherlands contains a high proportion of dissolved solid hydrocarbons; the crude oil is liquid as it flows from the well at some 70°C but becomes almost solid on cooling, owing to crystallisation of the solid compounds. Many crude oils from the USA, on the other hand, contain only a small proportion of solid hydrocarbons and remain liquid even at low temperature.

The simplest hydrocarbon is methane, a gas consisting of one carbon atom and four hydrogen atoms. The methane molecule can be represented as

$$\begin{array}{c} H \\ | \\ H-C-H \\ | \\ H \end{array} \quad \text{or} \quad CH_4$$

The carbon atom has four bonds that can unite with either one or more other carbon atoms (a property almost unique to carbon) or with atoms of other elements. A hydrogen atom has only one bond and can never unite with more than one other atom. The larger hydrocarbon molecules have two or more carbon atoms joined to one another as well as to hydrogen atoms. The carbon atoms may link together in a straight chain, a branched chain or a ring. Examples are:

$$\begin{array}{c} H \quad H \quad H \quad H \quad H \\ | \quad | \quad | \quad | \quad | \\ H-C-C-C-C-C-H \\ | \quad | \quad | \quad | \quad | \\ H \quad H \quad H \quad H \quad H \end{array} \quad \text{or} \quad CH_3-CH_2-CH_2-CH_2-CH_3$$

straight-chain hydrocarbon or "normal" compound

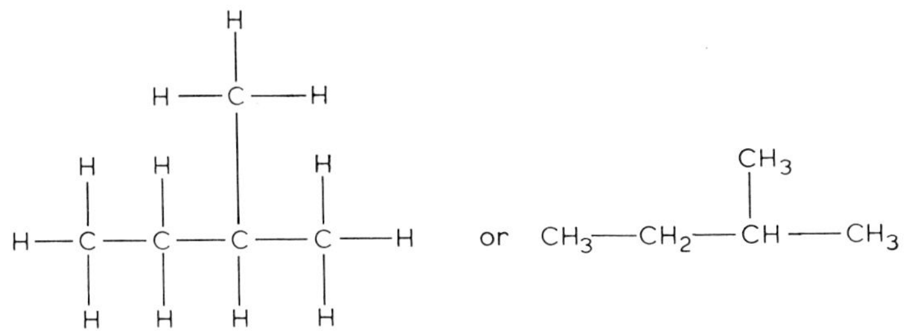

HYDROCARBONS

[Structural diagrams of a branched-chain hydrocarbon shown as expanded structural formula with H atoms and as condensed formula: CH₃—CH₂—CH—CH₃ with CH₃ branch]

branched-chain hydrocarbon or "iso" compound

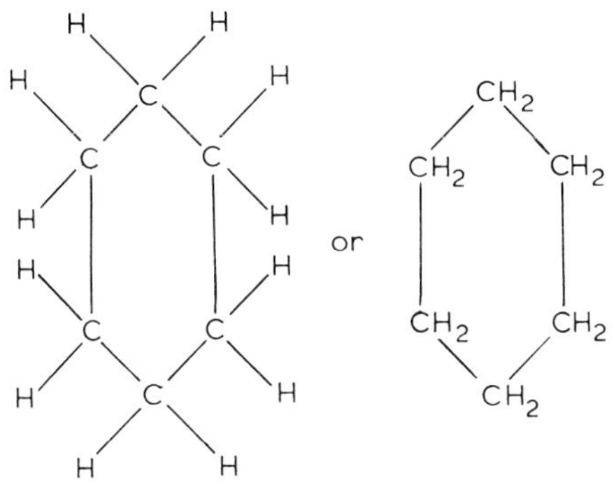

[Structural diagrams of a cyclic hydrocarbon shown as expanded structural formula with H atoms and as condensed formula with CH₂ groups arranged in a hexagonal ring]

ring or "cyclo" compound (rings of other sizes are possible)

From these three basic configurations a considerable number of hydrocarbons can be built up, especially since more complicated compounds may be formed by combinations of chains and rings, for example:

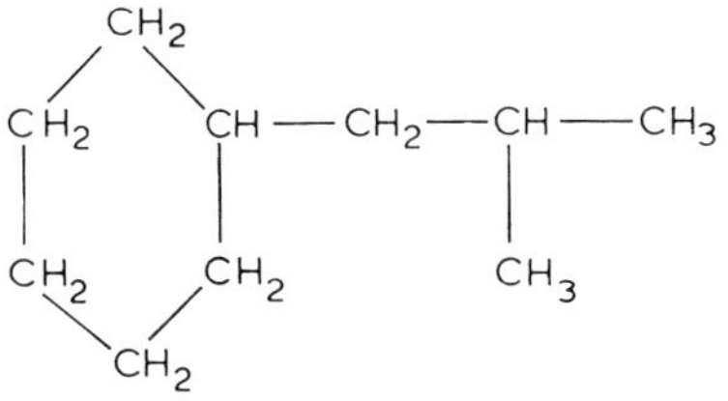

[Structural diagram showing a cyclohexane ring attached to a branched chain: CH—CH₂—CH—CH₃ with a CH₃ branch]

The number of hydrogen atoms associated with a given skeleton of carbon atoms may vary. When a chain or ring carries the full complement of hydrogen atoms, the hydrocarbon is said to be saturated, and such hydrocarbons are known as paraffins, paraffinic hydrocarbons or alkanes/cycloalkanes. Straight-chain structures are normal paraffins, branched-chain structures are isoparaffins, and ring-type structures are cycloparaffins or naphthenes. Thus for three hydrocarbons with five carbons atoms, all pentanes, we have, amongst others, the following structures:

(a) $CH_3-CH_2-CH_2-CH_2-CH_3$, normal pentane (C_5H_{12})

(b) 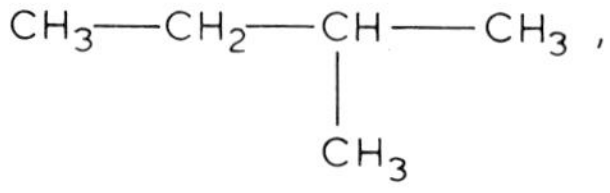 $CH_3-CH_2-CH(CH_3)-CH_3$, isopentane (C_5H_{12})

(c) 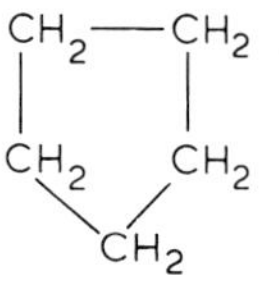 cyclopentane (C_5H_{10}) — ring of five CH_2 groups

Thus normal pentane, normally abbreviated to n-pentane, is a straight-chain, i.e. unbranched, paraffin, isopentane is branched and cyclopentane a ring compound.

When less than the full complement of hydrogen atoms is present in a hydrocarbon chain or ring, the hydrocarbon is said to be unsaturated. Unsaturated hydrocarbons are characterised by having two adjacent carbon atoms linked by two or three bonds instead of only one. These links are known as double bonds and triple bonds, respectively; they are not stronger than the single bond, but on the contrary surprisingly vulnerable, with the result that the unsaturated compounds are chemically more reactive than the saturates.

Straight- or branched-chain hydrocarbons with one double bond are called mono-olefins or alkenes, hydrocarbons with a double bond in a ring are cyclo-olefins, or cycloalkenes, and those with two double bonds in the structure diolefins or dienes. Hydrocarbons with a triple bond are called acetylenes or alkynes.

The simplest members of the olefin and acetylene series are ethylene and acetylene, and butadiene is the simplest diolefin:

$CH_2=CH_2$ $CH\equiv CH$ $CH_2=CH-CH=CH_2$
ethylene   acetylene   butadiene

Neither olefins nor acetylenes occur in crude oil or natural gas, but are produced by conversion processes in the refinery and are important raw materials for chemical syntheses.

Ring compounds containing one or more six-membered rings with three alternate double bonds form an important group known as aromatics because most of them have a characteristic smell.

The simplest member is benzene, C_6H_6, in which each carbon atom carries only one hydrogen atom:

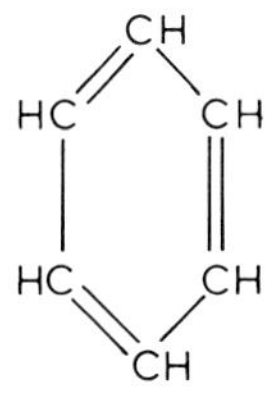

NON-HYDROCARBONS

```
      CH
    ⁄⁄  \
  HC    CH
  ‖      ‖
  HC    CH
    \  ⁄⁄
     CH
```
benzene

More complex molecules of the aromatic series are obtained by replacing one or more hydrogen atoms by hydrocarbon groups or by "condensing" one or more rings:

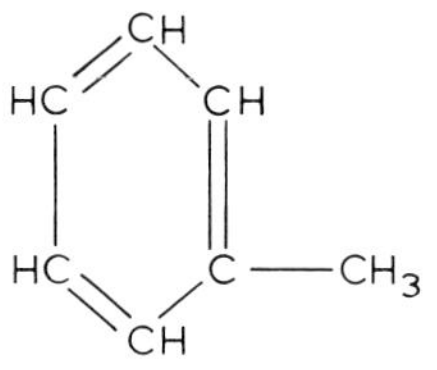

$C_6H_5CH_3$
toluene

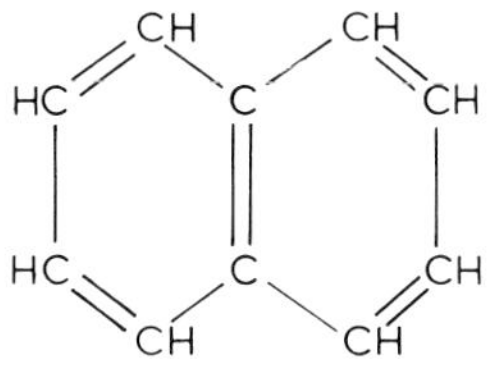

$C_{10}H_8$
naphthalene

From these few examples it will be obvious that there is no end to the number and complexity of hydrocarbon structures. By introducing other elements, in particular oxygen, nitrogen and sulphur, the number of possibilities based on a carbon skeleton (and thus the number of possible organic chemicals), increases tremendously.

NON-HYDROCARBONS

A brief reference has already been made to the non-hydrocarbons that may occur in crude oils and oil products. Although small in quantity, some of them have a considerable influence on product quality. In many cases they have noxious or harmful effects and must be removed, or converted to less harmful compounds, by refining processes. In a few cases their presence is beneficial and they should not be removed or converted.

The most important elements occurring in non-hydrocarbons are sulphur (S), nitrogen (N) or oxygen (O); in some crude oils there are small amounts of metal compounds, of vanadium (V), nickel (Ni), sodium (Na) or potassium (K) for example. An account of these compounds will help to explain the background of some of the refining and treating processes described in succeeding chapters.

Sulphur compounds

Many types of sulphur compounds occur in crude oils in widely varying amounts from less than 0.2 per cent by weight in some Pennsylvanian, Algerian and Russian crudes to over 6 per cent by weight in some Mexican and Middle East crudes.

A distinction is often made between corrosive and non-corrosive sulphur compounds. The corrosive ones are free sulphur, hydrogen sulphide and thiols (mercaptans) of low molecular weight. Moreover, they have an obnoxious smell.

Hydrogen sulphide, H_2S, has the structure H–S–H. If one of the hydrogen atoms is replaced by a hydrocarbon group, the compound is called a mercaptan or thiol, for example:

$$C_2H_5SH, \text{ ethanethiol}$$

The compounds are formed during the distillation of crude oils; they may cause severe corrosion of the processing units, and addition of chemicals, proper temperature control and the application of special alloys in plant equipment are required to control them.

The non-corrosive sulphur compounds are sulphides (thioethers), disulphides and thiophenes. If both of the two hydrogen atoms in hydrogen sulphide are replaced by hydrocarbon groups, the compound is called a sulphide or thioether, for example:

$$C_2H_5-S-C_2H_5, \text{ diethyl sulphide}$$

The disulphides are formed either from mercaptans by oxidation or from sulphides and sulphur:

$$C_2H_5-S-S-C_2H_5, \text{ diethyl disulphide}$$

Thiophenes are sulphur compounds with a ring structure containing five atoms:

$$\begin{array}{c} HC = CH \\ \| \quad \| \\ HC \quad CH \\ \diagdown S \diagup \end{array} \quad \text{or } C_4H_4S, \text{ thiophene}$$

The non-corrosive sulphur compounds, although not directly corrosive, may cause corrosion on decomposition at higher temperatures and therefore also require careful temperature control in processing units.

Apart from their unpleasant smell, both corrosive and non-corrosive sulphur compounds are undesirable in most products. In fuels, the sulphur burns to sulphur dioxide and sulphur trioxide; these oxides combine with the water formed by combustion to give sulphurous and sulphuric acids, which may cause serious corrosion in the colder parts of engines or furnaces. Furthermore, some sulphur compounds reduce the effect of anti-knock additives (tetraethyllead and tetramethyllead) on the octane rating of gasolines. Sulphur compounds in illuminating kerosine promote charring of the wick and cause a bluish white deposit on the lamp glass. In dry-cleaning solvents they may give a bad odour to cleaned goods and in paint thinners may affect the colour of the dried film.

Some natural gases have a high content of hydrogen sulphide; that from Lacq in France contains 15 per cent by volume, and in Canada there are wells producing natural gas with even 32 per cent by volume.

The lower thiols are insoluble in water, but soluble in hydrocarbons, and have an intolerable odour. They react with sodium and copper to form sodium and copper mercaptides and with oxygen to form disulphides.

Thioethers or sulphides are also insoluble in water, but soluble in hydrocarbons, and have an offensive odour. However, because of their relatively unreactive nature, drastic treatment is necessary for their removal. Disulphides are more reactive than thioethers, on account of the S–S linkage, and can readily be oxidised to compounds soluble in water. Thiophenes have a pleasant odour, comparable with that of benzene, and are relatively stable; they may even be beneficial.

Nitrogen compounds

Most crude oils contain less than 0.1 per cent by weight of nitrogen, but some from California, Japan and South America contain as much as 2 per cent by weight. The nitrogen compounds in the crude are complex and for the most part unidentified, but on distillation they give rise to nitrogen bases (compounds of pyridine, a six-membered nitrogen-containing ring) in the derived products.

Nitrogen bases often cause discoloration of heavy gasolines and kerosines, particularly when associated with phenols. In gasolines they may also cause engine fouling and in lubricating oils engine "lacquer". In heavy gas oil feedstocks for catalytic cracking they may reduce the activity of the catalyst by increasing coke deposits. Nitrogen bases can be removed by acid treatment and recovered by neutralisation of the acid extract.

Oxygen compounds

Some crude oils contain oxygen compounds. Their structure has not yet been established, but on distillation of the crudes the oxygen compounds decompose to form ring compounds with a carboxylic acid group (COOH), in the side chain, for example:

$$\begin{array}{c} CH_2 \\ CH_2 \quad CH_2 \\ | \quad\quad | \\ CH_2 \quad CH-CH_2-COOH \\ CH_2 \end{array}$$

These compounds are known as "naphthenic acids", large quantities having been originally found in distillation products of Russian naphthenic crudes. The carboxylic acid group(s) may, however, be attached to hydrocarbon groups other than naphthenes, and "petroleum acids" would be a more accurate term; however, "naphthenic acids" is generally accepted. Some of these acids are highly corrosive and special alloys have to be used in processing equipment.

Naphthenic acids are extracted from distillates by alkali treatment, either during distillation or afterwards, and are recovered by acidifying the extract. They are valuable by-products used in the manufacture of paint-driers, emulsifiers and cheap soaps.

Phenolic compounds occur in some crudes and are formed during cracking. They are oxygen compounds containing one or more OH groups, derived from aromatic hydrocarbons. The simplest members are phenol, the cresols and the xylenols, which are recovered during refining:

phenol

Other compounds

Several other elements occur in crude oils, either as inorganic or organic compounds, and remain in the ash on burning. They vary from crude to crude, but many crudes contain vanadium and nickel. Sodium and potassium are usually present, derived from saline water produced together with oil. Copper, zinc and iron are also found. These elements are generally of little account, but sometimes they are important e.g. vanadium is recovered as vanadium ashes from deposits on furnace walls, or from flue gases, when high vanadium fuels are burnt in refinery furnaces. Vanadium metal is an important component for the manufacture of special steels. Vanadium, iron and nickel in the feedstocks for catalytic cracking may spoil catalyst activity, and so the feedstocks have to be carefully distilled or redistilled to leave the metal compounds in the residue.

HYDROCARBON REACTIONS

Of the four main groups of hydrocarbons (paraffins, olefins, naphthenes and aromatics), the olefins are the most reactive and the paraffins the least. In the refining of crude oil and in the manufacture of petrochemicals, certain basic reactions play an important role. Some of them are also of interest in connection with the performance properties of oil products, e.g. in the deterioration of gasoline and lubricating oils through oxidation and polymerisation.

The following are the most important of these reactions:

Dehydrogenation — the elimination of hydrogen atoms from a molecule. A saturated hydrocarbon becomes unsaturated, and a chemical substance changes its type:

$$CH_3-CH_3 \longrightarrow CH_2{=}CH_2 + H_2$$
ethane → ethylene + hydrogen

$$\underset{\text{isopropyl alcohol}}{\begin{array}{c}CH_3\\|\\CHOH\\|\\CH_3\end{array}} \longrightarrow \underset{\text{acetone}}{\begin{array}{c}CH_3\\|\\C{=}O\\|\\CH_3\end{array}} + H_2 \text{ (hydrogen)}$$

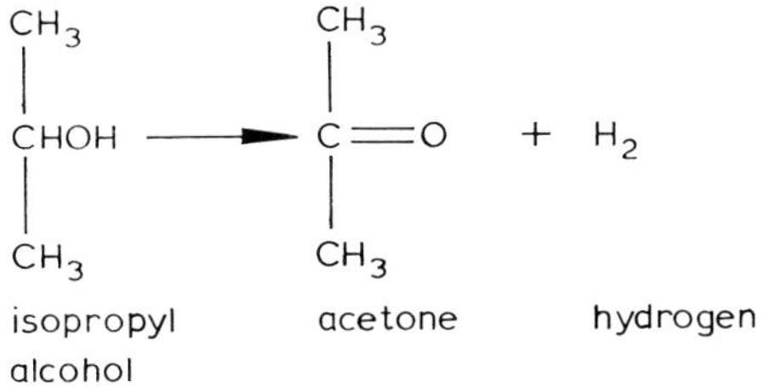

Hydrogenation — the reverse process to dehydrogenation; the filling up of the "free" places or double bonds in unsaturated structures by hydrogen atoms (addition):

$$CH_2{=}CH_2 + H_2 \longrightarrow CH_3-CH_3$$
ethylene + hydrogen → ethane

Cracking — disruption of the carbon–carbon bonds in large hydrocarbon molecules by heat, so that smaller molecules (both saturated and unsaturated) are obtained:

$$CH_3-CH_2-CH_2-CH_2-CH_2-CH_2-CH_2-CH_2-CH_2-CH_2-CH_2-CH_3 \longrightarrow$$
$C_{12}H_{26}$
dodecane

$$CH_3-CH_2-CH_2-CH_2-CH_3 + CH_2{=}CH-CH_3 + CH_2{=}CH-CH_2-CH_3$$
C_5H_{12} C_3H_6 C_4H_8
pentane   propylene   butylene

Pyrolysis — a severe form of thermal cracking; the disruption reaction is usually accompanied by a rearrangement of the fragments:

$$\underset{\text{propane}}{\underset{|}{\overset{|}{\underset{CH_3}{\underset{|}{CH_2}}}}\overset{CH_3}{\underset{|}{CH_2}}} \xrightarrow{800°C} \underset{\text{propylene}}{\underset{|}{\underset{CH_3}{\overset{CH_2}{\overset{||}{CH}}}}} + \underset{\text{ethane}}{\overset{CH_3}{\underset{|}{CH_3}}} + \underset{\text{ethylene}}{\overset{CH_2}{\underset{||}{CH_2}}} + \underset{\text{methane}}{CH_4} + \underset{\text{hydrogen}}{H_2}$$

$$\underset{\text{methane}}{CH_4} \xrightarrow{1200°C} \underset{\text{hydrogen}}{H_2} + \underset{\text{carbon}}{C} + \underset{\text{acetylene}}{CH \equiv CH}$$

Isomerisation — the rearrangement of the carbon skeleton of a molecule, conversion of a straight chain into a branched chain and the reverse:

$$\underset{\text{n-butane}}{CH_3-CH_2-CH_2-CH_3} \longrightarrow \underset{\text{isobutane}}{CH_3-\underset{\underset{CH_3}{|}}{CH}-CH_3}$$

Cyclisation — conversion of a chain into a ring molecule, hydrogen being lost:

$$\underset{\text{n-hexane}}{CH_3-CH_2-CH_2-CH_2-CH_2-CH_3} \longrightarrow \underset{\text{cyclohexane}}{\text{(cyclohexane ring)}} + \underset{\text{hydrogen}}{H_2}$$

Alkylation — the introduction of a straight- or branched-chain hydrocarbon group, into an aromatic or branched-chain hydrocarbon:

$$\underset{\text{benzene}}{C_6H_6} + \underset{\text{octene}}{C_8H_{16}} \longrightarrow \underset{\text{octylbenzene}}{C_6H_5-C_8H_{17}}$$

$$\underset{\text{isobutane}}{CH_3-\underset{\underset{CH_3}{|}}{\overset{\overset{CH_3}{|}}{CH}}} + \underset{\text{propylene}}{\overset{CH_3}{\underset{\underset{CH_2}{||}}{CH}}} \longrightarrow \underset{\text{isoheptane}}{CH_3-\underset{\underset{CH_3}{|}}{\overset{\overset{CH_3}{|}}{C}}-\underset{\underset{CH_3}{|}}{\overset{\overset{CH_3}{|}}{CH}}}$$

HYDROCARBON REACTIONS

Polymerisation and copolymerisation — the combination of a number of unsaturated molecules of the same or different compounds to form a single large molecule, called a polymer or homopolymer when it is built up from a number of identical monomers, and a copolymer when it is a combination of two or more different types:

$$n\ CH_2{=}CH_2 \longrightarrow CH_3{-}CH_2{-}CH_2{-}CH_2{-}CH_2{----}CH_3$$
$$\text{ethylene} \qquad\qquad \text{polyethylene}$$

Polymers are often solids (such as plastics and synthetic fibres), the properties of which depend largely on their molecular size.

Oxidation — the reaction of oxygen with a molecule that may or may not already contain oxygen. Oxidation may be partial, resulting in the incorporation of oxygen into the molecule or in the elimination of hydrogen from it, or it may be complete, forming carbon dioxide and water (combustion):

$$2\ \underset{\text{ethylene}}{\overset{CH_2}{\underset{CH_2}{\|}}} + \underset{\text{oxygen}}{O_2} \longrightarrow 2\ \underset{\text{ethylene oxide}}{CH_2\overset{O}{\overset{\triangle}{-}}CH_2} \qquad \text{partial oxidation}$$

$$\underset{\text{ethyl alcohol}}{2\ CH_3CH_2OH} + \underset{\text{oxygen}}{O_2} \longrightarrow \underset{\text{acetaldehyde}}{2\ CH_3{-}CHO} + \underset{\text{water}}{2\ H_2O} \qquad \text{partial oxidation}$$

$$\underset{\text{methane oxygen}}{CH_4 + 2\ O_2} \longrightarrow \underset{\substack{\text{carbon}\\\text{dioxide}}}{CO_2} + \underset{\text{water}}{2H_2O} \qquad \text{complete oxidation (combustion)}$$

Reduction — the reverse of oxidation: the proportion of oxygen to hydrogen in the molecule is decreased:

$$\underset{\text{acetaldehyde}}{CH_3{-}CHO} + \underset{\text{hydrogen}}{H_2} \longrightarrow \underset{\text{ethyl alcohol}}{CH_3CH_2OH}$$

Chlorination — in the reaction of a saturated hydrocarbon with chlorine one or more of the hydrogen atoms may be replaced by chlorine atoms with the formation of hydrochloric acid. The replacement of hydrogen by another atom in this way is called substitution:

$$\underset{\text{methane chlorine}}{CH_4 + Cl_2} \longrightarrow \underset{\substack{\text{methyl}\\\text{chloride}}}{CH_3Cl} + \underset{\substack{\text{hydrochloric}\\\text{acid}}}{HCl}$$

In the reaction of an unsaturated hydrocarbon with chlorine, two chlorine atoms are directly attached to the double bond. This is known as an addition reaction:

$$CH_2\!\!=\!\!CH_2 + Cl_2 \longrightarrow CH_2Cl\!-\!CH_2Cl$$
$$\text{ethylene} \quad\quad \text{chlorine} \quad\quad \text{dichloroethane}$$

Hydration — the addition of water to a double bond without breakdown of the molecular structure:

$$CH_2\!\!=\!\!CH_2 + H_2O \longrightarrow CH_3CH_2OH$$
$$\text{ethylene} \quad\quad \text{water} \quad\quad \text{ethyl alcohol}$$

Dehydration — the reverse process in the chemical field:

$$CH_3CH_2OH \longrightarrow CH_2\!\!=\!\!CH_2 + H_2O$$
$$\text{ethyl alcohol} \quad\quad \text{ethylene} \quad\quad \text{water}$$

However, in oil manufacturing the term is also used for simple drying of a product (elimination of dissolved or emulsified water).

Esterification — the reaction of an alcohol with an organic or mineral acid with elimination of water to form an ester:

$$C_2H_5OH + CH_3COOH \longrightarrow CH_3COOC_2H_5 + H_2O$$
$$\text{ethyl alcohol} \;\; \text{acetic acid} \quad\quad \text{ethyl acetate} \quad\quad \text{water}$$
$$\quad\quad\quad\quad\quad\quad\quad\quad\quad\quad \text{(ester)}$$

Hydrolysis — the decomposition of a molecular structure by the action of water. The hydrolysis of an ester results in the formation of an alcohol and an acid, and is the reverse of esterification:

$$CH_3COOC_2H_5 + H_2O \longrightarrow CH_3COOH + C_2H_5OH$$
$$\text{ethyl acetate} \quad\quad \text{water} \quad\quad \text{acetic acid} \quad\quad \text{ethyl alcohol}$$

Condensation — the coupling of organic molecules accompanied by the separation of water or some other simple substance, e.g. alcohol. A catalyst is usually required to promote the reaction:

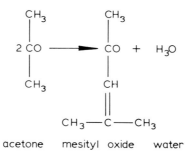

acetone   mesityl oxide   water

Sulphonation — the action of concentrated sulphuric acid on an aromatic hydrocarbon, e.g. benzene, to form a sulphonic acid. The hydrocarbon group in a sulphonic acid is directly linked to the sulphur atom:

benzene + sulphuric acid → benzene sulphonic acid + water

Sulphation — the reaction of an olefin with sulphuric acid. An ester is produced by addition of the sulphuric acid to the double bond and the hydrocarbon group is linked to the sulphur atom through an oxygen atom:

$C_6H_{13}-CH=CH_2 + H_2SO_4 \rightarrow$ mono-octyl sulphuric acid ester

octene + sulphuric acid → mono-octyl sulphuric acid ester

Hydrodesulphurisation — the elimination of sulphur from sulphur-containing chain molecules in crudes or distillates by the action of hydrogen under pressure over a catalyst:

$$C_8H_{17}-S-C_6H_{13} + 2H_2 \rightarrow C_8H_{18} + C_6H_{14} + H_2S$$
$$C_{16}H_{33}SH + H_2 \rightarrow C_{16}H_{34} + H_2S$$

Catalysis — the alteration of the rate of a chemical reaction by the presence of a "foreign" substance (catalyst) that remains unchanged at the end of the reaction, for instance hydrogenation using metallic platinum or nickel, and the cracking of a hydrocarbon using a silicate.

TYPES OF CRUDE OIL

Crude oils vary widely in appearance and consistency from country to country and from field to field. They range from yellowish brown, mobile liquids to black, viscous semi-solids. However, all crude oils consist essentially of hydrocarbons. Their differences are due to the different proportions of the various molecular types and sizes of hydrocarbons previously described.

One crude oil may contain mostly paraffins, another mostly naphthenes. Whether paraffinic or naphthenic, one may contain a large quantity of lower hydrocarbons and be mobile or contain a lot of dissolved gas; another may consist mainly of higher hydrocarbons and be highly viscous, with little or no dissolved gas.

The nature of the crude governs to a certain extent the nature of the products that can be manufactured from it and their suitability for special applications. A naphthenic crude will be more suitable for the production of asphaltic bitumen, a paraffinic crude for wax. A naphthenic crude, and even more so an aromatic one, will yield lubricating oils whose viscosities are rather sensitive to temperature. However, modern refining methods permit greater flexibility in their use of crudes to produce any desired type of product.

Crudes are usually classified into three groups, according to the nature of the hydrocarbons they contain.

Paraffin-Base Crude Oils

These contain paraffin wax (higher molecular weight paraffins which are solid at room temperature), but little or no asphaltic (bituminous) matter. They consist mainly of paraffinic hydrocarbons and usually give good yields of paraffin wax and high-grade lubricating oils.

Asphaltic-Base Crude Oils

These contain little or no paraffin wax, but asphaltic matter is usually present in large proportions. They consist mainly of naphthenes and yield lubricating oils whose viscosities are more sensitive to temperature than those from paraffin-base crudes, but which can be made equivalent to the latter by special refining methods. These crudes are now often referred to as naphthene-base crude oils.

Mixed-Base Crude Oils

These contain substantial proportions of both paraffin wax and asphaltic matter. Both paraffins and naphthenes are present, together with a certain proportion of aromatic hydrocarbons.

This classification is a rough-and-ready division into types and should not be used too strictly. Most crudes exhibit considerable overlapping of the types described and by far the majority are of the mixed base type.

Chapter 5

OIL PRODUCTS — MANUFACTURE

MANUFACTURING ACTIVITIES

A wide range of hydrocarbons occurs naturally in crude oil. Whilst crude oil may be utilised directly as an energy source (burnt as an under-boiler fuel), the full benefit of the different properties of the constituent hydrocarbons may be realised only if the constituents are separated. Physical separation of the constituent hydrocarbons has been the traditional function of an oil refinery and continues to be an important part of many refining activities. The advent of chemical conversion processes, by which the constituent hydrocarbons may be changed in structure, gave birth to the modern oil refinery and its role in making products of the quality and quantity demanded in the market from available crude oils — which either did not contain the required products or contained them in the wrong proportions.

To build today a sophisticated oil refinery would cost some one billion (10^9) US dollars. Of this sum, the cost of the equipment to carry out the physical separation and chemical conversion processes is only about one-third. Significant expenditure must also be made on handling systems, for the storage, transfer, blending and loading of large volumes of oil; on utilities systems to optimise energy production and consumption within the refinery; on process control systems and their associated computers; and on facilities to minimise adverse environmental effects to ensure the safety of personnel working within the refinery.

A medium-sized refinery, capable of processing some five million tons of crude oil annually, will typically have a permanent staff of close to five hundred people, and be providing additional employment for another two to three hundred contract labour. Thus, although the refining industry is highly capital-intensive, its effect on employment is significant. The trend has been to utilise increasingly skilled staff, particularly since the escalation of energy prices in the last decade emphasised the virtues of efficiency and economy.

The fifteen-fold increase in crude oil prices since 1972, with its attendant

Pernis refinery, The Netherlands. *Copyright Aerocamera–Bart Hofmeester*

disruptions of oil supply, has increased the complexity of refinery operations and forced a closer degree of coordination between Supply, Manufacturing and Marketing Functions in managing the business of acquiring, producing and selling oil products. Refineries have been required to become more flexible and more energy-efficient, but above all to ensure that each ton of crude oil received is processed as effectively as possible.

These considerations will be introduced in the following paragraphs on the main refinery activities, and will be dealt with in more detail in the succeeding sections.

Physical Separation Processes

Various separation techniques are applied in most crude oil and intermediate product processing steps. Common to all the techniques is that no change of molecular structure occurs during the operations and no new compounds are formed. Higher oil prices and greater differentials between oil products have emphasised the importance of maximum physical separation between products of different values. Separation can be according to molecular size or molecular type, and occasionally both size and type are distinguished. The main technique is:

Distillation — separation according to molecular size, making use of the difference in boiling point.

Other techniques are:

Absorption — separation according to size or type, making use of the difference in solubility in a liquid.

Solvent extraction — separation according to type, making use of the difference in miscibility with a third component.

Crystallisation — separation according to size or type, making use of the difference in melting point and solubility.

Adsorption — separation according to size or type, making use of the difference in adhesion to porous materials (gas/solid and liquid/solid systems).

Chemical Conversion Processes

Although all energy prices have increased over the last ten years, oil prices have shown the greatest increases both relatively and absolutely. As a result, there has been a tendency to substitute non-oil (notably coal) energy sources for their oil equivalents (particularly fuel oil), where practical. This process of substitution has accelerated the normal historical trends which affect the demand for different oil products, and has required refineries to make a markedly different product slate from that available from the distillation of crude oil. This has been achieved by the use of conversion processes, which involve a change in the size and structure of the hydrocarbon molecules. There are three main categories of conversion processes, involving:

Reduction of molecular size (cracking) — in which fuel oil components are converted into lighter, distillate products such as gas oil and gasoline. The main cracking processes are thermal cracking (including visbreaking and coking), catalytic cracking and hydrocracking. Investment in cracking has been at particularly high levels since the escalation of oil prices, and continued investment may be expected to enable refineries to maximise distillate manufacture.

Change of molecular structure without deliberate size change — for example, catalytic reforming and isomerisation (both of which are used to convert naturally occurring gasolines into products suitable for use in high-compression car engines).

Increase in molecular size — for example, polymerisation and alkylation (both of which convert gaseous hydrocarbons into liquids suitable for motor gasoline blending).

Treating and Subsidiary Processes

By a combination of physical and chemical processes, product streams may be purified and otherwise brought up to marketing specifications as to odour, colour, stability etc. Hydrotreating, for the removal of sulphur, is the major treating process in refineries. Subsidiary processes are applicable to "specialty" oil products, such as lubricants and bitumen, which are sold on the basis of their performance characteristics rather than their energy content. Often the subsidiary processes are carried out in separate installations and are not normally regarded as "refining" processes.

Control and Supervision of Refinery Processes

One of the fields of refinery operations to have undergone a quite dramatic development in the last twenty years is that of control and supervision of those operations. The change in the character of the refinery — for example, more complex plants involving more complex operations, and increasing integration of refinery units, requiring that all controls be concentrated in one control centre — and in addition the increasing cost of crude oil, made it vital to improve the quality of control, which meant that conventional instrumentation no longer sufficed.

Fortunately, new developments in electronics over the same period have made it possible to meet these changing requirements. Particularly, mention should be made of the micro-computer ("chip"), which has made a major contribution to the design of powerful instrumentation and control systems of great flexibility.

The two principal capabilities of the computer — to store large quantities of data and to perform calculations at high speed — have also allowed refineries to introduce sophisticated systems for supervision and scheduling.

All these developments are an ongoing process, with complete refinery control a prospect for the future.

Utilities

For the operation of the processing units, large quantities of heat, power, cooling water and compressed air are required, and the operation, efficiency and safety of the refinery depend to a large extent on the reliable functioning of these utility services.

In the past, utility plant design was mainly dictated by reliability considerations. Minimum energy consumption was secondary, owing to the availability of relatively cheap refinery fuel. Earlier refineries were characterized by process plants operated in isolation from each other. No use was made of the heat integration of process plants, and furnace efficiency was poor. Steam was generated in boiler houses spread over the refinery and used for mechanical drives for reasons of reliability. Power was imported from the public grid or was generated in the refinery power plant at moderate steam pressures with condensing steam turbines.

Over the last decade, more attention has been paid to efficient generation and the use of energy. This has resulted in the following developments:
- heat integration in process plants
- co-generation of steam and power in refinery power plants
- more efficient furnaces with recovery of heat from flue gases from steam production or for combustion-air preheating
- power generation from potential energy in process plants
- shift from inefficient steam turbines to electric motors
- utilisation of low-level heat for district heating
- speed control of electric motors
- use of energy models.

The necessity to make an optimum use of energy will be even more stringent for future refinery design.

The fuels available for steam and power generation depend very much on the type of process plant in the refinery and range from refinery fuel gas of varying calorific value via residue from refinery process plant to products like petroleum coke. Heavy residual fuel oil or petroleum coke with very high sulphur and metals content can no longer be fired in conventional ways because of environmental regulations regarding sulphur emission.

The electric power demand of refinery processes is generally high, and most conversion processes require high-level heat, while surplus heat at medium- and low-temperature levels becomes available.

The balancing of power and heat generation against power/heat consumption and low-level waste heat production in process plant often presents problems and may well result in unconventional schemes in the future (gasification, fluidized-bed combustion, etc.).

The higher degree of integration and sophistication makes demands on the reliability of the utilities system as a whole. However, the much higher efficiency of co-generation and combined cycles justifies this increase in complexity, requiring better controls and/or duplication of components in the system.

DISTILLATION

The first step in the manufacture of petroleum products is the separation of crude oil into the main fractions by distillation. This is the most important process in the refinery, because, in addition to its use for separation, it plays an important part in refining the products to marketing specifications.

A main distinguishing feature of the various petroleum products is their volatility, or ability to vaporise. This is associated with the size of the molecule; in compounds of a similar type, the larger the molecule, the lower the volatility. At ambient temperatures and pressure, gasoline is a liquid that vaporises readily, while kerosine and fuel oils are liquids requiring higher temperatures to vaporise them. Products such as paraffin wax, solid under normal conditions, require heating to a relatively high temperature before they liquefy and to still higher temperatures before they vaporise.

Volatility is related to the boiling point; a liquid with a low boiling point is more volatile than one with a higher boiling point. When a liquid, say water, is heated, the energy of its molecules increases and more molecules are able to pass through the surface of the liquid into the space above, i.e. more molecules pass into the vapour state. The pressure in the space above the surface, normally atmospheric pressure, tends to restrict the formation of vapour, but the temperature of the liquid determines the number of molecules leaving the surface of the liquid, and this in turn determines the vapour pressure of the liquid at that temperature.

When the vapour pressure is equal to or slightly higher than atmospheric pressure, vapour forms freely throughout the whole liquid, as is shown by the disturbance of the liquid surface and the formation of vapour bubbles in the liquid; the liquid is said to boil. The temperature at which a pure liquid boils is its boiling point and remains constant until all the liquid has evaporated, an important characteristic of a pure substance. The boiling point varies with pressure. At normal atmospheric pressure pure water boils at 100°C (212°F), ethyl alcohol at 78°C (172°F). Similarly, each of the individual hydrocarbons present in crude oil has its own characteristic boiling point. The boiling point is lowered by reducing the pressure in the space above the liquid (by creating a vacuum) and raised by increasing the pressure.

The heat transferred to the liquid in the process of boiling is retained in the vapour (latent heat of evaporation), and if this heat is removed, the vapour condenses back into the liquid state, giving off the heat of condensation. This is seen when steam (water vapour) from a kettle of boiling water strikes a cold surface.

Simple Distillation

The series of operations comprising boiling and condensation is known as distillation. A simple laboratory distillation apparatus is shown in Figure 5.1. The liquid is boiled in a flask or "still", the vapour is condensed in a tube or "condenser" surrounded by cold running water, and the distillate collected in a receiver.

In a mixture of several liquids of different boiling points, each component has its own characteristic vapour pressure, and the total vapour pressure above the liquid is the sum of the partial vapour pressures of the components. The mixture boils when the total vapour pressure is equal to the (external) pressure above the liquid.

When such a mixture is distilled, molecules of each component will vaporise, and the composition of the vapour phase will depend on the vapour pressures and the concentrations of the components in the liquid phase. Since the lower-boiling-point components have the higher vapour pressures, the distillate will at first be richer in these than in the higher-boiling-point components, whereas the liquid in the still will have a higher concentration of high-boiling-point components. As

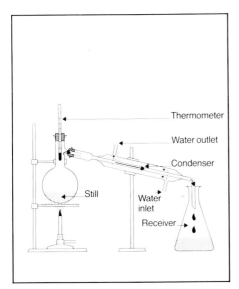

Figure 5.1 **Simple laboratory distillation apparatus**

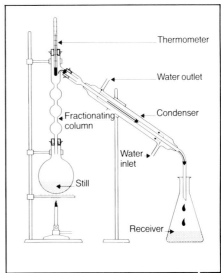

Figure 5.2 **Laboratory fractional distillation apparatus**

distillation proceeds, the composition of both distillate and residue will change progressively until all the liquid has been distilled into the receiver.

Boiling starts at a temperature that lies somewhere in the range of the boiling points of the components and depends on their ratio in the mixture. The initial boiling point (IBP) is defined as the temperature at which the first drop distils over. The temperature gradually increases during distillation, and the more volatile components distil over. The liquid becomes richer in higher-boiling-point components until the last drop of liquid evaporates at the highest temperature, the final boiling point (FBP).

Fractional Distillation

Using a simple distilling apparatus as described above, it is not possible to effect sharp separation between the components of a mixture in one distillation. By redistilling the first portion, a distillate richer in the more volatile components will be obtained, but the yield will be low, since part of the components always remains in the still. To effect a good separation it is necessary to modify the apparatus for continuous condensation and redistillation by inserting a still-head or "fractionating" column between still and condenser, as shown in Figure 5.2. Some of the vapour from the boiling liquid condenses as a liquid fraction in each bulb of the column. The condensation of further vapour from the still supplies heat, which re-evaporates the lighter or lower-boiling-point components from the liquid in the bulbs. These components condense in the next higher bulb, and so on up the column. As it becomes richer in the heavier, less volatile and higher-boiling-point components, the liquid in the bulbs flows back to the still. Thus there is a countercurrent flow of vapour and liquid, the vapour ascending the column and becoming lighter as the heavier components condense, and the liquid descending and becoming heavier as the lighter components re-evaporate. The vapour passing over the top into the condenser consists at first of the low-boiling components, and as these are removed the temperature of the liquid in the still increases steadily and higher-boiling components distil over. By changing the receiver at intervals, several different fractions are obtained. A fraction separated in this manner may consist of a relatively pure component from a simple mixture or a number of components from a complex mixture, depending on the composition of the mixture distilled and the type of apparatus. This process is called "fractional distillation".

Column Internals

We have seen above how fractional distillation requires a countercurrent flow of boiling liquid and condensing gas in good contact with each other. To do this on a large scale, a fractionating column is used, incorporating special contacting equipment known as "internals".

In oil refineries, the most commonly used distillation column internal is the tray. This is a horizontal plate covering the whole column cross-section, except for the "downcomers" — channels carrying liquid from one tray to the next (see Fig. 5.3a). The plate is perforated to allow passage of gas. In operation, liquid flows on the tray from a downcomer and into a highly turbulent froth made by bubbling and jetting of gas from the perforations. The froth is held on the tray for a short time by the weir, then flows over into another downcomer. Gas/liquid disengagement occurs in the downcomer and also in the vapour space above the froth. The column must be designed and operated with enough space to allow disengagement, otherwise it will fill up with froth, which cannot be separated. This undesirable condition, in which operation becomes impossible, is known as "flooding".

By placing trays one above another in the column a series of evaporation/condensation steps is produced, as in Figure 5.3a, by which continuous fractional distillation can be effected on a large scale. Previously, trays with bubble caps were used, but these have been largely discarded in favour of simple holes (sieve trays), or holes covered by discs of metal, which rise and fall with the gas flow rate (valve trays).

An internal sometimes preferred to trays, particularly in vacuum columns, is packing. A packed section (Fig. 5.3b) generally consists of a large number of

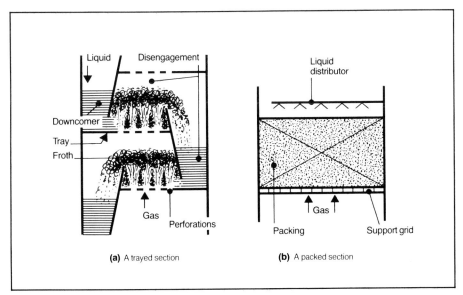

Figure 5.3 **Column internals**

Figure 5.4 **Pall rings**

small (approximately 50 mm dia.) rings piled up to a height of one or more metres, and resting on a support grid. "Pall" rings (Fig. 5.4) are nowadays commonly used, rather than the old "Raschig" type. Liquid is spread over the top of the packing and trickles down in films and rivulets which contact the countercurrently flowing gas. Again a number of evaporation/condensation steps are produced.

As well as facilitating distillation, trayed and packed columns perform absorption, stripping, washing and drying operations in the refinery. Other internals are also used, such as spray for direct-contact heat transfer, and auxiliary internals such as gas and liquid distributors, draw-off trays for removing side-streams, and demister mats catching fine droplets being carried up by the gas. All these internals must be carefully selected and designed for the required throughput and duty.

Distillation of Crude Oil

The products obtained by distillation of crude oil do not consist of single hydrocarbons, except in the case of simple gases such as ethane and propane. Each product fraction contains many hydrocarbon compounds boiling within a certain range and these can be broadly classified in order of decreasing volatility into gases, light distillates, middle distillates and residue.

The gases consist chiefly of methane, ethane, propane and butane. The first two are utilised as fuel or petrochemical feedstocks. Propane and butane may also be liquefied by compression and marketed as liquefied petroleum gas (LPG). Butane may to some extent be added to motor gasoline.

The light distillates comprise fractions which may be used directly in the blending of motor and aviation gasolines or as catalytic reforming and petrochemical feedstocks; these fractions are sometimes referred to as tops or naphtha.

The heavier, higher-boiling-point fractions in this range are the feedstocks for reforming processes and lighting, heating and jet engine kerosines.

Heavier distillates are used as gas oil and diesel fuel and also for blending with residual products in the preparation of furnace fuels.

The residue is used for the manufacture of lubricating oils, waxes, bitumen, feedstocks for cracking units and as fuel oil.

In the early days of refining, simple batch stills were used to produce illuminating oil (kerosine), the main product. Following the development of the internal combustion engine, the need for improved fractionation led to the use of simple fractionating columns corresponding, in principle, to the laboratory fractional distillation apparatus.

Demand for increased throughputs and higher-quality products then resulted in the development of continuous fractionation units. A simple continuous crude distillation unit is shown in Fig. 5.5. The crude oil feed first passes through a heat exchanger in countercurrent flow with the outgoing hot residue product. The preheated crude oil then enters the furnace, where it is heated to about 350°C depending on crude feedstock and products to be made; higher temperatures could lead to "cracking" and thermal decomposition. The hot vapour/liquid leaving the furnace enters the main fractionating column in the form of a mist

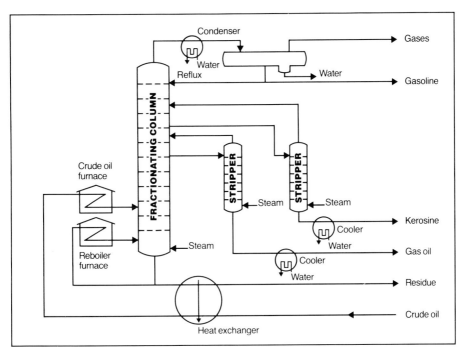

Figure 5.5 **Simple crude distilling unit**

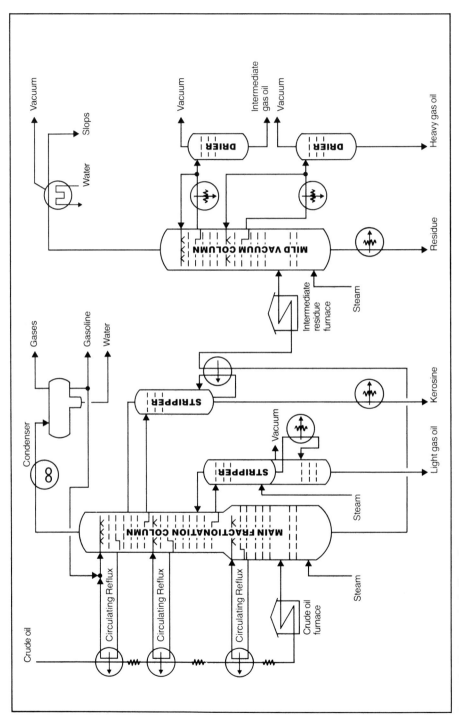

Figure 5.6 **Modern crude distilling unit**

which separates, the vapour passing upwards and the liquid downwards. This column normally operates slightly above atmospheric pressure and therefore the majority of hydrocarbons with atmospheric boiling points below 350°C are vaporised and those with higher boiling points remain as liquid and pass into the "stripping" section where any light components are vaporised by the action of steam. When steam is continuously blown into the bottom of the column, it contributes its own partial pressure, thus reducing the partial pressure of the hydrocarbons so that they boil at a lower temperature.

The vapour and steam pass up through the column; gas, gasoline and steam are removed from the top of the column. The gasoline vapour and steam are condensed in the condenser; the uncondensable gaseous hydrocarbons remain in the vapour phase and are thus separated from the gasoline. Water is removed from the system by settling and part of the gasoline is returned to the top of the column as "reflux", thus maintaining a downward stream of liquid. The temperature at the top of the column is controlled by varying the amount of "reflux" returned.

Other distillate products are removed from the column as liquid side-streams, the point of removal being chosen to give a product of the appropriate boiling range. The FBP of each product is controlled by the amount of product withdrawn, which in turn controls the amount of liquid flowing down the column below the liquid draw-off. In Figure 5.5 two such side-streams are shown, representing a kerosine and gas oil fraction. In order to improve the fractionation between the side-stream products and lighter fractions, each side-stream passes to a "stripper". Strippers are, in effect, small fractionating columns in which all components more volatile than those required in the product are vaporised with the aid of steam injected into the bottom of the stripper; the vapour stripped off is returned to the main column, and bottom products from the side-strippers are cooled before being sent to storage.

The quantity of individual distillate fractions of a given boiling range and the component distribution within these fractions may vary considerably (as therefore may the properties of these fractions too) between crudes from different sources.

The products of a modern crude distilling unit have to meet stringent specifications dictated by the market or downstream process units. Most distillate specifications include flash point, final boiling point and freezing/pour/cloud point (see Glossary). The flash point is determined by the proportion of light components in the fraction and the final boiling point and freezing/pour/cloud point by the heavier component "tail" in the fraction. Poor fractionation will increase the length of "tail" and therefore reduce the yield of specified product; good fractionation will maximise yield but requires taller, more expensive fractionating columns and higher fuel/energy input into these columns.

The modern crude distiller is designed to maximise yields of high-quality distillates whilst minimising the fuel/energy requirements by installing optimum heat-recovery systems. A typical modern 2-column distilling unit is shown in Figure 5.6. The crude feed passes through a train of heat exchangers, where it recovers heat from circulating reflux streams and the main product streams, entering the furnace with a high degree of preheat. The hot oil and vapour enter the first fractionating column at 300–350°C, depending on the feedstock and products to be made.

As the vapour passes up through the column, heat is removed, and it is partially condensed by circulating reflux streams, thus providing the liquid flow down the column and temperature control at various points in the column. The circulating reflux streams allow heat to be recovered from the column at higher temperature levels for feed preheat, thus improving the thermal efficiency of the system. Gas and gasoline are removed from the top of the column and cooled, water is separated, the gas is compressed and recombined with the gasoline and the mixture is then hydrotreated for sulphur removal before being debutanised and separated into specification products in a train of fractionating columns operating at higher pressures.

A kerosine fraction is withdrawn as a side-stream from the main column and passes through a stripper heated by a reboiler controlling the flash point of the kerosine product (to meet water content specifications, live steam for stripping is no longer used here).

A light gas oil fraction is also withdrawn as a side-stream lower down the column and then steam-stripped before being vacuum-dried.

The liquid leaving the bottom of the main fractionator after steam-stripping to remove the lighter hydrocarbons provides reboil heat for the kerosine stripper and then passes to a second furnace before entering a mild-vacuum column for further fractionation. Vacuum is obtained by means of steam ejectors or vacuum pumps which withdraw the non-condensable vapour from the top of the column. Circulating reflux streams withdraw heat from the column at various levels, condensing the hydrocarbon vapour rising up the column. Intermediate gas oil, heavy gas oil or waxy distillate side-streams are withdrawn from the column, stripped and dried under vacuum.

The "long residue" leaving the bottom of this column may be further processed in high-vacuum fractionators, or utilised as fuel oil, for bitumen manufacture, or as feedstock for cracking.

Extensive use is made of automatic control instruments to maintain steady operating conditions and product quality.

Vacuum Distillation

To recover additional distillates from long residue, distillation at reduced pressure and high temperature has to be applied. This vacuum distillation process has become an important chain in maximising the upgrading of crude oil. As distillates, vacuum gas oil, lubricating oils and/or conversion feedstocks are generally produced. The residue from vacuum distillation — short residue — can be used as feedstock for further upgrading, as bitumen feedstock or as fuel component.

The technology of vacuum distillation has developed considerably in recent decades. The main objectives have been to maximise the recovery of valuable distillates and to reduce the energy consumption of the units.

At the place where the heated feed is introduced in the vacuum column — called the flash zone — the temperature should be high and the pressure as low as possible to obtain maximum distillate yield. The flash temperature is restricted, however, in view of the cracking tendency of high-molecular-weight hydrocarbons. Vacuum is maintained with vacuum ejectors and lately also with liquid ring pumps.

In the older-type high-vacuum units the required low hydrocarbon partial pressure in the flash zone could not be achieved without the use of "lifting" steam. Those units are called "wet" units.

The latest development in vacuum distillation has been the vacuum flashers, in which no steam is required. These "dry" units operate at very low flash zone pressures and low pressure drops over the column internals (Fig. 5.7). For that reason the conventional reflux sections with fractionation trays have been replaced by low-pressure-drop spray sections. Cooled reflux is sprayed via a number of specially designed spray nozzles in the column countercurrent to the upflowing vapour. This spray of small droplets comes into close contact with the hot vapour, resulting in good heat and mass transfer between the liquid and vapour phase.

To achieve a low energy consumption, heat from the circulating refluxes and rundown streams is used to heat up the long residue feed. Surplus heat is used to raise medium-pressure and/or low-pressure steam or is given off to another process unit (heat integration).

The direct fuel consumption of a modern high-vacuum unit is approximately 1% on intake, depending on the quality of the feed. The steam consumption of the dry high-vacuum units is significantly lower than that of the "wet" units. They have become net producers of steam instead of steam consumers.

Three types of high-vacuum units for long residue upgrading have been developed for commercial application, viz.:
- Feed preparation units
- Luboil high-vacuum units
- High-vacuum units for bitumen production.

OIL PRODUCTS — MANUFACTURE

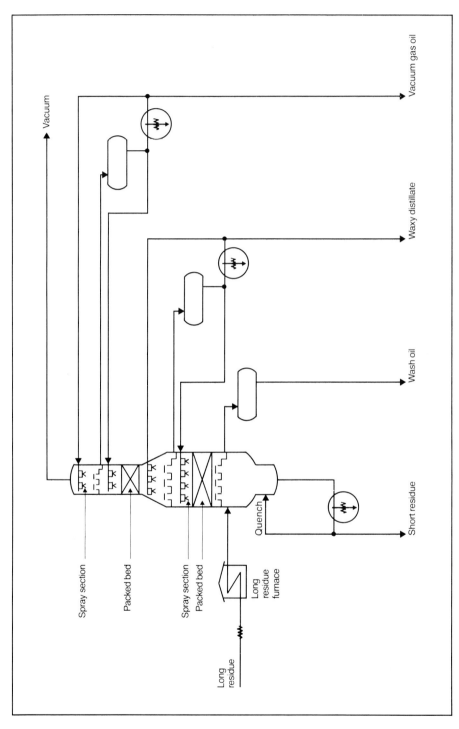

Figure 5.7 **Typical 'dry' high-vacuum unit**

DISTILLATION

Feed Preparation Units

These units make a major contribution to deep conversion upgrading ('cutting deep in the barrel"). They produce distillate feedstocks for further upgrading in catalytic crackers, hydrocrackers and thermal crackers.

To obtain an optimum waxy distillate quality a wash oil section is installed between feed flash zone and waxy distillate draw-off. The wash oil produced is used as fuel component or recycled to feed. The flashed residue (short residue) is cooled by heat exchange against long residue feed. A slipstream of this cooled short residue is returned to the bottom of the high-vacuum column as quench to minimise cracking.

Luboil High-Vacuum Units

Luboil high-vacuum units are specifically designed to produce high-quality distillate fractions for luboil manufacture. Special precautions are therefore taken to prevent thermal degradation of the distillates produced.

The units are of the "wet" type. Normally, three sharply fractionated distillates are produced (spindle oil, light machine oil and medium machine oil). Spindle oil and light machine oil are furthermore steam-stripped in dedicated strippers. The distillates are further processed to produce lubricating base oil. Short residue is normally used as feedstock for the solvent deasphalting process to produce deasphalted oil, an intermediate for bright stock (see Glossary) manufacture.

High-Vacuum Units for Bitumen Production

Special vacuum flashers have been designed to produce straight-run bitumen and/or feedstocks for bitumen blowing (see p. 325). In principle, these units are designed on the same basis as the previously discussed feed preparation units, which may also be used to provide feedstocks for bitumen manufacture.

Fractionators for Conversion Units

Visbreaker

In the fractionating section of a visbreaker (see p. 280) the soaker effluents are separated into the following fractions: gas (butane and lighter), gasoline, kerosine, gas oil and visbroken residue.

The gas is used as refinery gas after treating for removal of sulphur compounds. In view of a high content of unsaturates, components of the cracked gas may also be included in the feed to alkylation or polymerisation units. The gasoline is used, after further processing, as a blending component for motor gasoline manufacture or included in platformer feed. The kerosine is used directly as a low-viscosity fuel oil blending component. The gas oil is used directly or after hydrodesulphurisation as a component for diesel and domestic heating oil manufacture or as a fuel oil blending component. The visbroken residue is used in the manufacture of fuel oils.

In visbreaking soaker effluents at about 430°C are first separated in a cyclone. The bottom product (cyclone bottoms) is quenched in a circulating quench system and routed to a stripper column where the lightest components are removed by stripping with steam. The cyclone and stripper vapour products go to the fractionating column, where the rising vapour first passes through a wash oil section equipped with trays suitable for operating under coke-forming conditions. The fractionator bottom product (fractionator bottoms) is stripped with steam and combined with the stripped cyclone bottoms to form the combined visbroken residue product. After heat exchange with the short residue feed, the visbroken residue is further cooled in steam generators and coolers before being routed to storage.

In the upper fractionating part of the column, a gas oil side-stream is drawn off and steam-stripped in a side-stripper column for control of the gas oil flash point. A circulating reflux section below the gas oil draw-off controls the final boiling point of the gas oil product. Higher up the column a kerosine side-stream is similarly drawn off and steam-stripped in a side-stripper column. The vapour passing over the top of the fractionator column is cooled and partly condensed in overhead condensers (air- or water-cooled). The condensed liquid is partly returned to the column as reflux, the rest being sent on to further processing as "unstabilised" gasoline product. The uncondensed vapour goes to further processing as gas product.

Thermal Gas Oil Unit

The fractionating section of a thermal gas oil unit (TGU) (see p. 282) has many similarities with, and basically makes the same products as, the fractionating section of a visbreaker.

The cyclone overhead vapour from a TGU contains a much higher proportion of heavy ($350°C^+$) distillates than that in the visbreaker because of the processing of long residue a compared to short residue feed. This heavy distillate is condensed and drawn off in a circulating reflux section just above the wash oil section of the fractionating column. In the TGU, the fractionating column is often referred to as "combination tower". The heavy distillate drawn off at about 390°C and collected in the distillate surge drum is fed to the distillate cracking furnace. The effluents from this furnace are reintroduced below the wash oil section of the combination tower. A recycle (to extinction) of the heavy distillate fraction is thereby established. The upper part of the combination tower is similar to that of the fractionating column of a visbreaker.

LPG Recovery / Production

Recontacting

The fractionating columns of crude distillers, thermal, catalytic and hydrocracking units are usually operated at the lowest pressure possible in practice in order to minimise energy consumption. This will cause the major part of the propane and butane present in the column top product vapour to remain in the vapour phase after cooling in the overhead condensers.

In many cases it is desirable to recover propane and butane in liquid form (for the production of LPG) rather than to leave them in the fractionator gas product to become refinery fuel. For this purpose a "recontacting system" can be applied (see Fig. 5.8). The gas product is compressed to a pressure of 12 to 15 bar, mixed with the fractionator liquid top product (gasoline), cooled and routed to a flash vessel, where gas and liquid are again separated. Because of the higher pressure, a larger part of the propane and butane is recovered in the liquid phase in the recontacting flash compared with the flash in the overhead product accumulator. Components heavier than butane are also recovered from the fractionator gas top product with recontacting.

Following the recontacting step the "unstabilised" gasoline with a high content of butane and lighter components is separated (possibly after processing in a hydrotreater) in a debutaniser column into a gasoline bottom product with a very

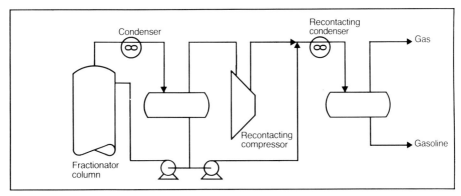

Figure 5.8 **Recontacting system**

low butane content and a top product consisting of butane and lighter components. The conditions in the debutaniser overhead accumulator are such that the top product will remain partly in the vapour phase. Here again, the recontacting principle (with a recontacting pressure of 25 to 30 bar) is applied in some cases to increase the recovery of propane and butane.

Similarly, recontacting can be applied in the product work-up section of platformers and in the overhead system of platformer stabilisers.

Chilling/Absorption

As alternatives to recontacting, "chilling" and "absorption" may be applied for recovery of propane and butane from vapour products that would otherwise become refinery fuel. In addition to vapour top products from debutaniser and stabiliser columns, this may include off-gases from various hydroprocessing units (hydrotreaters, hydrodesulphurisers, hydrocrackers).

Chilling in its simplest form, shown in Figure 5.9, comprises the cooling of the LPG-rich gases to a very low temperature (10 to 15°C). As cooling medium, use is made of "chilled cooling water", circulating via a chilling unit applying absorption, compression or steam jet systems for refrigeration. The liquid formed by the chilling is separated from the remaining vapour product in a simple flash vessel.

Absorption in its simplest form, shown in Figure 5.10, utilises a small packed column, in which the LPG-rich gases are brought into intimate contact with a "lean oil". As lean oil, a debutanised gasoline or naphtha can be used. After having absorbed butane and lighter components in the absorber column, the

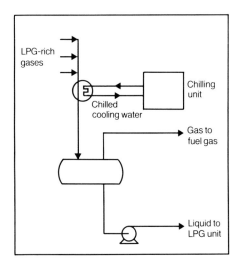

Figure 5.9 **Chilling LPG recovery system**

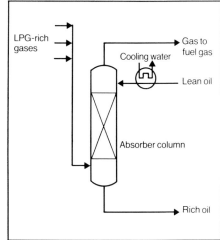

Figure 5.10 **Absorption LPG recovery system**

DISTILLATION

"rich oil" is reprocessed in a debutaniser, where butane and lighter components are recovered in the top product.

In more complex LPG recovery systems, chilling may be combined with either recontacting or absorption.

LPG Production

For the production of marketable LPG products from the various liquefied "butane and lighter" streams produced in debutanisers, stabilisers, chilling systems, etc., these streams are processed in an "LPG unit" (see Fig. 5.11).

This unit consists of a deethaniser and a depropaniser. The liquid feed is

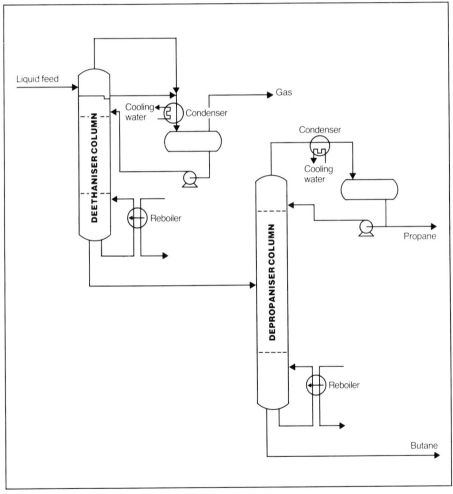

Figure 5.11 **LPG unit**

introduced on the top tray of the deethaniser column, which is heated by a reboiler. This column, operating at 25–30 bar, effects an asymmetrical separation whereby a very low ethane content of the bottom product is achieved while some heavier components are left in the top product, so that condensation can take place in a normal cooling-water/air-cooled condenser. With this system recoveries at about 85% propane and 95% butane can be attained.

The deethaniser bottom product is routed to the depropaniser column operating at about 16 bar. This column too is heated, effecting a symmetrical separation between propane and isobutane.

Catalytic Cracker Main Fractionator

The catalytic cracker reactor effluent product (see p. 288) is separated in the main fractionator in accordance with the same principles as in crude oil distillation. The main difference is that in this case the feed is a superheated vapour. The reactor effluent product, at a temperature of approximately 480–510°C, is desuperheated in the main fractionator bottom section to a sufficiently low level to prevent excessive coking. Desuperheating is accomplished by contacting with cooled bottom circulating reflux via grid trays.

The bottom product, often called slurry, is usually a fuel oil component. (Steam-)stripping is not required to meet flash point specification. Two sidestreams are taken from intermediate points in the column and stripped in stripping columns to produce light and heavy cycle oil. The heavy cycle oil is recycled to the catalytic cracker and/or used as fuel oil component. The light cycle oil may be used as gas oil and/or diesel fuel. The fractionator overhead vapour can only be partially condensed. The non-condensed vapour is recontacted (in two stages) at high pressure with a slipstream of the condensed liquid. The rest of the liquid is pumped as lean oil to the absorber. If all liquid is used for recontacting, debutaniser bottom product is used as lean oil.

The absorber separates the light dry gas fraction from the main fractionator overhead stream with the objective of high propane recovery from the gas. This recovery can be improved by applying chilled water as cooling medium in the absorber condenser system. The absorber bottom product is split in a debutaniser into LPG and full-range gasoline. The LPG is separated in the depropaniser into propane and butane. The full-range gasoline may be split in a gasoline splitter into various gasoline fractions.

Hydrodesulphuriser

The liquid product from a hydrodesulphuriser (see p. 306) contains small quantities of H_2S and low-boiling-point hydrocarbons dissolved from the hydrogen-rich gas and/or formed during the reactions. These light ends, including H_2S, are effectively removed by "stripping". The hydrodesulphurised product enters a stripper column (see Fig. 5.5) at the top and then flows downward over fractionation trays. At the bottom of the column some 10-15% of the product is vaporised because of steam injection. The created vapour flows upward, contacting the downward flowing liquid tray by tray, and, after leaving the column at the top, is partly condensed in the overhead condenser.

The uncondensed part of the vapour, comprising the stripped-off H_2S and low-boiling-point hydrocarbons, is typically sent to an amine treating unit for H_2S removal, after which it is utilised as fuel gas. From the condensed liquid, any free water is separated by gravity, while the remaining liquid is routed back to the column in combination with the feed.

The stripped, hydrodesulphurised product leaving the stripper column at the bottom contains water, as a result of the contact with the stripping steam. To remove the water, the stripped product is routed to a drier. Here the water is flashed off under partial vacuum, leaving a dry product.

If an absolutely dry product is required, like naphtha as platformer feed, the stripping is performed with reboiling instead of steam injection.

SOLVENT EXTRACTION

Separation by means of distillation is based on differences in boiling points and does not differentiate between chemical types, such as paraffins, aromatics and naphthenes. A fraction separated by distillation contains more or less of each type, depending on its boiling range and the crude oil used. However, the performance of a product in service depends on the chemical nature of the components as well as on their physical properties.

The presence of aromatic hydrocarbons in kerosine produces a smoky flame when burnt in a wick-type lamp and a more or less luminous flame in a jet engine. Formerly, sulphuric acid treatment was used to remove aromatic components and so improve the burning qualities. The treatment was expensive and the disposal of the spent acid and the constituents so removed presented a difficult problem. About 1907, a process was developed for the removal of aromatic hydrocarbons from kerosine with the aid of a solvent, liquid sulphur dioxide, which has advantages over acid treatment in that the undesired aromatic hydrocarbons can be recovered unchanged and the refining agent can be recovered and used again. The process, known as the Edeleanu process after the inventor, opened the field for the development of other solvent extraction processes, which are now used not only for the refining of kerosine, but also for the manufacture of high-grade lubricating oils, high-octane gasoline fractions and aromatics (feedstock for the chemical industry).

Principle of Solvent Extraction

The process is based on the use of a solvent in which one group of feed components, usually the aromatics, is preferentially dissolved. The choice of solvent depends primarily on its selectivity, i.e. its ability to distinguish between two (or more) chemically different groups of components. It must be cheap and readily available, resistant to chemical change during use, non-corrosive and cheap to handle. Its boiling point must be sufficiently different from that of the feed to be extracted that recovery of the solvent can be performed by simple distillation or flashing.

If such a solvent is thoroughly mixed with an oil fraction and the mixture is allowed to settle, two layers or phases are formed. One phase (the extract phase) will contain nearly all the solvent plus the dissolved components (e.g. the aromatics); the other (raffinate) phase will consist of undissolved components (e.g. the paraffins) and some solvent. After removal of the solvent, a raffinate and an extract are obtained which are chemically unchanged.

Each of the phases described above contains some of the components of the other phase. The quality of the raffinate could be improved by repeated extraction with fresh solvent. Such operation would result in a low yield of raffinate, as in each stage part of the paraffins is also lost. Moreover, large amounts of solvent would be required, which have to be recovered by distillation. The resultant costs would in general be so high as to make such a process unattractive. In practice, therefore, the extraction is performed in a number of stages through which feed and solvent flow countercurrently. In this way the raffinate is treated in the last stage with pure solvent; the feed is first contacted with solvent, which is partly loaded with soluble components. The selectivity of the solvent and its ability to dissolve certain groups of components are influenced by temperature. Each process is therefore operated at the most suitable temperature, the temperature range throughout the extraction system being carefully controlled.

Extraction Equipment

The mixing and separating can be performed in a series of mixer/settler tanks. This method involves high capital costs (for both the equipment and the large volume of solvent present in the settler) and requires a large ground area. Its use is therefore limited to systems requiring a small number of stages.

For large commercial units, a column-type extractor is used. The feed and the solvent enter near the bottom and the top, respectively, and flow countercurrently through the column. The raffinate phase is withdrawn from the top, the extract phase from the bottom.

In order to promote mixing, the column is equipped with packing material

SOLVENT EXTRACTION   259

(small cylinders, rings, saddles) or with perforated or slotted trays. An improved device for contacting feed and solvent, which has been developed by Shell, is the rotating disc contactor or RDC (see Fig. 5.12). The column is divided into a number of compartments by horizontal plates, each with a central hole (stator

Figure 5.12 **A rotating disc contactor.**

rings). Between the stator rings, rotor discs attached to a central shaft are rotated. By adjusting the rotor speed the degree of mixing can be controlled, thus maintaining the efficiency. The RDC is widely used. Other types of agitated column contactors also exist.

For systems which are very difficult to separate, sometimes centrifugal extraction is employed. In this type of extractor phase, separation is not obtained by gravity but by a (much higher) centrifugal force.

Solvent Extraction Processes

Solvent Extraction of Lubricating Oil

The lubricating oil fractions obtained from crude oils by vacuum distillation contain aromatic components, which are undesirable since they oxidise in engines to sludge-forming compounds and, moreover, have poor viscosity/temperature properties. In the manufacture of high-grade lubricating oils, therefore, it is necessary to remove these aromatic compounds, which can be done by solvent extraction. Insulating oils, white oils and medicinal oils made from lubricating oil distillates require a severe sulphuric acid treatment, but it is more economical to remove the aromatics by solvent extraction prior to acid treatment.

The Edeleanu process was first applied to lubricating oils in 1926, but sulphur dioxide has certain limitations as a solvent in lubricating oil extraction on account of low solubility of higher-boiling aromatics. Although, by using a blend of sulphur dioxide and benzene, some improvement can be effected, more suitable solvents such as furfural, phenol and N-methyl-2-pyrrolidone are used. All these

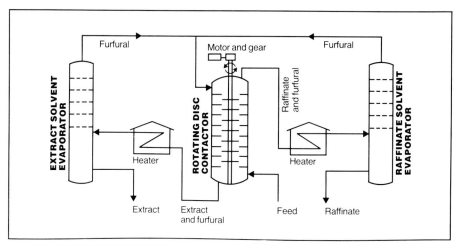

Figure 5.13 **Furfural extraction process**

solvents have a cyclic structure. Their boiling points are well below the boiling range of the lubricating oils, so that separation and recovery of the solvent by simple flashing is possible (furfural 162°C, phenol 181°C, N-methyl-2-pyrrolidone 206°C).

Furfural extraction is the most widely used extraction process for lubricating oils and is applied in many Shell refineries; solvent recovery is a relatively easy matter, although precautions must be taken to avoid oxidation and decomposition of the solvent.

In view of the high cost of solvent recovery and the low value of lubricating oil extract, it is essential that the maximum amount of refined oil should be produced with the minimum use of solvent. The highly efficient RDC is therefore normally used as extractor. Moreover, the temperatures in the extraction operation are carefully maintained to obtain maximum results. A diagram of a furfural extraction unit is given in Figure 5.13.

Solvent Deasphalting

Lubricating oil fractions are in general prepared by vacuum distillation, leaving a residue. From this residue the base material for high-quality residual lubricating oil is prepared by propane deasphalting. In this process, which is a single-solvent extraction process, the feed is treated countercurrently with liquid propane. The process operates at a pressure of 35–40 bar to keep the solvent a liquid at the extraction temperature. An RDC or a plate column is normally used, with the oil entering at the top and the propane at the bottom. The paraffinic compounds are preferentially dissolved in the propane and are withdrawn from the top, whilst the remaining, undissolved, asphaltenic compounds are withdrawn from the bottom. After distillation and recovery of the propane, a lubricating stock that is free from bituminous components is obtained. The deasphalted oil is extracted with furfural to obtain a lubricating oil.

The same process can also be carried out with butane or pentane as solvent. With these solvents, a higher yield of deasphalted oil can be obtained but in general with a lower quality. The deasphalted oils produced with butane or pentane as solvent are normally used as feedstock for conversion processes.

Solvent Extraction of Gasoline Fractions

Solvents like furfural are not suitable for this application, mainly because their boiling point is too close to, or even within, the boiling range of the feed, and they are therefore difficult to recover. Other solvents, often with a much higher boiling point, are employed.

The solvent at present most widely used is Sulfolane, a five-number cyclic sulphur compound with a boiling point of 282°C. On the basis of this solvent the Shell Sulfolane Extraction Process was developed. In this process liquid/liquid extraction is combined with extractive distillation, i.e. distillation in the presence of a selective solvent. The combination of these two process steps allows very sharp separations to be made, e.g. obtaining a raffinate with only minor amounts of aromatics while the extract is nearly free from paraffins. The mixture of extract and solvent is separated in a normal distillation column; the recovered solvent is recirculated to the extractor. A simplified scheme of a Sulfolane extraction unit is given in Figure 5.14.

The Sulfolane extraction process is mainly used for the manufacture of very pure, light aromatics (benzene, toluene, xylenes) which are employed as base material in the petrochemical industry. The feedstocks used are streams with a high aromatic content, such as catalytic reformate or the gasoline fraction obtained in the cracking process for ethylene manufacture. The process is also utilised for the preparation of low-aromatic-content hydrocarbon solvents or for separating gasoline, white spirit and kerosine fractions into a high- and a low-aromatic-content fraction.

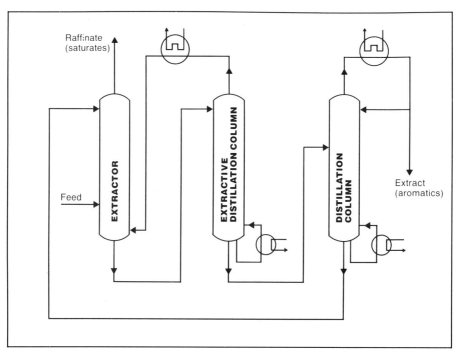

Figure 5.14 **Sulfolane extraction process**

Extractive Distillation of Gasoline Fractions

The process scheme is very similar to the one given in Figure 5.14 less the extractor. The feed is introduced halfway up a distillation column; the solvent at the top. In the extractive distillation step, the aromatics are taken up by the solvent, while the saturates appear as top product. The aromatics and the solvent are separated by distillation and the solvent returned to the top of the extractive distillation column. The process can only operate on fractions with a narrow boiling range and is less flexible than the combination of extraction/extractive distillation described above. The solvents most commonly used are phenol, Sulfolane and acetonitrile.

CRYSTALLISATION AND ADSORPTION

Crystallisation

Wax consists essentially of paraffin hydrocarbons of high molecular weight, which readily separate by crystallisation when an oil fraction containing them is cooled. Wax occurs in lubricating oils made from paraffinic crude oils and must be removed in order to avoid poor performance by congealing of the oils at low temperatures. This can be achieved in the dewaxing process, by means of which the waxy oils are separated into dewaxed lubricating base oils and slack waxes. The slack waxes can be worked up further to manufacture highly refined food-grade waxes by reduction of the oil content to a low level in the deoiling process, followed by thorough purification (see p. 314).

The process is essentially cooling of the wax oil to cause the wax to crystallise and then separation of the solid from the liquid by filtration or centrifuging. In older-type plants the oil is cooled in chillers, in which the viscous mixture of oil and wax is kept moving through the chiller by a worm conveyor. The chilled mixture is pumped through large filter presses in which the wax crystals are retained, allowing the wax-free oil to pass through. When full, the presses are opened and the cake of crude wax is removed. This type of process has three disadvantages. Firstly, it is discontinuous and a number of operators are required to handle the presses. Secondly, the high viscosity of the oil renders filtration difficult, especially with heavy oils. Thirdly, it cannot be used for residual oils on account of the presence of microcrystalline waxes, consisting of crystals of microscopic size. In order to overcome these difficulties, a diluent was used (a heavy gasoline or gas oil) to maintain fluidity. Wax, however, is appreciably soluble in such diluents and low dewaxing temperatures were required.

Solvent Dewaxing/Deoiling

Modern solvent dewaxing/deoiling processes were introduced to overcome these difficulties. By using a mixture of solvents it is possible to obtain a filtrate of low viscosity at chilling temperature and a good crystallisation of all types of wax in a form suitable for efficient filtration. The low viscosity makes it possible to replace discontinuous filters by more economic continuous rotary vacuum filters. The use of a combination of solvents, adjusted to different feedstocks, increases the flexibility of the process.

A diagram of the solvent dewaxing process is shown in Figure 5.15. Usually two solvents are used: toluene which dissolves the oil and maintains fluidity at low temperatures, and methyl ethyl ketone (MEK), which dissolves little wax at low temperatures and acts as a wax precipitating agent. Propane and a chlorinated hydrocarbon such as dichloromethane are sometimes used.

The process has three stages: mixing the waxy oil with the solvents and chilling; filtration of the chilled oil to separate the wax; recovery of the solvents and their recycling.

The waxy feed is mixed with the solvents and heated to ensure complete solution. The mixture is cooled down to filtration temperature (usually $-20°C$), first by heat exchange with cold, outgoing filtrate, followed by refrigeration with liquid ammonia or propane. The cooling is carried out in concentric pipes, with the process stream flowing through the inner pipe and coolant through the outer

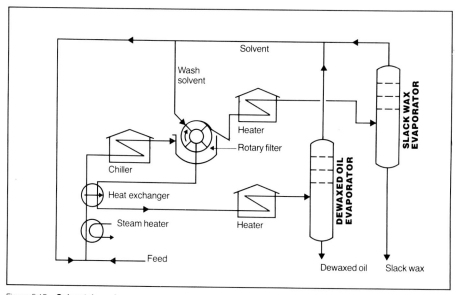

Figure 5.15 **Solvent dewaxing process**

CRYSTALLISATION AND ADSORPTION 265

pipe. The inner pipe wall is kept free of wax by a spring-loaded rotating scraper in order to keep the heat-transfer resistance and pressure drop sufficiently low. The chilled mixture flows to the rotary vacuum filter (Fig. 5.16), which is a cylindrical drum covered by cloth on coarse-mesh metal gauze or nylon grid. The drum rotates slowly on a horizontal axis, the lower portion passing through a cylindrical tank into which the oil and wax slurry is fed. The drum is divided into a number of segments by radial partitions from the centre to the circumference, and each segment is connected by a pipe to the end of the drum. As the drum rotates, these pipes make contact with stationary ports on the filter casing. As each segment passes through the wax, oil and solvent mixture, vacuum is applied to the interior of the segment through the appropriate port on the casing (A on Fig. 5.16) and the solution of oil in the solvent is drawn through the filter cloth into the drum, whilst the wax forms a cake on the surface and is carried round by rotation of the drum. Cold solvent is sprayed over the wax layer to wash the adhering oil from the wax cake on the filter. This oil and solvent leave the drum through the appropriate port (B) as the rotation continues. The wax cake is loosened from the filter surface by a slight pressure of inert gas, which is applied to the segment as it comes opposite the next port (C). The wax is removed by a

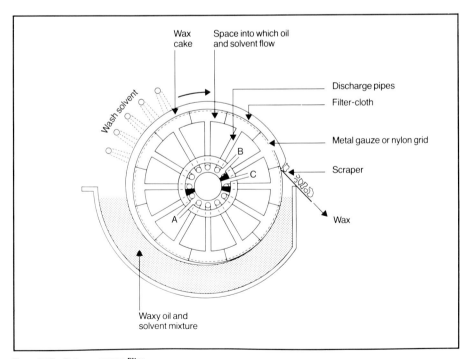

Figure 5.16 **Rotary vacuum filter**

scraper to a trough, along which it is moved by a worm conveyor. Finally, the solvents are separated from both wax and oil fractions by distillation and recycled to the process, while the "slack wax" and lubricating oil are run to storage.

The solvent deoiling process is essentially identical, apart from the higher filtration temperatures applied, namely in the range of −5 to 45°C.

Sweating Process

Another physical separation process where use is made of the difference in melting point is the "sweating process" for removing the lower-melting-point components and most of the oil from slack wax. This process consists in gradually increasing the temperature of the solid slack wax so that at first the oil, together with the lowest-melting-point waxes, drains away from the cake. As the temperature is increased, waxes with still higher melting points become liquid and drain away. By control of the temperature of sweating, oil and lower-melting-point waxes, referred to as "foots" oil, are removed, leaving a mass of paraffin wax with the required melting point. In this way fully refined paraffin waxes with less than 0.5% wt. oil are obtained. Sweating may be stopped at an earlier stage to produce "scale wax", with an oil content up to 2% wt.

The best sweating equipment for the manufacture of high-quality waxes is the "vertical-tube sweating stove". It consists of a cylindrical vessel containing a bundle of vertical tubes one inch in diameter and a horizontal perforated plate in the bottom. Firstly, the bottom section is filled with water until the water layer is level with the perforated plate. Then the molten wax is charged to the vessel and floats on the water layer. The charge is allowed to solidify by circulation of cooling water through the tubes, after which the water is drained off, leaving the solid wax cake resting on the perforated plate. On raising the temperature of the vessel by circulation of hot water through the tubes, the sweat oil drips through the perforated plate, whereas the solid wax remains on it. After sweating, the wax is melted by injection of live steam into the vessel.

The advantages of the vertical-tube sweating stove are its low price, simple construction, easy operation and low energy consumption. However, its application is limited to the manufacture of waxes of low molecular weight, since for heavier types the oil would be occluded in the interstices between the wax crystals and, hence, not drain away during sweating.

Adsorption

Solid adsorbents such as molecular sieves, aluminas and active carbon have increased considerably in scope of application in the last ten years, particularly in the natural gas and LPG industries. Their use will be illustrated by a very common application, namely virtually complete water removal from natural gas prior to liquefaction.

In Figure 5.17, the process line-up is shown, in this case with two adsorbent beds, one adsorbing and one in regeneration. Typically, a cycle time of eight hours is used, corresponding to one operating shift. In the adsorption mode the natural gas is dried to less than 0.5 ppm (weight) of water at about 50 bar pressure, and at the end the adsorbent will contain about 10% of its own weight of water.

The bed is regenerated by bringing it to about 300°C using dried product gas, compressed and heated in a fired heater. The regeneration gas is then cooled, compressed and, after separation of condensed water, returned to the feed of the unit. Typically, a single liquefaction module has a feed of 300 MMScf/d of

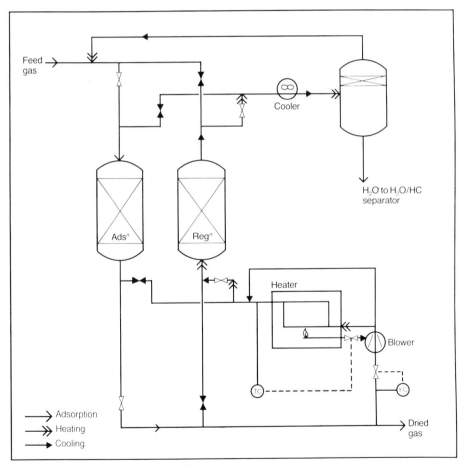

Figure 5.17 **Two-bed natural gas drying unit (molecular sieves)**

natural gas and a single adsorption bed may contain 20 tons of molecular sieves, which cost about $4 per kg. In practice, more than 2 beds may be used, depending on optimisation studies. Automatic timing and switching equipment is applied, requiring a minimum of operator attention.

Such "temperature swing" regenerable adsorption cycles have cycle times of eight hours or more, and are limited by the amount of adsorbent needed to applications with up to, say, 500 tons of acid gas contaminant per year when competing with solvent processes. Pressure swing adsorption (PSA) achieves regeneration by pressure reduction and purging with considerably reduced cycle times. This increases the removal of contaminant per bed enormously compared with temperature swing. Such techniques are used for CO_2 removal from hydrogen in synthesis of hydrogen via steam reforming.

Aluminas may also be used for drying of gas or LPG, but are more reactive and may give side-effects such as hydrolysis of carbonyl sulphide to H_2S in LPGs, putting the product off specification. They are commonly used for HCl removal from refinery gases or LPGs on a non-regenerative basis.

Activated carbon has a great variety of applications, often on a non-regenerable basis, for cleaning up gas and waste water streams. Sulphur compounds, elemental sulphur, heavy hydrocarbons, phenols, colours, pesticides, etc. can be removed. Active carbon has limited regeneration capacity and is usually applied on a disposable basis. Typical cost is $3–4 per kg.

REFORMING

Introduction

Reforming processes were developed for the purpose of converting low-octane heavy gasoline fractions (naphthas) into product with a higher ignition quality, in terms of octane number, for blending into motor and aviation gasoline. This conversion involves subjecting the compounds of the naphtha to complex chemical reactions, at high temperature and pressure, with the aim of producing primarily aromatics and to a lesser extent also isoparaffins.

In these processes byproducts are also formed, mainly as a result of breakdown reactions yielding liquefied petroleum gas (propane, butane), light hydrocarbon gases and hydrogen. Originally, the thermal route was extensively followed until well after World War II. Since 1950, however, catalytic reforming has rapidly taken over because the yields and qualities of catalytic reformate were substantially better.

Catalytic Reforming

Historical

Several forms of catalytic reforming processes were developed from 1935 onwards, but none of the early processes found wide acceptance, mainly because of high investment and operating costs. A major breakthrough in catalyst technology occurred in 1949, when Universal Oil Products (UOP) Company introduced the first platinum-on-alumina catalyst. This heralded the era of catalytic reforming and the decline of thermal reforming.

In the following years, quite a number of catalytic reforming processes for the upgrading of naphthas were announced. These reforming processes can be classified in two main groups, depending on the type of catalyst used, i.e. catalyst containing the metal platinum as one of the main active ingredients and catalyst containing other catalytically active metals or metal oxides. The active metals are finely dispersed on a carrier, usually a highly porous form of aluminium oxide. It is believed that the most widely used reforming processes utilise a catalyst containing 0.2–0.8% wt. platinum with or without modifiers or activators, which can be metals as well, and around 1% wt. halogen.

There are four basic unit types, non-, semi-, fully and continuously regenerative.

The capacity of catalytic reformers as of January 1, 1980 in non-Communist areas is estimated to be about 380 million tons per annum, which is nearly 13% wt. of crude intake. The catalytic reformers in Shell refineries are based on UOP technology and use UOP catalysts containing platinum. The majority of the reformers are of the semi-regenerative type and the remainder consist of fully regenerative units (in the USA and Canada). Since 1977, five continuously regenerative units have also been in operation. The latter type of unit is expected to replace a substantial part of the semi-regenerative unit capacity in the medium or longer term.

Chemistry

The main reactions occurring in catalytic reforming processes are the following:
  (a) dehydrogenation of naphthenes, yielding aromatics and hydrogen;
  (b) dehydro-isomerisation of alkyl cyclopentanes to aromatics and hydrogen;
  (c) isomerisation of paraffins and aromatics;
  (d) dehydrocyclisation of paraffins to aromatics and hydrogen;
  (e) hydrocracking of paraffins and naphthenes to lighter, saturated paraffins at the expense of hydrogen.

The above reactions take place concurrently and to a large extent also sequentially. The majority of these reactions involve the conversion of paraffins and naphthenes and result in an increase in octane number and a net production of hydrogen. Characteristic of the total effect of these reactions is the high endothermicity, which requires the continuous supply of process heat to maintain reaction temperatures in the catalyst beds.

The reactions take place at the surface of the catalyst and are very much dependent, amongst other factors, on the right combination of interactions between platinum, its modifiers or activators, the halogen and the catalyst carrier. During the operating life of the catalyst, the absolute and relative reaction rates are influenced negatively by disturbing factors like gradual coke deposition, poisons (permanent and temporary), and deterioration of the physical characteristics of the catalyst.

Operating conditions

The reforming reactions proceed at economic rates in the temperature region of 450–530°C. To limit the catalyst performance decline rate due to coke deposition, reactor pressures in the range of 10–40 bar are required, leading to hydrogen partial pressures in the 5–35 bar range. The high pressure is also important to limit equipment size. The lower operating pressures are applied mainly in the fully and continuously regenerative types of processes. These two types are economically very attractive, since the lower pressures entail higher liquid and hydrogen yields.

Catalysts

Shell's catalytic reformers use UOP-manufactured catalysts containing platinum or platinum combined with a second catalyst ("bimetallic catalyst").

A survey of the type of process and catalyst used is given in Table 5.1.

In the bimetallic catalyst types the platinum is combined with small quantities of a second metal. The best-known second metal is rhenium. The advantage of bimetallic over monometallic catalysts is their higher stability under reforming

Table 5.1. **Catalysts used for reforming processes**

Type	Catalyst types	
	Monometallic	Bimetallic
Semi-regenerative	R-10, R-11, R-12	R-16, R-18, R-22, R-50
Fully regenerative	R-9, R-55	
Continuously regenerative	–	R-22, R-32

conditions. A disadvantage is the higher sensitivity towards poisons, process upsets and susceptibility to non-optimum regenerations.

If the catalyst performance after a certain period of operation, which is in the range of 1 week to 1 year, declines to an uneconomic level, the performance of the catalyst can usually be restored to fresh or nearly fresh levels by (*in-situ*) regeneration techniques. These techniques, which have become very complicated, consist nowadays of essentially three parts, i.e. a carbon-burn step to remove carbonaceous deposits, a metal redispersion step and a metal oxide reduction step. The majority of the catalyst batches attain an overall life in the 4–7 years' range.

Equipment

A description of the three types of processes and equipment used in Shell refineries is given below. A catalytic reformer comprises a reactor section and a product-recovery section. More or less standard is a feed-preparation section in which, by a combination of hydrotreatment and distillation, the feedstock is prepared to specification.

Semi-Regenerative Process. A simplified flow scheme of a semi-regenerative reformer is shown in Figure 5.18. The desulphurised straight-run gasoline is first distilled in the feed-preparation section (not shown). This is necessary to adjust the initial and final boiling points and to remove dissolved oxygen and water. A fraction boiling between 40 and 180°C is removed from the column, mixed with hydrogen (recycle gas) and passed through heat exchangers in countercurrent with the hot product from the reactors. It is then raised to reactor temperature in a heater. The vaporised feed now passes in succession through a series of three or four reactors consisting of cylindrical steel vessels, filled with the pelleted catalyst, through which a centrally placed perforated pipe runs. The oil vapour enters at the top of the reactor casing and so passes into the annular space between the reactor wall and the catalyst. It then flows through the bed of catalyst towards the centre pipe, passing through the perforations and out at the bottom. Heat is absorbed during the reactions, so that the vapour has to be reheated. This is done by passing it through heaters between the reactors, in order to maintain the correct temperature in each reactor. The effluent from the last reactor passes to a gas separator and thence to the product fractionating section. Here the lighter hydrocarbons (LPG) are removed and the reformed bottom product, called "platformate", can be split into a light and heavy grade, if so desired, for gasoline blending purposes. The gas from the reactor effluent contains the hydrogen set free by the reforming reactions; it is recycled after the removal of the heavier hydrocarbons in the gas separator. This recycling is necessary to ensure a

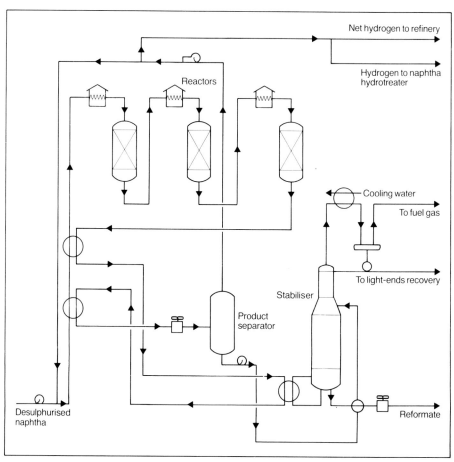

Figure 5.18 **Semi-regenerative reformer**

sufficient hydrogen partial pressure in the reactors in order to suppress coke formation on the catalyst.

Platformate Research octane numbers (see p. 397) which can be achieved in this process are usually in the range of 85–100, depending on an optimisation between feedstock quality, gasoline qualities and quantities required on the one hand and on the other the operating conditions required to achieve a certain planned cycle length (6 months to one year). The catalyst can be regenerated *in situ* at the end of an operating cycle. Often the catalyst inventory can be regenerated 5–10 times before its activity falls below the economic minimum, whereupon it is removed and replaced. Platinum and other promoters or activators can be recovered from the spent catalyst for re-use.

Unit capacities vary from 250 to 3500 t/d naphtha intake. Main product yields are in the range of 75–90% wt. of platformate and 1–3% wt. of hydrogen, depending on feed quality, operating conditions and degree of deterioration of the catalyst.

Fully Regenerative Process. The fully regenerative reformer or cyclic reformer has long been in commercial use and has specific advantages over semi-regenerative units in high-octane operations with poor-quality feedstocks. The lower octane ceiling of semi-regenerative units is the direct result of the long periods between regenerations, with a resultant higher average coke level on the catalyst.

To attain the higher octanes while maintaining high platformate and hydrogen yields, lower reforming pressures are required. The significance of this becomes even more marked as octane severity increases. Currently, the cyclic units are operated at pressures as low as 8–10 bar. Coke lay-down rates at these low pressures and high octane severities (100–104 range) applied are so high that the catalyst in individual reactors becomes exhausted in less than 1 week up to 1 month. Consequently, a line-up system is used that allows each reactor to be taken out individually for regeneration. The schematic flow diagram of such a unit and its regeneration system is shown in Figure 5.19. Unit capacities range typically between 300 and 3000 t/d.

Continuously Regenerative Process. Whereas in the semi-regenerative and cyclic processes fixed catalyst beds in individual reactors are used, in the novel continuously regenerative process moving beds of catalyst are applied. This technology, developed by UOP in the late sixties, makes it possible to reform severities as severe as those applied in the cyclic process, but avoids the drawbacks of the latter, i.e. cyclic operational disturbances and operation of very complex line/valve systems. A schematic flow diagram is shown in Figure 5.20.

The reactor section, as regards process flow, is similar to semi-regenerative platformers. Four reactors are usually employed, owing to the enhancement of endothermic reactions at low pressure.

To effect gravity flow of catalyst from one reactor bed to the next the reactors are in a stacked position rather than in a side-by-side configuration as in the other two types of processes. The reactors are housed in one pressure-containing shell. The catalysts beds are contained in an annular space between screens. Catalyst flow between reactors is via multiple (6–12 typically) equally spaced catalyst transfer pipes. A continuous slipstream of catalyst is withdrawn from the bottom of the last reactor (via the lock hopper 1 system), regenerated in the regeneration section and re-injected into the top of reactor 1 via the lock hopper 2 system. The catalyst flow rate for the entire system is set by the flow through the regeneration column. This flow is regulated by the catalyst flow-control hopper, which transfers small loads of catalyst on a continual, timed basis from the regeneration column to the surge hopper. The other transfers of catalyst, by the two lock hopper systems, are on level demand.

274   OIL PRODUCTS — MANUFACTURE

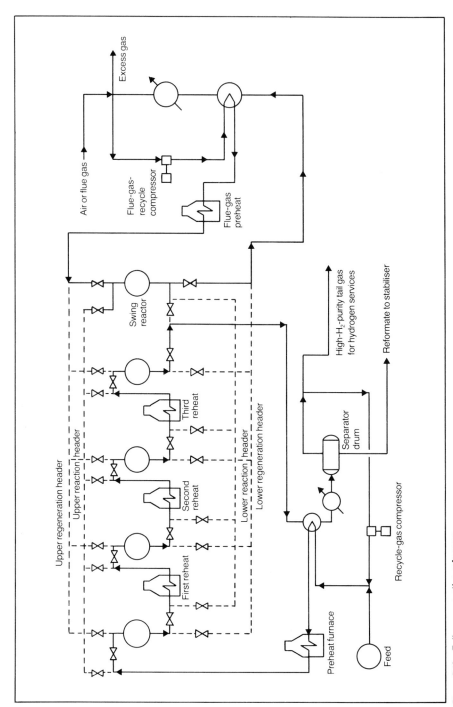

Figure 5.19 **Fully regenerative reformer**

REFORMING

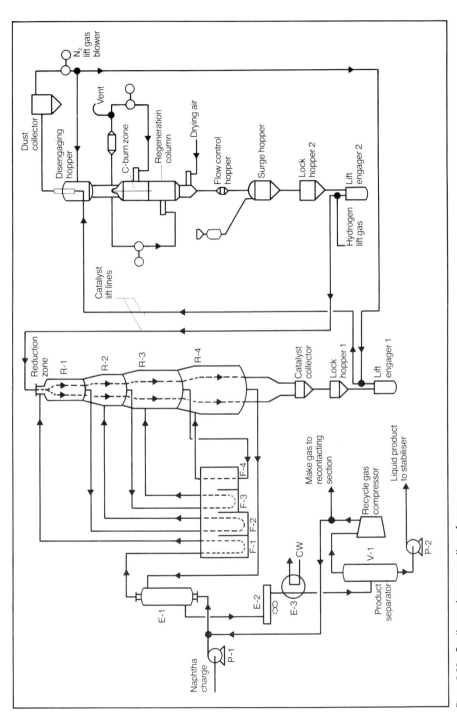

Figure 5.20 **Continuously regenerative reformer**

In the regeneration column, the catalyst is regenerated on a continuous basis. However, the final step of the regeneration, i.e. reduction of the oxidised platinum and second metal, takes place in the top of the first reactor.

Design capacities of units for this process vary from 1000 to 4500 t/d; operating pressures are in the 9–11 bar range; and design reformate octane number in the 95–104 range.

ISOMERISATION

Isomerisation is the transformation of one molecular structure into another (isomer) whose component atoms are the same but are arranged in a different geometrical structure. Since structural isomers may differ greatly in physical and chemical properties, isomerisation offers the possibility of converting less desirable compounds into isomers with desirable properties. The main field of applications of isomerisation are:

(1) Isomerisation of normal butane into isobutane

$$CH_3-CH_2-CH_2-CH_3 \longrightarrow CH_3-\underset{\underset{CH_3}{|}}{\overset{\overset{CH_3}{|}}{C}}-H$$

normal butane   isobutane

Since isobutane is an essential component for the manufacture of alkylate, this form of isomerisation is closely linked with alkylation.

(2) Isomerisation of pentanes and hexanes into higher-branched isomers.

$$CH_3-CH_2-CH_2-CH_2-CH_3 \longrightarrow CH_3-\underset{\underset{CH_3}{|}}{\overset{\overset{H}{|}}{C}}-CH_2-CH_3$$

normal pentane   2-methylbutane

$$CH_3-CH_2-CH_2-CH_2-CH_2-CH_3 \longrightarrow CH_3-\underset{\underset{CH_3}{|}}{\overset{\overset{CH_3}{|}}{C}}-CH_2-CH_3$$

normal hexane   2,2-dimethylbutane

Since branched isomers have a higher antiknock quality than the corresponding linear paraffins, this form of isomerisation is important for the production of motor fuels.

In addition to the above applications, isomerisation is applied for the conversion of *ortho*-xylene and *meta*-xylene into *para*-xylene, used for the manufacture of polyester fibres.

Isomerisation of low-molecular-weight paraffins has been commercially applied for many years. After extensive laboratory work had been carried out during the 1930s, World War II prompted the development of the laboratory processes into full-scale commercial units in order to meet the demand for isobutane necessary for the manufacture of large amounts of alkylate (see section on Alkylation, p. 300). While the first butane isomerisation unit went on stream in late 1941, by the end of the war nearly 40 butane isomerisation units were in operation in the USA and the Caribbean. Two pentane and two light naphtha isomerisation units also came on stream towards the end of the war to provide an additional source of blending stock for aviation gasoline.

Though butane isomerisation has maintained its importance, present-day interest in isomerisation is especially focussed on the upgrading of fractions containing C_5 and C_6 for use as motor gasoline components. This application has been prompted by the world drive to remove the lead additives gradually from motor gasoline in order to reduce air pollution. The octane loss caused by the removal or reduction of the lead antiknock additives can be compensated for by isomerisation of the pentane/hexane paraffin fraction of the gasoline.

Isomerisation technology has also substantially improved. In order to achieve the low temperatures necessary to obtain an acceptable yield of isomers, the isomerisation reactions have to be carried out in the presence of a catalyst. Catalyst systems used in the early units were based on aluminium chloride in some form. These catalyst systems, however, had the drawback of being highly corrosive and difficult to handle. In recent years, catalysts of a different type have come into use. These are solid catalysts consisting of a support having an acidic carrier and a hydrogenation function, frequently a noble metal. Modern isomerisation units utilise these dual-function catalysts and operate in the vapour phase and the presence of hydrogen. For these reasons, these processes are called hydro-isomerisation processes.

The first hydro-isomerisation unit was introduced in 1953 by UOP, followed in 1965 by the first BP one, while in 1970 the first Shell hydro-isomerisation (Hysomer) unit was started up. At present the following hydro-isomerisation processes are commercially available:

UOP Butamer   for butane isomerisation
UOP Penex   for pentane/hexane isomerisation

BP C$_4$ isomerisation   for butane isomerisation
BP C$_5$/C$_6$ isomerisation   for pentane/hexane isomerisation
Shell Hysomer   for pentane/hexane isomerisation

All these processes take place in the vapour phase on a fixed bed of catalyst containing platinum on a solid carrier.

As an example, the Shell Hysomer process will be briefly described. The flow scheme is shown in Figure 5.21. The liquid feedstock, a pentane/hexane mixture, is combined with the recycle gas/fresh gas mixture. The resultant combined reactor feed is routed to a feed/effluent heat exchanger, where it is heated and completely vaporised by the effluent of the reactor. The vaporised combined reactor feed is further heated to the desired reactor inlet temperature in the reactor charge heater. The hot charge enters the Hysomer reactor at the top and flows downwards through the catalyst bed, where a portion of normal and mono-branched paraffins is converted into higher-branched (higher-octane) compounds. Temperature rise from heat of reaction release is controlled by a cold-quench gas injection into the reactor. Reactor effluent is cooled and subsequently separated in the product separator into two streams: a liquid product (isomerate) and a recycle gas stream returning to the recycle gas compressor.

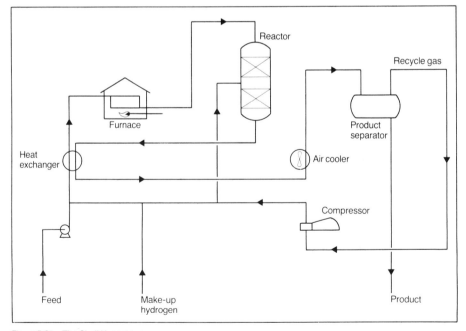

Figure 5.21 **The Shell Hysomer process**

The catalyst is a dual-function catalyst consisting of platinum on a zeolite basis, highly stable and regenerable.

Temperatures and pressures vary in a range of 230–285°C and 13–30 bar, C_5/C_6 content in product relative to that in feed is 97% or better, and octane upgrading ranges between 8 and 10 points, depending on feedstock quality.

The Hysomer process can be integrated with a catalytic reformer, resulting in substantial equipment savings, or with iso–normal separation processes which allow a complete conversion of pentane/hexane mixtures into isoparaffin mixtures. An interesting application in this field is the total isomerisation process (TIP), in which the isomerisation is completely integrated with the Union Carbide molecular sieve separation process.

THERMAL CRACKING

Thermal cracking is the oldest and, in a way, the simplest cracking process. It basically aims at the reduction of molecular size by application of heat without any additional sophistication such as a catalyst. At a temperature level of 450–500°C the larger hydrocarbon molecules become unstable and tend to break spontaneously into smaller molecules of all possible sizes and types. By allowing a particular feedstock to remain under cracking conditions for a certain time, the desired degree of cracking (conversion) can be achieved. Hence, temperature and time (residence time) are important process variables.

Obviously, the cracking conditions to be applied and the amount and type of cracked products will depend largely on the type of feedstock. In practice, the feedstock for thermal cracking is a mixture of complex heavy hydrocarbon molecules left over from atmospheric and/or vacuum distillation of crude. The nature of these heavy, high-molecular-weight fractions is extremely complex and much fundamental research has been carried out on their behaviour under thermal cracking conditions. However, a complete and satisfactory explanation of the reactions that take place cannot be given, except for relatively simple and well-defined types of products. For instance, long-chain paraffinic hydrocarbon molecules break down into a number of smaller ones by rupture of a carbon-to-carbon bond. (The smaller molecules so formed may break down further.) When this occurs, the number of hydrogen atoms present in the parent molecule is insufficient to provide the full complement for each carbon atom, so that olefins or "unsaturated" compounds are formed.

$$CH_3-CH_2-CH_2-CH_2-CH_2-CH_2-CH_2-CH_3 \rightarrow CH_2=CH-CH_3 + CH_3-CH_2-CH_2-CH_2-CH_3$$

The rupturing can take place in a variety of ways; usually a "free radical" mechanism for the bond rupture is assumed.

However, paraffinic hydrocarbons are usually only a small part of the heavy petroleum residues, the rest being cyclic hydrocarbons, either aromatic or naphthenic in character. In these, the rupture takes place in the paraffinic side-chain and not in the ring. Other side-reactions also take place. In particular, the condensation and polymerisation reactions of the olefins and of the aromatics are of considerable practical importance, since they can lead to undesirable product properties, such as an increase in the sludge or tar content. Hence, in practice, it is very difficult to assess the crackability of various feedstocks without plant trials.

The final products consist of gases, light hydrocarbons in the gasoline and gas oil range and heavier products. By selection of the type of unit, feedstock and operating conditions, the yields and quality of the various products can, within limits, be controlled to meet market requirements.

When thermal cracking was introduced in the refineries some 60 years ago, its main purpose was the production of gasoline. The units were relatively small (even applying batch processing!), were inefficient and had a very high fuel consumption. However, in the twenties and thirties a tremendous increase in thermal cracking capacity took place, largely in the version of the famous Dubbs process. Nevertheless, thermal cracking lost ground quickly to catalytic cracking (which produces gasoline of higher octane number) for processing heavy distillates with the onset of the latter process during World War II. Since then and up to the present day, thermal cracking has mostly been applied for other purposes: cracking long residue to middle distillates (gas oil), short residue for viscosity reduction (visbreaking), short residue to produce bitumen for briquetting, wax to olefins for the manufacture of chemicals, naphtha to ethylene gas (also for the manufacture of chemicals), selected feedstocks to coke for use as fuel or for the manufacture of electrodes.

In modern oil refineries there are three major applications of the thermal cracking process:
 (1) visbreaking,
 (2) thermal gas oil production,
 (3) coking.

Visbreaking

Visbreaking (i.e. viscosity reduction or "breaking") is an important application of thermal cracking because it reduces the viscosity of residues very substantially, thereby lessening the diluent requirements and the amount of fuel produced in a refinery.

Figure 5.22 shows a typical layout of a visbreaker processing short residue. The feed, after appropriate preheat, is sent to a furnace for heating to the cracking

THERMAL CRACKING

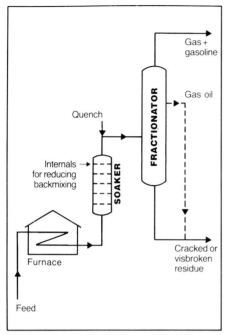

Figure 5.22 **Shell soaker visbreaking process**

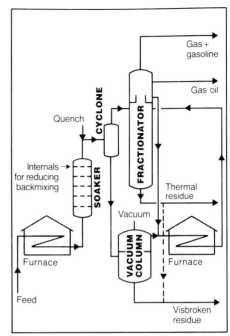

Figure 5.23 **Two-stage thermal cracking process (thermal gas oil unit)**

temperature, 450–460°C. The cracking takes place to a small extent in the furnace and largely in a soaker (reaction chamber) just downstream of the furnace. At the soaker outlet, the temperature is lower than at the furnace outlet/soaker inlet because the cracking reaction is endothermic. The products are quenched at the soaker outlet to stop the cracking reaction; after that the products enter the fractionator at a temperature level of 300–400°C and from here onward the processing is similar to any normal distillation process. The products are separated into gas, gasoline, gas oil and residue. Often gas oil is left in the residue or blended back into the residue. The residue so obtained has a far lower viscosity than the feed (visbreaking).

The (upflow) soaker provides for a prolonged residence time and therefore permits a lower cracking temperature than if the soaker was not used. This is advantageous as regards cost of furnace and fuel. Modern soakers are equipped with internals so as to reduce back-mixing effects, thus maximising the viscosity reduction. Since only one cracking stage is involved, this layout is also named one-stage cracking.

The cracking temperature applied is about 440–450°C at a pressure of 5–10 barg in the soaker. The fractionator can be operated at 2–5 barg as convenient.

Thermal Gas Oil Production

This is a more elaborate and sophisticated application of thermal cracking than visbreaking. Its chief aim is not only to reduce viscosity of the feedstock but also to produce and recover a maximum amount of gas oil. Altogether, it can mean that the viscosity of the residue (excluding gas oil) run down from the unit can be higher than that of the feed.

A typical flow scheme of this type of unit is shown in Figure 5.23. The first part of the unit is similar to a visbreaking unit. The visbroken residue is vacuum-flashed to recover heavy distillates, which are then recracked, together with heavy distillate recovered from the fractionator, in a second furnace under more severe cracking conditions (temperature 500°C; pressure 20–25 barg). More severe conditions are necessary because the feedstock has a smaller molecular size and is therefore more refractory than the larger residue molecules in the first stage. This layout is also referred to as two-stage cracking.

A view of Shell's largest thermal cracker, 11 000 tonnes daily, at Singapore.

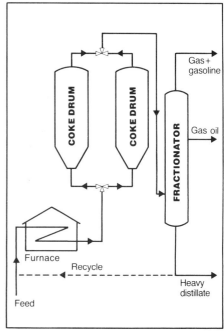

Figure 5.24 **Delayed coking process**
(Auxiliary systems like hydraulic cutting, coke crusher and transport, calciner, blow-down, etc are not shown)

Delayed Coking

This is an even more severe thermal cracking application than the previous one. The goal is to make a maximum of cracking products — distillates — whereby the heavy residue becomes so impoverished in hydrogen that it forms coke. The term "delayed" is intended to indicate that the coke formation does not take place in the furnace (which would lead to a plant shutdown) but in the large coke drums after the furnace. These drums are filled/emptied batchwise (once every 24 hours), though all the rest of the plant operates continuously. Figure 5.24 shows a simplified flow scheme of a delayed coker.

A plant usually has two coke drums, which have adequate capacity for one day's coke production (500–1500 m^2). The process conditions in the coke drum are 450–500°C and 2–3 bar. Only one coke drum is on-line; the other is off-line, being emptied or standing by.

Only the vapour passes from the top of the coke drums to the fractionator, where the products are separated into the desired fractions. The residue remains in the coke drum to crack further until only the coke is left. Often the heaviest part of the fractionator products is recycled to feed.

Product Quality

Thermal cracking products — the distillates — are not suitable for commercial use as produced in a unit; they require further refinement or treatment in order to improve their quality. Formerly, wet treating processes, for example treatment with caustic or other extraction medium, were applied to remove or "sweeten" the obnoxious sulphur products, but nowadays the modern catalytic hydrotreatment is employed almost without exception, both for the gasoline and for gas oil range products. Of course, the gases too have to be desulphurised before being used as fuel gas within the refinery.

The residual products from thermal cracking are normally not treated any further, except for coke, which may be calcined if the specifications require it to be so treated. The cracked residue is normally disposed of as refinery or commercial fuel. Here a very important aspect of the process is the stability of the cracked residues or of the final fuels after blending with suitable diluents. Residues contain asphaltenes, which are colloidally dispersed uniformly in the oil in a natural way. During the cracking, the character of the asphaltenes as well as of the oil changes, and if the cracking is too severe the natural balance of the colloidal system can be affected to the extent that part of the asphaltenes precipitates in the equipment or in the storage tanks, forming sludge. If the sludge formation is excessive, i.e. above a certain specified limit, the product (fuel) is considered to be unstable.

Plant Operation / Decoking

A practical aspect of operation of thermal cracking units is that, in spite of good design and operating practice, furnaces, and sometimes also other equipment, gradually do coke up, so that the unit has to be shut down and decoked. Furnaces can be decoked by "turbining" (using special rotary tools to remove coke from inside furnace pipes) or by steam–air decoking. In the latter case, the coke is burnt off in a carefully controlled decoking process in which air and steam are passed through the tubes at elevated temperatures. Air serves to burn coke, whereas the steam serves to keep the burning temperatures low so that they do not exceed the maximum tolerable temperature. Other coked equipment is usually cleaned by hydrojetting techniques. Owing to these unavoidable stops for decoking, the on-stream time, i.e. on-stream days per annum, for thermal cracking units is slightly shorter than for most other oil processes.

CATALYTIC CRACKING

Introduction

Catalytic cracking is a process for the conversion of heavy hydrocarbon fractions mainly into high-quality gasoline and fuel oil components, which are lighter, less viscous and thus more valuable than the feedstock.

The feedstock that is upgraded in this process would otherwise serve as a heavy fuel oil component. Heavy fuel oils are among the least valuable products obtained from crude oil, and they are also the easiest to find substitutes for. Hence, with increasing cost of crude oil there is an increasing incentive for application of catalytic cracking. As a consequence, considerable development of all aspects of this relatively old process is taking place.

The conversion, or cracking, predominantly takes place in the vapour phase in the presence of a catalyst. This catalyst, which is available as pellets or as a powder, has the ability to enhance the rate of cracking reactions and selectively to promote certain types of reactions. This results in products and product properties that are characteristic of the catalytic cracking process, e.g. formation of relatively large quantities of olefins, iso-components, and aromatics. These components contribute significantly towards the high octane number of the gasoline.

Part of the feedstock is converted into gas consisting of a mixture of hydrocarbons with four or fewer carbon atoms per molecule, hydrogen sulphide, and a small percentage of hydrogen. This gas may be processed in various ways. Usually it is separated in a gas separation (see p. 253) unit into a C_2 (i.e. ethane) and lighter fraction, a propane/propylene fraction, and a butane/butylene fraction.

CATALYTIC CRACKING

The C_2 and lighter fraction may be either used as refinery fuel or sold as town gas after treating to remove the hydrogen sulphide (see p. 256). The treated propane/propylene and butane/butylene fractions may be sold as such as LPG (liquefied petroleum gas) or they may be used as feedstock for polymerisation (see p. 303) and for the chemical industry. The propane and butanes which pass unchanged through the polymerisation and chemical processes are available for sale as LPG.

Alkylation (see p. 300) of the isobutane with light olefins is another means of converting part of the cracked gases into valuable high-octane gasoline components. Furthermore, part of the butanes is blended into the finished gasoline to satisfy volatility requirements. The gasoline obtained in catalytic cracking must be treated to remove contaminants (see p. 314). It has a Research octane number of about 90.

The rather aromatic light gas oil produced is partly blended to diesel and gas oils, if necessary after hydrodesulphurisation (see p. 307), and partly to fuel. The heavy gas oils, which are very aromatic, are used for fuel blending or burnt in the refinery as refinery fuel.

During the cracking reactions some heavy material, known as "coke", is deposited on the catalyst. This reduces its catalytic activity and regeneration is required. Regeneration is accomplished by burning off the coke, after which the catalyst activity is restored. Because of this phenomenon, there are three steps in the cycle of the catalytic cracking process that can be distinguished in the various commercial applications of the process, viz. a cracking step in which the reactions take place and a stripping step to eliminate hydrocarbons adsorbed on the catalyst before the third step, regeneration, in which coke is burnt off the catalyst.

The feedstock for catalytic cracking has traditionally been, and still is, normally obtained from vacuum distillation (see p. 248) or solvent deasphalting (see p. 261). In this way, the feed is virtually free of asphaltic materials and metals. Asphaltic material tends to cause excessive coke formation. The metals, which deposit on the catalyst and cannot be removed from it, have their own undesirable catalytic effect which leads to increased formation of coke and light gases at the cost of gasoline.

Although catalysts are now being developed that can tolerate some metals deposition and thereby facilitate processing of residual material containing moderate amounts of asphaltic material and metals, a catalytic cracker complex usually includes a vacuum distilling unit for feed preparation. It, furthermore, includes distillation columns for separation of the products, and treaters for the final products.

The Houdry and the Thermofor Catalytic Cracking Processes

To meet the need for high-octane gasoline, catalytic cracking was already being carried out on a commercial scale in 1916. Aluminium chloride was used as a catalyst, but, owing to the high cost of the catalyst and the difficulty of recovering it, the process could not compete economically with the thermal cracking processes that were being developed at that time. A considerable effort was therefore mounted to develop a more suitable catalyst. This work concentrated on the treating of naturally occurring clays. By 1930 the Frenchman Houdry succeeded in obtaining a catalyst in this way. He moved to the USA and in 1936 the first commercial unit went on stream using the Houdry fixed-bed process. In this process, three separate vessels were used, each containing a bed of pelleted catalyst. After the catalyst in one vessel had served for the promotion of the cracking reaction for a certain length of time, thereby being deactivated by the coke deposited on it, it was stripped and purged of hydrocarbons by blowing steam through the catalyst bed. The catalyst was then regenerated by burning off the coke, which was effected by introducing air into the vessel. This cyclic type of process, the cycle being of the order of half an hour, had inherent disadvantages, and eventually a moving-bed process was developed. In this process the catalyst is still used in the form of pellets or beads with a diameter of the order of 5 mm. However, the catalyst is continuously transported from the vessel where the reaction takes place (reactor) to the vessel where the regeneration takes place (regenerator). The Thermofor kiln, which had been used in refineries for other purposes, was adapted for use as a regenerator, and the resultant process was therefore called Thermofor Catalytic Cracking (TCC). The first TCC came into operation in 1943. The reactor and the regenerator were located alongside each other and catalyst pellets were carried by means of two bucket elevators: spent catalyst from the bottom of the reactor to the top of the regenerator and regenerated catalyst from the bottom of the regenerator to the top of the reactor. Later the transport of catalyst via buckets was replaced by transport by means of a high-velocity gas flow. In yet a later stage the reactor was located on top of the regenerator, whereby only one catalyst lift pipe is required, the overall structure being very high (approximately 100 m). A modern gas-lift moving-bed unit is depicted in Figure 5.25.

Regenerated catalyst enters the catalyst-engaging vessel and is carried upwards by a stream of gas to the catalyst-disengaging vessel, from which it flows downwards to the reactor via the seal leg. In the reactor, the catalyst is contacted with hot feed (approximately 400°C), and the reaction takes place. The catalyst is then stripped with steam before flowing to the regenerator via another seal leg. In the regenerator the catalyst is regenerated as the coke is burnt off. A major part of the heat from the coke burning is removed via cooling coils. The heat which can be transported with the catalyst from the regenerator to the reactor is sufficient only to vaporise and crack the feed. A furnace must always be included to heat the feed.

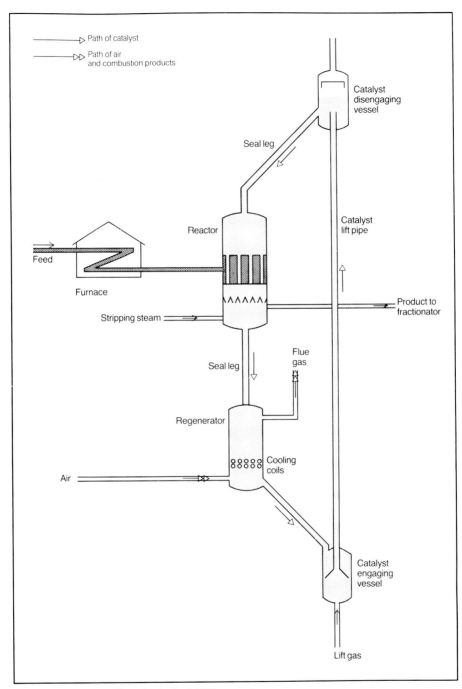

Figure 5.25 **Modern gas lift moving-bed unit (TCC process)**

The Fluidised Catalytic Cracking Process

While the search was going on for suitable cracking catalysts based on natural clays, some companies concentrated their efforts on developing a fully synthetic catalyst. This resulted in the synthetic amorphous silica–alumina catalyst, which was commonly used until 1960, when it was slightly modified by incorporation of some crystalline materials (zeolite catalyst). When the success of the Houdry fixed-bed process was announced in the late 1930s, the companies that had developed the synthetic catalyst decided to try to develop a process using finely powdered catalyst. Subsequent work finally led to the development of the fluidised bed catalytic cracking (FCC) process, which has become the most important catalytic cracking process.

Originally, the finely powdered catalyst was obtained by grinding the catalyst material, but nowadays it is produced by spray-drying a slurry of silica gel and aluminium hydroxide in a stream of hot flue gases. Under the right conditions, the catalyst is obtained in the form of small spheres with particles in the range of 1–50 microns.

When gas is passed through a bed of powdered catalyst at a suitable velocity (0.1–0.7 m/s), the catalyst and the gas form a system that behaves like a liquid, i.e. it can flow from one vessel to another under the influence of a hydrostatic pressure. If the gas velocity is too low, the powder does not fluidise and it behaves like a solid. If the velocity is too high, the powder will just be carried away with the gas.

When the catalyst is properly fluidised, it can be continuously transported from a reactor vessel, where the cracking reactions take place and where it is fluidised by the hydrocarbon vapour, to a regenerator vessel, where it is fluidised by the air and the products of combustion, and then back to the reactor. In this way the process is truly continuous.

The first FCC unit went on stream in Standard Oil of New Jersey's refinery in Baton Rouge, Louisiana, in May 1942. Since that time, many companies have developed their own FCC process and there are numerous varieties in unit configurations.

The Modern Fluidised Catalytic Cracking Process

Description

Figure 5.26 shows a schematic diagram of a modern FCC unit. Hot feed, together with some steam, is introduced at the bottom of the riser via special distribution nozzles. Here it meets a stream of hot regenerated catalyst from the regenerator flowing down the inclined regenerator standpipe. The oil is heated and vaporised

CATALYTIC CRACKING

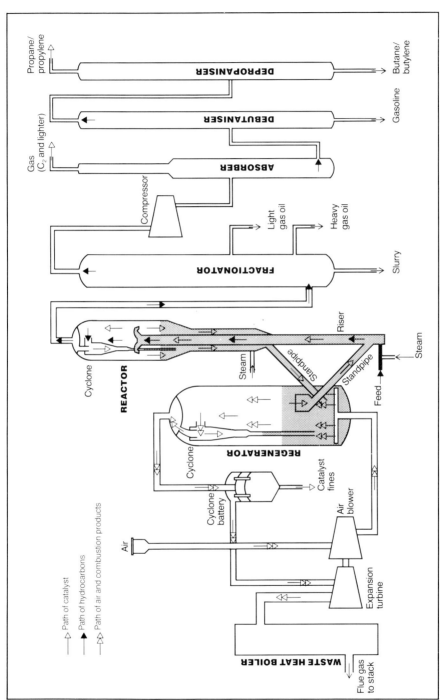

Figure 5.26 **Modern type FCC unit**

by the hot catalyst and the cracking reactions commence. The vapour, initially formed by vaporisation and successively by cracking, carries the catalyst up the riser at 10–20 m/s in a "dilute phase". At the outlet of the riser the catalyst and hydrocarbons are quickly separated in a special device.

The catalyst (now partly deactivated by deposited coke) and the vapour then enter the reactor. The vapour passes overhead via a cyclone separator for removal of entrained catalyst before it enters the fractionator and further downstream equipment for product separation (for description see p. 256).

The catalyst then descends into the stripper where entrained hydrocarbons are removed by injection of steam, before it flows via the inclined stripper standpipe into the fluidised catalyst bed in the regenerator.

Air is supplied to the regenerator by an air blower and distributed throughout the catalyst bed. The coke deposited is burnt off and the regenerated catalyst passes down the regenerator standpipe to the bottom of the riser, where it joins the fresh feed and the cycle recommences.

The flue gas (the combustion products) leaving the regenerator catalyst bed entrains catalyst particles. In particular, it entrains "fines", a fine dust formed by mechanical rubbing of catalyst particles taking place in the catalyst bed. Before leaving the regenerator, the flue gas therefore passes through cyclone separators where the bulk of this entrained catalyst is collected and returned to the catalyst bed.

To minimise energy consumption, the air blower in a modern-type FCC unit is driven by an expansion turbine. In this expansion turbine, the current of flue gas at a pressure of about 2 barg drives a wheel by striking impellers fitted on this wheel. The power is then transferred to the air blower via a common shaft. This system is usually referred to as a "power recovery system". To reduce the wear caused by the impact of catalyst particles on the impellers (erosion), the flue gas must be virtually free of catalyst particles. The flue gas is therefore passed through a vessel containing a whole battery of small, highly efficient cyclone separators, where the remaining catalyst fines are collected for disposal.

Before being disposed of via a stack, the flue gas is passed through a waste heat boiler, where its remaining heat is recovered by steam generation.

In the version of the FCC process described here, the heat released by burning the coke in the regenerator is just sufficient to supply the heat required in the riser to heat up, vaporise, and crack the hydrocarbon feed. The units where this balance occurs are called "heat-balanced" units. Some feeds cause excessive amounts of coke to be deposited on the catalyst, i.e. much more than is required for burning in the regenerator and to have a "heat-balanced" unit. In such cases, heat must be removed from the regenerator, e.g. by passing water through coils in the regenerator bed to generate steam. Some feeds cause so little coke to be deposited on the catalyst that heat has to be supplied to the system. This is done

CATALYTIC CRACKING

by preheating the hydrocarbon feed in a furnace before contacting it with the catalyst.

The very early FCC units had both cooling of the regenerator to remove heat and a preheat furnace for the feed to provide heat.

From this it may easily be deduced that, especially for a "heat-balanced" unit, circulation of the correct amount of catalyst is of the utmost importance.

Main Characteristics

The main characteristics of any version of a modern FCC unit, in contrast with earlier versions, are:
- a special device in the bottom of the riser to enhance contacting of catalyst and hydrocarbon feed;
- the cracking takes place during a short time (2–4 seconds) in a riser ("short-contact-time riser") at high temperature (500–540°C at riser outlet);
- the catalyst used is so active that a special device for quick separation of catalyst and hydrocarbons at the outlet of the riser is required to avoid undesirable cracking after the mixture has left the riser. Since no cracking in the reactor is required or desirable, the "reactor" no longer functions as a reactor; it merely serves as a holding vessel for the cyclones;
- the regeneration takes place at 680–720°C. With the use of special catalysts, all carbon monoxide (CO) in the flue gas is combusted to carbon dioxide (CO_2) in the regenerator (complete CO combustion regeneration). In older units, some CO leaves the regenerator and to recover the heat of combustion the CO is burnt in a special boiler (CO boiler). In modern units without CO in the flue gas, recovery of sensible heat takes place in the waste heat boiler;
- modern FCC units include a power recovery system for driving the air blower, whereas in older units this was accomplished with steam turbine drives or electric motors.

A modern FCC unit is depicted in Figure 5.27. From left to right the following main equipment can be distinguished: two large storage vessels for catalyst (fresh and equilibrium); partly covered by these vessels is the fractionator, then comes the riser with the reactor on top (highest point about 45 m above ground level), the stripper underneath the reactor, and the standpipe to the regenerator; the last vessel just above ground level is the vessel containing the cyclone battery.

Feedstocks and Catalysts

With the low selectivity of the early-generation catalysts, i.e. the amorphous silica–alumina catalyst, a feedstock virtually free of asphaltic material and metals was required to avoid excessive coke production. This combination of feed and

Figure 5.27 **A modern fluidised catalytic cracker unit.**

catalyst yielded 30–35% wt. gasoline (on feed) with a Research octane number of 92–94. With the introduction of some crystalline zeolitic material in the amorphous silica–alumina catalyst (starting around 1960), a considerable improvement in cracking activity and selectivity resulted. At a coke production corresponding to heat-balanced operation, less gas and considerably more gasoline is obtained (about 50% wt. on fresh feed). However, the gasoline has a lower Research octane number (90–92) and the light gas oil is even more aromatic and therefore has poorer ignition qualities than the corresponding products obtained with earlier catalysts.

With the earlier catalysts, having lower activity and lower selectivity, it was often the practice to recrack part of the heavy gas oil by recycling this fraction to the reactor riser (recycle operation). This led to higher gasoline yield and lower production of fuel oil components. Recycling to extinction tended to form too much coke and gas. With the zeolite catalyst, having a higher selectivity and higher activity, recycle operation for the above purposes is no longer necessary.

A typical yield structure obtained with a modern FCC unit, when cracking predominantly vacuum distillate feedstock of a Middle East origin over a modern catalyst and without recycling heavy gas oil, will be:

	% wt. component on fresh feed
C_2 and lighter	2
C_3	5
C_4	9
Gasoline	40–50
Light gas oil	30–20
Heavy gas oil	9
Coke	5

In practice, a considerable variation will be encountered, depending on factors such as feed quality, catalyst, operating conditions and unit configuration.

The rising cost of crude oil and therefore also of vacuum distillate feed has greatly increased the incentive for cracking cheaper and usually inferior feedstocks. In this context, heavier distillates obtained by deeper-vacuum flashing and by flashing of thermally cracked residues are of interest, as are residual materials of suitable quality, i.e. reasonably low in asphaltenes and metals content.

The quality of these feedstocks can be improved by subjecting them to hydrotreatment (see p. 306). In such treatment, asphaltenes and other components of high coke-making propensity will be partly converted and the metals content reduced, such that a reasonable yield of valuable products is obtained in the catalytic cracking process. Hydrotreatment of normal feedstocks also results in a better yield of valuable products, mainly higher gasoline yield.

The development of increasingly selective catalysts allows of the processing of increasing amounts of inferior feedstocks. Catalysts are available that can tolerate such high metals content (mainly nickel and vanadium) that some residue can be included in the FCC feedstock. An increased portion of inferior material in the feed can be handled when heat removal from the regenerator is applied. Furthermore, the effect of metals can also be mitigated by application of passivation techniques, i.e. addition of a metal to the catalyst, which suppresses part of the adverse effects of nickel on the catalyst.

In addition to improving metals tolerance, catalyst development is also aimed at:
- developing catalysts that improve the octane number of the gasoline or the ignition qualities of the light gas oil;
- developing catalysts that will contribute to reduced emission of components containing sulphur and nitrogen as well as reduced emission of particulate matter (catalyst dust) and carbon monoxide from the FCC regenerator.

The catalyst in the FCC suffers from attrition (wear by mechanical rubbing) and, although very efficient cyclones are applied, some catalyst is lost from the system. Furthermore, the catalyst activity also suffers from the high temperatures at which the catalyst is applied, which leads to permanent catalyst deactivation. A certain daily addition of fresh catalyst is therefore required to maintain the right quantity and quality of catalyst in the unit. Additions of the order of 300–900 kg per 1000 tons of intake are normal.

HYDROCRACKING

The need for gasoline of a higher quality than that obtainable by catalytic cracking (p. 284) led to the development of the hydrocracking process. The history of the process goes back to the later 1920s, when a plant for the commercial hydrogenation of brown coal was commissioned at Leuna in Germany. Tungsten sulphide was used as a catalyst in this one-stage unit, in which high reaction pressures, 200–300 bar, were applied. The catalyst displayed a very high hydrogenation activity: the aromatic feedstock, coal and heavy fractions of oil, containing sulphur, nitrogen and oxygen, was virtually completely converted into paraffins/isoparaffins. The result of the Leuna plant — loss of octane number from aromatics hydrogenation — indicated that a two-stage process was to be preferred: a first stage for the hydrogenation of impurities in the feedstock, notably the nitrogen compounds, followed by a hydrocracking step. In 1939, ICI developed a second-stage catalyst for a plant that contributed largely to Britain's supply of aviation gasoline in the subsequent years.

During World War II, two-stage processes were applied on a limited scale in Germany, Britain and the USA. In Britain, the feedstocks were creosote from coal tar and gas oil from petroleum. In the USA, Standard Oil of New Jersey operated a plant at Baton Rouge, La. *, producing gasoline from a Venezuelan kerosine/

---

* Now owned by Exxon, USA.

light gas oil fraction. Operating conditions in those units were comparable: approximate reaction temperature 400°C and reaction pressures of 200-300 bar.

After the war, commercial hydrocracking was stopped because the process was too expensive. Hydrocracking research, however, continued intensively. By the end of the 1950s, the process had become economic, for which a number of reasons can be identified.

The development of improved catalysts made it possible to operate the process at considerably lower pressure, viz. 70-150 bar.

This in turn resulted in a reduction in equipment wall thickness, whereas, simultaneously, advances were made in mechanical engineering, especially in the field of reactor design. These factors, together with the availability of relatively low-cost hydrogen from the budding steam reforming process, brought hydrocracking back on the refinery scene. The first units of the second generation were built in the USA to meet the demand for conversion of surplus fuel oil (cycle oil from fluid catalytic cracking) in the gasoline-oriented refineries.

Hydrocracking is now a well-established process, which is offered by many licensors. Shell has developed three basic configurations, which are described below.

Basis for the Choice of Conversion Route

Refiners are continuously faced with trends towards increased conversion, better product qualities and more rapidly changing product patterns. Various processes are available that can meet the requirements to a greater or less degree: coking, visbreaking/thermal cracking, catalytic cracking and hydrocracking.

The type of processes applied and the complexity of refineries in various parts of the world are determined to a great extent by the product distribution required. As a consequence, the relative importance of the above processes in traditionally fuel-oil-dominated refineries such as those in Western Europe will be quite different from those of gasoline-oriented refineries in, for instance, the United States.

An important aspect of the coking, thermal and catalytic cracking processes is that they operate at low pressure. This gives advantages in the fields of capital cost, metallurgy and engineering.

A particular feature of the hydrocracking process, as compared with its alternatives, is its flexibility with respect to product outturn and the high quality of its products. In areas where a quantitative imbalance exists of light products, middle distillates and fuel, hydrocracking is a most suitable process for correction. Moreover, the hydrocracker does not yield a coke or pitch byproduct: the entire feedstock is converted into the required product range, an important consideration in a situation of limited crude oil availability. The development of the low-pressure catalytic reforming process (p. 269), which produces a relatively cheap, high-quality hydrogen, has contributed substantially to the economic viability of hydrocracking. On the whole, hydrocracking can handle a wider range of feedstocks than catalytic cracking, although the latter process has seen some recent catalyst developments which narrowed the gap. There are also examples where hydrocracking is complementary rather than alternative to the other conversion processes; as an example, cycle oils which cannot be recycled to extinction in the catalytic cracker can be processed in the hydrocracker.

Notwithstanding many extensive comparisons between the various processes, experience shows that generalisations with respect to the optimum conversion route still cannot be made.

Process Description

All hydrocracking processes are characterised by the fact that in a catalytic operation under relatively high hydrogen pressure a heavy oil fraction is treated to give products of lower molecular weight.

Hydrocracking covers widely different fields, ranging from C_3/C_4 production from naphtha, on the one hand, to luboil manufacture from deasphalted oils, on the other.

Most hydrocrackers use fixed beds of catalyst with downflow of reactants. The H-Oil process developed by Hydrocarbon Research Corp. and Cities Service R & D employs an ebullient bed reactor in which the beds of particulate catalyst are maintained in an ebullient or fluidised condition in upflowing reactants.

When the processing severity in a hydrocracker is increased, the first reactions occurring lead to the saturation of any olefinic material present in the feedstock. Next come the reactions of desulphurisation, denitrogenation and de-oxygenation. These reactions constitute a treating step during which, in most cases, only limited cracking takes place. When the severity is increased further, hydrocracking reactions are initiated. They proceed at various rates, with the formation of intermediate products (e.g. saturation of aromatics), which are subsequently cracked into lighter products.

Configurations

When the treating step is combined with the cracking reactions to occur in one reactor, the process is called a single-stage process. In this simplest of the hydrocracker configurations, the lay-out of the reactor section generally resembles that of a hydrotreating unit (p. 308). This configuration will find application in cases where only a moderate degree of conversion (say, less than 50%) is required. It may also be considered if full conversion, but with a limited reduction in

HYDROCRACKING

molecular weight, is aimed at. An example is the production of middle distillates from a heavy distillate oil.

The catalyst used in a single-stage process comprises a hydrogenation function in combination with a strong cracking function. The hydrogenation function is provided by sulphided metals such as cobalt, molybdenum and nickel. An acidic support, usually alumina, attends to the cracking function. Nitrogen compounds and ammonia produced by hydrogenation interfere with the acidic activity of the catalyst.

In the cases where high/full conversion is required, the reaction temperatures and run lengths of interest in commercial operation can no longer be adhered to. It becomes necessary to switch to a multi-stage process, in which the cracking reactions mainly take place in an added reactor.

With regard to the adverse effect of ammonia and nitrogen compounds on catalyst acidity, two versions of the multi-stage hydrocracker have been developed.

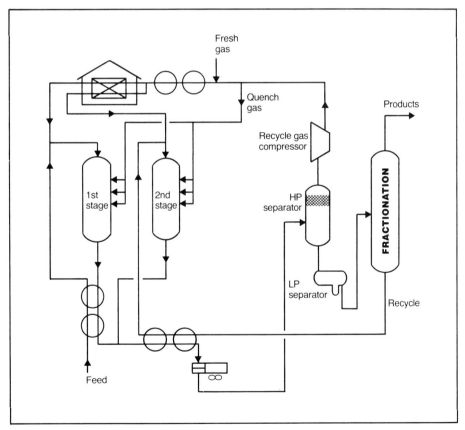

Figure 5.28 **Two-stage hydrocracker**

In one type, these undesirable compounds are removed from the unconverted hydrocarbons before the latter are charged to the cracking reactor. This type is called the two-stage process. The other variety is often referred to as a series-flow hydrocracker. This type uses a catalyst with an increased tolerance towards nitrogen, both as ammonia and in the organic form.

A diagram of the reactor section of a two-stage process is given as Figure 5.28. Fresh feed is pre-heated by heat exchange with effluent from the first reactor. It is combined with part of a hot fresh gas/recycle gas mixture and passes through a first reactor for the desulphurisation/denitrogenation step. These reactions, as well as those of hydrocracking, which occurs to a limited extent in the first reactor, are exothermic. The catalyst inventory is therefore divided among a number of fixed beds. Reaction temperatures are controlled by introducing part of the recycle gas as a quench medium between the beds. The ensuing liquid is fractionated to remove the product made in the first reactor. Unconverted

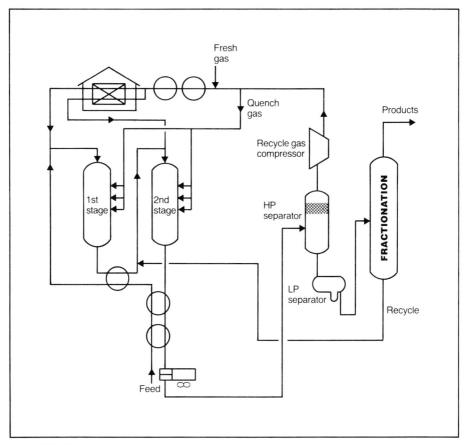

Figure 5.29 **Series-flow hydrocracker**

material, with a low nitrogen content and free of ammonia, is taken as a bottom stream from the fractionation section. After heat exchange with reactor effluent and mixing with heated recycle gas, it is sent to the second reactor. Here most of the hydrocracking reactions occur. Strongly acidic catalysts with a relatively low hydrogenation activity (metal sulphides on, for example, amorphous silica–alumina) are usually applied. As in the first reactor, the exothermicity of the process is controlled by using recycle gas as quench medium between the catalyst beds. Effluent from the second reactor is cooled and joins first-stage effluent for separation from recycle gas and fractionation. The part of the second reactor feed that has remained unconverted is recycled to the reactor. Feedstock is thereby totally converted to the product boiling range. Figure 5.28 depicts a two-stage hydrocracker.

The series-flow configuration is shown as Figure 5.29. The principal difference from Figure 5.28 is the elimination of first-stage cooling and gas/liquid separation and the interstage ammonia removal step. The effluent from the first stage is mixed with more recycle gas and routed direct to the inlet of the second reactor. In contrast with the amorphous catalyst of the two-stage process, the second

Figure 5.30 **A hydrocracker.**

reactor in series flow generally has a zeolitic catalyst, based on crystalline silica–alumina. As in the two-stage process, material not converted to the product boiling range is recycled from the fractionation section.

Both two-stage and series-flow hydrocracking are flexible processes: they may yield, in one mode of operation, only naphtha and lighter products and, in a different mode, only gas oil and lighter products. In the naphtha mode both configurations have comparable yield patterns. In modes for heavier products, kerosine and especially gas oil, the two-stage process is more selective because product made in the first reactor is removed from the second reactor feed. In series-flow operation this product is partly overcracked into lighter product in the second reactor.

ALKYLATION

Alkylation is the introduction of an alkyl group into a molecule. Though a number of alkylation reactions are carried out commercially, the alkylation reaction most commonly used in the oil industry is that of a saturated branched-chain hydrocarbon, isobutane, with light olefins, usually mixtures of propylene and butylene. The product of this reaction, alkylate, is a liquid consisting of a mixture of isoparaffins and is characterised by excellent antiknock quality. A typical example of the alkylation reaction is

$$CH_3-\underset{\underset{CH_3}{|}}{\overset{\overset{CH_3}{|}}{CH}} + CH_2=CH-CH_2-CH_3 \longrightarrow CH_3-\underset{\underset{CH_3}{|}}{\overset{\overset{CH_3}{|}}{C}}-CH_2-\underset{}{\overset{\overset{CH_3}{|}}{CH}}-CH_3$$

$$\text{isobutane} \qquad \text{1-butylene} \qquad \text{isooctane}$$

Alkylation can be effected by heat alone but high pressures and temperatures would be required; this has led to the development of catalytic processes allowing of mild reaction conditions. The first commercial catalytic alkylation units were constructed in 1938 and mainly used sulphuric acid as a catalyst. Many oil companies contributed to these early developments in the years before World War II.

The demand for alkylate for aviation gasoline during that war required unprecedented quantities of this component and prompted a rapid development of both alkylation capacity and technology. In 1942, the first commercial hydrofluoric acid alkylation unit was put into operation by Phillips Petroleum Company (Fig. 5.31). Subsequently, Universal Oil Products also developed and built hydrofluoric acid alkylation units. By the end of the war, the alkylation capacity

ALKYLATION

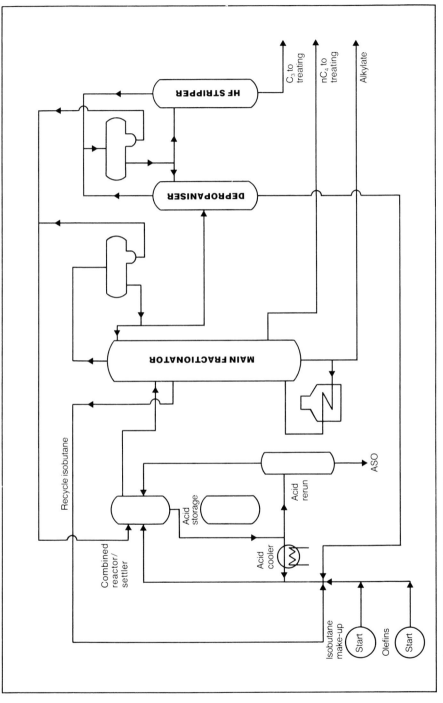

Figure 5.31 **The Phillips HF acid alkylation process**

in the USA alone was approximately 22,000 tons of alkylate a day. Of the latter quantity, about half was manufactured using the sulphuric acid process and the other half by means of the hydrofluoric acid process.

The main features of the HF process compared with the H_2SO_4 version are that spent HF can be regenerated economically, thus avoiding the appreciable quantities of spent sulphuric acid produced in the competing process, and that reactor temperatures can be somewhat higher, eliminating the need to install refrigeration equipment, as is employed in the sulphuric acid process.

At the present time, the predominant use of alkylation process in the oil industry is for the manufacture of motor gasoline components, and world alkylation capacity is equally split between the sulphuric acid and hydrofluoric acid versions of the process.

Olefins used as alkylation feedstock are produced by the catalytic cracking processes. Though butylenes are the preferred feedstock, since they produce an alkylate consisting mainly of isooctane with excellent antiknock characteristics, propylene and amylenes may also be used as feedstocks. Alkylation capacity in refineries is usually limited by the availability of isobutane and often, depending on the refinery's configuration, butane isomerisation facilities must be added if all the light olefin fraction from a catalytic cracking unit has to be alkylated.

The properties which make alkylate an excellent gasoline blending component are high octane number, good response to lead addition and small difference between octane numbers determined according to the Motor Method and the Research Method (see p. 407). Apart from its contribution to the quality of gasoline, alkylation also plays an important role in the quantities of gasoline which refineries can produce, since alkylation upgrades LPG-type components (gases) into liquid gasoline products.

The importance of the alkylation process in present refinery operation is increasing, since the reduction of lead levels in gasoline will increase the demand for high-octane blending components like alkylate.

A short description of the Phillips hydrofluoric acid alkylation unit will now be given.

Dried olefin feed together with recycle and make-up isobutane is charged to a reactor/settler system (exclusive Phillips design) where the hydrocarbon feed is highly dispersed into a moving bed of HF catalyst. Upon leaving the reaction zone, the reactor effluent flows upwards to a settler where the acid separates from the hydrocarbons and, by means of gravity flow, returns through an acid cooler to the reactor zone. Reactor temperatures between 25 and 45°C are used.

The hydrocarbon phase, consisting of propane, recycle isobutane, normal butane and alkylate, flows upwards through the settler and is charged to the main fractionator, the bottom product of which is motor alkylate.

A few trays above the bottom of the main fractionator, normal butane is removed as a vapour side-draw, condensed, treated and sent to storage.

Main fractionator overhead, consisting mainly of propane, isobutane and HF, is charged to a depropaniser. The depropaniser overhead product propane is passed through the HF stripper for HF removal, then treated to eliminate traces of HF and alkyl fluorides and routed to LPG propane storage.

The depropaniser bottoms represent a portion of the recycle isobutane, the main part of the recycle isobutane being removed as a vapour side-draw below the feed tray, condensed, cooled and returned to the reaction zone.

To keep the acidity of the catalyst at a constant level, a slipstream of acid is continuously pumped from the acid cooler to an acid regeneration column where by distillation acid-soluble oils (formed as by-products from undesired reactions) are rejected together with minor quantities of water that may enter with the feed. The regenerated acid vapour top product returns to the acid settler, where it is condensed, while the acid-soluble oils and water are routed to a disposal system.

The catalyst of this process, hydrofluoric acid, is toxic and very corrosive and special precautions have to be taken in handling it. Reactor temperatures between 25 and 45°C are used.

POLYMERISATION

Polymerisation is the combination of small molecules of the same compound to form a larger molecule while maintaining the original atomic arrangement of the basic molecule. The product obtained by polymerisation from the basic molecule, the monomer, is called polymer. Product formed by combination of two, three or four of the same monomer are called dimers, trimers and tetramers, respectively. For example, two molecules of the unsaturated hydrocarbon isobutylene can combine to form a new molecule, di-isobutylene

$$CH_2{=}\underset{\underset{CH_3}{|}}{\overset{\overset{CH_3}{|}}{C}} \quad + \quad \underset{\underset{CH_3}{|}}{\overset{\overset{CH_3}{|}}{C}}{=}CH_2 \quad \longrightarrow \quad CH_3{-}\underset{\underset{CH_3}{|}}{\overset{\overset{CH_3}{|}}{C}}{-}CH_2{-}\underset{\underset{CH_2}{\|}}{\overset{\overset{CH_3}{|}}{C}}$$

isobutylene   isobutylene   di-isobutylene

Polymerisation processes were developed in the early 1930s in order to make use of the light olefinic gases derived from the cracking processes to manufacture liquid products in the boiling range of gasoline. Originally, attention was paid to purely thermal polymerisation, but the resultant low olefin conversion, together with the requirement of high temperature and pressure, led to the development of catalytic polymerisation, and around 1935 catalytic polymerisation plants came into commercial operation. Shell Development Company's sulphuric acid and Universal Oil Products' phosphoric acid processes were used for polymerisation of propylenes and butylenes.

While sulphuric acid polymerisation is practically obsolete, the phosphoric acid process has maintained its importance for the manufacture of motor gasoline components where isobutane for the production of alkylate is either unavailable or expensive. The gasoline liquids produced from polymerisation, though characterised by a high octane number, are not as good a gasoline blending stock as alkylate, since the latter responds better to lead addition and has a lower spread between Research and Motor octane numbers. In modern refinery operation, however, polymerisation should be seen as supplementing the alkylation process, since it can convert just enough of the olefins to balance the alkylation isobutane supply.

Apart from its use for the production of motor gasoline components, the (UOP) phosphoric acid polymerisation process can also be employed to produce diesel and jet fuels and finds increasing application in the chemical industry, where olefins are used as building blocks for the manufacture of plastics, resins, alcohols and detergents.

A recent development in the field of catalytic polymerisation has been the introduction by the Institut Français du Pétrole of the DIMERSOL process for the dimerisation of propylene and/or n-butylenes for production of high-octane gasoline or C_6 to C_8 olefins for the chemical industry. This process is characterised by a liquid phase reaction performed at low pressure and ambient temperature in the presence of a soluble catalyst.

As an example of polymerisation, the phosphoric acid polymerisation process will be briefly described. The catalyst used in this process is phosphoric acid on pellets of kieselguhr, a naturally occurring silica, used as a carrier. The flow scheme is shown in Figure 5.32. The feedstock is a propane/propylene and/or a butane/butylene fraction. As a result of exothermic polymerisation reactions taking place in the reactor, heat is liberated in proportion to the olefinic content of the feed. To achieve an optimum olefinic concentration in the reactor feed, a recycle stream (mainly propane/butane) is therefore introduced together with the feed prior to being heated to the required reactor inlet temperature through feed/effluent heat exchange. The temperature in the reactor is controlled by introducing a liquid hydrocarbon stream as a quench between the reactor beds. The effluent from the reactor passes through a flash drum providing the required recycle and is routed to a stabiliser where the poly gasoline is separated as a stabilised product having a desired vapour pressure. Typical operating conditions of a polymerisation unit are 40–80 bar and 190–230°C (depending on feed quality and type of product required), and a conversion of typically 85–95% is achieved.

POLYMERISATION

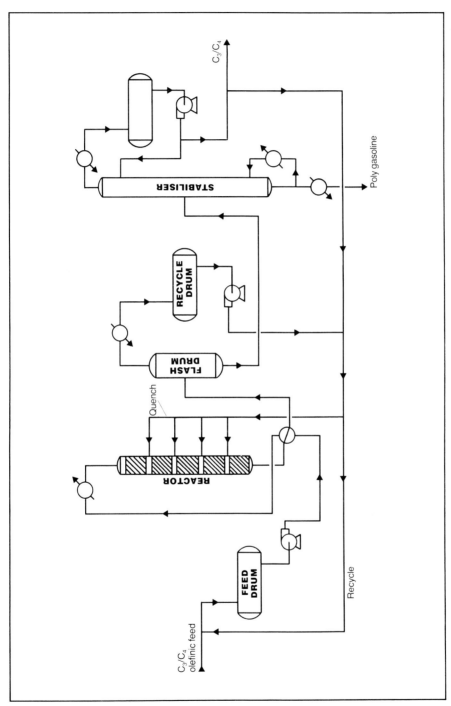

Figure 5.32 **Polymerisation process for motor fuel production**

HYDROTREATING

Until the end of World War II there was little incentive for the oil industry to pay significant attention to improving product quality by hydrogen treatment. Since that time, however, a number of factors have led to the development of increasingly sophisticated and flexible hydrotreating processes. During the postwar period, the increased production of high-sulphur crudes made more stringent demands on the product blending flexibility of refineries, and the marketing specifications for the products became tighter, largely owing to environmental considerations. This situation resulted in the necessity for higher sulphur removal capability in the refineries.

At the same time, the availability of relatively cheap hydrogen as a by-product of catalytic reforming gave additional momentum to the development of the sulphur removal process by hydrogen treatment. In this treatment the sulphur compounds are removed by converting them into hydrogen sulphide by reaction with hydrogen in the presence of a catalyst. This results in high liquid product yields, since only the sulphur is removed. Furthermore, the hydrogen sulphide produced can be easily removed from the product stream, for example by an amine wash. In this way, the hydrogen sulphide is recovered as a highly concentrated stream and can be converted into elemental sulphur by the "Claus" process (see Gas Treating and Sulphur Recovery, p. 322).

Hydrodesulphurisation has been extensively used commercially for treating naphtha as feedstock for catalytic reformers to meet the very stringent sulphur specification of less than 1 ppm wt to protect the platinum catalyst. It has also been widely applied for removal of sulphur compounds from kerosines and gas oils to make them suitable as blending components. In cases where products from catalytic or thermal cracker operations are present in the feedstock, saturation of olefins to improve thermal and storage stability, and to a limited extent denitrification, can be achieved using this process.

In recent years, the world energy situation has encouraged the development and application of more effective conversion processes, whereby the crudes processed by the refineries can be upgraded to give larger quantities of lighter, more valuable distillates and less fuel oil. As a consequence, the quantity of conversion or cracked products within the refineries is increasing relative to straight-run products. Since these conversion products are characterised by high olefin, aromatics and nitrogen contents, often in addition to sulphur, blending to market specifications becomes increasingly difficult. For this reason the hydrodesulphurisation process has evolved to its current status as a hydrotreating process with, by application of suitable catalysts and operating severity, the capability of reducing these undesirable quality features to an acceptable level.

Hydrodesulphurisation / Hydrotreating of Distillates

Figure 5.33 is a diagram of the Shell Hydrotreating Process for distillate fractions. An impression of a commercial unit is given in Figure 5.34. Two basic processes are applied, the liquid phase (or trickle flow) process for kerosine and heavier straight-run and cracked distillates up to vacuum gas oil, and the vapour phase process for light straight-run and cracked fractions. Both processes employ the same basic line-up: the feedstock is mixed with hydrogen-rich make-up gas and recycle gas. The mixture is heated by heat exchange with reactor effluent and by a furnace and enters a reactor loaded with catalyst. In the reactor, the sulphur and nitrogen compounds present in the feedstock are converted into hydrogen sulphide and ammonia, respectively; the olefins present are saturated with hydrogen and part of the aromatics will be hydrogenated. The reactor operates at temperatures in the range of 300–380°C and at a pressure of 40–60 bar. The reaction products leave the reactor and, after having been cooled to a low temperature, typically 40–50°C, enter a liquid/gas separation stage. The hydrogen-rich gas from the high-pressure separation is recycled to combine with the feedstock, and the low-pressure off-gas stream rich in hydrogen sulphide is sent to a gas-treating unit, where hydrogen sulphide is removed. The clean gas is then suitable as fuel for the refinery furnaces. The liquid stream is the product from hydrotreating. It is normally sent to a stripping column, where H_2S and other undesirable components are removed, and finally, in cases where steam is used for stripping, the product is sent to a vacuum drier for removal of water.

The catalyst used is normally cobalt and molybdenum finely distributed on alumina extrudates. Nowadays, with the emphasis on energy conservation and the more stringent treating requirements for the products from conversion processes, a high-activity cobalt and molybdenum catalyst is normally applied. This is able to achieve higher treating levels at lower reactor temperatures.

Other catalysts have also been developed for applications where denitrification is the predominant reaction required or where high saturation of olefins is necessary. In such cases a nickel/molybdenum-containing catalyst is used. Another interesting application of the hydrotreating process is the pretreatment of feedstock for catalytic cracking units. By utilisation of a suitable hydrogenation-promoting catalyst for conversion of aromatics and nitrogen in potential feedstocks, and selection of severe operating conditions, hydrogen is taken up by the aromatics molecules. The increased hydrogen content of the feedstock obtained by this treatment leads to significant conversion advantages in subsequent catalytic cracking, and higher yield of light products can be achieved.

308 OIL PRODUCTS — MANUFACTURE

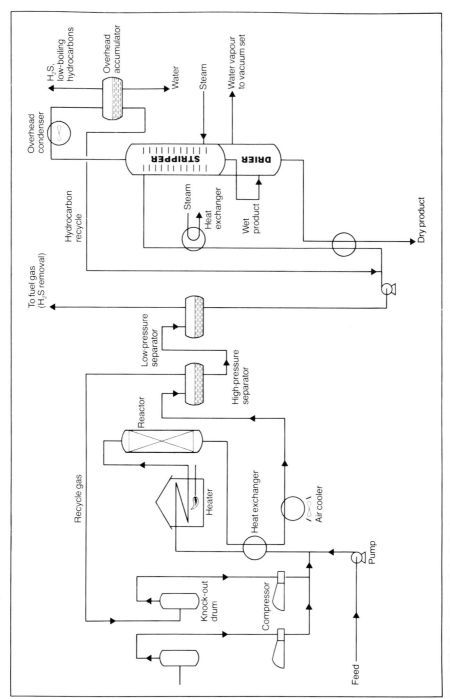

Figure 5.33 **Typical line-up of a hydrotreater**

Figure 5.34 **A hydrotreater.**

Hydrotreating of Pyrolysis Gasoline

Pyrolysis gasoline (pygas) is a by-product from the manufacture of ethylene by steam cracking of hydrocarbon fractions such as naphtha or gas oil. Since its yield is typically some 70–80% of the ethylene production, it is a contributory factor in the economics of ethylene manufacture. Pygas is normally produced from the ethylene plant fractionator as a C_5-180/205°C cut.

Traditionally, the outlet for pygas has been into mogas blending, a suitable route in view of its high octane number. Only small proportions, however, can be blended untreated owing to the unacceptable odour, colour and gum-forming tendencies of this material. The quality of pygas can be satisfactorily improved by hydrotreating, whereby conversion of di-olefins into mono-olefins provides an acceptable product for mogas blending.

An alternative objective is the recovery of chemicals such as benzene, toluene and xylenes (BTX) and production of feedstocks for isomerisation units. For this reason two different processes, first-stage and second-stage pygas hydrotreating, are applied, either separately or integrated.

The purpose of the first-stage hydrotreater (FSHT) is selectively to hydrogenate di-olefins to mono-olefins and thus improve mogas quality, or to produce feedstock for a second-stage unit. The process operates under trickle-flow conditions employing high pressures (approximately 60 bar) and low reactor temperatures (80–130°C). A specially developed nickel-containing catalyst is used. Liquid product recycle is applied to limit the temperature rise within the reactor. In the FSHT no desulphurisation occurs.

A second-stage hydrotreater (SSHT) is used if the objective is to produce feedstock for BTX recovery. In this case the product from an FSHT is further treated in an SSHT to saturate the olefins present and produce an almost completely desulphurised product meeting the stringent olefin and sulphur specifications. Such an SSHT is normally integrated with an FSHT, and a simplified flow diagram of a two-stage integrated unit is given in Figure 5.35. The SSHT reactor operates at a higher temperature (230–280°C) and a pressure of 45–65 bar. A nickel/molybdenum catalyst on an extrudate carrier is generally used.

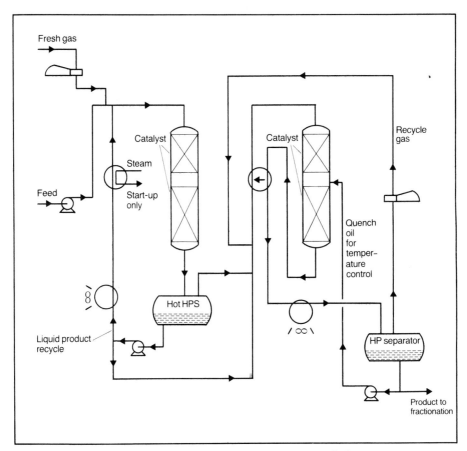

Figure 5.35 **Typical flow scheme of an integrated first- and second-stage pygas hydrotreater**

Again liquid recycle may be applied. Cool high-pressure separator liquid may also be injected into the reactor between the catalyst beds to control the temperature rise.

Smoke Point Improvement of Kerosine

The smoke point is an important burning property of kerosine. Kerosine may be broadly classified as paraffins, naphthenes and aromatics. Of these, paraffins have the least tendency towards smoke formation and aromatics the greatest. Naphthenes are intermediate between the other two. Hence a good-quality kerosine must contain a sufficiently high proportion of paraffinic hydrocarbons relative to aromatics.

By hydrotreatment of a kerosine fraction aromatics present can be converted into naphthenes, thereby producing an improvement of burning characteristics. For this purpose the Smoke Point Improvement (SPI) process can be applied. This process closely resembles the conventional hydrotreating unit in line-up; however, an aromatics hydrogenation catalyst consisting of a noble metal on a special carrier is used. The reactor operates at pressures in the range of 50–70 bar and temperatures of 260–320°C. To restrict temperature rise due to the highly exothermic aromatics conversion reactions, quench oil is applied between the catalyst beds. The catalyst used is very sensitive to traces of sulphur and nitrogen in the feedstock, and therefore a pretreatment is normally applied in a conventional hydrotreater before the kerosine is introduced into the SPI unit.

Hydrodesulphurisation of Residual Fractions

Residual petroleum fractions of Middle East origin contain 3–5% sulphur. Fuel oil composed of such fractions therefore yields upon combustion in, for example, power stations sulphur dioxide, which is emitted into the atmosphere. To reduce or eliminate this pollution, flue-gas desulphurisation can be applied. Another solution is hydrodesulphurisation (HDS) of residual fractions prior to combustion.

As a catalytic hydroprocess, residue HDS is a later development of hydrotreating and HDS of distillate fractions and of hydrocracking of vacuum distillate into lighter fractions. It was introduced in the late sixties and early seventies. Most applications are currently found in Japan.

Unlike the feedstocks in the other catalytic hydroprocesses, residual fractions contain asphaltenes. Asphaltenes are complex and large molecules with the hetero-atoms sulphur, nitrogen, vanadium and nickel built in a matrix of aromatic structures which have a low hydrogen-to-carbon ratio. Vacuum residues of Middle East origin typically contain 100–200 ppm (weight) of metals, whereas distillate fractions are virtually free from vanadium and nickel. The hydrogen-to-carbon ratio is around 1.5, as opposed to 1.9 for distillates.

The line-up of a residue HDS unit resembles that of a conventional hydrodesulphuriser. The feed, atmospheric or vacuum residue, is mixed with hydrogen, heated to reaction temperature, and passed through trickle-flow reactors. Here hydrogen and sulphur-containing species react to give hydrogen sulphide and desulphurised product. The catalyst that enables this reaction to proceed gradually deactivates and is therefore replaced, say, every 6 or 12 months. Upon cooling, the reactor effluent is separated into a gas phase composed of hydrogen sulphide and excess hydrogen, and a liquid phase of desulphurised product. The reaction product hydrogen sulphide is absorbed into an amine solution, while the hydrogen is recycled. Concentrated hydrogen sulphide leaving the regenerative amine absorption unit is eventually converted into elemental sulphur, a marketable by-product.

The desulphurisation of residual fractions containing metals and asphaltenes differs from conventional desulphurisation. The hydrogen partial pressure needed to process these fractions is considerably higher, up to 150 bar, while the lower reactivity of the sulphur-bearing species calls for substantially larger reactor volumes for a given throughput. The large high-pressure reactors contribute significantly to the costs of residue HDS. The catalysts applied are of a special nature.

The catalyst is gradually deactivated by the deposition of metal sulphide formed from the metal compounds present in the feed. This deposition of metal sulphide occurs mainly at the periphery of the catalyst particles and eventually leads to complete plugging of the catalyst pore mouths. Then the sulphur-containing species can no longer penetrate the interior of the catalyst, and the desulphurisation comes to a halt. Narrower pores plug earlier, but have a larger catalytic surface area for desulphurisation, and *vice versa*. A tailor-made catalyst with an optimum combination of metal tolerance and desulphurisation activity can therefore be manufactured by adjusting the pore diameter. The use of tailor-made catalysts in a multiple reactor system is illustrated in Figure 5.36. In the tail-end of the reactor system, the oil still contains relatively low concentrations of metals and a narrow-pore catalyst with emphasis on desulphurisation activity rather than on metal tolerance is optimal. Conversely, the front-end catalyst "sees" high metal concentrations and should therefore be of the wide-pore type, tolerant to metals. Proper applications of tailor-made catalysts in multiple-catalyst systems can lead to substantial reductions in total catalyst volume for a given duty, and, hence, to an appreciably lower capital expenditure.

A further development to residue HDS is residue hydroconversion. One of the applications of this process is to convert residual fractions into feedstocks, which can be further processed in conventional crackers to yield lighter products. In this

HYDROTREATING

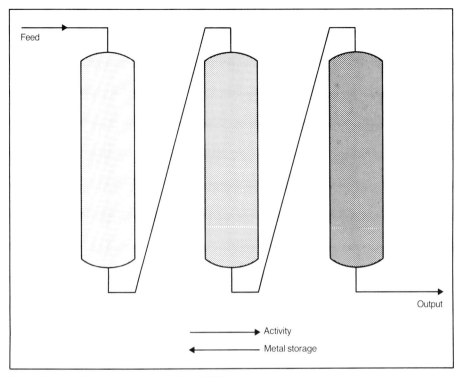

Figure 5.36 **Multiple-catalyst system for demetallising**

way, the production of "white" transportation fuels, gasoline, kerosine and diesel, from a barrel of oil is maximised at the expense of residual fuel oil. Consistently with this, the latter is replaced by other energy sources, such as coal. A second application of residue hydroconversion is the upgrading of tar sands bitumens and heavy oil, as occur in vast amounts in for instance Canada and Venezuela, into pumpable synthetic crude, which can be further processed in conventional refineries. In particular, the hydroconversion of the metal-rich and hydrogen-deficient unconventional feedstocks requires special techniques, such as withdrawal of catalyst saturated with metals and addition of fresh and regenerated catalyst during the run.

Hydrofinishing of Lube Base Oils

Catalytic hydrotreatment of lubricating oils was originally introduced in the USA by several oil companies (Esso, Gulf, Shell Oil) and was applied for the first time in Europe by BP in their Dunkirk refinery in 1959. The reasons for its application were the unattractive aspects of the alternative treating route of acid and clay treating (pp. 323–325). The latter steps are very costly: chemical consumption cost is high and labour requirements considerable. Moreover, the disposal of acid sludge and spent clay gives rise to serious problems.

In the hydrofinishing process the feedstock is contacted with hydrogen over a catalyst at elevated pressure and temperature (typically 100–125 bar and 300–375°C). Feedstock, nitrogen and sulphur compounds are partially converted into ammonia and hydrogen sulphide; aromatic compounds are partially hydrogenated to naphthenes. The layout of a lube base oils hydrofinishing plant is very similar to that of a gas oil hydrodesulphuriser (p. 307). The hydrofinishing route can be applied to a large range of feedstocks. Examples are

- treatment of low-viscosity-index naphthenic (LVIN) oils such as cutting oils and low-grade machine lubricants to improve colour and colour stability, and to reduce their polycyclic aromatics content;
- treatment of medium-viscosity-index naphthenic (MVIN) oils to improve colour, colour stability, oxidation stability and various other specifications related to special applications such as transformer oil, refrigerator oil, hydraulic fluids etc.

Wax Hydrofinishing

Paraffin waxes and microcrystalline waxes (p. 458) require a finishing step to remove coloured, odoriferous and unstable components. In addition, the finishing step for wax grades used in or on food, or in blends for food packaging purposes, must reduce any trace amounts of potentially carcinogenic hydrocarbons to below extremely low levels.

The traditional finishing routes, activated earth (or an equivalent adsorbent) for microcrystalline waxes, sulphuric acid/activated earth for paraffin waxes, are increasingly being replaced by hydrotreatment. The problem of acid sludge and spent earth disposal is thus eliminated. Moreover, hydrofinishing, with a typical temperature range and pressure of 300–350°C and 125 bar, yields waxes with properties superior to those of conventionally treated waxes: apart from a very good colour and undetectable odour and taste (or nearly so), polycyclic aromatic hydrocarbons are almost completely removed.

GASOLINE TREATING

Gasoline produced by simple distillation of crude oil or via conversion processes such as thermal or catalytic cracking contains a large variety of undesirable impurities which are different both in quality and quantity, depending on the manufacturing process applied. The quantity varies between a few to some thousands of parts per million, and their influence on the properties of the

GASOLINE TREATING

gasoline may be quite dissimilar. Examples of such impurities which are found in gasoline fractions and in particular in catalytically cracked material are:
- sulphur compounds such as hydrogen sulphide and mercaptans (thiols);
- oxygen compounds such as alkyl phenols, thiophenols and organic acids;
- nitrogen compounds;
- unsaturated hydrocarbons.

The effect of these constituents can be classified as follows:
- bad smell: mainly caused by H_2S (toxic!) and mercaptans;
- corrosivity: generally caused by organic acids, and, more specifically for copper parts of the engine, caused by H_2S and certain mercaptans;
- reduction of octane number and susceptibility to lead caused by mercaptans;
- formation of gum, which leads to damage to and possible blockage of the engine caused by thiophenols, alkyl phenols, nitrogen compounds and unsaturated hydrocarbons.

In view of the above adverse properties of these impurities in the raw gasoline, their partial or complete removal is required, while the conversion of a harmful constituent into a less harmful one may also give the desired improvement. The two main process principles which are applied for the reduction of mercaptans in gasoline and in particular of catalytically cracked gasoline are mercaptan extraction and mercaptan oxidation to disulphides, or "sweetening".

Both processes have in common the fact that other undesirable contaminants are removed in a caustic prewash. Only in exceptional cases (excessive mercaptan content) is the mercaptan extraction process — sometimes combined with sweetening — applied for gasoline treatment. For this reason only the sweetening process will be briefly discussed in more detail (see Fig. 5.37).

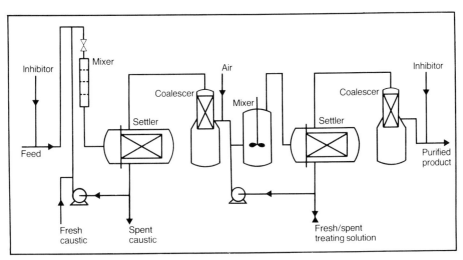

Figure 5.37 **Gasoline sweetening process**

The gasoline entering the treating unit is first contacted with a weak aqueous caustic solution of some 5% wt. NaOH. Both phases are intimately contacted in a globe valve followed by a plate mixer in which the mixture is redispersed by forcing it through a set of perforated plates. Separation takes place in a settler usually fitted with a set of parallel plates to enhance the separation of both phases. The separated caustic solution is then returned to the mixing devices, while entrained caustic is separated from the gasoline leaving the settler by passing this stream through a coalescer, usually containing a polypropylene wool packing. In the prewash, the organic acids, H_2S and part of the alkyl phenols, thiophenols and mercaptans are removed.

The pretreated gasoline is then contacted with a 15% wt. aqueous caustic soda solution and at the same time air is injected into the mixture. In a propeller mixer an intimate and prolonged contact between the three phases is established during which the mercaptans and thiophenols are oxidized to disulphides which remain in the gasoline. In order to enhance the oxidation of mercaptans, use can be made of compounds that increase the solubility of mercaptans in the caustic solution (Shell Air/Solutiser or Air/Caustic process) or of an oxidation catalyst (the UOP Merox process). The latter process has almost completely replaced the other processes.

During the treatment as described above, alkyl phenols are only incompletely removed, while nitrogen compounds are not removed at all, but in practice this does not cause gum problems as long as the thiophenols are removed in the process. Moreover, to combat gum problems caused by unsaturated hydrocarbons, an anti-oxidant (e.g. Topanol A) is injected both in the feed and the product line of the treater.

KEROSINE TREATING

Kerosines and jet fuels as produced in crude distilling units generally do not meet product specifications, since they usually contain impurities which render them unsuitable for their specific use. Of the two products, the kerosine produced for domestic use such as illumination, cooking or as a solvent requires a rather rigorous treatment, for which purpose hydrotreatment is almost exclusively used, sometimes in combination with some finishing process. However, for jet fuels, where in most cases desulphurisation is not required, it is usually sufficient to apply the much simpler and cheaper sweetening process.

Examples of impurities which adversely affect the properties of jet fuels and which quantitatively depend on the type of crude processed are sulphur compounds such as hydrogen sulphide and mercaptans and, moreover, oxygen compounds such as naphthenic acids. The effect of these impurities can be

KEROSINE TREATING

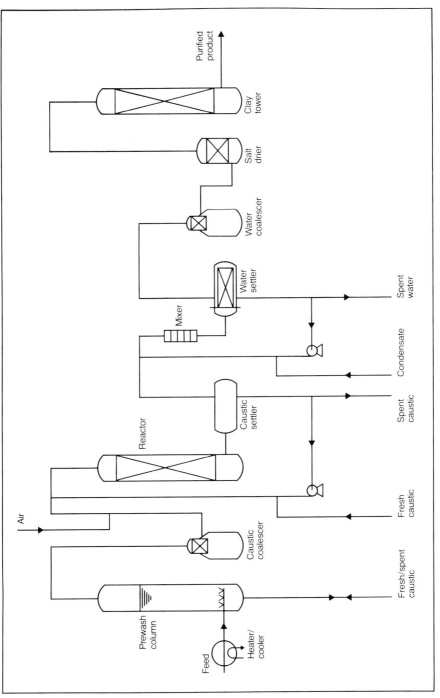

Figure 5.38 **Kerosine Fixed Bed Merox sweetening unit**

classified as follows:
- bad smell, caused by H_2S (toxic!) and mercaptans;
- corrosivity, generally caused by naphthenic acids and, more specifically for silver-plated and copper parts, caused by H_2S and certain mercaptans;
- stabilising effect on water dispersed in the jet fuel, caused by naphthenic acids.

In the sweetening process, the mercaptans are almost completely oxidised to disulphides, while in a caustic prewash the H_2S and naphthenic acids are completely removed.

Since, however, the mercaptans present in kerosine fractions are more refractory than those found in gasoline fractions, UOP have developed a specific sweetening process for the former streams making use of a more active catalyst impregnated on charcoal. This process, which is known as the UOP Fixed Bed Merox Sweetening Process, is briefly described below and shown in Figure 5.38.

The kerosine entering the unit is first contacted in a contacting column with an aqueous 2% wt. caustic soda solution. In this stage the H_2S and naphthenic acids are removed. In a coalescer the entrained caustic is removed, after which the kerosine enters the reactor. In the feed to the reactor the oxidation air is injected. The reactor contains a bed of activated charcoal, which has been impregnated with the catalyst and periodically wetted with a 15% wt. caustic soda solution. In the caustic settler, the entrained caustic is removed and the product is then water-washed, dried with rock salt and passed over a clay tower, the latter to remove traces of impurities which may be present owing to non-optimum treatment.

GAS TREATING AND SULPHUR RECOVERY

Types of Gases and Their Contaminants

Crude oils contain up to 4% wt. sulphur, and the gas streams produced on refining or converting a crude fraction contain significant quantities of hydrogen sulphide (H_2S). This highly poisonous and corrosive compound must in general be removed. LPGs (propane and butane) also contain carbonyl sulphide and mercaptans, and these too may have to be removed, depending on the sales specifications set for the product.

Gasification of residual fuel oils (or, in the future, of coals) may be applied to make hydrogen, methanol or ammonia from the raw gas (carbon monoxide and hydrogen) or to produce a low-sulphur gaseous fuel. The gas from the gasifier has to be treated to remove H_2S and carbonyl sulphide (sometimes to very low levels) and in many applications large quantities of CO_2 have to be removed.

Natural gas streams may contain sulphur in the form of H_2S, carbonyl sulphides or mercaptans, and significant amounts of CO_2 may also be present. Because of the great variety of combination of contaminants in naturally occurring gases, these projects usually pose the most problems to the process designer. Typical specifications for natural gas for domestic use are 4 ppm (vol.) H_2S, and about 100 ppm (vol.) for other sulphur components. The CO_2 specification may be set by calorific value of product, or by secondary processing, such as cryogenic plants. Complete liquefaction, for example, requires a CO_2 specification of about 100 ppm vol.

The water content should often be reduced to avoid formation of hydrates, which will plug the pipeline.

Gas-Treating Processes

The removal of the contaminants discussed above is almost always carried out by absorption in regenerable solvents. Occasional exceptions, for small amounts of contaminant, are adsorbents (see p. 266) or caustic soda on a disposable basis.

The regenerable solvents can be classified as chemical, physical, and mixtures of physical and chemical. The choice of solvent depends on pressure and type of feed gas, amount and combination of contaminants in the feed, and treated gas specification.

Another relevant factor in choice of solvent is the composition of the sulphur-rich stream removed when the solvent is regenerated, in cases in which a process for recovering sulphur from this stream is required. The "Claus" process is by far the most common sulphur recovery process, with limitations on the amount of CO_2 and hydrocarbons in its feed.

There may therefore be specifications on selectivity of the treating process for the absorption of H_2S as opposed to absorption of CO_2 or hydrocarbons.

Chemical absorption

The two most important contaminants we are concerned with, H_2S and CO_2, are both acidic in aqueous solution, and early in the development of gas-treating processes a weakly basic water-soluble solvent was sought which would react reversibly with these components. Many alkanolamines have the correct combination of properties, and already in the early thirties di-ethanolamine (DEA) gained popularity, under the name "Girbotol process". Other alkanolamines such as mono-ethanolamine (MEA) and di-isopropanolamine (DIPA) have also found wide application.

A typical high-pressure gas-treating process scheme, with sulphur recovery, is shown in Figure 5.39. The H_2S and all or part of the CO_2 are removed by countercurrently contacting the gas with DEA solution in a column with trays.

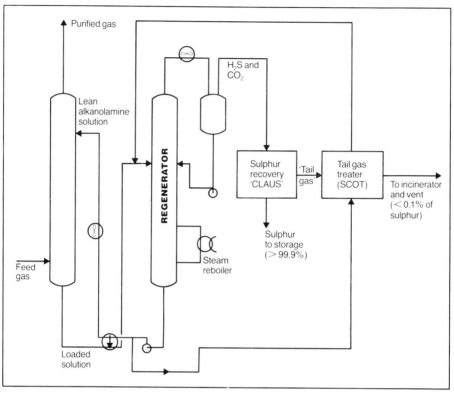

Figure 5.39 Alkanolamine process for removal of H_2S (and CO_2) for gases

The amount of solution and number of trays is chosen to meet the specification on H_2S and, if relevant, CO_2. The DEA solution leaving the absorber is let down in pressure to allow dissolved and entrained hydrocarbons to escape. These gases are usually sent to the fuel gas system. After picking up heat from the hot regenerated solvent, the DEA solution enters the regenerator, where it is contacted countercurrently with steam. The solvent is raised to its boiling point of about 110°C and stripped by the steam, and the regenerated solvent, after giving up heat to the loaded solvent, is cooled to about 40°C before re-entering the absorber. Typically, the solvent inventory is circulated 50 times per hour between regenerator and absorber. A single regenerator may serve several gas absorbers and also LPG extractors (see below).

The steam in the H_2S/CO_2 stream from the regenerator is largely condensed by cooling to 40°C and the water is returned to the top of the regenerator.

The H_2S/CO_2 gases are then fed to the sulphur recovery unit, shown in this case as two processes in series to remove 99.9% of the sulphur (see below for further details).

Chemical absorption processes consume appreciable amounts of steam, and, in an effort to reduce costs, there is a move towards higher concentrations of amine and higher H_2S/CO_2 loadings. This tends to go with higher degradation and corrosion rates, so a compromise must be found. Among the various alkanolamines, those which are most selective towards H_2S in the presence of CO_2 are gaining ground, since the sulphur recovery unit becomes considerably smaller if the CO_2 content in the feed to this unit is low, and also the overall treating costs are much reduced. DEA has a poor selectivity and is losing some ground for this reason.

It should be noted that deep removal of carbonyl sulphide from gases is not possible with chemical solvents, owing to the slow reaction rate, and mercaptans are hardly removed at all.

Physical Solvents

The solubility of H_2S, CO_2 and carbonyl sulphide is much higher than that of methane, carbon monoxide and hydrogen in many liquids — methanol for example. Since the heat of solution is much lower than the heat of reaction in chemical solvents, desorption can be achieved by pressure reduction, possibly combined with moderate heating or inert gas stripping.

Such solvents are only attractive, however, when sufficiently high loadings of contaminant in the solvent can be achieved, avoiding excess solvent requirements. Such high loadings are possible when the partial pressure of the contaminant is very high (typically above 10 bar) or when refrigeration is used to cool the solvent to increase solubility.

Methanol, under the trade name "Rectisol", is widely applied for removal of H_2S, CO_2 and COS from synthesis gas. The lean Rectisol is cooled to $-40°C$. Very deep sulphur removal is achieved and the solvent is selective for H_2S. Such units are complex, with high capital costs.

The application of other physical solvents for natural gas treating is limited to feeds with low concentrations of propane and heavier hydrocarbons, since the solubility of these components is too high.

Mixed Physical / Chemical Solvents

The Sulfinol solvent, developed in the early sixties, combines many of the attractive properties of physical and chemical solvents. It is a mixture of Sulfolane (see p. 262), an alkanolamine, and water. It has found wide application in treating natural and syngases. Its main features are deep removal of H_2S, carbonyl sulphide, mercaptans and, when required, CO_2. Operating costs are in general significantly lower than for purely chemical processes.

LPG Treating

LPGs may contain H_2S, COS and mercaptans, the amount depending on origin. The H_2S is usually removed by applying chemical absorption solvents (see above) in a packed column.

COS may be removed if sufficient residence time is created in a mixer–settler also using a chemical solvent, with a propeller mixer as contacting device. A solvent regeneration system combined with gas absorbers can be used.

For mercaptan removal, alkali hydroxide solution is generally used, also in a packed column. The caustic soda may be regenerated by oxidising the mercaptides to disulphides, which are then separated from the solution by contacting with gasoline.

Sulphur Recovery and Tail Gas Treating

The "Claus" process is by far the most widely applied means of sulphur recovery, and is based on partial combustion of H_2S to SO_2 (at 1200–1400°C) and the further reaction of H_2S and SO_2 to form elemental sulphur in accordance with

$$2H_2S + SO_2 \rightleftharpoons 3S + 2H_2O$$

Equilibrium limits sulphur recovery to about 95%. The cost of the process and sulphur emission increases with CO_2 concentrations in the feed, and excessive hydrocarbons in the feed can result in black and unsaleable sulphur. These set the two main requirements on the upstream treating process.

The process is exothermic, and if little CO_2 is present the combined gas treating and sulphur recovery units are at least self-sufficient in terms of energy.

The so-called "tail gas" from a sulphur recovery unit is the mixture of sulphur gases (SO_2, H_2S etc.) remaining after condensing the sulphur, and is usually incinerated to SO_2 and vented. More stringent environmental requirements have led to the development of several processes for treating the tail gases to increase the overall sulphur recovery to as high as 99.9%. Such high recoveries can be achieved in Shell's "SCOT" process, where all sulphur compounds are converted in a catalyst bed into H_2S, which is re-absorbed using a selective alkanolamine solvent and recycled to the Claus unit. Selectivity is essential to avoid CO_2 build-up. The process is particularly attractive when the tail gas solvent requirements can be integrated in an elegant way with the gas treating process upstream of the Claus.

TREATING OF BASE OILS

Sulphuric acid was one of the first chemicals applied to petroleum. Thanks to its availability and the versatility of its action, treatment with this acid, sometimes followed by treatment with a suitable type of clay, for many years remained the most important refining process for mineral oils.

However, the process has several drawbacks. It is therefore gradually being phased out and replaced by hydrogen treatment processes. Partly for historical reasons, the various aspects of acid/clay treatment are highlighted, followed by its disadvantages in comparison with hydrogen treatment.

Sulphuric Acid Refining

The process is carried out in agitators in which the mixture of oil and acid is agitated either by air-blowing or by pump-mixing. Subsequently, the acid sludge formed is separated either by gravity in a settling tank or by centrifuging. Since sulphuric acid acts both as a chemical and as a solvent, its action on petroleum is highly complex and incompletely known.

Sulphuric acid treating results in partial or complete removal of unsaturated hydrocarbons, sulphur, nitrogen and oxygen compounds, and resinous and asphaltic substances, thus improving oxidation stability, demulsibility, carbon residue, neutralisation number, colour and odour of the oil.

A number of process variables determine the effectiveness of a sulphuric acid treatment of mineral oils.

1. Nature of Oil Treated

The reactivity of hydrocarbons with respect to sulphuric acid decreases from olefins to aromatics, naphthenes and paraffins in the order given.

2. Acid Strength

Since at lower acid concentrations the solvent power of the sulphuric acid diminishes rapidly, acid of 98% strength is normally applied. For very thorough refining (e.g. for medicinal and transformer oils), oleum (fuming sulphuric acid) has to be used to remove constituents which do not dissolve in acid of 98% strength. The sulphonic acids formed require a separate removal step (washing with an alcoholic caustic soda solution).

3. Treating Temperatures

An increase in temperature increases the chemical reactivity of the acid and may promote undesirable sulphonation, but the effect is not the same as that brought about by increasing the acid strength. Furthermore, oxidation of the oil by the acid increases rapidly with the increase in temperature. The optimum treating temperature varies between 35 and 70°C, depending on the nature and viscosity of the base oil, the strength of the acid used and the type of product to be manufactured.

4. Quantity of Acid/Oleum

The severity of the acid treatment can be varied by changing the acid dosage, but the results are not necessarily the same as those obtained by changing the acid strength or the treating temperature. For economic reasons the quantity of acid should be kept as low as possible, but it varies within wide limits because of the differences in specifications for the final oils (lubricating oils 0.5–4% wt. of acid, transformer oils 5–15% wt. of oleum, medicinal white oils up to some 10 treatments of 5% wt. of oleum each).

5. Contact Time

The mixing time is closely related to the degree of agitation, the viscosity of the mineral oil, the treating temperature, the quantity and strength of the acid used and the number of portions in which the acid is applied. The contact time may vary from 0.5–1 hour for lubricating oils to 1–3 hours for transformer oils and to more than 24 hours in total for medicinal white oils, while the relevant settling times for separating the acid sludge are even longer.

Clay Treating

A final clay treatment is sometimes applied to extracted mineral oil fractions to improve their colour, stability (turbine base oils) and, for transformer oils, their electrical properties, by removing traces of unstable compounds still present.

Acid-refined mineral oils require a clay/lime treatment mainly to absorb reaction products and traces of acid sludge.

The oil is preheated to the required temperature in an agitator and thoroughly mixed with the clay/lime, either by air-blowing or by pump-mixing. After the necessary contact time, the spent clay/lime is removed from the oil by filtration in a suitable type of filter (e.g. "Sweetland"). Any fines still present are subsequently removed by a plate and frame ("blotter") press.

At present, mostly activated clays are used, which have undergone a special acid treatment to increase their activity as an adsorbent.

The effectiveness of a clay treatment depends not only on the type of clay used but also on the following process variables:

1. Treating Temperature

Depending on the grade to be produced, this temperature may vary between 70–80°C for lubricating and transformer oils and 80–90°C for white oils.

2. Quantity of Clay/Lime

Clay/lime treatment of acid-refined oils is in most cases carried out in 2–4 steps. The total percentage of clay/lime varies between 0.5 and 3.5% wt., depending on base material and grade to be produced.

3. Contact Time

The contact time of the clay/lime with acid-refined oil is about 0.5–1 hour for lubricating and transformer oils, and can amount to 18 hours or more for white oils.

Comparison of Acid/Clay Refining with Hydrogen Treatment

Acid/clay refining is a cumbersome batch process which is expensive on account of its poor yield, high chemical consumption and manpower requirements. However, its main disadvantage is the problem caused by the disposal of the acid sludge and spent clay which, in view of the increasing severity of environmental legislation, is becoming more and more expensive, if not impossible. While free dumping of these materials is out of the question, burning causes considerable air pollution.

Hydrogen treatment can overcome these disadvantages. It involves a continuous operation in which the mineral oil is submitted to a catalytic hydrogen treatment at elevated temperature and high pressure, in order to convert unstable compounds, thus yielding superior products at lower cost than the acid/clay/lime process. Further details of this treatment may be found under Hydrotreating.

BITUMEN BLOWING

Asphaltic bitumen — normally called "bitumen" — is obtained by vacuum distillation or vacuum flashing of an atmospheric residue (p. 251). This is "straight-run" bitumen, the properties of which are described on p. 464. Bitumen is also obtained by precipitation from residual fractions by propane or butane (solvent deasphalting, p. 261).

The bitumens thus obtained have properties which derive from the type of crude oil processed and from the mode of operation in the vacuum unit or in solvent deasphalting unit. The grade of the bitumen (p. 464) depends on the amount of volatile material that remains in the product: the smaller the amount of volatiles, the harder the residual bitumen.

In most cases, the direct refinery bitumen output does not meet the market requirements. Authorities and/or industrial users have formulated a variety of bitumen grades with often stringent quality specifications, such as narrow ranges for penetration and softening point. These special grades are manufactured by blowing air through hot liquid bitumen in a bitumen blowing unit.

What reactions take place when a certain bitumen is blown to grade? Bitumen

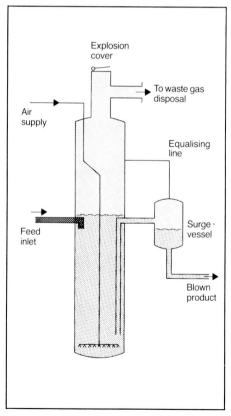

Figure 5.40 **Blowing column**

may be regarded as a colloidal system of highly condensed aromatic particles (asphaltenes) suspended in a continuous oil phase. By blowing, the asphaltenes are partially dehydrogenated (oxidised) and form larger chains of asphaltenic molecules via polymerisation and condensation mechanisms. Blowing will yield a harder and more brittle bitumen (lower penetration, higher softening point), not by stripping off lighter components but by changing the asphaltenes phase of the bitumen. Too soft a blowing feedstock cannot be blown to an on-specification harder grade.

The blowing process is carried out continuously in a blowing column (see Fig. 5.40). The liquid level in the blowing column is kept constant by means of an internal draw-off pipe. This makes it possible to set the air-to-feed ratio (and thus the product quality) by controlling both air supply and feed supply rate. The blowing feed (approximately 210°C), entering the column just below the liquid level, flows downward in the column and then upward through the draw-off pipe. Air is blown through the molten mass (280–300°C) via an air distributor in the bottom of the column. The bitumen and air flow are countercurrent, so that air low in oxygen meets the fresh feed first. This, together with the mixing effect of the air bubbles jetting through the molten mass, will minimise the temperature effects of the exothermic oxidation reactions: local overheating and cracking of bituminous material. The blown bitumen is withdrawn continuously from the surge vessel under level control and pumped to storage through feed/product heat exchangers.

ENERGY MANAGEMENT IN REFINERIES

Introduction

Energy use depends strongly on the type of refining activities and the type of process applied. A simple hydroskimming refinery uses only 4% of its oil intake as refinery fuel, whereas a complex refinery, with cracking and lubricating oil facilities, may utilise as much as 11%.

An average refinery consumes about 300,000 tons per year of refinery fuel at a cost of about 50 million US dollars per year. This expenditure on refinery fuel is almost 50% of the total refinery operating costs.

Because of these high costs, refineries have since 1960 put much effort into reducing refinery fuel consumption. These efforts were intensified after the large increases in oil prices in 1973. Today's average refinery uses only 80% of the energy that a similar refinery (with the same throughput and product package) would have used in 1972.

Principles of Energy Saving

Improving energy efficiency is technically not difficult: it is basically an economic, organisational and also, to a certain extent, political problem. The technological principles of energy saving have been known for a long time. Energy saving costs money: capital for new or improved equipment, labour costs for the design and installation of the improved equipment, labour costs for efficient operation and maintenance of the equipment.

Energy costs have increased drastically in the past 10 years, but so have the costs of equipment and labour and also the costs of borrowing money. New equipment not economically attractive in 1972 is often still not economically attractive today. The fact that refineries, nevertheless, have achieved a 20% saving in refinery fuel is to a large extent due to human effort.

A refinery in the Far East for instance had considerable success with the following efforts:
- The refinery manager visibly showed an interest in energy conservation and set targets for the next twelve months.
- Plant managers were made accountable for their energy use; monthly reporting systems were introduced and an energy team was appointed.
- More meters were installed and meters were calibrated more often; greater attention was paid to training and motivation, results being published in the house journal.

The result was a 30% energy saving in six years.

Experience shows that the following principles should receive close attention in all efforts to reduce refinery energy consumption:

(1) Fire furnaces and boilers efficiently; install oxygen meters in flue gas ducts.
(2) Repair leaks quickly.
(3) Be careful with draining and venting.
(4) Keep heat exchangers clean.
(5) Optimise stripping steam rates, reboiler heat, recycle and solvent circulation rates.
(6) Use "best" available catalyst.
(7) Test meters frequently.
(8) Avoid intermediate storage; do not cool intermediates.
(9) Improve insulation and maintain existing insulation well.
(10) Raise steam at the highest possible pressure, condense steam at the lowest possible pressure and use the pressure differential for power generation.

An analysis of the 20% saving achieved since 1972 reveals the following. Of this 20%, about

ENERGY MANAGEMENT IN REFINERIES

8% was due to better firing and improvements of furnaces,
5% to less flaring and less leakage,
3% to less steam consumption in process plants,
2% to less steam consumption in the tank farm,
1.6% to less steam for auxiliaries,
0.4% to better firing of boilers,
20% total.

At least half the saving was due to human effort such as better operation and maintenance and somewhat less than half to capital expenditure or new equipment.

Equipment or systems that receive great attention in the literature but which (for economic reasons) are as yet only rarely applied in the daily practice of oil refining are: heat pumps, organic Rankine cycles, recovery of power from pressure, variable-speed electric drives.

Energy and Temperature Levels

The law of conservation of energy teaches that energy is not used as such but converted into other (less valuable) forms. For a refinery, this means that oil or gas is converted into heat at a high temperature level and discarded to cooling water or to the atmosphere at a much lower temperature.

In a typical oil refinery, heat demand will be greatest in the temperature range of 300 to 500°C; heat is usually discarded at a temperature around 150°C.

Although flame temperatures are normally around 1500°C, the range between 1500 and 1000°C is never used, mainly because of the high costs of the exotic materials of construction needed in this temperature range. Some use is being made here and there of the 1000 to 500°C range, for example in gas turbines for power generation. The temperature range between 100 and 150°C is rarely used because of the very bulky and therefore expensive equipment needed (e.g. organic Rankine cycles).

City District Heating

However, heat in the 100–150°C range is eminently suitable for heating domestic or commercial buildings. Several refineries have therefore recently been connected to city district heating systems. For example, a medium-sized refinery in Gothenburg, Sweden, supplies heat to the city equivalent to 35,000 tons of fuel per year. Other cities, however, are hesitant to install such systems because of the high capital costs and the relatively low earning power of such projects.

Combined Heat and Power Generation

The total energy needed for a refinery can be roughly subdivided as follows:
63% heat from process furnaces,
25% steam, half for turbines to drive pumps and compressors, half for heating,
12% electricity, mainly for electric motors to drive pumps.

This opens possibilities for combined heat and power generation. During the last twenty-five years a number of high-pressure boilers have been installed which simultaneously generate electric power and medium-pressure steam by expansion of high-pressure steam through a turbo-alternator. The ratio of steam to electricity is about 10 tons of steam per MWh of electricity. Originally, this combined generation of steam and electricity was a very successful development, but for a modern refinery the ratio of steam to electricity produced is too high. Better suited to today's refinery needs is a gas turbine/alternator with a waste-heat boiler that generates 2.5 tons of medium-pressure steam per MWh of electricity at a higher efficiency. In future years, we may expect the introduction of gas turbines, which, instead of generating steam, use their waste heat directly to heat up process streams, e.g. the crude to the crude distilling column.

Other comparable techniques are the use of back-pressure turbines (medium- to low-pressure steam) and waste-heat boilers using the heat of hot flue gases from furnaces. These techniques were applied on a large scale in the period 1960 to 1970, sometimes to the extent that some refineries nowadays are faced with surplus steam for which they have no outlet.

Choice of Fuels

To lower the energy bill of a refinery, efforts should be made to reduce not only the amount but also the value of the material used for refinery fuel. The present-day tendency is to use refinery fuel gas from which valuable components such as propane and butane have been removed and to burn residues with the highest possible viscosity (the higher the viscosity the lower the value of the fuel).

Organisation of Energy Management

Large or complex refineries will often have an efficiency or energy manager who forms part of the management team. He will have a group of people reporting to him called technical auditors, whose task it is to collect and analyse data, to compare the refinery's performance with that of other refineries on the basis of an index system, and to formulate recommendations for improvement. Furthermore, they will assist operating departments in their day-to-day work with advice, especially on furnace firing and steam consumption. About one man per 100,000 tons of fuel consumed per year could be a good yardstick.

Computers can be of assistance in reducing energy consumption, e.g. with process supervision systems and with linear programming models for utilities. These will be very useful, provided sufficient skilled manpower is available for updating, analysing and interpreting the results.

PROCESS CONTROL AND SYSTEMS TECHNOLOGY

Control of refinery operations includes a broad range of activities at various levels. At the lower end, we find control devices, automatic controllers, which keep process conditions such as temperatures, pressures and flow rates at their desired values. At the other end, there are the computer-generated production reports as an aid for refinery managers. The complement of techniques to assist operators, technologists and managers in doing their daily job has gone through an evolutionary development of which we have not yet seen the end. At all levels, the role of digital computers is become increasingly important, new concepts in plant operation are breaking through, and the division of labour between man and machine is being re-evaluated.

Some of these developments will be highlighted below.

Process Control — New Concepts

The change in the technology of refinery processes in the course of time is clearly reflected in the drastic changes in operational tools. In the early days of oil refining, the operator was quite able to keep his "still" in hand with only a wheel key, a thermometer and a pressure gauge; we see the same man now in an air-conditioned, centralized control room looking at colour TV screens on which he finds the requisite information for supervising the whole refinery. A number of factors have led to this. For example:
- more complex plants require more complex control systems to ensure stable and safe operation;
- integration of refinery units — i.e. the product of one unit being the feed for the next unit without intermediate storage, or integration via heat exchangers to save energy — puts more stringent demands on control and requires concentration of all controls in one control centre;
- more complex operation and more advanced automatic control strategies call for a carefully designed man–process interface and support by computational facilities.

The above requirements are difficult to meet with conventional pneumatic or electronic instrumentation. In the years between 1960 and 1980 both users and instrument manufacturers put much effort into realizing new concepts in automation. New developments in electronics, not least the advent of the micro-computer ("chip"), have enabled the design of powerful, flexible instrumentation and control systems.

A number of these systems have been in use for some time for applications such as control of batch reactor sequence and in-line blending, and the number of applications will grow rapidly. Complete refinery control is now following. More elaborate control strategies are possible and, because they work digitally, the controllers are more accurate and stable. Presentation of process information to the operator is predominantly on VDUs (Visual Display Units) and is highly selective. At any given moment, the operator sees a small part of the total available information. But this is not a severe limitation, because tests have shown that he is using only 20% of the available information during 80% of the time.

Strange as it may seem, there is a return to decentralised control in some instances, but this holds only for the hardware. The new instrumentation systems — "distributed control systems", as they are often called — allow the micro-computer-based controllers to be located near the units. Several of these unmanned systems are connected via "data channels" (coaxial cables) to the central control room, where the actual control and supervision take place.

The availability of the supervisory computer provides almost unrestricted computational and data storage facilities.

The stored data can be used to supply trend records on the VDUs replacing the chart pen recorders, and to provide computer graphics showing plant sections with live data or actual line-up of complex manifold systems. The computational facilities can be used for more advanced control structures which:
(1) take into account the dynamic interaction between plant sections: multi-variable control,
(2) prevent the exceeding of critical limits: constraint control, or
(3) use mathematical models (steady state or dynamic) to maximise yield of products and/or minimise energy consumption: optimising control.

Supervision Systems

Refinery processes often lead a dynamic life: it seldom happens that a unit is running in the same "mode of operation" — i.e. the same feedstock, the same product pattern, hence the same process conditions — for a number of consecutive days. This fact complicates the operational task, which is to manufacture, as economically as possible, the necessary quantities of gas, gasoline, kerosine, gas oil, etc. If for a certain feedstock the desired mode of operation is known — i.e. the associated optimum process conditions in terms of temperatures, pressures, flow rates etc. are known — one can rely on the control system to keep the unit on these "target values" during the run length of that mode. In addition, a supervision program in the computer could regularly check whether the unit is still on target and could inform the operator if deviations exceed certain limits. The problem is, however, to know these optimum process conditions for all possible modes of operation. Two different routes are being followed in actual process supervision systems to provide these target values and to guide operations:

(1) mode library,
(2) target calculation.

If a mode of operation is likely to be repeated often in the course of time, then it is attractive to establish once and for all optimum targets and to store these in the computer memory for later retrieval. In this way, mode supervision systems for primary processing have been in use with fifty or more modes in the library. It has been found possible to let the computer interpolate between two modes to accommodate desired modifications in product pattern or to handle known mixtures of feedstocks.

If an actual operation differs too much from one of the "library modes" — e.g. because the feedstock is unknown (new crude oil, or unknown mixture of different crude oils) or because for secondary processing catalyst ageing occurs — the target could be calculated. Then one needs technological models of the various units, such as distillation, platforming, thermal cracking; such a model is a set of relations between operational key variables. An effective way of process supervision is to use the two methods in a complementary manner: from the library one selects a mode that is nearest to the desired operation for the first best guess; in a second step the model is then used to trim the operation of the unit to remove the (small) differences between actual and desired operation.

Once a technological model of the unit(s) is available in the process computer, it can be used for other purposes as well, for example constraint checks, debottlenecking, optimisation, generation of data for computers higher up in the hierarchy, scheduling, and training of operators and/or technologists.

Apart from process supervision, the computer provides the necessary paperwork: hourly logs, daily balances, and all kinds of other reports to render efficient process management possible.

Another area well suited to computer-aided supervision is that of oil movements operations, which includes all housekeeping tasks in relation to the flow of feedstock and products to and from storage tanks, to and from processing units, into and out of the refinery, and in blending systems. All these activities are being managed from an oil movement control room; for the layout of that room the same concepts can be applied as outlined above.

The computer regularly scans tank levels, temperatures and, via a data link with the laboratory, product quality data. The system allows manual entry and retrieval of data and instructions and performs automatic reporting, e.g. sequential logs of daily operations, stock reports, tank balances, alarm summary. For tank or in-line blending it executes tasks such as scanning flowmeters and tank levels and monitoring blend ratios and qualities. Other important aspects are validity status checking of suction and discharge manifold valve positions.

If in a refinery all or most of the processing units and utility systems are provided with a computer system, it is quite feasible to send concentrated sets of data to a central refinery computer system. Here a complete refinery data base can be formed, with information ranging from crude oil arrivals via the operation of distilling and conversion units to the blending of final products. This is the best location to compare actual operation and refinery targets and to generate feedback data to refinery scheduling and crude oil acquisition. Here reports can be prepared to enable managers at various levels to improve their decision-making.

Scheduling and Programming Business Operations

The abilities of computers to store large quantities of information and to perform rapid calculations are ideally suited to problems of scheduling. Examples are the routing of road tankers and the scheduling of deliveries for heating fuels by recording the normal rate of use by customers and estimating likely variations due to changes in temperature.

In many Shell Operating Companies, extensive use is made of a technique known as linear programming for determining the best plant configuration and process profile for the refining operation in the longer term.

In the shorter term, the same models are used to program the refinery by determining the most economical way of converting the crude oil that is available into the products that are required.

The complement of units in a refinery complex is the outcome of a historical development during which, from time to time, estimated future market requirements raised the need for modifications and extensions. The flexibility in selection of raw materials, cut points for distilling and the freedom in operating conditions for the secondary units are the refinery manager's means of varying product yields and qualities.

Computers are increasingly used in the day-to-day scheduling of refinery operations to match in an optimum way the feedstocks, the mode of operation of all units, the blends and the product offtakes, such that no shortages or tank overflows occur. Extensive use is being made of "conversational" computer programs in which the computer performs calculations and invites the scheduler to evaluate the results and, based on this evaluation, to decide how the program should continue. The effect of a scheduler's decision on future tank levels can be made visible on display screens.

In the construction of new facilities such as refinery expansion or oil/gas production platforms, models using the critical path technique often form the basis for management control of the operation. Sophisticated programs permit the network information relating to costs, resources required and expected timings to be represented in many different ways in order to facilitate the scheduling of the construction activities and the effective management control of the project.

Computers do not in themselves solve all these problems, but competitiveness does depend in part on the skill of managers and computer personnel in using the computer as an aid in decision-making.

PROTECTING THE ENVIRONMENT

When considering a refinery's performance in protecting the environment, local conditions such as topography, e.g. distance from residential areas, meteorology, plant age and capacity, type of processing units, etc. are of overriding importance.

Cost-effective pollution control begins before a plant is built. Problems can be minimised or even solved before they arise if the plant is well designed, properly operated and well maintained.

Potential impacts on the environment should be identified and, if need be, attended to in the earliest possible stage of development of a new project.

Gaseous Effluents

The main gaseous effluents of a refinery are: combustion gases from fuel burning, tail gases from sulphur recovery units, fugitive hydrocarbon emissions and combustion gases from flares.

With the exception of fugitive hydrocarbon emissions emanating from valves, pump seals, flanges, drainage system, tanks, etc., the gaseous effluents mentioned above are vented to atmosphere via stacks high enough to ensure adequate dispersion before ground level is reached.

The burning of fuel forms the major contribution to the total atmospheric emission of a refinery. Sulphur dioxide (SO_2), nitrogen oxides (NO_x) and particulates are the main contaminants in these combustion gases.

The quantity of SO_2 in the stack gases depends on the sulphur content of the refinery fuel pool and the capacity and efficiency of the sulphur recovery unit and typically amounts to, say, 2 tons per 1,000 tons of crude throughput. The contribution of the sulphur recovery unit to the total SO_2 emission of a refinery may normally range from 10 to 50%.

Sulphur recovery efficiency may be boosted from, say, 95% to more than 99% by installing a Shell Claus Off-Gas Treating (SCOT) plant downstream of the sulphur recovery unit.

The fuel pool of a refinery may consist of gas, which in most cases is desulphurised, liquid fuel, mostly residue, and, if the refinery has a catalytic cracker, coke burned off from the catalyst. The last component normally forms only a small part of the fuel pool. The sulphur content of a refinery's residue is determined by its crude diet and its processing scheme. Desulphurisation of a refinery's residual fuel or flue gas scrubbing as a means of reducing the SO_2 emission is seldom economic. Replacement of high-sulphur residue by desulphurised distillate is an alternative that would normally result in a heavy economic penalty and a higher sulphur content of the refinery's product package.

Without reducing the actual emission, the building of high stacks is a cost-effective means of reducing ground level concentration of air contaminants, including SO_2, in the surroundings of a refinery.

NO_x is another constituent of combustion gases. A typical average emission figure for a refinery is 8 kg of NO_x, expressed as NO_2, per ton of fuel burnt. The extent of atmospheric nitrogen fixation in the flame during combustion is among other things related to burner and furnace/boiler design parameters. Also, the chemically bound nitrogen in the fuel is partly converted to NO_x.

Little commercial experience is available on the reduction of NO_x emissions from stationary sources. It is expected that changes in combustion techniques rather than flue gas treating may effect a reduction in NO_x emission from stationary sources.

Airborne particulate emissions in a refinery originate from the burning of liquid fuel, and, if present, the operation of the catalytic cracker. Emissions of soot/black smoke from stacks and flares may occasionally occur during abnormal operating conditions. The continuous particulate emission, normally a few tons per day, seldom poses a real problem. If need be, catalyst fines emission from a catalytic cracker can be reduced by installing extra cyclone arrangements and/or electrostatic precipitators. Careful operation, regular maintenance and, if need be, special burner design may minimise particulate emission from liquid fuel burning.

As stated, hydrocarbon emissions originate from a number of small diffuse sources in a refinery. Though difficult to quantify, experience indicates an emission ranging from 0.04% to 0.4% of the crude intake. These hydrocarbon emissions often contain odorous compounds like H_2S and thiols, which may give rise to an objectionable smell in the vicinity. Control has to be sought in pollution-conscious design and operation, and good maintenance and housekeeping. Emissions from the storage of volatile products and crudes can be controlled by the use of floating roof tanks.

Hydrocarbon vapour relief streams from refinery equipment are collected in a closed system leading to a flare of adequate height, where the vapour released is burnt. Streams rich in H_2S are collected in a separate flare system. The flares have steam injection facilities to reduce smoke formation. Part of the flare gas may be re-compressed and used as refinery fuel gas. When the refinery has a continuous gas surplus a ground flare may be installed to reduce the nuisance of a continuous flame at the top of the elevated flare stack. Flaring rates typically range from negligible quantities to, say, 0.5% of the throughput.

Aqueous Effluents

Introduction

In any refinery, there are a large number of identifiable waste water streams with various levels of contamination. The most contaminated streams are the condensates from the processing units and the water drained from the storage tanks; they contain volatile/malodorous compounds, hydrocarbons, dissolved organics and suspended solids (SS). Some other streams are contaminated with oil and SS only (for instance, rain water, ballast water), some streams are normally oil-free but can be contaminated by oil in the event of mis-operation or equipment failure, and finally some streams are always oil-free.

Over the years, the oil refineries have developed a number of measures to limit the impact of their activities on neighbouring surface waters, more particularly because since the early sixties a large proportion of the new refineries have been built inland and have been discharging their effluent into fresh-water lakes or rivers.

In general, water pollution abatement in refineries involves:
(1) in-plant control measures aimed at minimising the pollution at source,
(2) drainage systems designed to prevent complications in treatment units,
(3) end-of-pipe treatment units.

In-Plant Controls

To limit pollution at source, modern refineries are generally designed for minimum water consumption, for instance by using air and/or recirculating water for cooling instead of once-through cooling water, by selecting processes that do not produce as much waste water, and by applying a high level of water re-use; equipment and operating procedures have also been improved to reduce the content of impurities in the waste water.

Drainage Systems

In modern refineries, the categories of waste water described in the introduction are collected in separate sewer systems. This ensures that the most contaminated streams are not unduly diluted with relatively clean water, which would unnecessarily increase the size of the final treatment units and would also impair their efficiency. With this system of segregated sewers, it is possible to submit the most contaminated streams to the most thorough treatment. Accordingly, the water from the processing units containing volatile, malodorous impurities such as phenols, hydrogen sulphide, thiols, ammonia and light hydrocarbons is collected in a closed sewer system and is countercurrently stripped with steam before joining other contaminated streams. Often the stripped condensates are re-utilised as wash water in crude oil desalters.

End-of-Pipe Treatment Units

The sequence of end-of-pipe units normally applied to treat the most contaminated stream (process waste water) is shown as a block diagram (Fig. 5.41). It includes the following steps:

(1) Oil Separation and Equalisation. The primary treatment involves the removal by gravity separation of oil droplets (dispersions) and coarse articulate matter. The oil interceptor is often of the corrugated-plate type, in which laminar flow is assured. Buffer tanks may also be installed to avoid wide variations in the water flow or in the concentration of impurities.

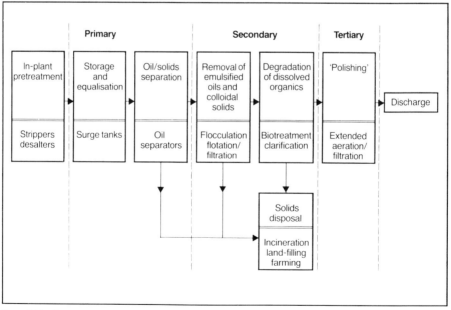

Figure 5.41 **Flow scheme for treatment of process water**

(2) Removal of Emulsified Oil and Colloidal Solids (secondary treatment). Waste water that has passed through a gravity separator can still contain appreciable amounts of oil, mostly in the form of very small droplets (emulsion). These must be removed in order to meet the effluent discharge specifications and to protect the biotreater (which is the next treatment step) against excessive ingress of oil.

This can be accomplished by addition of chemicals under controlled conditions in a flocculation unit (see Fig. 5.42). The flocculants can produce any of the following effects:
- destabilisation of the colloids/emulsions, allowing them to coalesce;
- generation of a dense precipitate, which adsorbs and enmeshes the contaminants;
- formation of bridges between the particles/droplets, which then agglomerate.

The resultant aggregate can then be separated from the water by sedimentation, filtration or air flotation.

(3) Removal of Dissolved Organic Material by Biological Oxidation. In a biological treatment unit, the waste water is contacted with a high concentration of active micro-organisms under intensive aeration. 10–50% of the degradable

Figure 5.42 **A flocculation basin.**

organics are converted into insoluble material (biomass), the balance being carbon dioxide and water.

The activity of the micro-organisms is sensitive to the environment and in particular to the presence of an excess of oil, or suspended solids, which is why deep deoiling is performed prior to biological treatment.

For refinery effluent treatment, the completely mixed activated sludge process is often used.

(4) Polishing/Tertiary Treatment. After-treatment (polishing) of the biotreater effluent is generally limited to an after-aeration step prior to final discharge. When the treated effluent is to be reused as make-up for a circulating cooling system, a filtration step is advisable.

(5) Sludge Dewatering and Disposal. Several sludges are produced in the treatment sequence for refinery waste waters:
- bottom sludges in the oil/water separators,
- scum and sludge in the flocculation unit,
- excess biological sludge in the biotreater.

They are characterised by high water content and low calorific value. After collection, homogenisation and primary dewatering in a thickener, the sludges can be mechanically dewatered, e.g. in centrifuges, and incinerated, e.g. in a fluid bed type of incinerator.

At present fuel prices, incineration is an expensive process, and therefore methods are now being developed for alternative forms of sludge disposal. For instance, the sludge from a refinery biotreater is of a quality that can be used for land-fill and disposal on soil with a minimum of environmental effects.

Oil Spills

In contrast to aqueous effluents, oil spills are not a continuous discharge from a refinery but a consequence of accidents inside or outside the fence. They are thus unpredictable in quantity and location. Oil spills within a refinery (e.g. a tank overflow) can usually be contained by preventive measures, the most important being a good sewer system. Those originating from outside (e.g. ships, mooring buoys, jetties) can usually not be contained by permanent structures. To minimise the impact of these outside spills, speedy and well-coordinated action is required. This is possible only if levels of responsibility, sequence of actions, internal and external alert procedures and materials and equipment are provided for by an oil spill contingency plan.

Noise

Control of noise from refinery equipment is necessary to protect the ears of the workers in the refinery and to avoid annoyance in neighbouring residential areas, if any.

The field of industrial noise control is still relatively young; there is no international consensus on criteria for acceptability of noise or on details of noise measurement procedures. Local regulations should be consulted.

It is commonly accepted that extended exposure to noise levels above 90 dB(A) may eventually cause damage to the ears of workers to a varying extent. Noise levels below 80 dB(A) are generally regarded as safe in this respect.

The range of variation in noise level considered acceptable in the neighbourhood is quite wide, i.e. between 35 and 65 dB(A). These levels are low enough to avoid any direct health effect, but, as the refinery noise may still be clearly audible and recognisable, it may cause annoyance. Investigations have shown that people living near a refinery or similar industry generally accept the steady noise — if it is not too loud — but complain about unusual noises such as from flares or sirens. Maximum noise levels are often specified by authorities when a permit is granted for construction or extension of a refinery. British Standard 4142 could be used for guidance.

The use of decibels, abbreviated dB, is a traditional peculiarity of acoustics and noise control. The term easily confuses a layman. Additional confusion is caused by using the same term to describe different concepts. The type of decibel most commonly used, and quoted above, represents the logarithmic scale for sound pressure. Above 120 decibels "sound pressure level" or "noise level" the experience of noise is painful to the ears. Zero decibels sound pressure level is just about the lower limit of audibility. A second type of decibel in common use represents the logarithmic scale for sound power, i.e. the acoustic energy that equipment produces per second. Decibels sound power level can be used to characterise and rate equipment with respect to the noise it generates. It is also a starting point for various types of calculations, such as the noise levels in residential areas near a refinery. Sound pressure levels at a certain distance from a source can be calculated from the sound power level and *vice versa*. Refinery noise consists of components of various frequencies. The human ear is able to hear frequencies between 20 Hz and 20,000 Hz (1 Hz (Hertz) = 1 cycle per second). The sensitivity of the ear varies with frequency. It is most sensitive in the middle of the range, 1000-4000 Hz. Sound level meters used for noise measurements have what is called an A filter that approximates this property of the ear. When the A filter has been used the resultant sound level is quoted as dB(A).

In the design of a new refinery, or a refinery extension, noise control should be taken into account from a very early stage: buildings and tank farms may be used to form barriers to the noise that is produced mainly in the processing area. This should be borne in mind when the general lay-out of the plant is decided on. Another aspect at this stage is that noise-control equipment such as silencers, acoustic insulation on pipes and acoustic enclosures is fairly voluminous. Sufficient space should be reserved. At a later stage, it will be necessary to determine the maximum acceptable noise levels for individual items of equipment, so that the equipment suppliers can incorporate the necessary noise-control measures in their equipment design. After start-up of the plant, it is customary to check on the noise levels in the plant and take corrective action where necessary.

The emission of noise from a refinery is usually very steady; most equipment operates 24 hours a day. In the vicinity, the noise from the refinery fluctuates by a few dB(A) over intervals of a few minutes. Fluctuations over large intervals (days or weeks) may be as large as 20–25 dB(A) at a distance of, say, 1000 m. Both types of fluctuations are caused by changes in atmospheric conditions. Wind direction, wind and temperature gradients with elevation and turbulence in the air influence the propagation of noise. In particular, the effect of wind direction is quite pronounced. In a certain position, the refinery noise may be clearly audible and even dominating when the receiver is downwind of the refinery; yet the refinery may be inaudible against the background of other noises when the wind is in the opposite direction. As noise propagates from the various sources in the refinery to the neighbourhood it also changes its character. High-frequency components (hissing, sharp sounds) that may be present near the equipment gradually disappear with increasing distance; only the low-frequency components (rumbles) remain. Noise from a refinery, other industry and busy roads all becomes alike at large distances.

SAFETY

The approach to safety is regarded as part of the effort to conserve the environment to protect the health of all concerned. This applies to all activities, including the safety of products delivered. The protection of equipment is a further factor of importance.

In order to illustrate the above, an oil-refining project will be discussed from its conception until well into the operational phase. The following phases may be distinguished: conception, design, procurement, construction, commissioning, operation and maintenance. In practice, these phases overlap, but they will be discussed separately, with the emphasis on safety aspects.

Conception

In this phase an important issue is that the project should fit into its environment, also in the longer term. The project may be an extension or modernisation of an existing installation or a greenfield project. Especially in the latter case, the infrastructure for supply of feedstocks and delivery of products has to be agreed with the local authorities, due consideration being given to the proximity of, for instance, public housing and other industrial installations, in connection with potential hazards from the latter.

In the selection of the process, safety aspects are considered together with the economics of the project.

Design

After selection of the process, a process design is made in which reliability aspects and on-stream maintenance possibilities are included. A layout study is started, often utilising block models. A number of rules govern the layout. Most of these rules are based on safety considerations, for instance to avoid escalation of an incident and to enable proper maintenance. The latter makes a major contribution to the overall safety of the plant. Fire hazards are taken into consideration with respect to safe distances of individual pieces of equipment and the piping layout. Prevailing wind directions are also taken into account with a view to potential gas release.

In the process design, the extreme pressures and temperatures are determined which can occur during normal and emergency operation, such as utility failures.

When the project has been further defined by process engineering, flow schemes or piping and instrumentation diagrams (PIDs), safety reviews are made by a team of all disciplines concerned. Use is made of scaled piping models, which are studied for operability, safety and maintenance.

As part of the process design, the safeguarding by instrumentation and control, overpressure protection and relief systems are important safety aspects. The fail-safe concept is generally applied and emergency instrumentation systems are designed for on-stream testing. Corrosion aspects are reviewed, with the possible implications for start-up/shutdown procedures and limitations for process conditions.

Especially in the case of a new process, a multi-disciplinary team carries out a systematic analysis of potential hazards, which may result in adaptation of the design and additional research.

Fireproofing, fire protection (fixed and mobile) and fire detection are incorporated in respect of the fire hazards, and gas detection systems on the basis of potential release of explosive or toxic gas. Blastproofing of buildings is reviewed in the light of potential explosion hazards. Emergency shutdown systems are examined.

In the engineering design, it is ensured that materials are selected that can cope with the possible occurrence of extreme process conditions and that the applicable engineering codes and standards are used. Generally speaking, company design engineering practice and company-accepted international or national codes are used, whichever are the more stringent.

Procurement

Before equipment and materials are ordered, a final check is made to ensure that they conform to code requirements.

Construction

If a plant is built by a contractor, a company construction team should monitor progress in the field to ensure that good engineering practice and the appropriate codes and standards are being followed. It is also important to establish that the proper materials are used.

Commissioning

The personnel who are to man the plant are phased in in good time, so that they can study the project and plant, peruse the manufacturers' and operating manuals, be trained, and make preparations for start-up. The local operating crew is normally supported by a start-up team and expertise from within the company or outside, as required.

A final check is made that the plant conforms to PIDs and the construction is examined on proper assembly, accessibility, escape routes and maintenance aspects. Documentation is handed over by the contractor and a filling system set up.

Personnel protection facilities are reviewed and checked, and emergency procedures agreed. Rules are laid down for work permits and for accessibility of the plant to personnel of various disciplines. Detailed commissioning procedures are prepared and reviewed for safety. Before start-up all safety features are checked with regard to proper setting and operation.

Operation and Maintenance

The many checks conducted during the design and construction of the plant to ensure that a safe plant is built will be to no avail if it is not operated and maintained correctly once it has been brought on stream. To ensure that the basis of the safety design philosophy is properly understood, these aspects are explained in the design and operating manuals, and the information is used for training the operational staff. In this way, personnel are informed of the hazards existing in the plant, how to recognise them, and the corrective action to be taken in the event of the occurrence of a hazardous situation.

In modern processing units, much of the equipment is safeguarded by automatic systems. It is essential that these devices function properly on the infrequent occasions that they are required, and for this purpose they are checked regularly to ensure that they do in fact operate. Some other safety features are not as complicated or sophisticated, for instance the drainage systems and fire-fighting water mains. Nevertheless, reliance has to be placed in them, and so they too are subjected to regular checks to determine that they still possess the required capacity.

Whilst in operation the plant will undergo wear and tear in a number of ways, such as thinning of pipe and vessel walls as a result of corrosion and erosion, mechanical wear and ageing of various kinds. If undetected, deterioration of this type may lead to leaks and hence unsafe situations. The plant is therefore inspected at regular intervals. Many measurements can be made with the plant in operation by using radiographic and ultrasonic techniques, but for other checks it is necessary actually to enter the equipment. For this purpose the plant has to be stopped, and this is done in accordance with a schedule agreed between the inspection staff and the manager of the plant. The object of these inspections is to predict when the wear will reach intolerable limits so that maintenance or replacement can be planned properly and in good time.

In a refinery, it is not uncommon, over the years, for feedstocks to change and operating conditions to be altered from those envisaged when the units were first designed. Before such changes are made, the new conditions are carefully checked against the specifications of the equipment to ensure that they will not be outside the equipment limits.

It can be seen that safety for refinery plant is an integral concept resulting from many detailed checks. In operation, it is essential that this integrity is not compromised, and hence it needs to be thoroughly understood by operators and regularly examined by inspection and maintenance staffs.

STATIC ELECTRICITY IN PETROLEUM LIQUIDS

In general, when two substances of different material come into very close contact, by mere pressure or by friction, "charge separation" will occur, i.e. the one substance will preferentially collect positive, the other negative electric charges.

Precisely this same process takes place when oil products flow through pipelines. In the beginning, electric charges of either sign are evenly distributed so that the oil and the pipeline are neutral, but after some time the liquid will have obtained an excess of charges of one sign, while the counterpart of these charges of the other sign has flowed to the pipeline wall.

This charging of petroleum liquids is directly related to the velocity at which the liquid flows through a pipeline. Still, this charging is comparatively moderate. Really high charging, actually encountered in practice, occurs for example in microfilters and/or when the liquid is contaminated with water.

The use of microfilters aims at removal of very fine contaminants of the fuel. As the name suggests, the passage of liquid through the filter is very narrow, and thus the friction between liquid and filter is intensive and the generation of electric charge abundant.

Contamination of petroleum liquid with water is another source of intense electrification, because water is virtually insoluble in oil products. The water therefore tends to be dispersed as very fine droplets, and a charging mechanism is set up as described previously, but now at each surface dividing water droplet and the oil.

The electric charges that have been generated as described above themselves cause an electric voltage between the liquid and the pipe wall. This voltage tends to drive the charges back to the wall, and the degree to which the latter succeeds is directly related to an important property of the liquid, viz. its electric conductivity. Thus an equilibrium will eventually be reached such that the amount of charge added by charge separation equals the amount of charge transported back to the wall.

Unfortunately, in this respect most refined petroleum fractions have extremely low conductivities, so that electric charges may be retained in such liquids for quite long periods. Nevertheless, given sufficient time the liquid may be able to relax its charge to a harmless level.

So far, nothing has occurred that is in itself actually dangerous. The fluid in the pipeline, it is true, may have become electrically charged and, being a good electrical insulator, will retain its charges to a certain extent. But even if, for one reason or another, the charge density had become extremely high, so that a spark would occur from the inside of the liquid to the wall of the pipe, such a spark would do no harm. This is so because liquid oil is uninflammable.

Inflammability of an oil product requires evolution of a certain amount of vapour and mixing with air (oxygen). The evolution of vapour may have occurred naturally, i.e. by the ambient temperature, or artificially, i.e. by some kind of application of heat. Furthermore, the mixture of vapour and air must be within a certain range of concentration in order to be inflammable. The limits of this range are generally called the lower explosivity limit (LEL) and the upper explosivity limit (UEL). Summarising, mixtures of oil vapour and air below the LEL or above the UEL are uninflammable, as is the oil liquid.

Motor gasoline is an example of a relatively volatile oil fraction. When standing in a storage tank, it will develop so much vapour that the resultant vapour/air mixture is far above the UEL: it is called "overrich". Gas oil, on the other hand, is a relatively non-volatile fraction. It will give off so little vapour that the resultant vapour/air mixture is below the LEL: this mixture is called "too lean" to be inflammable.

An example of a liquid giving an inflammable vapour/air mixture (under ordinary climatic conditions) is jet fuel, a light kerosine fraction (naphtha).

A dangerous situation, it will now be appreciated, may occur if naphtha contaminated with water is pumped through a pipeline at such a velocity that severe electric charging occurs, and if this liquid is pumped into a tank where the water settles out, thus further increasing the electric charge density. Meanwhile, the naphtha gives off a vapour, rendering the vapour space in the tank explosive. The potential danger in this instance is that the electric charge may collect on the liquid surface and from there spark towards the tank's roof or part of its structure. Such a spark may have sufficient energy to ignite the vapour/air mixture.

The remedies for the hazards of static electricity follow from the above:
(a) Limit the charge generation by
- restriction of flow velocity,
- removal of microfilters,
- guarding against presence of water.
b) Prevent the charge from coming into a tank by
- increasing the conductivity of the liquid (additive),
- permitting the liquid to relax its charges (relaxation chamber).
(b) Prevent the occurrence of an explosive atmosphere by
- use of floating roof tanks,
- use of floating covers in fixed roof tanks,
- making the mixture in vapour space too lean (inert gas, air blowing),
- making the vapour space over-rich (use of fuel gas).

Each of these remedies has appropriate applications. The choice of remedy will normally be determined by economic considerations, design practice and good housekeeping.

Chapter 6

MARKETING OF OIL PRODUCTS

MARKETING ORGANISATION

The marketing organisation of an oil company provides the essential interface between the consumer and the rest of the company. It is responsible for providing a bridge between the two, for ensuring that the consumers' essential interests and needs are correctly represented to the company and that the company's requirements are met satisfactorily in its dealings with a multiplicity of subjects and problems ranging from fuels and lubricants to bitumen or wax. Representation must embrace knowledge of the full range of products and the performance requirements ranging from those of the private motorist to the industrial consumer. A general representative may be expected to deal with safety, finance, distribution, training, investment, selling and to act as a business counsellor where needed.

Management within marketing must deal with an even wider range of circumstances, which can be expected at different times to include many diverse and sometimes conflicting problems: relations with trade unions and dealer organisations; optimising the use of available resources including staff, investment and working capital, product research and development; safety; product quality, supply and distribution; and selling methods, including advertising; questions of profitability and cash flow.

Oil industry management has been required to deal with many changes in the last forty years. Immediately after World War II oil was the fuel that permitted rapid rehabilitation and expansion. To meet the steady growing requirements, marketers had to provide ever-increasing facilities with more service stations, more depots, more vehicles; supply sources and capabilities were constantly expanded throughout the world to keep up with demand. This cheap oil fueled the post-war recovery of the 1960s and '70s.

This situation, which had continued for more than 20 years, changed suddenly in 1973 when the price of oil was increased fourfold and severe shortages in supply were experienced as a result of political and military actions. The era of cheap oil had finished and it was realised that relatively scarce oil would have to be used more selectively.

Oil industry marketers know that their basic product has become more politically sensitive. They know that unbridled imports of crude oil supply are capable of distorting the balance of payments of even the strongest countries, producing very serious economic setbacks. They know that some sources of crude oil supply are very unstable and may be cut off at any moment or that the price level may be raised overnight to levels that are untenable in the longer term. They know that they are dealing with a natural resource which is finite and limited in the longer term. Responsible marketers are, therefore, reacting accordingly and progressively concentrating their marketing efforts on oil-specific markets where there is no real practical alternative to oil pending the development of synthetics, which will be a slow and expensive process. Obvious examples are transportation and lubrication. The good marketer of oil products will make every effort to help and advise his customers on how to conserve energy and thus get the best value for money from the oil products that he buys. For under-boiler firing, in power generation, where cheaper and more plentiful energy sources can be used, oil is steadily being replaced. It should be remembered, however, that inter fuel price competitiveness depends on producer governments, on the development of refineries owned by governments in both producer and consuming countries, and to a considerable degree on the taxation policies of the consuming country. All too often customers in a consuming country are prone to complain about the high price of oil products in the market whilst forgetting that these products are frequently used by governments as major revenue raisers.

The job of the oil marketer is becoming more complex all the time. He is subject to the vicissitudes of circumstances that are largely outside his control. Any year may start with a shortage of supply and high costs due to political or military circumstances affecting important supply sources. As political tension eases, supplies return to normal and costs fall, but in the period of shortage customers have taken alternative measures, so a product surplus develops and market prices weaken as the result of intense competition. All this can occur within a very short period of time, producing considerable confusion in the minds of the public. The marketer must balance his options carefully, taking care not to over-commit his future capability in times of difficulty and at the same time not to reduce his ability to make an adequate return on his investments in periods of surplus product and low market prices.

As was explained in some detail in the first chapter, traditionally the major oil companies have been organised on an integrated basis whereby they have accepted the responsibility of exploring for crude oil, developing producing fields, transporting by pipeline or ship to their refineries, manufacturing finished products and marketing/distributing to the end user either directly or through a system of agents or dealers. However, in recent years this system has begun to change with the arrival of increasing producer government control of exploration and production, producer governments seeking to widen their range of customers, consumer governments seeking to secure oil supplies through deals with producer governments, the development of refineries owned by governments in both producer and consuming countries, and the growth of independent marketers who now have greater access to refined product. These factors are causing the marketing functions within integrated oil companies to become more flexible in their attitude to supply sources. The constant pressure on margins is also causing these marketing companies to improve further their marketing and distribution capabilities, including the provision of a wider range of products and services for consumers, advances in performance of products such as lubricants of value to consumers, and greater cost effectiveness in distribution and delivery with automated systems.

Organisation

Oil products reach the end consumer through a variety of different channels of trade and distribution routes, depending on the trading size and marketing patterns of the country concerned and the individual purchasing habits of the consumer. The patterns of delivery and the way the oil company and market segments being supplied are organised are other relevant factors. Typical market sectors include:

- Automotive Retail
- Domestic Heating
- Agriculture
- Aviation
- Marine
- Commercial Road and Rail Transport
- Manufacturing and Processing Industry
- Civil Engineering and Construction
- Power Generation

The marketing company may be organised so that it has a division to cover each of these sectors or groups of sectors. A degree of product co-ordination is also needed, or it may be organised according to the type of oil product supplied. For large customers it will probably sell direct, but where volumes are small and the selling points numerous and widely separated it is probable that dealers or agents will be employed for cost-effectiveness.

Sometimes an oil marketing company may concentrate on one product only, such as LPG (liquefied petroleum gas), bitumen, waxes or motor gasoline. In the

USA, for example, there are many thousands of fuel jobbers, and some of the most prominent and successful marketers of LPG are companies that concentrate on this product alone. Single product orientation tends to be appropriate to the special products, such as LPG, bitumen and lubricants. Integrated companies with oil refining and marketing facilities usually market a wide range of products, but problems of representation in marketing are the same.

Service Stations

The most noticeable oil marketing activity to the man in the street is the automotive retail service station. Most major oil companies sell through this channel under their own brand, and a large proportion of their total business depends on their success or failure in this sector. For many years the number of sites developed grew rapidly to keep pace with the development of motor car ownership. However, in many of the more developed countries maturity has been reached and many of the smaller sites have been forced to close by the pressure of competition from larger, more efficient, better situated sites able to sell profitably with lower prices and usually through self-service systems. These pressures are

Figure 6.1 **A modern service station.**

expected to continue in many markets in the future and may well intensify with the introduction of automated direct debiting systems and equipment, leading to the extensive introduction of unmanned sites.

A common misconception is that service stations are staffed and operated by employees of the oil company under whose brand name the station operates. In some cases this may be true, but many of the stations are operated by independent businessmen under a franchise.

Product Handling and Distribution

It is essential that the marketing organisation of an oil company provides customers with products they require as efficiently as possible. Transport by road, rail, water and pipeline has to be chosen on its merits to suit particular circumstances. In spite of the wide range of products handled by a large company, great care has to be taken to ensure that contamination does not occur, and to this end adequate testing facilities form part of the operation. Because of the wide variety of lubricants needed, lubricating oil blending plants have to be strategically located and, where appropriate, automated warehousing used to reduce handling and distribution costs. Because of the relatively high value of oil products there is inevitably some conflict between maintaining adequate stocks and not carrying excessive inventories. All of these various factors place a premium on the whole distribution network. Thus, in spite of relatively low growth in some areas, significant investment has still to be made in order to achieve lower costs.

Product Quality / Technical Service

Technical aspects of oil products, their distribution and the equipment in which they are used are an essential part of the marketing activity of any oil company. The needs, present and future, must first be determined by the marketer and action taken to ensure that products with the correct technical characteristics are available in adequate quantities in the market place at all times. The following pages cover some of the more basic technical subjects involved in marketing oil products.

The Automotive Retail Market

Superficially retail service stations have changed but little in the post-war years. Larger forecourts and the presence of self-service pumps are perhaps the most conspicuous changes. In reality, however, the whole concept of automotive retailing has undergone a massive change in the developed world where gasoline markets are now expanding slowly if at all.

In the early days of most markets gasoline was sold by a number of retailers e.g. provision stores, hardware stores and automobile agents as just another product line. As market demand grew, filling stations with off-the-road driveways and a little building serving as oil store and office appeared. Later came service stations offering initially lubrication and washing, and then vehicle repairs also. Later still the office was converted to a salesroom for tyres, batteries, accessories and sometimes a much wider range of products unrelated to the car. Then developments took place during a steady growth in the demand for gasoline. Today, however in the absence of market growth, a drastic re-appraisal of facilities is taking place, with the emphasis on the most cost-effective competitive operations.

In order to reduce distribution costs it is desirable for a service station to have a maximum possible throughput of fuel. For this reason there has been a steady trend towards fewer service stations, each with a large number of pumps featuring self-service and situated in areas where a large throughput of vehicles can be expected. At the same time the services offered by filling stations have altered through the years. Whereas in the past, in addition to fuel and lubricants, vehicle service and repairs formed a major part of the operations, they are now a smaller part of the whole. On the other hand, car accessories and various other merchandise continue to be offered at service stations as these products provide not only convenience to customers but are able to contribute to the overheads of operating the station.

Although distribution costs fall if a minimum number of different fuels are offered, modern trends are contrary to this form of saving. In addition to two grades of gasoline, premium and regular, non-leaded gasoline is offered in some countries. Increasingly there is a need to offer gas oil as automotive diesel cars are used in increasing numbers, while in some markets there has been extensive development of LPG as a fuel. As LPG is stored and supplied under pressure this calls for different types of equipment from that normally used in gasoline and diesel dispensers, but it is the challenge to provide safe equipment that has been faced and overcome by the industry. In other markets, supply of fuel containing ethyl alcohol is a requirement and in some cases ethyl alcohol alone is used as a fuel in either anhydrous or hydrated form. In addition to the problems of handling a multiplicity of fuels high standards of housekeeping are obviously essential. This is particularly the case with the alcohol-containing fuels in contact with water, giving rise to technical problems.

The efficiency of the automotive retail network is being steadily increased by application of electronics and data transmission networks. The electronic dispensing pumps which transmit data to the cash point within service stations are increasingly evident. However, the data so acquired can also be transmitted back to a central point so that optimum refuelling schedules can be arranged for the

Figure 6.2 **An alcohol dispensing pump.**

stations and the behaviour of the network over time and in response to price changes and other factors can be monitored rapidly.

Already in some countries the ubiquitous credit card is being supplanted by electronic funds transfer in which the customer's bank account is immediately debited with the cost of the services provided. This has obvious cash flow advantages for the retail network and helps to improve its competitiveness.

Generally speaking, automotive retail sites are in prime positions and they have a steady flow of customers. Because of the intrinsic value of the sites there is an increasing tendency, particularly in city centres, to use the "air space" of the sites for additional activities which may or may not be automotive-related.

It is clear that in the future the automotive retail sector will be increasingly taxing in its demands on management and equipment of increasing complexity will be needed to minimise overall operating costs and provide the customer with excellent service.

Aviation

It was more than forty years after the birth of the oil industry in Pennsylvania, USA, that the aviation transport industry made its first hesitant steps with the historic flight of the Wright brothers in 1903, also in North America. The pre-condition for powered flight was the combination of lightweight engines with lightweight fuels; and as air transport developed lightweight fuels became more and more important. Long-range powered flight, by aircraft capable of carrying an economic payload of passengers or freight, requires the lightest possible airframe and engine and the lightest possible fuel load. With the exception of the coal-based synthetic fuels produced by Germany during World War II, the fuel requirement has always been entirely met by petroleum fuels. The sole fuel for the first forty years was aviation gasoline (AVGAS), but since the introduction of the gas turbine engine in the 1940s, aviation turbine fuel (ATF) has always been increasingly used.

Like ballooning which preceded it, powered flight was little more than a sporting activity until 1914; commercial flying developed only slowly through the 1920s and '30s. Nevertheless, from the outset it was conceived of as essentially a long-range, international mode of transport, providing a new era of shortened journey times for passengers and rapid transmission of mails. Air transport was synonymous with high speed, and speed has always made its presence felt in all phases of the air transport business. The second, and even more important characteristic of the business, is its great commitment to safety. As a result its safety record is remarkably good, so much so that it is safer to fly in a commercial aircraft than it is to be a pedestrian. Considerations of safety (like speed) have dominated every aspect of aviation, including the supply of high quality fuels, lubricants and other special products such as hydraulic fluids and greases.

Expansion of air transport since 1950 has been rapid. Much of it is in passenger-carrying, which increased at an average annual rate of 12 per cent up to 1973 and has continued to grow at a somewhat slower rate since then. In spite of this vast growth, air travel still accounts for little more than 5 per cent of total passenger transport mileage, although it uses some 15% of transportation fuel. The expansion of air travel has not been fully reflected in fuel consumption because great strides have been made in the fuel efficiency of aircraft. Nevertheless, the rate of increase in demand for aviation fuels and lubricants has been high and is expected to continue despite further marked improvements in fuel efficiency. The economic pressure to achieve ever higher levels of fuel efficiency has become intense as fuel prices have escalated since 1973/74.

The consequences of the pressures to control the cost of fares and freight rates, and to sustain growth in traffic, have had important effects on the fuel supply business. The De Havilland Comet I, which entered airline service in 1952 and was the first jet airliner, had a seating capacity of only 36, compared with 68 seats in the Douglas DC-6 piston-engined airliner that was contemporary with the Comet. However, the Comet's fuel capacity was greater than the 9,800 litres of the DC-6 but was only a small fraction of the full fuel load of a modern Boeing 747 (Jumbo) of nearly 180,000 litres. The need to load fuel in such quantities into aircraft of much greater size has necessitated the development of entirely new fuelling systems and equipment. In three decades, fuelling rates have increased from about 250 litres per minute to over 5,000 litres per minute. Multi-company storage depots at major airfields have developed into considerable tank farms, which in some cases are supplied by pipelines from refineries. At the other end of the range of size of the aviation business are the airfields of the private flying clubs, at which only AVGAS is dispensed (usually in very small quantities and by one oil company only). To cover the full range worldwide a major international company may operate on something like a thousand airfields in nearly eighty countries.

To market aviation fuels and to service airline supply contracts is a complex business that requires considerable investment, organisation and rapid communications. An airline will usually contract with a company for fuel to be supplied at a number of airports in various countries. Major oil groups are able to arrange for

Figure 6.3 An 'Oxford' 18m³ capacity airfield fueller.

the airline to be fuelled in various countries by the local company that markets aviation fuel at the various airports involved. The contracting group will make all arrangements for the airline at the airports to be visited and will ultimately present the airline with an invoice for all fuel supplied.

A very important contributor to speed and safety in commercial aviation has been the R & D (research and development) for military aviation. Although the supply of product to air forces is a modest proportion of total aviation product sales, military business is nevertheless an important part of the aviation business and continues to provide valuable opportunities for advanced product and service developments which often "spin off" into civil applications.

Products for both commercial and military use are very closely controlled by mandatory specifications, which are produced by a variety of bodies through cooperation between industrial, operator, military and governmental interests. Aviation products for military use are covered both by national specifications and by international bodies such as NATO, whilst commercial products are covered by a combination of IATA (International Airline Transport Authority) and aircraft and engine builders' specifications and/or approvals. These differences in the quality specifications for military and civil applications are not the only factors that differentiate between the way that aviation product sales are handled for these two types of customer. In civil aviation, at all levels from private flying clubs to major airline operation, an "into-aircraft" fuelling service is generally the principal characteristic of an oil company's aviation operation. There is an important exception in North America, where the aircraft fuelling service is carried out by independent airfield fuelling service companies rather than by the oil company supplying the fuel. In so far as lubricants and other special products are concerned, only engine oil replenishment is carried out as part of the servicing between flights and the remainder of the products are delivered to airline stores for use by maintenance crews during overhaul. Air forces naturally receive delivery into their own bulk tankage and carry out all aircraft fuelling and lubrication services.

Great importance is attached to the airfield delivery service by airline and other customers. Fuelling has to be carried out very rapidly as part of a complex operation of servicing between flights, and it must be done with full regard to the safety of the personnel and the airliner (which represents a large investment). This requires meticulous planning by the oil companies and the application of product quality control systems from the refinery to the aircraft tanks. Fuelling equipment needs to be maintained and operated to high standards by well trained crews. The development of mutual confidence between customers and suppliers over years of trouble-free operation is a very important aspect of the aviation products business. The major companies, who have great experience of this market, support the airfield service with a high degree of technical expertise in all aspects of oil products application. This sometimes involves an oil company in cooperation with aircraft engine and airline companies in the design and development of new aircraft and of airfield facilities. For example, in the case of the Concorde supersonic airliner, Shell's Thornton and Amsterdam Research Laboratories played a very active part in the fuel system development. The technical service offered to customers not only covers familiar aspects of science and technology, such as metallurgy and high-temperature oxidation of hydrocarbons, but it sometimes extends to less expected topics such as microbiological growth in aircraft fuel tanks and the phenomenon of static electricity formation and dissipation in fuel pumping. Basic research carried out over many years by Shell research workers in the Netherlands and the UK produced the anti-static additive ASA-3, which is now present in most supplies as a safety precaution against the possibility of an accidental fire developing from a spark discharge, which could occur when fuelling at very high rates.

The delivery of fuel into an aircraft's tank may either be from a mobile fueller or an underground hydrant system. Modern fuellers are fitted with pumps, filtration equipment, meters and hoses. They often tow a trailer. Hydrant systems pump fuel from the tank farm through filter/water separators to underground pipelines, which terminate at a series of fixed hydrant valves set conveniently in the aircraft parking apron. Mobile hydrant dispensers are used to link the hydrant system through flexible hoses to the aircraft's tanks. Each grade of aviation fuel is provided with a separate pump and pipeline to avoid contamination. The system incorporates safety devices and is usually fully automatic in operation.

Some degree of simplification has occurred in major airports since AVGAS sales passed their peak in 1965. Although there still were some piston-engined freighters and small feeder line aircraft in operation in 1980, they were becoming comparatively scarce in commercial service and AVGAS was becoming increasingly a light aircraft fuel. During the 1980s ATF will be the dominant fuel supplied at major airfields, mainly in the form of kerosine-based Jet A-1 rather than wide-cut gasolines such as Jet B. Wide-cut fuels such as JP-4 are usually specified for military use.

Domestic Heating

The increased prosperity that has been enjoyed since the 1940s by all the developed countries with temperate or cold climates coincided with oil and natural gas becoming the cheapest fuels. Thus the new automatic domestic heating systems that were introduced were in most cases fuelled by either oil or natural gas. In countries like Canada, the United States and the Netherlands (and later the UK), where natural gas was indigenous, a large part of the increased heating demand was satisfied by that fuel. Elsewhere heating oil greatly increased in popularity; for example, in Western Europe domestic heating oil consumption increased six-fold between 1960 and 1970.

The oil heating market boom of the 1960s was halted by the 1973/74 price escalation, which in most areas of the world changed the price relationship with competing energy sources. Conversions to oil in existing buildings ceased, and new buildings were constructed almost entirely with gas or coal heating systems, where available. The sudden price escalation, which quickly spread to other energy sources, focused attention in many countries on the poor quality of building thermal design in existing buildings and the low standards set in many national building regulations. Standards of insulation were rapidly changed for new construction, and in many countries financial incentives were provided to

Figure 6.4 **A modern domestic boiler.**

improve the insulation of existing premises. In some countries legal limits were introduced to control maximum temperatures in buildings. The net effect was a sharp fall in demand, and very little growth in this market can be expected in coming years. The average annual increase in total energy demand for space heating will be very much smaller than hitherto because of the improvements that are being made to insulation and by other methods of improving efficiency. Oil is likely to obtain only a small share of that total demand increase.

The changed scene in this market sector presents new challenges and new opportunities. Over the next several decades there will be a substantial opportunity available in the sale, installation and servicing of retrofit insulation and efficient energy management and control systems in buildings. Another area in which oil companies already have some experience is in total energy and CHP (combined heat and power) systems. Large-scale (city-wide) and medium- or small-scale (neighbourhood or group) systems are capable of making much of the waste heat from the generation of electrical power (the heat of nearly two gallons out of every three gallons of fuel burnt is wasted in conventional power generation) usable for space and heating. Such schemes are capital-intensive but are likely to increase in economic attractiveness as the real price of energy increases in future. It is unlikely that oil will be burnt in the largest schemes, which involve major power stations; but in some medium and small schemes, where natural gas is not available, oil-fuelled diesel engines, gas turbines or Stirling engines are the preferred prime movers. Another important potential contributor to space heating (with or without air-conditioning) is the heat pump. A heat pump draws energy from the atmosphere, the ground or a nearby water supply and raises its temperature to the level required for space heating without directly transferring heat from the combustion of fuel to the air space. The heat pump operates on the same principle as a refrigerator but in the reverse direction, providing heat rather than cold. Historically, most heat pumps have been powered by electricity, and for every unit of electricity input there can be an output of energy for space heating of approximately two units equivalent of electricity. In this way many of the losses incurred in conventional power station electricity generation are in effect recovered by the heat pump; consequently, the overall thermal efficiency in the conversion of heat input to power station boilers and heat into the building room spaces increases from around thirty per cent to more than fifty per cent. Gas- and oil-fired refrigerators are already familiar, and in the same way heat pumps can be developed to use gas or oil instead of electricity. In this way the roundabout and costly route of converting heat into electricity, only to return it finally to heat, is avoided. The ratio of energy output to input (coefficient of performance) will not be as high for gas or oil heat pumps as for electric heat pumps but they will be much more economic and thermally efficient overall. The other energy sources that are likely to play an increasing part in many countries in space and water heating of buildings are direct solar radiation, urban waste, geothermal energy, and biomass and wind or water power in rural areas.

In the past the choices of energy source have been relatively simple. A wider range of possibilities now exists which, with the higher cost of all forms of energy, makes the choice of the economic optimum much more complicated. In most instances it will be a mixture of sources that will give the best solution rather than one source, as was the case in the past.

In developing the market for oil heating, oil companies produced a simple solution to meet a comparatively simple market need. That solution involved participation in equipment and appliance development, systems design and installation and after-sales technical service. The new market situation, although much more complex, finds oil companies well experienced and prepared to play a prominent part in multi-energy source schemes. In this way such oil as is used for the purpose of space and water heating will be used responsibly and efficiently.

Marine

The displacement of sailing ships by steam-propelled ships was already well under way before the beginning of the oil industry, but it was not until the beginning of the twentieth century that the very considerable advantages of oil fuel were first recognised and exploited in ships. Compared with coal, oil fuel required much less manpower in the engine-room and took up less storage space. With its higher calorific value it permitted a greater operational range and duration and the bunkering operation became much easier and cleaner. Thus the significant price advantage that coal enjoyed in the early years of the oil industry did not prevent the widespread adoption of oil fuel for warships and its considerable use in passenger liners. After the 1914/18 war oil fuel steadily replaced coal in cargo vessels powered by steam. The predominant product used was residual fuel oil (sometimes referred to as "heavy bunker fuel"), which was burnt in boilers to raise steam.

The advent of marine diesel engines, which burned distillate fuels and not residual fuel oil, also occurred in the early years of this century, but the displacement of steam by diesel did not occur to any significant extent until the 1940s. Until the late 1940s, the relatively small price difference between distillates and heavy fuels gave very little incentive to change to heavier fuels. When at that time residual fuels became significantly cheaper than distillates this price factor, together with the development of anti-wear cylinder lubricants, stimulated the conversion of diesel ships to the use of residual fuel oils. In due course diesels, burning the cheaper fuel, replaced steam turbines in all but the biggest ships. By the mid-1970s well over 90 per cent of merchant vessels were diesel-powered.

A third potential prime mover for marine service appeared during the 1950s.

Gas turbines, burning distillate oils, are mainly used in naval vessels, but it is interesting to note that one of the first applications of this type of power unit to a ship was in the Shell tanker "Auris" in 1950. Altough it has a high power/space ratio compared with a diesel engine, a gas turbine is unable to equal its fuel economy in marine service. With the cost of fuel having risen in shipping (as in aviation) to 50 per cent or more of total operating costs, the power unit with the best fuel economy is bound to predominate in merchant vessels, and so the marine gas turbine is restricted to naval vessels.

Escalation of fuel prices following the 1973/74 oil crisis not only had a short-term effect in interrupting the steady growth of oil consumption in this mode of transport, but also had a long-term effect by introducing more uncertainty into the future pattern of marine engine development and choice of fuel. The possibilities of nuclear propulsion were revived, but it seems highly improbable, in the face of growing environmental pressures, that this can make any significant impact on the marine market during this century. Modern technology in the handling and burning of coal, either in fluidised-bed continuous-combustion systems or as micro-slurries in big diesel engines, is at an advanced stage. Although there are problems remaining to be resolved, coal burning could return to a limited degree before the end of the century. The combination of "sailing ships", with diesel-electric auxiliary power for the purpose of manoeuvring and to make optimum use of the available winds, is attracting interest in shipping circles, particularly for ships of 20,000 dwt or less on selected routes. In these special circumstances a 50 per cent saving in fuel, compared with a similar conventional motor vessel, is estimated. The ships will have high capital cost and crewing will be difficult. Until real energy costs are substantially higher their adoption is likely to be very slow. However, in the long run it is possible that there will again be a place in the shipping market for "sail".

So whilst there may well be competition from coal, sail and nuclear in the long-term future, the marine energy market can be expected to depend on oil for at least the next two decades, almost regardless of what happens to fuel prices. Meanwhile, the main pressure in the market will be for improved fuel economy, which can be made through improvements in ship design as well as in engine development. It must also be remembered that in sea transport power requirements and fuel consumption increase very rapidly with speed, and so there is likely to be a continued reduction of average cruising speeds, despite the unwelcome increase in passage time.

Shipping has lost most of its passenger-carrying role (except for cruising) to the international airlines and is now predominantly a cargo transport mode. As such the growth of demand for shipping services closely follows the development of trade. In international trading, transportation of oil (both crude and product) accounts for about half the total cargo ton-miles, as compared with about 20 per cent for grain, ores and coal together. Consumption of oil in all its many applications therefore has a dominant influence on international trading and on marine fuel consumption. Even though the increasing world population can be expected to have an influence on trade and hence on shipping, "plateauing" in the production of oil will inevitably limit the growth of international oil trading and marine fuel consumption in the years to come, but it will still remain significant.

The marketing of marine fuels and lubricants requires the establishment of a worldwide network of several hundred port bunkering installations to meet the requirements of ocean-going ships and coastal vessels. Additionally, there is a secondary network of outlets exclusively distributing distillate marine diesel fuels to barge traffic on inland waterways, short-journey passenger and car ferries, fishing fleets and motor pleasure boats.

The main bunkering installations stock all grades of residual and distillate marine fuels and lubricants, with the heavy bunker grades being stored in heated tanks at a temperature suitable for pumping. Tanks may have capacities as large as 12,000 tonnes, and the largest pumps can deliver fuels at rates of up to 1,500 tonnes per hour. In addition to a steam-raising boiler house for heating the tanks, an installation is equipped with hose gear capable of matching the oil inlet designs of a wide variety of vessels.

Ships' bunkering operations may be carried out during loading or unloading of cargo where wharves are equipped with permanent bunkering lines. Alternatively, fuels may be delivered from barges moored alongside or from road or rail tankers. Some ports have special bunkering berths or, where ships cannot berth alongside a wharf, bunkering may take place through floating lines or through submarine pipelines leading to a mooring buoy. The cost of delaying a ship is large, particularly if a tide is missed. Consequently, shipping lines require a prompt and reliable service from their suppliers.

Availability of a complete range of ships' lubricants can be as important as the speedy delivery of fuel. Breakdowns or the need for premature overhaul or repair of ships' machinery can be very costly and may be dangerous. Chief engineers of ships usually pay close attention to the performance in service of the wide variety of lubricants, greases and hydraulic fluids which a modern ship consumes, often at a rate of several hundred tonnes each year. They attach a good deal of importance to the specialist technical service which those major companies which market a marine product range provide in all major ports, particularly when expert advice is available from staff who combine product knowledge with shipboard experience. Major integrated oil companies are themselves operators of tanker fleets and are therefore able to combine these two types of expertise in the service of their customers.

Manufacturing and Process Industries, Commercial Road and Rail Transporters and Civil Engineering Industry

The Private Road, Domestic Heating, Agricultural, Aviation and Marine market sectors, which have been discussed, all represent fairly homogeneous groupings of consumers calling for specific ways of conducting business. The remaining categories — Manufacturing and Processing Industries, Commercial Road and Rail Transporters and the Civil Engineering Industry — are very much less distinctive groupings but have in common the fact that representation in business is usually directly with the consumer rather than through dealers or agents. There is, however, as much difference in the requirements of individual customers for products and services as there is between the different categories.

Additional factors are the wide variation in the size of consumers and the manner in which different customers prefer to arrange their purchases. Some may offer to purchase on a long-term contract or by annual tender, whilst others will wish to buy on a spot basis as the need arises. Quite often the way in which fuels or bitumen are bought differs from that for lubricants by the same customer. This may mean that more than one representative is required to serve one customer; in another case only one representative may be needed to cover the full range of products sales and services. If a successful relationship is to be maintained with all such customers it is vitally important that representatives are well trained in all aspects of products and their application. Good business performance in this very mixed market sector calls for an effective product range backed up by a professional, commercial, and technical representation from within a flexible organisation. Customers may be widely dispersed throughout every marketing territory. Representation therefore has to be divided geographically as well as by a degree of product specialisation. Some complication exists with client companies with multiple locations, away from the head office, where negotiations or service have to be carried out.

This big and diverse sector may also have a spread of small consumers, which leads to indirect selling through agents' distributors in addition to direct selling by company representatives.

The task of the market manager is somewhat akin to that of the composer of music and the conductor of an orchestra. His problem is to achieve a worthwhile balance of business using the resources available and choosing from the myriad of opportunities that may exist. This involves a massively detailed knowledge of all potential customers' needs, product by product, and associated commercial and technical matters, to be able to determine how best to deploy resources in satisfying such needs. Within the framework of the chosen mix, representatives or intermediaries can be allocated to particular clients whose custom is to be solicited. Naturally, business proposals can differ greatly in their attractiveness.

Road and Rail Transport

Road and Rail Transport is one of the most price-sensitive parts of this market, particularly since the fuel has become such a high proportion of the total operating costs both of road vehicle fleets and of diesel-powered railway engines. As they are the principal carriers of inland freight, consumption grows at much the same rate as the growth of trade. Although many thousands of small commercial vehicle operators purchase product on a spot basis and through the retail network, in many countries both bus and lorry operations are increasingly concentrated in big fleets, and railways are by nature big company operations. Much of this business is consequently put out to competitive periodic tender. Successful bidding requires a great deal of study and analysis. Though price is a critical factor, customers are particularly concerned about lubricant quality and technical service; and a reputation for prompt delivery of large volumes of fuel over what is often a very big territory can also be of considerable significance when bidding is close.

Road vehicle fuel consumption is potentially a factor subject to considerable variation and fleet economics are considerably dependent upon the achievement of a high level of ton-miles per gallon. Big fleets employ expert engineers who have a prime responsibility for this aspect of performance. They can be considerably assisted by the sophisticated laboratory and other technical services that an oil company can offer as they strive to improve both fuel economy and reduce maintenance costs. Medium size and smaller operators will generally have somewhat less in-house expertise of this kind, and technical services designed to meet their particular circumstances are a powerful competitive sales aid.

Manufacturing and Processing Industries

There is a very significant variation between the individual industries in the amount of energy consumption per unit of output, which reflects the diversity of products manufactured and the processes that are employed. Industries which produce basic materials (metals, ceramics, cement, paper, chemicals, glass and bricks) are by far the most energy-intensive. They account in many countries for nearly three-quarters of the total fuel consumption of manufacturing industry, and almost all are composed of a relatively small number of very big plants. Being so energy-intensive, they need to be able to replace oil by other fuels whenever it is both feasible and price relationships indicate that change is necessary. There has been a greater economic incentive to increase energy efficiency since the 1973/74 price rise. The less energy-intensive finished product fabrication industries use a considerable amount of electricity for machine operation and other fuels are mainly used for steam raising and space heating.

Figure 6.5 'Package' oil-fired boilers.

When the power and heat requirements are in such proportions that a CHP (combined heat and power) system is feasible, with generators driven by either diesel engines or gas turbines, oil may remain the preferred fuel because of the very high overall thermal efficiency of energy use, which in some instances may be in excess of 80 percent. Thus whilst manufacturing and processing industry is a declining market outlet for oil fuel, there are still many situations in which the consumers' needs for "energy efficiency" go hand in hand with that for fuel.

The market for lubricants in manufacturing and processing industry is completely different, because there are no significant substitutes for petroleum-based-lubricants. As manufacturing processes become more capital-intensive, reliability of operation without involuntary shutdowns becomes increasingly important. Also, given the need for energy economy, the best lubrication practice can significantly contribute to diminishing losses through friction. One estimate suggested that, in 1980 in the UK, savings of the order of £0.5 billion each year in energy costs were potentially achievable through better lubrication practices. Thus, consumption of petroleum-based lubricating oils, unlike oil fuels, can be expected to reflect industrial activity, but the amounts used will also be affected by advances in technology.

Figure 6.6 **500MW oil-fired power station boiler.**

Companies with products of good performance backed by an expert technical service to meet the needs of a diverse and sophisticated sector will continue to be in demand. Wherever there are demonstrable benefits from high performance, products command a premium. A company will usually concentrate on its proprietary branded product range. However, many big industrial customers publish product specifications which suppliers are required to meet, and in some countries the sale of such special grades can be a significant proportion of the total.

In addition to lubricants, a wide range of other refinery products is sold to manufacturing industry. They include white oils used in toilet and cosmetic products, as plasticisers and processing aids in the plastics industry, as rubber extenders, and as carriers in agricultural sprays. Consumption can be large. For example, the quantity of oil used as an extender in a tyre factory amounts to nearly half the quantity of rubber used. White oils are very strictly controlled in regard to health hazards because of their potential contact with human beings. Synthetic fibres being processed into textiles need a range of textile fibre oils that are emulsifiable and can be removed in the later processing stages. In the engineering industry there is a need for cutting oils, heat treatment oils, drawing

Figure 6.7 **4 tonne/hr burner firing a water tube boiler.**

oils and heat transfer oils. The rapid growth of the use of electricity has brought with it requirements for electrical insulating oils for transformers, cables, switches and capacitors. Petroleum products also make a great contribution to the whole range of permanent and temporary protectives that are used on both finished and intermediary goods. Included are solvents used in paints and metallic coatings, as well as removable (temporary) protective compounds capable of providing protection for periods of a few weeks to several years in hostile environments.

Civil Engineering and Construction

Like manufacturing and processing industry, this market sector is diverse in regard to both size of clients and scale of projects with which they are concerned. There are giant consortia, some of which are multinational in composition (engaged in the construction of major cities, bridges, canals and transport systems), which are frequently surrounded by a multitude of small sub-contractors, all of whom require fuels, lubricants and construction materials. At the other end of the spectrum are small jobbing builders and road maintenance contractors. Their needs differ in many ways, commercially and technically. In many countries a considerable part of road construction and maintenance work is sub-contracted by either central or local government departments. The civil engineering and construction sector is a major user of fuels, lubricant and bitumen. Servicing this industry requires very considerable organisational flexibility and an adaptable delivery service operating wherever the construction project may be. With big schemes demand can be very large over periods of months and years, but it can rise and fall dramatically during the course of construction and in response to adverse weather conditions. On some major contracts the supplier of fuels may undertake to install temporary on-site storage, whereas the smallest end of the industry spectrum may be supplied through an intermediary. On technical matters, service has often to be made available both to an authority and to a contractor.

Power Generation

One of the largest markets for petroleum products lies in thermal power generation. Electricity utilities are the main customers: in some countries these are private enterprise concerns, in others they are operated by the state. By far the largest volume of product consumed is fuel oil for steam generation. When diesel generators are used in utilities they are usually operated on a heavy fuel, but for standby generation distillates are used. Significant quantities of kerosine are used in gas turbines for peak shaving as the rapid start-up and response of these units makes them ideal for the purpose.

Generally speaking, products are sold on specification against tender and it is a highly competitive market. Lubricants for both gas and steam turbines, and of course diesel engines, are needed but again are usually sold by tender against the specification.

Agriculture

A principal feature of oil product supply to agriculture that is common to all supplies in rural areas arises out of the dispersion of population. In the countries that are most advanced in agricultural mechanisation, rural populations have fallen from 80 per cent or more of the total to less than 20 per cent. The small numbers and their geographical spread encourage the linking of marketing of farm supplies to the general domestic and commercial heating business even though many of the farm products are different. For much the same reason it is common for an oil company to wholesale its products for an area of farming country to a specialist distributor, who may well trade in a range of other supplies to farmers. In this way direct contact with the final user is lost but the business is in the hands of specialists who have intimate knowledge of their customers' needs and problems.

As almost every farming process has become mechanised, a large capital investment is required by each individual farmer if he wishes to obtain every piece of specialised equipment. Many of these pieces of equipment are required only for short periods, e.g. for planting or fertilising or harvesting. As a consequence there have developed in many countries agricultural contracting companies who rent machines to different farmers in turn. Alternatively, a similar sharing of machine time is achieved where farmer cooperatives exist. Such contractors or cooperatives may prefer to trade directly with a major oil marketer, maintaining their own storage of product and control of its use, rather than rely upon the services provided through the company's authorised agricultural market distributor.

A special feature of trading with farmers derives from the nature of many agricultural processes in which the farmer has a considerable cash outflow at seedtime followed by a steady cash outflow during the gestation period, before harvesting brings in what may be, for some crops, a whole year's cash inflow. The specialist agricultural distributor has a business that is finely tuned to the rhythms of the countryside, bringing some relief to the primary product supplier. The experience and the financial soundness of authorised dealers in an oil marketing company's products is of great importance in their initial appointment.

In addition to the farmer and the agricultural contractors, there is a rural network of specialist agricultural machinery distributors and repair companies who are themselves potential customers, particularly for lubricants. Most machinery has a specified recommendation list of approved proprietary products from which a choice has to be made. That particular choice may well rest with the machinery distributor or even with individual mechanics working in the company. Suitable literature is made available to assist in that choice.

Despite the revolution that farming has undergone in this century, many of its attitudes and traditions have survived. Farmers continue to attend markets and agricultural shows regularly. It is there that they keep in touch with developments in technology. An oil company that is active in this sector of the oil business must ensure adequate and fully competent representation on these occasions because farmers are a highly professional and critical section of the oil-using public.

Mention has already been made of the growing importance of alcohol fuels for petrol engines and of vegetable oils for diesels. Although in most countries diesel engines now predominate as prime movers for agricultural machines, that situation could change in some countries when vegetable-derived fuels become more readily available. In regard to these alternative fuels, oil marketing companies are contributing their know-how and experience in fuels technology and lubrication and in the distribution and handling of such products.

Special Product Businesses

The foregoing sections have reviewed the various main categories of market into which all oil products are sold, and the different principal characteristics of each have been described. Some of those differences have been between the various classes of product that are involved, such as fuels, lubricants, waxes, bitumen, LPG and process oils. There are important other differences in the marketing of LPG, lubricants, waxes and bitumens that need to be emphasised. Because of this they are sometimes referred to in the oil industry as "special product businesses." By comparison with "main products" (liquid fuels) they account for no more than about 8 per cent of the total volume of sales in the world outside the Communist areas. In 1980 LPG consumption was about 90 million tonnes, about half of which was in the USA. Bitumen sales amounted to about 57 million tonnes in that year, and 25 million tonnes of lubricants (including greases and process oils) and 1 million tonnes of wax were also sold.

Most major integrated oil companies trade in the full range of special products in some countries, but not in all countries. Furthermore, there are many refineries throughout the world operated by integrated companies, independent companies and governments which have no special facilities for the production of special products. Consequently, marketing divisions of some oil companies may have to purchase all their supplies of these products either from overseas associates or from other companies. Whilst major integrated companies enjoy a large share of the total markets for LPG, lubricants, bitumen and wax, in each of these businesses there are important independent companies who specialise in one or other of these products. These independent companies mainly confine their operations to their home country and only the biggest are themselves refiners of the products that they sell.

Waxes Business

Business in paraffin and microcrystalline waxes has some of the characteristics of other commodity businesses. In particular, a considerable proportion of the trade to final consumers is handled by specialist companies. Nevertheless, oil-refining companies which produce a range of basic wax grades also participate in marketing some of their output. This is done in a variety of ways. One important method is through participation in a subsidiary company which produces and sells a finished wax product, such as candles. The development of waxed paper and board packaging of foodstuffs has created opportunities for product innovation involving costly and sophisticated research. Oil companies have been prominent in the development of this technology and thus of market-branched wax formulations which are recommended by major packaging machinery manufacturers.

In general, there are three distinct roles for oil companies to play in the wax business. There is participation in the international trade in wax, of which a substantial proportion of production is sold to specialist resellers and formulators. Secondly, there is investment in wax-processing companies, and thirdly, there is direct participation in markets requiring premium quality products which involve a high level of scientific and technological input.

Bitumen Business

Bitumen is a basic engineering material, used either as a binder in road asphalt or incorporated as a component in a wide range of formulated industrial adhesives, paints, waterproofing materials, sealants and roofing felts. It therefore needs to be marketed in ways that are similar to other products with which it is either competing or to which it is complementary.

Although there is a small international trade in bitumen, most is marketed locally as a technical performance product. There are usually three types of outlet within the road construction and maintenance industry and the importance of each varies greatly between different countries. The three outlets are government agencies, quarrying companies which operate asphalt mixing plants, and road construction and maintenance contractors. For major construction projects binder will be sold to asphalt mix producers, who will frequently demand provision of on-site bitumen heated storage facilities from their bitumen suppliers. The supply of construction sites (either for roads or for other civil engineering projects) often involves long distance road haulage by special heated bulk bitumen vehicles. Some other products used in road construction, such as tack coat, are sold to construction and maintenance contractors. Sales to government agencies are usually of products used for routine maintenance programmes in which contracts are let annually. The purchasing decisions made by these three categories of customer are often influenced by their senior engineers, and successful marketing of bitumen usually involves specialist technical representation and a supporting laboratory service. Since the technical standards and regulations which control the materials used in the construction industry have considerable significance for the suppliers of bitumen, it is important for the specialist bitumen staff of those suppliers to participate in the work of the bodies which produce the standards.

The premium that bitumen is able to command over its alternative use as a fuel is limited by competition with cement concrete prices in road construction. Usually the premium is of sufficient magnitude for bitumen to be the preferred product despite the extra costs of storage, distribution and selling that are usually incurred. It is not possible to move in and out of the bitumen business with alacrity because special investments in refinery plant as well as marketing plant are involved, and also because the quality of bitumen must obtain recognition as being reliable in meeting the expected performance over the years to come. This reliability can only be perceived through many years of good past performance.

DISTRIBUTION AND STORAGE OF OIL PRODUCTS

One of the main functions of an organisation marketing oil products is to distribute the products from their source at the refinery to their destination at the point of sale. The details of a distribution system naturally vary with circumstances, but the general pattern is from refinery to installations, from installations to depots, and from depots to customers or retail outlets. Nevertheless, where more convenient, customers or retail outlets may be supplied direct from refinery or installation.

Bulk distribution from refineries to installations can be made by tanker (generally the cheapest means of transport), by trunk pipeline, rail or road. The operation of tankers and trunk pipelines is often entrusted to separate companies. The policy of siting refineries near areas of consumption has considerably reduced the amount of product distribution by tanker from refineries to distant installations, although distribution by tanker is still the normal method of supply to installations in marketing areas where refineries do not exist. Since the mid-1970s there has been an increase in the amount of refinery capacity in the vicinity of oilfields, and consequently there is developing some revival of ocean transport of finished oil products.

The local marketing companies distribute products from installations and depots by a combination of road and rail transport, sometimes supplemented by product pipelines and by coastal and inland water transport. The pattern of distribution depends on the size and topography of the marketing area, e.g. larger areas will require more installations and depots, and countries rich in waterways will use more water transport.

An installation is a main centre of distribution equipped with all necessary facilities for receiving supplies from the refinery and storing, blending and issuing them in smaller quantities within a marketing area. Installations are supplied by ocean-going or coastal tankers, or by pipeline, rail or road. Usually an installation supplies customers and retail outlets within a convenient distance by direct deliveries, but supplies more distant points via depots.

Depots are secondary centres of distribution which receive supplies from an installation by road, rail, water or pipeline, store them and issue them in smaller quantities throughout a limited local marketing area.

Planning a Distribution System

The total investment in an oil product distribution system is somewhat greater than the investment required for the refining of products. Much of the cost of marketing a product is attributable to distribution costs; it is therefore important to have an efficient and economical distribution system.

As a distribution system develops, many problems require solution; the size and location of individual installations, depots and retail outlets, the methods of transport to be used in each link of the distribution chain, the design of each item of plant and equipment, and the techniques and methods to be followed. These problems cannot be viewed in isolation, but must be considered against the background of the system as a whole. In addition the planner must look ahead and try to foresee the changes likely to take place over the next five, ten or even twenty years.

Planning is influenced by many factors; the geography and climate of the marketing area, the distribution of population and industry, the relative prosperity of different districts, the available means of communication, variations in consumption, the location of sources of supply, the activities of competitors, the availability of finance and so on. Many of these factors are constantly changing and the distribution system must be flexible enough to change with them.

Transport

Product Pipelines

Product pipelines are used only where there is a large, concentrated market area to be fed from a refinery. The high capital cost of a pipeline can then be justified by savings in transport cost over conventional means of transport. Typically the cost of transport by pipeline is about one quarter of the cost of a comparable movement by rail and an even smaller fraction of the cost of road transport. Pipelines can also compete economically with all types of inland barge movements.

An example of a product pipeline is the Trapil line connecting four separate refineries near Le Havre with some thirty depots in the Paris area. A 10 in.-diameter line was commissioned in 1953, when the traffic through the line was 300,000 tonnes per annum. By 1961 the annual traffic had reached 2,000,000 tonnes and a second line of 12 in. diameter was being installed. Later a third 20 in.-diameter line was commissioned running parallel to other lines and meanwhile the traffic has grown to about 3,500,000 tonnes per annum divided between some forty different products (taking into account the diversity of origin and characteristics). This pipeline has not eliminated conventional water transport on the Seine, but has shown the two methods of transportation to be complementary; water transport alone could not have dealt with the tremendous increase in product demand.

Water Transport

Water transport is comparatively cheap and where geographical conditions permit, is widely used for distributing products. In coastal waters small tankers are used, varying in capacity from 500 to 6,000 tons. Such vessels supply ports that are inaccessible to larger tankers, or where it is more economic to supply in comparatively small lots.

On inland waterways and estuaries, barges of 50–1,500 tons capacity are used in similar ways. They may be self-propelled, differing from small tankers only in details of construction, or "dumb" barges without propelling machinery, towed or pushed by a tug, often in trains of several barges. Such barge trains can amount in total capacity to 25,000 tons.

Self-propelled and dumb barges of 150–1,000 tons capacity are also used in harbours and roadsteads for supplying bunkers to vessels that cannot come alongside a wharf or jetty. Bunker craft are equipped with hose handling gear or flow booms and pumps capable of delivering oil at rates up to 400 tons an hour. These craft also usually carry small bulk stocks of lubricating oil for ship's machinery and of gas oil for ship's galleys and auxiliaries.

Road Transport

Road transport plays a very large part in the distribution of petroleum products. By far the greater proportion of petroleum products is delivered to the customer in bulk tank lorries but packed trucks are also used mainly for lubricating oils.

Bulk lorries range in total capacity from 1,000 to 12,000 UK gal. (4,550 to 54,550 litres), the size used depending on restrictions imposed by road conditions, legislation and the nature of the distribution network. Every effort is made to use the maximum size of vehicle as the unit cost of delivery decreases as the capacity increases.

Modern bulk lorries are designed either as general-purpose vehicles or as special-purpose vehicles. The general-purpose vehicle is usually divided into compartments as in many countries the quantity of gasoline that can be carried in a single compartment is limited by law and ranges from 800 to 3000 UK gal. (3,650 to 9,100 litres). The tanks are fitted with sumps to ensure complete draining, and in the case of mild steel tanks the internal surfaces are lined with epoxy resin to ensure cleanliness. Tanks are increasingly being made of aluminium alloy as this enables more product to be carried for a given vehicle weight, and in some countries the aluminium tank has almost superseded the conventional mild steel variety. However, aluminium needs special techniques to repair, and in areas where no such repair facilities are available, mild steel, although heavier, is still the best material.

Tanks made of polyester resin, reinforced with glass fibre, are being increasingly used, especially for the middle range of distillates, as they require little maintenance and are light and strong. Their more general use has so far been restricted by their thermal susceptibility; they cannot be used at temperatures above 120°C. Where vehicles are intended for use with return freights, stainless-steel tanks are sometimes used because they can be cleaned easily, but the use of stainless steel is restricted by its high cost.

Bulk lorries are usually filled through open manholes on top of the tank, with quantities either metered in or filled to a fixed ullage level. Discharge is generally through tank bottom connexions with a flexible hose from the tank outlet pipe to the inlet of the receiving tank, and deliveries are made where possible by full compartment parcels and more preferably by full tank loads. When products have to be discharged to a level higher than the lorry tank, or when more viscous products such as heavy fuel oils are being delivered, the lorries can be fitted with discharge pumps.

For deliveries of domestic heating oils to houses, or of small parcels of products to dealers, lorries have meters fitted in the outlet system to measure quantities delivered.

The use of large capacity vehicles has led to more rapid filling and discharge arrangement so as to get the maximum number of trips from a vehicle. Discharge by gravity has been accelerated by use of larger outlet pipelines and, where this is not possible, by battery electric pumping. With heavy traffic congestion in some cities, deliveries are frequently made at night, the driver controlling the delivery in the absence of the customer, and this has led to more intensive use of vehicles.

Special-purpose vehicles are used where general-purpose vehicles are not suitable, e.g. bitumen vehicles which can also be used for fuel oil and for sulphur, provided the tanks are internally lined with aluminium. These products need to be heated, and the tanks are therefore equipped with flame tubes or steam coils and are lagged with non-conductive material. Discharge is usually by air pressure, the most rapid and simple method, but pump discharge is used in some countries as this enables a less robust tank to be used.

LPG also requires specially designed vehicles. Stronger tanks are required, almost double the weight of similarly sized mild steel tanks for non-pressurised products, but high-tensile steel tanks are being introduced to reduce the tank weight. LPG tanks are not fitted with compartments and the liquid is discharged to customer's storage via a meter and by means of a pump driven from the vehicle engine. Bridging delivery vehicles are discharged by static pumps or compressors. No insulation is provided for LPG consumer tanks. Safety relief valves on the tanks are set to a pressure in excess of any that may be encountered under extremes of climatic conditions in the country in which the vehicle operates.

Other products, such as ethylene, require to be refrigerated as well as compressed before they are transported, and tanks similar to LPG tanks are used with up to five inches thickness of insulation, which limits the temperature rise to about $0.5°C$ ($1°F$) per hour. Special fittings are required for refrigerated products due to the risk of moisture freezing on them and rendering them inoperative.

Rail Transport

In many countries, the railways are still the mainstay of the internal distribution system. Products are distributed in bulk by rail wagons from refineries or ocean terminals to installations or depots, or in some cases direct to customers.

Bridging of large quantities of products from refineries or ports to installations is now being undertaken in some countries by liner trains made up entirely of bulk tank wagons, and these operate at high speeds with rapid turn-round at each end achieved by large capacity loading and discharge arrangements. Although freight tariffs for liner trains are much lower than for single wagons, much higher capital investment is required at loading and discharge points and liner trains made up of bulk tank wagons have to be equipped with brakes and running gear designed to operate at high speeds.

Installations and Depots

Both installations and depots have the same essential functions: to receive products in large quantities, to store them and to issue them in smaller quantities. In general, therefore, the nature of the plant and equipment provided to carry out these functions is much the same at both; installations, however, operate on a much larger scale than depots, and consequently the size and range of their facilities are generally greater. Moreover, individual installations and depots vary greatly in size and scope; some handle only "white" products, some only "black" products, some both. Again, some handle products only in bulk, others partly in bulk and partly packed, while depots may handle only packed products. The actual plant and equipment required in any particular case are determined by the range of products handled, the volume of the trade and the nature of the operations carried out.

Both installations and depots require amenities for staff and labour, ranging from the usual washrooms, canteens, etc., customary in any factory, to living quarters and services for a complete community in an isolated location.

The organisation of an installation must be such as to ensure that the demands of the market are promptly and efficiently met, stocks of products are maintained at an adequate but not excessive level, the quality of products is up to specification, losses are eliminated as far as possible, plant and equipment are maintained in good condition, safety measures are observed and costs kept as low as possible.

The manager of an installation is in contact not only with his own staff and the local branch office of his company but also with labour unions and staff associations, the officers of tankers, the officials of transport organisations, contractors, and numerous local authorities such as police, fire, public health, factory inspectorate, customs, weights and measures, harbour board, etc. In addition he often becomes virtually the unofficial mayor of the installation community. The duties of the superintendent of a depot are similar, though of course on a smaller scale.

Discharging Facilities

Where installations or depots receive their bulk supplies of products by tanker or barge, the cargoes are discharged through one or more pipelines leading to the manifold or hose exchange. If the vessels can come alongside there is nothing unusual about these pipelines, but if the vessels have to moor offshore it is necessary to provide either a submarine pipeline from shore to the mooring point or a floating pipeline that can be launched and towed into position when required. The connexion between a vessel and the pipeline is made by means of flexible hoses.

Tanker discharge lines are usually fairly long and have to be reasonably large (with capacities up to 2000 tons/hour) to reduce discharge time and to speed tanker turn-round. It would therefore be very expensive to provide a separate line for each product and generally only two lines are required, for black and white products respectively. Segregation of individual white oils is usually ensured by pumping water between successive grades. This causes no difficulty as the water rapidly settles to the bottom of the shore tank and is periodically drawn off. Black oils cannot be handled in this way as they form fairly stable emulsions with water. They are therefore generally pumped between products and the small amount of down-grading due to mixing at the interface is accepted. Whenever possible, tanker discharge lines are left full of product; otherwise, white oil lines are cleared with water and black oil lines with compressed air.

Some products, such as bitumen and LPG, require special handling and have their own separate discharge lines. Other products, such as lubricating oils, need special care to prevent contamination and may require separate lines, either one for each grade or one for each group of compatible grades. Lubricating oil lines are always cleared with compressed air and never with water.

When installations or depots are supplied by road or rail the discharge lines are generally short and fairly small and segregation is ensured by providing a separate line for each product.

Pumps

Pumps are used for all movements of oil through installations and depots. Reciprocating, duplex, double-acting pumps (pumps with two cylinders and using both sides of the piston), were at one time the most widely used, but are being replaced to an increasing extent by centrifugal pumps, which have the advantage of a smooth instead of a pulsating flow, together with simple installation and control. For heavy products, however, rotary, positive-displacement pumps, which have a very fine clearance between impeller and casing, are used and can efficiently handle liquids with viscosities up to about 3,500 sec Redwood I (4,000 SSU, 113° Engler) at pumping temperature. With products of still higher viscosity it is generally more economic to reduce viscosity by heating than to use more powerful pumps.

The capacity of each pump depends on the service required; for filling packages it depends on the capacity of the filling machine; for filling bulk lorries and rail tank wagons a filling rate of about 15 minutes per vehicle is generally aimed at. The individual pumps are usually capable of dealing with two filling points at a time; when larger throughputs are required two or more pumps are used in parallel. Pumps are often operated by remote control and when two or more are used in parallel they may be arranged to start or shut down automatically in sequence according to the demand at the filling points. For filling barges and small coastal tankers, and for bunkering, pumping rates up to 500 tons/hour or more may be needed. Bunkering pumps are generally controlled by an operator, who receives instructions by telephone from the bunkering point, but remote and automatic controls are also used to an increasing extent.

Storage Tanks

Since the earliest days of the industry cylindrical tanks have been used for the bulk storage of crude petroleum and its products. Within the Royal Dutch/Shell Group of Companies the types and sizes of tanks have long been standardised. Such tanks are of all-steel construction with butt-welded shells and lap-welded bottoms and roofs. For most purposes vertical tanks are preferred and are constructed in capacities up to 100,000 m^3 (22 million UK gal.). Horizontal tanks are also used, especially when tanks have to be buried, and vary in capacity from 50 to 260 m^3 (11 to 57 thousand UK gal.).

For storing non-volatile, high flash-point products, such as gas oil, lubricating oil, fuel oil, tanks are operated at atmospheric pressure, but for storing volatile, low-flash products, such as gasoline, it is necessary to maintain a slight pressure in the vapour space of the tank to reduce evaporation losses or, alternatively, to eliminate the vapour space by the use of a floating roof or plastic blanket.

The fixed roofs of standard tanks are conical and self-supporting, i.e. there are no internal columns supporting the roof. The roof consists of thin steel plates welded together at the edges, resting on a supporting steel framework and attached to the tank only at the top of the shell. This ensures that in the event of an explosion in the tank the roof sheets will blow off at the periphery, thus acting as a safety valve and avoiding damage to the roof framing or tank shell.

Floating roofs consist either of a single deck supported by pontoons or of a double deck over the whole surface of the tank. The roof floats on the surface of the liquid and rises or falls with the level of the product in the tank.

Standard tanks are not suitable for the storage of LPG owing to the very high pressure required to keep it liquid, about 85 $lb/in.^2$ (6 kg/cm^2) for butane and 250 $lb/in.^2$ (18 kg/cm^2) for propane. Pressure vessels are usually either long, heavily built, small-diameter horizontal tanks with rounded ends, or spheres.

Unlike other steel structures, storage tanks do not usually require concrete or masonry foundations. Except on the very poorest soils, tanks are usually erected on a simple foundation of consolidated rubble covered with a layer of sand about 10 in. thick and finished off with a 2 in. layer of sand–bitumen mix, which seals the foundation against weather erosion and protects the underside of the tank bottom against corrosion. The foundation raises the bottom of the tank 30 in. above ground level. On poor soils it may be necessary to limit the load on the foundation by using a shallower tank of greater diameter. Where this is not practicable a concrete raft foundation supported on piles may be required.

Tank Fittings and Accessories

Dip Hatches. All storage tanks are provided with various fittings according to the products stored. Every tank has one or more dip hatches on the roof through which the height of liquid in the tank can be measured and samples extracted. On floating roofs and non-pressure fixed roofs the dip-hatch is merely a hole with a hinged lid, but on pressure roofs a gas-tight fitting must be used to permit gauging and sampling without loss of internal pressure. The use of remote-reading automatic gauging devices is becoming more common, but they supplement rather than replace dip-hatches since "dipping" remains the most accurate method of measurement.

Vents and Manholes. The roof of every tank is also provided with one or more vents to permit air to escape when the tank is being filled and to enter when it is being emptied; otherwise the tank might be damaged by an excessive difference between internal and external pressure.

Steam Coils. Tanks used for the storage of heavy oils and bitumen are provided with steam coils to keep the tank contents warm enough to be easily pumped. The coils usually consist of rows of steam pipes, connected at alternate ends by hairpin bends, situated in a horizontal plane a few inches from the bottom of the tank.

Compound or Fire Walls. Storage tanks are usually surrounded by oil-retaining walls known as "bund" or "compound" walls. Tanks containing high-flash-point products need only walls high enough to prevent leaking oil from draining on the adjacent land, but tanks containing low-flash-point products or hot products must have walls high enough to enclose a volume sufficient to contain any oil likely to leak or boil over should a tank catch fire. Usually there are local regulations fixing the size of compound in relation to the capacity of the enclosed tanks. For protection against normal risks a compound capable of holding the contents of the largest tank plus 10% of the capacity of the other tanks in the enclosed space is considered sufficient.

Bulk Filling

One of the main operations at nearly every installation and depot is the loading of products in bulk into small craft, rail tank wagons, or bulk lorries for deliveries to depots, retail outlets or customers. The delivery of bunkers to vessels lying alongside is also a bulk filling operation. To ensure segregation it is usual to lay a separate pipeline for each product from the manifold or hose exchange to each bulk filling point.

For filling small craft and for bunkering, the filling point at the jetty is connected to the pipeline aboard the vessel by a flexible hose. Filling is usually to an ullage mark, although increasing use is being made of meters.

Rail tank wagons usually come in on a siding, and the filling lines for the various products run parallel to the tracks with branches fitted with control valves at intervals corresponding roughly to the length of a tank wagon. If the tank wagons are filled from the bottom, the end of the appropriate branch is connected to the bottom inlet of the tank by a flexible hose. If the wagons are filled from the top, the branches are carried to the top of a platform running parallel to the rail track and level with the tops of the tank wagons. Filling is done either by lowering flexible hoses or articulated filling arms connected to the branches into the open manholes of the tanks or by connecting flexible hoses to filling tubes in manhole lids. Filling is usually to a fixed mark inside the tank, although meters may also be used.

Bulk lorries are usually similarly loaded from overhead. The lorries run either underneath a gantry carrying the filling lines or alongside a filling platform similar to that used for rail tank wagons. Filling is generally through articulated filling arms of flexible hoses lowered into open manholes of the lorry tanks or through hoses connected to filling tubes in the manhole lids; bottom loading is sometimes carried out through flexible hoses with quick acting, self-sealing coupling. Lorries are often filled to an ullage mark, though meters, often with preset stop valves, are increasingly used.

Drums and Small Packages

Bulk storage and transport are the most economical means of handling oil products but the ultimate delivery to the customer is frequently required in relatively small quantities contained in drums or small packages. To retain the advantages of bulk handling as far as possible, the filling of packages is carried out in the last stage of the distribution, i.e. at the installation or depot. Many different sizes and types of packages are used, especially for lubricants and special products, but by far the commonest packages for all products outside the USA are the 46 UK gal. (209 litre) drum and the 4 UK gal. (18 litre) tin.

Drums. Drums are generally made of mild steel. "Non-returnable" drums are made of light, 18-gauge material; the "returnable" drums from heavier, 14-gauge material. The former, in spite of their name, can usually be re-used a good many times before being scrapped; the latter, naturally, have a much longer life. Drums are usually painted externally and may be lacquered or metallised internally. Returnable drums are sometimes galvanised both internally and externally.

Drums may be filled by volume or by weight. Even the smallest drum-filling plant incorporates some form of simple mechanical conveyor system to facilitate handling of drums both before and after filling. In plants with a large number of filling points, comprehensive conveyor systems are employed, comprising both gravity and powered conveyors, in order to move both empty and full drums as efficiently as possible with the minimum of man-handling.

Drums inevitably get knocked about in use and need periodical reconditioning. The work is done by outside contractors where possible, but often has to be done at the installation.

Tins. The 4 UK gal. tin, made of tin plate, is approximately 10 in. square by 14 in. high. It has been the traditional package of the petroleum industry for some 60 years and is still widely used in many parts of the world, especially in the Middle East, Africa, India, Pakistan and the Far East, where the tin itself has a considerable re-sale value. Even today, with increasing bulk distribution, millions of tins are used each year. At one time kerosine was distributed almost entirely in tins, and tins are still used mainly for kerosine and to a much smaller extent for other products.

Tins may be filled either by weight or by volume by means of semi-automatic filling machines. The filling hole may be closed by a screw cap, an expanded cap, or a press cap, applied by a capping machine.

Many other types of small tins, especially lithographical tins, are used for marketing lubricants and various special products. For the most part they are packed and distributed in cases or cartons. The packages are normally filled at installations by means of semi-automatic or fully automatic high-speed filling machines. A typical modern high-speed filling line comprises automatic machines for filling and seaming, carton packing, glueing, sealing and palletising. The machines are linked by gravity and powered conveyors and elevators to provide a fully automatic flow from start to finish.

Plastic Packages. Of recent years the use of thermoplastic resins as packaging materials has developed rapidly. Conventional packaging materials, glass, tinplate and mild steel, are being replaced by synthetic materials such as PVC (polyvinyl chloride), polyethylene and polystyrene in the form of tubes, bottles, jerricans and even drums. These new packages have the advantages of lightness, corrosion resistance and good impact strength. Plastic packages, mainly high-density polyethylene, are used for a wide variety of products including lubricating oils, kerosine, detergents and other chemical products.

Storage of Packages. Practically every installation and depot handles part of its throughput in packages, some in very large quantities. The movement and storage of these packages can easily require much labour and storage space. To save labour the maximum use is made of mechanical handling devices, e.g. roller conveyors, hoists, drum trucks, tractor–trailer trains and fork-lift trucks. Fork-lift trucks are generally used in conjunction with pallets, i.e. rectangular platforms on each of which a number of drums or small packages are stacked and handled as a unit. In some of the biggest and most modern warehouses Robo-trailers have replaced manually operated handling vehicles and are entirely computer-controlled. Small packages are stored under cover, but drums are generally stacked in the open. Whether stored inside or outside, packages are stacked as high as safety and convenience permit in order to make best possible use of available space.

Storage and Handling of Special Products

To an increasing extent products such as LPG, lubricants and bitumen, which at one time were invariably filled into packages at refineries and so distributed, are now distributed in bulk. Installations may therefore require facilities for receiving and storing these products in bulk and for blending and filling them for distribution to the market, either packed or in bulk.

Additives are increasingly blended into products at installations, and facilities for this purpose are usually required. Anti-knock additives (TEL/TML, see p. 397) are more usually added at the refinery.

Safe Operating Practices

Although installations and depots may appear to be dangerous places because of the type and large quantities of products handled, rigorous observation of the Institute of Petroleum's Code of Safe Practice has achieved an enviable safety record.

The dangerous feature of volatile oil is the inflammability of vapours. These vapours, however, burn only if mixed with air in the correct proportion and if ignited by a flame, a spark, or incandescent metal. A spark with sufficient energy for ignition can be caused not only by an electric circuit but by static electric charges generated within the product and released on the liquid surface during or shortly after transfer of the product from one container to another. The safety measures at installations and depots are based on the elimination of all possible sources of ignition from all areas where dangerous concentrations of inflammable vapours are at all likely.

The emphasis is thus on the prevention of fires rather than on their extinction, but since fires may nevertheless occur, each installation and depot is provided with "first aid" fire fighting equipment of various sizes at strategic points for dealing with small outbreaks. Fire extinguishers used in installations and depots may contain foam, soda/acid, carbon dioxide, dry chemical or vaporising liquid, each type being suitable for a particular purpose. Facilities are generally provided for producing large quantities of foam for extinguishing larger fires in storage tanks.

Areas surrounding possible sources of inflammable vapour are designated "restricted areas", the size of the area depending on the type of products handled. Within restricted areas all possible sources of ignition are eliminated by rigid rules against smoking, carrying matches, or lighting fires, by the use of flameproof or intrinsically safe electrical equipment, engines and machinery, and by the exclusion of all non-flameproof equipment and vehicles.

Where toxic or corrosive chemicals are handled, suitable safe operating practices are applied, including, for example, the provision of suitable protective clothing.

OIL PRODUCTS APPLICATION, SPECIFICATION AND TESTING

Each individual product of an oil industry (whether it be the output of a single process plant or a blend of components produced by several different processes) is required by the customer to perform a defined function, e.g. as a fuel, lubricant, adhesive, construction material, waterproofing agent, corrosion protective, electrical insulant, etc. In order to perform satisfactorily in service, each product requires certain characteristics which can be quantitatively described. For example, it may need to be sufficiently fluid between minus 10°C and plus 35°C to be pumpable. That characteristic then has to be translated into one or more test measurements. In this example an upper limit of viscosity at a defined temperature might ensure fluidity under cold conditions, and a vapour pressure measurement might guarantee that flow would not cease through vapour lock at high operating temperatures. A combination of all the technical requirements of an application provides the basis on which a fuel can be described or specified in terms of standard test methods. Other test methods are needed to protect the integrity of the product during transportation, storage, and handling and environmental considerations may impose some further special constraints. Levels of quality above these basic requirements are a matter of commercial judgement. The total mixture of requirements translated into standard test methods becomes a specification that defines the range within which a particular product must fall when it is delivered to customers if it is to provide the level of customer satisfaction at which the supplier aims.

In this section the principal applications of oil products will be described, with particular reference to the required performance characteristics that have to be satisfied in a marketable product.

About 85 per cent of the world's oil supply is used as fuel, either in engines to produce power for transportation or in continuous combustion applications to produce heat for use as such or translation into other power media. The world

APPLICATION, SPECIFICATION AND TESTING

still is dependent for most of its power and heat on the burning of wood and fossil fuels.

All fuels need oxygen for burning; the more intimate the mixture of air and fuel, the more complete and efficient the combustion. Gaseous fuels, and liquid fuels that vaporise easily, mix readily with air, while liquid fuels that do not vaporise easily can be atomised into minute droplets which disperse rapidly throughout the air. Intimate mixing of the fuel with air permits the extremely rapid burning that takes place in the internal combustion engine. Solid fuels cannot be so easily dispersed in air, and consequently are less easily burnt quickly and efficiently. Petroleum fuels thus have an inherent advantage over solid fuels as regards ease of combustion. They also have the advantage of being almost completely combustible (with the formation of relatively little ash), and of being easily stored and handled.

Petroleum fuels comprise both gaseous fuels — natural gas and LPG (liquefied petroleum gas) — and liquid fuels. The latter are normally classified according to their volatility into gasolines, kerosines, gas oils, and diesel fuels, and residual fuel oils. Kerosines, gas oils and distillate diesel fuels together are sometimes referred to as "middle distillates".

Natural gas, LPGs and gasolines can each form an explosive mixture with air at ambient temperatures, and so special precautions have to be taken in handling and storage; strict control over ignition sources must be observed in their vicinity for the prevention of explosions and fires. Great care must also be taken to see that other heavier fuels are not contaminated with gasolines.

Motor Gasoline

The transport industry is a major consumer of petroleum products, and its development has been closely associated with that of the petroleum industry. The internal combustion engine is by far the predominant prime mover, and its requirements for fuels and lubricants have become more exacting.

There are two types of IC engine: those operating on intermittent combustion (almost entirely reciprocating piston engines, the exception being the rotary (Wankel) engine) and those operating on continuous combustion (gas turbines). Piston engines may in turn be divided into two classes: spark-ignition engines and compression-ignition engines. Although gas turbines may possibly be used some time in the future for automotive purposes — and experimental models have been operated — they are not yet established and will not be further considered in this chapter. Aviation and industrial gas turbines are described on p. 401 and p. 418; should automotive use develop, the general requirements are likely to be similar.

The Spark-Ignition Engine

The Engine. The spark-ignition engine is used extensively in passenger cars, trucks, motor cycles and outboard motors. It is also found in light trucks and older designs of agricultural tractors and aeroplanes. In the spark-ignition engine a charge of partly vaporised and partly atomised fuel is drawn with air into a cylinder, where it is compressed by the forward motion of the piston. The compressed mixture of fuel and air is then ignited by a spark, and the resulting combustion develops a pressure that forces the piston back and provides the driving power. The gases, having done their work, are exhausted from the cylinder and a fresh charge is introduced. Although this cycle of operations can be accomplished during 2 strokes of the piston most gasoline engines operate on the more efficient 4-stroke cycle (Fig. 6.8). The predominant use of gasoline engines is for road transport, and the aim of engine designers has always been to create a unit that is light, flexible and inexpensive to build. In the post-war period when gasoline was cheap there was no particular incentive to achieve maximum economy, neither was there any great concern about the gases emitted from the exhaust, which did in fact contain unburnt hydrocarbons, carbon monoxide and nitrogen oxides, as well as the residues from lead anti-knock additives. By the mid-1960s it had become clear that a major contributor to the "smog", a disturbing atmospheric phenomenon in the Los Angeles area, was a photo-chemical reaction involving unburnt hydrocarbons and nitrogen oxides. In this particular geographical situation, under some weather conditions little air movement

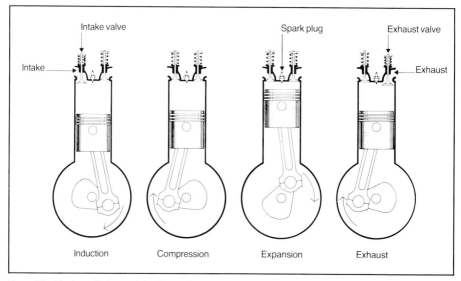

Figure 6.8 **The four-stroke spark-ignition cycle**

occurred with a consequent build-up of combustion products leading to smog formation. The American automobile industry was then required to produce vehicles which minimised this type of atmospheric pollution, and increasingly stringent performance standards were set to be met progressively. Although the initial limits could be met by modifications to gasoline engines, subsequent lower limits were largely achieved by the use of an oxidation catalyst in the exhaust system of the engine. For these catalysts to be effective, engines were set up to run with a slight excess of fuel, producing relatively high levels of carbon monoxide and hydrocarbons, which could be oxidised or "burnt" efficiently by the catalyst. This led to a loss in car fuel economy, although subsequent development of the so-called "3-way catalyst" wherein carbon monoxide, hydrocarbon and nitrogen oxides emissions are simultaneously reduced went a long way to restore this loss. To function effectively the 3-way catalyst requires precise control of the fuel–air mixture at the chemically correct value for complete combustion, which has led to the development of electronically controlled carburettors (see p. 390).

The post-October 1973 increase in prices and uncertainty of oil supplies observed at the time then brought about a quest for greatly improved fuel economy, and in the case of the United States a maximum fuel consumption was specified for the fleet of cars produced by each manufacturer. In order to achieve both acceptable emission standards and improved efficiency a massive re-design of engines has had to take place. A common feature of new engine designs is the trend towards higher compression ratios in order to improve thermal efficiency. Raising the compression ratio of engines normally increases their tendency to "knock" or "detonate". The ability of a fuel to resist knock is measured by its "octane number". Gasolines are rated by comparison with blends of isooctane or normal heptane. Isooctane is defined as having an anti-knock rating of 100 — normal heptane 0. Thus a gasoline with anti-knock properties similar to a blend of 97% isooctane and 3% normal heptane would be said to have an octane number of 97. Through improved control of mixture movement in the combustion chamber designers are finding ways of increasing compression ratio but not raising the octane requirement of the engine. Lean-burn operation is also attractive for reducing fuel consumption, but exhaust emission control requirements may constrain what can be done. For example, to meet the stringent emission legislation in USA and Japan extensive use is made of 3-way catalyst systems. However, these systems have to operate at an air–fuel ratio for the individual fuel which is less than optimum for best-fuel economy. Thus engine design often has to compromise between emissions and economy. To improve economy further the designer can optimise engine size and vehicle gearing, but then he has to compromise between economy and performance. This has led to some using two sparking plugs and others a separate pre-chamber or specially shaped combustion chamber. If the engine can be kept light, then the rest of the car can be lighter too and this too brings about a significant economy of fuel consumption. This has led to increased interest in supercharging engines, i.e. the mixture of fuel and air is compressed before introduction into the cylinders. Supercharging may be achieved either by a centrifugal blower (similar to that in a vacuum cleaner) driven by a small turbine in the exhaust system of the engine or alternatively the compressor may be a positive displacement (Roots blower), driven by the engine itself. It will be appreciated that, with all this development, detailed design of car engines in the next few years is likely to show considerable variation on that which had become the norm until the 1960s.

Carburation. The mixture of the fuel and air needed for the engine is traditionally provided by means of a carburettor supplied by a fuel pump connected to the fuel tank. Air is drawn through the carburettor by the suction of the piston and carries fuel with it partly as vapour, partly as liquid droplets into the inlet manifold, through the inlet valves into the cylinders. By the early 1960s in order to achieve improved performance from a given size of engine, "fuel injection" had been developed in which individual supplies of fuel were injected close to the inlet valves of individual cylinders. This type of system offers advantages in improved mixture distribution and is used to achieve improved performance in some modern engines but is significantly more expensive than a carburettor. Both carburettors and fuel injection systems now offer much more accurate control of mixture strength tailored to provide the required compromise between emissions, economy and performance. Electronic control systems are being applied and of particular interest are systems with a closed-loop control of mixture strength, e.g. the use of an exhaust oxygen detector to control the mixture precisely for the operation 3-way catalysts.

The Combustion Process. When the mixture of fuel and air is compressed in the cylinder and then ignited by the spark, the resulting flame travels rapidly from the spark plug through the compressed gas. The combustion causes a further increase in pressure, and the unburnt mixture beyond the advancing flame front is rapidly compressed, whereby its temperature is raised. It should ideally continue to burn smoothly until all the fuel is burnt, and thus provide a steady thrust to the piston.

Knocking. As already noted under certain conditions, however, the temperature of the unburnt mixture may rise to a point at which it self-ignites, and a very rapid burning — an explosion — of the residual mixture occurs. The effect is like a hammer blow on the cylinder, and the resulting vibrations give rise to a sharp metallic ping known as "detonation" or "knocking". This sequence of events obviously increases the mechanical and thermal stresses on the engine, and is therefore to be avoided.

Knocking depends partly on the design and operating conditions of the engine and partly on the type of fuel. It is promoted by high temperatures and pressures during combustion, and any engine factors resulting in these conditions tend to

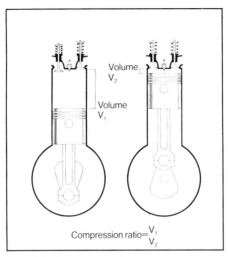

Figure 6.9 **Compression ratio**

promote knock. Relevant factors include spark timing, throttle opening, intake air and coolant temperature and compression ratio. Compression ratio is the ratio of the volume of air and fuel when the piston is at the lower end of the cylinder to the volume when the mixture is compressed and the piston is in its topmost position (Fig. 6.9). It affects the combustion performance of the fuel more than any other single factor, and increasing the compression ratio increases the tendency to knock. The higher the compression ratio the higher is the thermal efficiency of the engine, i.e. the ratio of power obtained from the engine to the maximum power available in the fuel, but the limiting compression ratio that can be utilised is set by the knock characteristics of the fuel. Increasing the octane number of the fuel permits the use of a higher compression ratio, and gives more power, higher efficiency and lower fuel consumption. However, no advantage is gained by increasing the octane number beyond that necessary to give knock-free performance.

Deposits. During the running of an engine, deposits derived from the fuel, the lubricant and atmospheric dust, form within the combustion chamber, and may have an important effect on the combustion process and the efficiency of the engine. An appreciable proportion of the deposits is derived from the additives that are incorporated in engine fuels and lubricants to improve their performance in various ways.

The deposits decrease the volume of the combustion space, thus in effect increasing the compression ratio; they prevent the loss of heat by conduction, and increase the thermal capacity, thus causing higher cylinder temperatures; they provide local hot spots that encourage abnormal ignition. The first two of these effects lead to what is known as "octane requirement increase" (ORI), i.e. the engine demands a fuel of higher octane number to give the same performance as previously. ORI is common to all types of spark-ignition engine and reaches an equilibrium level after about five thousand miles of operation, the level varying between two and five octane numbers for fuels with octane numbers of 90 or more. With lower octane fuels the variation may be much greater.

There are several types of abnormal ignition. Combustion should be initiated only by the spark from the plug; sometimes, however, a hot component in the cylinder—a hot plug electrode or a glowing deposit—may fire the mixture, a phenomenon known as surface ignition; pre-ignition if it occurs before the normal plug spark, post-ignition if it occurs afterwards. Depending on when and how surface ignition is initiated, it can give rise to various noises described as "wild ping" and "rumble". Pre-ignition is very dangerous and if sustained will very rapidly lead to engine failure.

Combustion chamber deposits may also lead to spark fouling and exhaust valve failure. Spark plug fouling is largely due to deposits rich in lead derived from the anti-knock additives in the fuel and is particularly likely to occur under low-temperature operating conditions. The deposits can cause misfiring by reducing the electrical resistance of the insulator and can also corrode the electrodes and reduce the life of the plug.

Exhaust valve failures can result from deposits on the valve stem or seat, which interfere with correct operation and cause valve burning. On the other hand, lead deposits in fact act as valve seat lubricants, and if lead-free gasoline is used, special valve seat materials must be used to prevent erosion.

Cylinder Wear and Corrosion. Mechanical wear of the cylinder by the reciprocating motion of the piston with its rings in contact with the cylinder wall is largely prevented by a film of lubricating oil. When the engine is running there is usually a sufficient supply of oil, and wear is negligible, but when starting there is a short period before circulation of the oil is established, during which some abrasive wear takes place.

Cylinder wear can also be caused by corrosion. All petroleum fuels contain low concentrations of sulphur compounds, and leaded motor gasolines contain "scavengers", both of which form acidic gases on combustion. Other acids are formed by fixation of atmospheric nitrogen during combustion. Water is also formed when petroleum fuels are burnt. When the engine is hot, the water passes out of the exhaust pipe as steam, and the corrosive gases go with it. When the engine is cool, liquid water condenses on the cylinder walls, and the corrosive gases dissolve in it to form acids which can attack the surface of the cylinder, particularly when starting from cold, before the lubricant has had an opportunity to establish a protective film. Corrosion of this type is thus more evident when the vehicle is used for a number of short journeys with much stopping and starting, and the engine does not reach its normal working temperature. However, it is less evident now than formerly, owing to the more effective thermostatic control of engine temperatures and the use of more resistant materials.

Mechanical Wear. Although wear outside the engine cylinder has given problems in the past, particularly with cams and tappets, the combination of modern lubricants and appropriate metallurgy has largely eliminated this problem.

Fuels for Spark-ignition Engines

The spark-ignition engine requires a fuel that will readily evaporate in the air stream drawn into the carburettor, be readily ignited in the cylinder by a spark, and burn smoothly without knocking or other operational difficulties. Petroleum gases, town gas, gasoline, kerosine, benzole, and lower alcohols can and have all been used, but gasoline is the preferred fuel in almost universal use. The performance of the engine depends very much on the quality of the gasoline, the most important properties in this respect being knock characteristics (or anti-knock value), volatility and stability.

Volatility. The volatility of a gasoline affects engine performance in several ways. If it is too low, insufficient vapour may be drawn into the cylinder to allow easy starting from cold, and warm-up will be slow. On the other hand, too high a volatility is apt to cause carburettor icing or vapour lock. A balance must be struck between these extremes. Volatility is assessed by a laboratory distillation test and by a Reid vapour pressure determination (see Glossary).

Cold Starting. A mixture of air and gasoline vapour is inflammable under engine conditions if the air/fuel ratio by weight lies between $6:1$ and $30:1$. A mixture containing just sufficient air to burn all the fuel present is referred to as "stoichiometric". However, to operate lean of stoichiometric $(14:1)$ great care has to be taken in the design of the mixture preparation and combustion system. Unless sufficient fuel vaporises to give at least one part of vapour to twenty parts of air the engine will not fire. Easy starting at low temperatures depends on the more volatile fractions of the gasoline; the higher the quantity of gasoline that evaporates by 70°C in the standard distillation test (E70) the easier will the engine start. The E70 value can vary from a minimum of 10 in summer up to a maximum of 45 in winter.

To aid cold starting, an excess of fuel is required to compensate for the smaller proportion that will evaporate at low temperatures, and thus produce an inflammable mixture in the cylinder. This excess may be achieved by use of a "choke" in the carburettor. If such an excess of fuel, or too volatile a fuel, were delivered when the engine was warm, evaporation would be so great that too much fuel vapour would be produced, the fuel/air mixture might be too rich to fire, and hot starting would be difficult.

Mixture Distribution, Driveability and Warm-up. Under cold conditions only about 10% of the gasoline may vaporise, as it is sprayed into the airstream to the engine cylinders. The fine droplets formed will follow the airstream with the vapour but the bulk of the fuel is deposited on the manifold walls and flows to the cylinders as a liquid. Thus a hot-spot (exhaust- or water-heated) is essential to promote sufficient evaporation of the fuel so that the engine will operate properly without excessive fuel enrichment. During the warm-up period the engine may exhibit such malfunctions as hesitation, stumble or even stalling if the enrichment is inadequate or the fuel too involatile. The fuel property that best describes fuel performance under these conditions is the percentage evaporated at up to 100°C in the standard distillation test, the so-called E100 value, which is controlled in

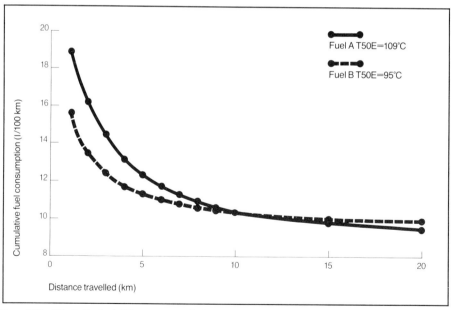

Figure 6.10 **Effect of fuel volatility on warm-up fuel economy for manual choke Morris Marina 1800**

the range 42–70% according to the season of the year. The gasoline's final boiling point is also limited.

The influence of fuel volatility on fuel economy is quite significant when short journeys correspond to the major use of the car. Figure 6.10 shows the effect of fuel volatility on warm-up fuel economy for a car with a manually operated choke. The effects are less marked with an automatic choke, although not if it is adapted to suit the particular fuel used. For short journeys less fuel is used with a more rather than a less volatile fuel. On the other hand, once the engine has warmed up, the less volatile fuel, which will probably have the higher specific gravity, will give better fuel economy.

Carburettor Icing. For efficient engine operation a large proportion of the fuel should be vaporised before it enters the cylinder. Heat is required to vaporise a liquid, and the gasoline in the carburettor takes this heat from the surrounding air and metal, thereby cooling them. If the air entering the carburettor is already cold and damp owing to the prevailing weather, the extra cooling due to gasoline evaporation may reduce the temperature sufficiently for ice to form. This ice may so interfere with the normal flow of fuel and air through the carburettor that fuel consumption can increase, or the engine stall. The more volatile the gasoline, the greater is the tendency to ice formation. The icing tendency of a gasoline correlates approximately to its E100 value, so there is a conflict between good non-icing performance, which requires an involatile fuel, and good warm-up performance, which requires high volatility. With most modern cars icing is not a serious problem with normal fuels; in many areas, however, anti-icing additives are incorporated into the gasoline in winter. These act either by depressing the freezing point of water or by lessening adhesion of the ice crystals to the internal surfaces of the carburettor.

Atmospheric conditions most conducive to icing are a temperature of about 4.5°C (40°F) and a relative humidity about 90%, but it can occur over the range -5.5 to $+10°C$ (20–50°F) at humidities over 60%.

Most modern engines now have heated air intakes to the carburettor. Control of air intake temperature permits more accurate control of mixture strength and allows more rapid warm-up of the engine. It has the incidental benefit of virtually eliminating the carburettor icing-problem.

Hot Fuel Handling. The maximum permissible volatility of a gasoline is governed by its liability to cause hot-fuel handling problems. The fuel pump in the average car can supply much more fuel than the engine requires. If the fuel becomes hot, however, and evolves a lot of vapour, the fuel pump supplies a mixture of liquid and vapour, and the total fuel supply may be insufficient for normal running and the pump is said to be vapour-locked. The most critical condition for vapour lock is when the car is accelerated following a hot soak period prior to which the car was driven fast in hot weather. The driver may experience a lack of response or roughness during acceleration and in severe cases a complete stall. A related problem again experienced after a long hot run is hot starting. Alternatively, a very rapid evolution of vapour may occur in the fuel pipe between pump and carburettor—usually when the car is standing hot after a long run—and force the liquid gasoline from the float chamber into the inlet manifold. Too rich a mixture, above the upper limit of inflammability, may then enter the cylinder, ignition fails to occur, and the engine will not restart.

Vapour lock tendency is greatly influenced by the design of the fuel system, but for a given car it depends on fuel volatility, atmospheric temperature and pressure. The effect of altitude can be marked since atmospheric pressure decreases with altitude, and the evolution of vapour from a liquid increases as pressure decreases. Increase in altitude thus causes the fuel to give off more vapour, and thereby encourages vapour lock.

The risk of vapour lock is reduced to a minimum by control of fuel volatility to suit the season and the locality. For many years the Reid vapour pressure (RVP) was used as the criterion, but this does not altogether correlate with practice. The best control criterion is the temperature required to evolve a given amount of gasoline vapour, usually expressed as a ratio to liquid volume. This however, is difficult to measure, and various other controls are normally used, based on a combination of RVP and distillation characteristics.

Mixture Distribution. If the gasoline is completely vaporised before entering the cylinder, the air and fuel are more or less evenly distributed among the cylinders. If a substantial proportion of the fuel remains as liquid droplets, uniform distribution is not so readily achieved and some cylinders undoubtedly receive a richer mixture than others, and general efficiency suffers; hence the advantage of fuel injection.

A limitation on the final boiling point of the gasoline helps to overcome this problem since it is the higher boiling parts of the fuel that are most difficult to vaporise. Heating the manifold also assists in the evaporation of the droplets and is achieved by use of heat from the exhaust or engine coolant.

Knock Rating. The tendency of a gasoline to knock, its "knock rating", or anti-knock value is best expressed as an "octane number", defined as the percentage volume of isooctane in a blend with normal heptane that matches the gasoline in knock characteristics in a standard engine run under standard conditions. Since in practice gasolines vary in their measured octane number with engine operating conditions and indeed with type of engine, it is desirable to measure the octane number under different conditions. Specifically, one set of conditions produces the research octane number (RON), and another more severe set of conditions the motor octane number MON. Additionally, it is important to measure the RON of that proportion of the fuel boiling below 100°C. The octane quality of the lower boiling fractions (up to 100°C) of the gasoline is of particular importance under accelerating conditions when fuel fractionation may occur in the inlet manifold if the lower boiling fractions evaporate, leaving heavier components as liquid. If the octane quality of the lower boiling fraction is lower than that of the gasoline as a whole, such fractionation may lead to knock.

The octane requirements and fuel preferences of car models can be characterised by the use of fuel standards in which the three laboratory octane numbers are varied independently. If, in addition, the distribution of car models in a market is known, then it is possible to calculate the proportion of that market that is satisfied by a fuel of given octane quality. The aim of the refiner is not normally to satisfy every car in the fleet but a specified high percentage. Although the octane number of a gasoline may appear on the pump from which it is distributed, usually the research octane number, this is not itself an adequate criterion of gasoline performance in practical engines.

The addition of certain metallic compounds to gasolines materially increases their anti-knock value. Many such compounds have been investigated, but tetraethyllead (TEL) and tetramethyllead (TML) are the only anti-knock additives in widespread use. Present-day lead contents are typically in the range 0.15–0.6 g Pb/l and result in increases in research octane numbers of about 2–6 units. In recent years there has been a considerable move to reduce or eliminate lead from gasoline. In the USA and in Japan this was essential for cars fitted with the catalytic type of emission controls. More recently there has been pressure for removal of lead on the grounds that it has had an adverse effect on population exposed to high lead concentrations. The precise effect of lead on people and the extent to which lead from gasoline engines contributes to lead found in people's bodies is a matter of debate. Nevertheless, legislation calling for a reduction to about 0.15 g Pb/l is being introduced in a number of countries, and others plan to use unleaded fuel. Inevitably, reducing lead content will increase crude oil consumption. Either the octane requirement of an engine must be reduced by reducing its compression ratio with attendant loss in efficiency, or more crude must be used to make the gasoline of equal octane number. Figure 6.11 illustrates the crude needed to produce a given amount of gasoline at a particular octane number at 3 different levels of lead, assuming that the engine and gasoline are well matched. Thus, as in the case of hydrocarbon and nitrogen oxide emissions, there is a conflict between cleaning-up the exhaust of an engine and achieving the most efficient use of crude oil.

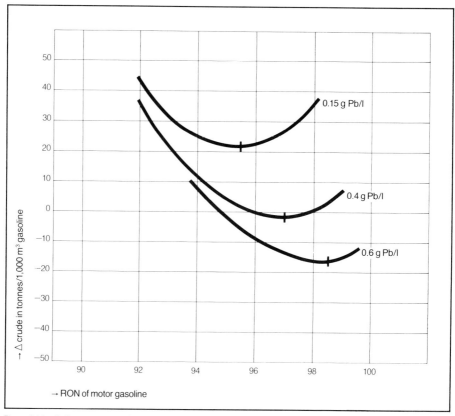

Figure 6.11 **Optimum octane of the integrated refinery/car system** (Car efficiency parameter at 1.6% volume per unit octane increase). Basis 96 RON mogas-pool at 0.5 g Pb/l; complex refinery

Other Gasoline Additives. In addition to anti-knock additives, other additives are used in gasolines. Many gasolines contain components which oxidise easily giving rise to gum formation and deposits in the engine unless an anti-oxidant is used.

Carburettor icing has already been referred to and can be ameliorated by the addition of suitable additives, such as dipropylene glycol or isopropyl alcohol.

A new car has its carburettor designed and adjusted to provide the optimum air/fuel ratio intended by the car manufacturer. However, over a period of time deposits build up in the carburettor, and the air/fuel ratio departs from the design condition. These deposits indeed are found throughout the whole of the inlet system but can be considerably reduced by suitable detergent additives, such as polyisobutene amine with a mineral oil carrier. This effectively preserves new car performance for much longer with attendant improvements in economy and exhaust emissions.

Alternative Fuels and Components. Alcohols, particularly ethanol, have a long history of use as a gasoline supplement. Interest in alternatives or extenders for hydrocarbons as fuels for internal combustion engines generally is now growing for a variety of reasons. These include the escalating cost of hydrocarbons causing balance of payment problems, fears of disruptions in supply caused by crisis in producing areas, strategic or political considerations and of course, legislation and other pressures to reduce lead content.

The main classes of components under consideration are alcohols, especially methanol, ethanol and tertiary-butyl alcohol (TBA) and ethers, especially methyl tertiary-butyl ether (MTBE). Methanol and ethanol can be used as neat fuels as well as blending components. All of these cost much more per octane number than lead components, but reduce undesirable exhaust emissions. MTBE is perhaps technically the best component, although the most expensive, as it has high octane quality, good volatility and is free from side effects. All the alcohols, whilst having high blending octane numbers and being cheaper than MTBE suffer from other drawbacks. Principal of these are increased vapour pressure leading to vapour lock problems, poorer driveability performance, loss of fuel economy and poor water tolerance. The last is most serious for methanol, although all alcohol–gasoline blends tend to absorb water, and at low temperatures the alcohol and water can separate out from the gasoline in storage tanks to form a separate phase. At low alcohol concentrations in gasoline, however, this is not a serious problem.

Ethanol has been used as a gasoline component for a number of years in various countries. Recently these have included the USA where a 10% ethanol blend is marketed as "Gasohol" and Brazil where a 20% blend is sold. Also in Brazil neat ethanol is being sold as an automotive fuel, and specially manufactured vehicles are available to use it. The high price of ethanol renders it uncompetitive with gasoline on the open market, but within a given economy there are foreign exchange advantages. These may be negated if an existing refining industry produces surplus gasoline in order to meet demand for other fuels such as diesel fuel.

Methanol is produced mainly from natural gas, although it can be produced at greater cost from a variety of other feedstocks, notably coal. Its use at low concentrations (below 5%) in gasoline is growing, especially in Germany, and various trials have been run at concentrations up to 15%. At these levels (even at 3%) use of a higher alcohol cosolvent is necessary, for example TBA, to improve the blends' water tolerance and prevent phase separation as well as reduce volatility. Methanol can also be used as a neat fuel, and several test fleets are operating on a trial basis, notably in the USA and Germany, but neither the fuels nor the special vehicles are commercially available at present.

Although cheaper than ethanol, methanol is still more expensive on an energy basis than conventional gasoline, although this situation may well change in the future and there could be opportunities to use on a limited scale cheap surplus methanol from chemical operations.

Kerosine. A special grade of kerosine called tractor vaporising oil (TVO) used to be supplied for spark-ignition-engined agricultural tractors. This had a much lower octane quality than gasoline and hence the engines needed low compression ratios. Moreover, because of its low volatility the engines had to be started and warmed up on gasoline and then switched to TVO, which was vaporised in an exhaust-heated vaporiser. Nowadays most tractors are diesel-engined and hardly any TVO is produced.

Liquefied Petroleum Gas (LPG). In the USA, Japan and some parts of Europe, LPG is being used as an automotive fuel. It has a high octane number and gives clean, complete combustion with even mixture distribution. The use made of LPG in road transport varies considerably from country to country; for instance, in Europe it varies from minimal in West Germany and UK to over 10 per cent of vehicle miles covered in both Italy and the Netherlands. The differences result from local availability of refinery LPG and alternative approaches to fiscal structure for automotive fuels which, in some cases, favour LPG. Rapidly increasing supplies of LPG extracted from natural gas liquid have served to stimulate further interest in these materials as automotive fuels. This is because increased application of LPG, together with the options to use locally produced alcohols, provide a means of helping to reduce the dependence of transportation on crude oil. From a technical viewpoint LPG represents an ideal fuel for the Otto cycle engine, but, being very volatile, it is less than ideal from a fuel handling viewpoint. It has a high octane number and gives clean, complete combustion with even mixture distribution in more efficient engines tuned to leaner air/fuel ratios. In most instances, simplicity of conversion of the gasoline engine to dual-fuel gasoline/LPG makes this a preferred route at present. However, recent developments have shown that high compression ratio dedicated LPG engines offer improved performance and very high efficiency.

Aviation Gasoline and Aviation Turbine Fuel

The development of gas turbines for aircraft propulsion has reduced the use of aviation piston engines to light aircraft and some helicopter applications. The conditions under which such aircraft operate are in some respects less severe than those experienced by the earlier generation of civil and military aircraft for which aviation gasolines were developed. Furthermore, there has been virtually no piston engine development that has created any significant fuel quality requirement greater than that which existed in the generation of engines originally developed for military purposes in the 1940s, subsequently adapted for civil use. There has been a significant reduction in the number of types of engine in use, and it has therefore been possible to rationalise the numbers of fuel grades supplied. Consequently, some of the grades of fuel originally available have been discontinued.

Aviation Gas Turbine Engines

Since the first flight of an aircraft powered by a gas turbine ("jet") engine during the 1940s, development of this type of engine has been rapid and still continues. However, unlike the spark-ignition piston engine, the increased power and improved fuel economy of the gas turbine engine does not depend upon a progressive improvement in fuel quality. Nevertheless, the fuel specification originally drawn up in the 1940s has been changed over the intervening years to reflect the changing needs of the various types of engine (e.g. the introduction of tests to determine the resistance of the fuel to thermal degradation, and the corrosion of metallic silver parts found in some engine fuel supply systems), and of fuel handling systems (e.g. the introduction of tests to determine more precisely the fuel/water separation characteristics).

Anti-knock quality is of no significance in a gas turbine engine, where combustion is continuous, rather than intermittent as in a piston engine. Consequently, there is no opportunity for vaporised fuel and air to form in a pocket ahead of the flame and cause self-detonation. In the gas turbine, fuel is sprayed at high pressure through atomisers into a continuously flowing stream of compressed air, which swirls and circulates within the combustion chamber in such a way as to create a stable ball of flame. Figure 6.12 illustrates a modern gas turbine aircraft engine. The primary air entering the front end of the combustion chamber is only a fraction of the total air flow, but is sufficient to provide a mixture of about 15 parts air to 1 part fuel in the flame zone of the combustion chamber. The remainder of the air is used to cool the thin heat-resistant metal of the combustion chamber, and is injected through secondary air holes into the very hot stream of combustion gases, in order to cool these gases to a temperature that is acceptable to the materials of the nozzle guide vanes and turbine blades.

The forward thrust of the jet is entirely dependent upon the energy released when the fuel burns in the compressed air stream. In older engine designs, the energy released was used entirely to accelerate the combustion gas through the jet pipe, thus producing thrust. Modern engines utilise an additional turbine stage to drive a large fan, which accelerates considerable quantities of air around the engine, thus "by-passing" the combustion chamber. Because the heat is released at a constant pressure, there is an instantaneous increase in the volume of the air

Figure 6.12 Olympus aircraft gas turbine (Rolls-Royce).

stream, which causes the stream to accelerate through the nozzle and turbine blades into the jet pipe. The amount of engine thrust is controlled by varying the flow of fuel to the combustion chamber. As the amount of fuel is reduced, the speed of rotation of the turbine and compressor fall, thereby reducing the airflow through the engine. Conversely, when the fuel flow is increased, the turbine and compressor speed up, airflow is increased, and more thrust is achieved.

Because little of the oxygen in the secondary air is burnt in the combustion chamber, there is a substantial proportion of oxygen in the jet gases. In some military aircraft and Concorde more fuel is injected into the jet stream and burnt, giving the gases further acceleration and enhanced thrust. This is known as "afterburning" or "reheat". It is a rather inefficient way of using fuel and is used only for short bursts of power at take-off, in combat, or to accelerate to supersonic speeds in Concorde.

Aviation Turbine Fuels

As in other forms of internal combustion engines, the thermal efficiency depends upon the compression ratio (i.e. the ratio of the pressure in the combustion chamber to the ambient air pressure). However, in gas turbines a further major restraint on efficiency is the ability of the turbine to withstand high temperatures. The full potential of the high temperatures generated in combustion cannot be realised because of the necessity to cool the combustion gases with secondary air to temperatures that the nozzle, turbine blades and disc materials can withstand. Improved thermal efficiency is sought through increased compression ratio, associated with higher turbine temperatures. Changes in compression ratio can influence the rate of carbon formation, particularly with higher boiling point aromatic compounds. If such carbon is present in the combustion system, it can lead to high thermal radiation, which can damage the thin combustion chamber. Also, deposition of carbon on the fuel nozzles may cause distortions of the atomiser spray pattern. Any disturbance of spray pattern is likely to reduce efficiency of combustion (which is generally maintained at a high level of 98% or more) and seriously affect temperature distribution in the stream of combustion gases, thus causing damage to the nozzles and turbine. Although careful combustion chamber design is the principal means of avoiding these problems, certain properties of the fuel are also controlled. Aromatics are the main source of the carbon particles and therefore have been controlled by specification.

Excessive sulphur in fuel can have an adverse effect on carbon deposition under some conditions, and it is therefore important to control sulphur content.

As the jet engine has been developed, the interval between maintenance has been greatly increased, with obvious economic advantage to airline operators. It is therefore important that deposition tendencies of fuels should be controlled at extremely low levels, in order that these intervals can be retained and, if possible, improved upon.

Combustion Characteristics. Potentially, as was remarked by Sir Frank Whittle when he first developed his jet engine, gas turbines are very catholic in their taste for fuels. Nevertheless, combustion system development proved to be a major obstacle in the early days, and Sir Frank was later to make a generous acknowledgement to Mr Lubbock and his colleagues in Shell for the assistance he received in making this important part of his engine work well. Under ideal conditions almost any petroleum distillate fuel can be made to burn efficiently in a gas turbine. However, conditions in aviation service are far from ideal and there are certain important considerations that must be fully satisfied. It is, for example, essential that under all the conditions that may be experienced, either on the ground or in the air, an instant engine start can be obtained. In this respect a fairly volatile distillate is preferable. Another important combustion consideration is that there should not be a loss of flame ("flame out") during flight manoeuvres.

One of the characteristics of a gas turbine engine is that engine speed changes do not occur as rapidly in response to throttle movements as in a piston engine. Consequently, under conditions of acceleration or deceleration there is a transitory richening or weakening of the fuel/air mixture strength in the flame zone. All fuels have rich and weak mixture limits beyond which combustion cannot be sustained and it is important that at no time should the condition in the combustion chamber of an aviation gas turbine engine exceed these limits for the particular fuel used. These various combustion considerations (combustion efficiency, deposit formation, flame stability and ignition) are primarily a function of the adequacy of fuel atomisation and of fuel volatility. The degree of atomisation depends upon fuel viscosity as well as fuel pressure. It is therefore important in specifying an aviation turbine fuel (ATF) that controls be included both for volatility (by a distillation test) and viscosity. Because in aviation practice fuels are exposed to extremely low temperatures, it is necessary for the viscosity at low temperature to be controlled.

Aircraft Fuel System Performance Characteristics. The fuel properties that are required in order to ensure consistent performance of aircraft fuel systems are as important as those needed for good combustion. Operation at very low temperatures requires that the fuel should have a low freezing point, i.e. the temperature at which wax crystals separate from the liquid phase. Were the freezing point to be above the temperatures to which fuel is exposed in aircraft fuel systems, clogging of filters accompanied by interruption of fuel flow and loss of engine power would result. Jet aircraft cruise at high altitudes in order to achieve fuel economy and as a consequence exceptionally low temperatures can be experienced and fuel freezing point has to be controlled. This requirement becomes increasingly difficult to meet as the choice of fuel is moved from the gasoline range through kerosine into middle distillates.

Wax is not the only danger associated with low fuel system temperatures; any free water that may appear in the system could also form ice crystals and clog filters. Therefore it is essential that in distribution and in the airfield supply systems, aviation turbine fuels must be kept free of any water that may be taken up into suspension. To ensure that all necessary precautions taken have been effective it is common practice to draw samples of fuel during the refuelling operation and subject them to a test to determine the approximate free water content (e.g. the Shell Water Detector Test, where a sample of the fuel is drawn through a chemically-treated filter pad, which changes colour from yellow to green if any free water is present in the fuel).

The high-pressure fuel pumps are sensitive to the viscosity of the fuel. If fuel is used which has inadequate viscosity characteristics to provide a sufficient lubrication for the heavily loaded sliding pump surfaces, very rapid wear occurs. On the other hand, high viscosity at low temperature could cause pump performance and fuel flow problems, and this is another reason for controlling the maximum viscosity of kerosine type fuels.

The introduction of supersonic aircraft has put greater emphasis on the thermal stability of ATF. At supersonic speeds, even when flying in the very rarified upper atmosphere and in extremely cold temperatures, much frictional

APPLICATION, SPECIFICATION AND TESTING   405

heat is generated. In order to avoid high metal skin and structure temperatures, the fuel has to be used as a "heat sink" or cooling medium. Fulfilment of this role requires a complex heat transfer system. It is of paramount importance that in performing this duty the fuel is not degraded to such an extent that it precipitates gum or coke-like deposits, which would reduce the rate of heat transfer from metal to liquid and also be liable to clog filters and fuel atomisers. Standard potential gum tests were found to be inadequate for this purpose and so a special test for thermal stability was introduced. The use of anti-oxidant and metal-deactivator additives in certain fuels can be beneficial with respect to thermal stability.

Fuels of aviation turbine engine systems must be free from acidity and from potentially corrosive sulphur compounds to prevent damage to sensitive metal parts.

Airline operation requires a rapid turn-around between flights. This puts a premium on the rate of refuelling of aircraft which carry a heavy fuel load. Although the generation of static electricity in the pumping of the fuels has long been recognised and precautions are taken to bond all components to avoid a build-up of static charges, it is nevertheless advantageous at high pumping rates to control the conductivity of the fuel itself. Such conductivity control can be achieved by the addition of an anti-static additive, e.g. Shell ASA-3 or DuPont Stadis 450.

To obtain satisfactory control of the engine, it is necessary that fuel density should be kept within certain limits. In the operation of long range aircraft, weight of fuel may be a critical factor limiting the pay load that can be carried, and slight variations in density can have a significant effect. The energy content of a fuel, which is purchased on a volume basis, is a function of its density, which is of prime importance to the user.

The most important reason for controlling the volatility of ATF is the loss of fuel, caused by evaporation from unpressurised fuel tanks at high altitude. Low temperatures at high altitude help to reduce evaporation, but when fuels are used as a heat sink, the problem is more serious. A control on the front end volatility ensures satisfactory cold starting whilst minimising evaporation losses.

Additives in ATF. Mention has already been made of the use of anti-oxidants, metal deactivators and anti-static additives. An anti-icing additive is sometimes also added if the fuel system is not equipped with heaters. Its purpose is to ensure that dissolved water in the fuel, which tends to come out of solution as fuel temperature falls, does not form ice in the fuel system.

Grades of Aviation Turbine Fuel and their Specifications. The considerations of combustion and fuel system performance discussed above suggested to the pioneers of aviation gas turbine engines that a "kerosine" rather than a "gasoline" type fuel would be most advantageous. An additional potential advantage was that in the event of a crash there would be less likelihood of fire, due to the higher flash point of kerosine fuel. This type of fuel (commonly known as AVTUR or Jet A-1) is currently used by most of the worlds' civil airlines. However, in the interests of greater availability, a wide-cut type of fuel (referred to as AVTAG or JP-4) was developed for military use, and a civil version of this fuel is now included by IATA (the International Aviation Transport Authority) as Jet B in its guidance to member airlines.

Aviation Piston Engines

The principles of operation of aviation and automotive spark-ignition piston engines are the same. The need to minimise weight in aircraft has, however, forced designers to produce aircraft engines which can obtain the maximum power for the minimum weight of engine without sacrificing fuel consumption. Because maximum power is needed only for a relatively short time, namely at take-off, the desired result can be achieved by "supercharging" or "boosting" a comparatively small capacity engine. This is done either with a mechanically driven supercharger or with a "turbo-blower", which utilises some of the energy that is present in the engine exhaust to drive a small turbo-compressor. The effect of this device is to compress the charge that is supplied to the cylinders, where it is able to produce much more power than would be possible in a "naturally aspirated" engine of the same size. The degree of supercharging is referred to as "boost" pressure.

Aviation piston engines may be water-cooled or air-cooled, with in-line or radial cylinder configurations. Because of their simplicity and lightness, in-line air-cooled engines are usually favoured for light aircraft. In the past, powerful supercharged engines were used in larger passenger aircraft in which a high boost pressure has a similar effect to a high engine compression ratio on gasoline octane rating requirement. Even though it has been possible to produce fuels which are superior in their anti-knock quality to isooctane (in that they have a performance number in excess of 100), the boost pressures used were sufficiently high for it to be usually necessary for aviation piston engines to have compression ratios lower than are used in automotive gasoline engines, e.g. 7:1 instead of 9:1. However, few of these older aircraft remain.

Aviation Gasoline

Octane Rating. Aircraft cruise using a weak fuel/air mixture, but when maximum power is required at take off, a richer mixture is necessary. Therefore, aviation gasolines are required to meet certain octane number specifications, which reflect these different modes of operation, and, in a similar way to the determination of octane number of motor gasolines, CFR engines (see Glossary) are used to test aviation gasolines.

The two engine test procedures used in this connection are (a) the motor method (using the same engine as for motor gasoline octane ratings, but with the results reported as Lean Mixture Rating Aviation Method, in terms of octane number (ON) for gasoline below 100 ON and performance number (PN) for gasoline above 100 ON), and (b) the supercharge method in which the engine is run with a constant compression ratio and "knock" is induced by increasing boost pressure. The mixture strength is set to produce maximum power for the sample being rated and a comparison is made with isooctane/n-heptane reference fuels. When top quality fuels that are superior to isooctane are being rated, the comparison is with reference fuels prepared from isooctane plus increasing quantities of TEL (tetra-ethyl-lead).

The motor method gives a lower rating than the supercharge method, and aviation gasolines with supercharge octane numbers of 85 and over are usually characterised by both performance numbers (e.g. 115/145 grade). The grade of aviation gasoline known as 100LL, with ratings of 100/130 but with a low lead content, is becoming the most widely used aviation gasoline, being suitable for all aircraft engines previously rated for 100/130 (high lead) gasoline, as well as those rated for 80/87 grade. The only other grade currently available is 115/145, which has a high lead content and is used in certain military aircraft, albeit in ever decreasing quantities.

Most aviation gasoline contains TEL anti-knock additive, and a great deal of octane enhancement is gained because of the isoparaffinic constitution of aviation gasoline.

The use of leaded fuels has influenced engine design in several respects. Exhaust valves that operate at high temperatures are made from metals that are resistant to attack from lead salts that may deposit upon their surfaces. Spark plug performance is also at risk from lead salts under the severe temperature conditions of high-powered engines; plug failure at take-off, when power needs are greatest, could be disastrous. Hence a special additive (ethylene dibromide) is used to scavenge the lead products of combustion.

Volatility. Unlike motor gasolines, which have different volatility specifications according to geographical location and from summer to winter, aviation gasolines require a universally applicable specification because with a single fuel fill an aircraft may be exposed to a wide variety of climatic conditions as well as the variation of temperatures and pressures that exist over a range of altitudes. The occurrence of vapour lock or carburettor icing in a car engine may be no more than an inconvenience, whereas in an aircraft engine it could be disastrous. Vapour pressure is, therefore, controlled at a level below that which is acceptable in motor gasoline.

Gum. In aviation gasoline the same considerations apply as in motor gasoline with regard to existent and potential gum, but for storage reasons both specifications are set at lower levels for aviation gasoline.

Freezing Point. The priority given to operational safety in aircraft is reflected by the use of a control on aviation gasoline freezing point. In flight, fuel may be exposed at high altitude to exceptionally low temperatures and there must be no possibility of it becoming solid at any point in the fuel system. Even though aviation gasoline has a comparatively low end point, and like all gasolines has a very low freezing point in consequence, minus 60°C is specified as the maximum freezing point that is acceptable.

Water and Solid Content. It is extremely important that fuel delivered into an aircraft's tanks should be free of contamination by solids and water. Water is particularly dangerous since it could form ice on filters at even moderately low temperatures experienced in flight, with resulting engine failure. Aviation gasoline does not normally come into contact with water in the storage and distribution system, but if this should happen by some mischance it is desirable that the water should not become emulsified with the fuel and should on standing easily and quickly separate from it. A water reaction test is consequently included in aviation gasoline specifications. Since water is able to form in the bottom of aircraft tanks in small quantities as a result of the condensation of humid air on cold tank walls, compliance with the water separation specification ensures that any such adventitious water, which has not been drained from the tanks during routine inspections, will not be likely to cause trouble.

Domestic (Illuminating) Kerosine

The introduction of kerosine as an illuminant during the nineteenth century was the first extensive application of any petroleum product except naturally-occurring bitumen. Crude oil was regarded mainly as a source of kerosine, to supplement the animal and vegetable oils then in use. The development of clean and efficient appliances for lighting, cooking and heating, together with the relative safety of handling kerosine in small volume packaging, has ensured the continuance of a large market for this product in the domestic sector. Alongside this demand, the development in the 1940s of the gas turbine engine for aircraft relied upon the availability of an exceptionally clean fuel of high calorific value and thermal stability, for which purpose kerosine was the best choice. The worldwide offtake for aviation use is now slightly greater than that for the domestic market, although of course there are large regional variations. For many years kerosine and similar products were also used in tractors fitted with spark-ignition engines (Tractor Vaporising Oil). This application has virtually disappeared, because of the introduction of diesel-engined tractors burning gas oil.

The major source of kerosine in the refinery is still the atmospheric distillation of crude oil, during which a cut boiling between 140° and 300°C is taken. The boiling range thus overlaps both the gasoline and gas oil volatilities and so there is competition for the hydrocarbons in this range. However, components from thermal cracking, hydrocracking and coking, suitably treated, increasingly need to be incorporated into the kerosine pool in order to meet the market demand. For domestic use, smoke point, sulphur content, flash point, and volatility are the major constraining requirements.

Kerosine Lamps

Kerosine lamps fall into three basic groups; those using a luminous-flame wick burner, those using a blue-flame wick burner with an incandescent mantle, and those using a pressure-jet burner with an incandescent mantle.

Luminous-flame wick burners are the oldest, simplest and least efficient of the three types, and examples exist from the very primitive to the most elaborate. They tend to be smoky and odorous, and to require frequent cleaning of the wick to remove char formed by thermal decomposition of the fuel and the deposition of any contaminants present in the fuel.

Blue-flame wick burners with an incandescent mantle are significantly more efficient at producing light than the luminous-flame type and are generally free from smoke or smell.

Pressure-jet burners are about twice as efficient at producing light as blue-flame wick burners but are somewhat noisy. The fuel is finely atomised by being forced through a very small orifice (jet) under pressure, and is vaporised by the reflected heat of the flame, by contact with the incandescent mantle, and by the pressure drop through the orifice.

Kerosine Cookers

Kerosine cookers, ranging from small single burners to complete cooking ranges, can employ either pressure or vaporising burners. Many have been developed for the preparation of traditional basic foodstuffs in those areas of the world where kerosine has been the major fuel available to supplement or replace scarce charcoal. The smaller kerosine stoves (Primus type) are now largely being replaced by LPG appliances, but in many areas of less developed countries they are still the staple cooking method.

Figure 6.13 **A kerosine burner.**

In cost, but not convenience, kerosine cookers compare favourably with gas appliances, and, similarly, kerosine refrigerators based on the absorption principle are still widely used in tropical areas.

Kerosine Burners for Heating

For small space heating applications, flueless appliances are often used, and these are of the long-drum or short-drum type. For the large applications, particularly where water heating is included, the kerosine appliance may be connected to a flued boiler. These units utilise pot burners, wall-flame or pressure-jet burners.

Long-drum Burners are so called on account of the long drum or chimney mounted over them to induce the air required for combustion. They may produce luminous (white) or non-luminous (blue) flames according to the design of the burner. In the luminous-flame type the kerosine is supplied via a cotton wick, and air, induced by the chimney, is supplied to both sides of the wick via perforated galleries. If a circular wick is used, air is supplied via a perforated gallery to the outside and via a central tube to the inside of the wick. The quantity of air is sufficient to maintain a smokeless white flame and the burning rate is controlled by varying the exposed area of the wick. In the blue-flame burners the kerosine is again supplied via a cotton wick, usually circular, and induced air is supplied in almost the same way as before. However, the design of the gallery and flame spreader is such that the air is supplied at a greater speed than in the luminous-flame type and a non-luminous blue flame results.

Short-drum Burners consist of two concentric perforated steel cylinders or "shells" with a circular wick between the bases of the cylinder. When the flame is first lit, kerosine vaporises close to the wick and heats the shells. As the rate of vaporisation increases the kerosine vapour burns over the whole height of the shells. Air is introduced internally through the central air tube, and hence through the perforations of the inner shell into the combustion zone, and externally between the air shield and the outer shell, and thence through the perforations of the outer shell into the combustion zone, producing a multitude of small blue flames within the combustion zone. The air is induced by the chimney effect of the hot burning gases between the shells and no long chimney is needed. Short-drum burners may have adjustable cotton wicks fed by capillary action from the reservoir or a fixed asbestos or glass fibre kindler set in an annular trough fed by gravity from a constant-level device. The burning rate is controlled in the capillary types by adjusting the wick and in the annular trough types by a fuel regulator and shut-off valve or alternatively by raising or lowering the burner and trough relative to the constant-level device.

Pot Burners, which are very simple yet effective burners, consist of an open pot with perforated sides, in the bottom of which a thin layer of kerosine is maintained by gravity feed from a reservoir. The fuel is ignited by means of a spill or piece of rag soaked in the fuel. The flame, which generally burns above the pot, provides sufficient heat to vaporise the oil. Air is obtained through the holes in the side of the pot and may be provided by a natural draught from a flue or by a forced draught from a low-power fan.

Wall-flame Burners are of the vaporising type. Oil is thrown in coarse droplets on a flame ring from a rotating oil distributor driven by a small electric motor. The fuel is vaporised from the flame ring and ignited by a high-tension electric spark. Combustion air is supplied by a fan driven by the same electric motor, and flame grills are used to ensure efficient combustion.

Pressure-jet Burners rely on fuel being atomised through an orifice and ignition of the resulting droplets. External ignition to the droplets is applied by means of either an electric spark of or tiny luminous pilot flame. Air is supplied at a controlled rate by means of a fan. These burners are basically the same units as those used for burning gas oil, but either for traditional or climatic reasons, kerosine is often the preferred fuel. The burners are always enclosed in a flued boiler unit with heat exchanger, and these are the basic unit for the modern central heating system for single dwellings.

Properties of Domestic Kerosine

Good combustion requires the intimate mixing of fuel and air and kerosine is sufficiently volatile in most ambient temperatures to produce a vapour that can be ignited, at worst with only a small amount of heat addition. As a result, kerosine appliances can often operate independently of any other power source, and this enables them to be a useful supplement when only small heat load requirements need to be met. Although kerosine competes with LPG in many markets, the low cost of a simple kerosine appliance, combined with very small, frequent purchases of kerosine, make kerosine use attractive in the less developed countries.

For appliances that rely on natural vaporisation to operate, safe use depends largely on the fuel being controlled for a combination of smoke point and volatility. The smoke point defines the maximum flame height (in mm) that a fuel will burn in a standard lamp without smoking, and in practice of course this defines the maximum flame volume available for heat release, under natural vaporising conditions. If the fuel is too volatile, vapour will be fed to the flame at a higher rate than can be burnt within the flame volume, and the flame will stretch beyond the smoke point seeking additional air for combustion. In extreme cases this leads to "flaring" with the implications of ignition of surrounding material or even explosion of the fuel container. If the fuel is not volatile enough it will not have sufficient air for complete combustion and may firstly smoke, and secondly lay down carbonaceous deposits on the wick or perforated plate surfaces.

For safety in storage and handling, the kerosine is controlled to a minimum flash point, which defines the minimum temperature at which the fuel will ignite from an external ignition source, again under standard conditions.

To ensure safety when used in confined areas, kerosine supplied for unflued appliances is normally controlled to an extremely low sulphur content. This ensures that the sulphur compound concentration in the flue gases does not reach a toxicologically hazardous level.

To minimise maintenance, the char value is also defined. This ensures that the frequency of cleaning the wick or plate is kept to reasonable intervals. Other properties relating to the cleanliness and corrosivity of the fuel are included in domestic kerosine specifications to check for good housekeeping in the distribution system and to prevent staining or corrosion of common household materials.

Gas Oils and Distillate Diesel Fuels

These products form the largest group derived from the refining of crude oil. The products range from only slightly less volatile than kerosine, to the heaviest components available from distillation of crude oil. Although free from residual components, they frequently contain components from thermal cracking, catalytic cracking, hydrocracking, or coking processes.

The use of these fuels is primarily either for fuelling high or medium speed diesel engines, or for domestic, commercial or small industrial continuous combustion applications. The terminology used does not, however, reflect the true end use accurately, and often is solely an indication of fiscal structure. By definition, gas oils have a distillation requirement which in general is of the order of not more than 65 per cent by volume recovered at 240°C, and not less than 85 per cent by volume recovered at 350°C. Diesel fuels may contain higher boiling components, often from vacuum distillation operations.

The term "gas oil" comes from the earliest use, which was to enrich water-gas (a low calorific value gas obtained by reacting steam with coke) in the manufacture of town gas.

Diesel Engines

Diesel engines, and particularly the high-speed variety which are commonly used for automotive purposes (in trucks, buses, railcars, taxis and more recently passenger cars), are increasingly the principal users of gas oil. In this type of reciprocating piston engine the fuel is injected with precise timing into the cylinder as a very finely atomised mist, in order to ensure rapid mixing with the hot compressed air in the combustion space. The engine has no spark plug, but instead relies upon the self-ignition of the fuel/air mixture. This can only occur if the increase in air temperature caused by compression in the cylinder is sufficiently great. Hence a diesel engine must have a very high compression ratio, usually more than 13:1. This high ratio ensures high efficiency and a good fuel consumption. The fuel is not pre-mixed with air, in contrast to the gasoline engine, and so it is possible to control the power output simply by changing the flow of fuel from the fuel pump and injectors. This enables the diesel engine to produce much better fuel economy than a gasoline engine at light load. On the other hand, the diesel engine is not able to generate as much power as a gasoline engine of comparable size because mixing of fuel and air is never complete, and so not all the air can be used. The maximum power that a diesel engine can produce is limited by the onset of black smoke in the exhaust as the fuel supply is increased. This occurs despite the fact that oxygen is still available in the chamber, because mixing is inadequate.

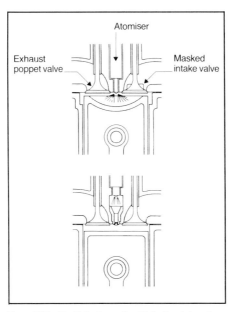

Figure 6.14 **Fuel injectors: direct injection (above) and indirect injection (below)**

Like the spark-ignition engine the diesel engine can be designed to operate on both two-stroke and four-stroke cycles, and both normally-aspirated and supercharged engines are in use. A diesel engine is not so easy to operate at very high speeds as a gasoline engine because of the mixing problem described above, and so even the so-called high-speed automotive diesel engine does not usually exceed 3,000 rpm. This again somewhat restricts the power output for a given size of engine.

There are basically two types of diesel combustion system (Fig. 6.14). The one at the top of the diagram is an open chamber (or a direct injection engine), whilst the one below is of pre-chamber (or indirect injection) configuration. The former are commonly used in slow- and medium-speed engines, whilst pre-chamber designs are most prominent in high-speed engines.

Fuels for Diesel Engines

Although it has a very high compression ratio, a diesel engine does not suffer from "knock" such as occurs in spark-ignition engines (see p. 390). Basically this is because in the diesel engine fuel and air are not pre-mixed and are not ignited from only one location (the spark plug). Consequently, octane number is of no significance in a diesel fuel.

However, a diesel engine does require a fuel with a good "ignition quality" if it is to perform satisfactorily. Firstly, it must be possible to start the engine when it is cranked under the coldest conditions that are experienced in service. Secondly, when ignition occurs combustion should proceed steadily and not too rapidly. If combustion is too fast there is a very rapid rate of pressure rise in the cylinder. This makes the engine noisy ("diesel knock") and puts the engine under severe mechanical stress. The combustion rate depends upon the length of the very brief interval (about half a millisecond) between the arrival of the fuel in the combustion chamber and the occurrence of self-ignition. The shorter this "ignition delay" is, the better. With a longer ignition delay more of the fuel will have arrived in the chamber prior to ignition and the fuel mist will tend to burn explosively. Ignition delay is longest when the engine is idling (hence the tendency to idle noisily) and is shortest at full power. The third requirement for good ignition quality is associated with the maximum amount of fuel that can be injected, before the exhaust begins to produce black smoke. A good quality fuel will permit rather more fuel to be injected before the onset of smoke, and the engine will therefore have a higher "smoke-limited power output".

The volatility of diesel fuel has some effect on cold starting and on smoke-limited power output, but the hydrocarbon composition (paraffins, olefins, naphthenes, aromatics) also has an important influence on ignition quality. Generally, the types of hydrocarbon that have high octane numbers and perform well in gasoline engines have poor diesel ignition quality. A diesel fuel which is predominately paraffinic is superior to one that is aromatic. Diesel fuel ignition quality is expressed as "cetane number", and the test sample is matched with reference fuel mixtures of cetane and alpha-methylnaphthalene in a standard diesel engine test. A good commercial fuel for high-speed diesel engines will have an ignition delay equal to a reference fuel mixture containing about 45 to 50 per cent by volume of cetane (45 to 50 cetane number), whereas inferior fuels may have a cetane number of 40 or less.

If an engine that has been designed to operate on a high cetane number fuel is run on a low cetane number fuel it will be much more difficult to start, it will be noisy (particularly at low loads and when idling) and it will tend to emit excessive black smoke at full throttle. In general, high-speed engines are more sensitive to ignition quality (cetane number) and volatility than medium-speed or low-speed engines. Automotive engines, which run at relatively high speeds, are normally supplied with the most volatile and highest cetane number gas oils (automotive diesel fuel).

In addition to volatility and ignition quality, cold flow, viscosity and sulphur content are important operational parameters.

Wax crystals start to come out from gas oils at a temperature that is dependent upon the boiling range and the paraffinicity of the product. Unfortunately, the best components in terms of ignition quality are those containing a high proportion of n-paraffin molecules, and it is these that come out of solution first as wax crystals. All diesel engines have the injectors protected by a filter, and in the case of high-speed small automotive diesel engines, this filter is of very fine mesh. It is obviously important that the filter is not blocked by wax crystals under the most severe ambient conditions expected, as otherwise fuel flow would cease and the engine would be stopped. The temperature at which wax crystals first appear during cooling is known as the cloud point of the fuel, and this is the historical method of specifying the low temperature operability of diesel fuels. However, fuel still continues to flow significantly below this temperature, even through very fine filters, and if the filters are located in a warm position, the rate of wax re-dissolution, combined with modification of the wax crystals by additives, can maintain the fuel flow down to even lower operating temperatures. In order to reflect this, most areas of the world outside North America use the Cold Filter Plugging Point (CFPP) as the criterion of low temperature operability. This is specified for the fuel at a level that is related to historical meteorological data. The lowest temperature at which fuel will flow is known as the pour point.

Because of the extremely fine tolerances and very high pressures in diesel injectors, fuel viscosity is critical. Too high a viscosity results in poor atomisation and thus incomplete or poor combustion, and too low viscosity leads to a lack of lubricity and thus to injector pump wear, and even injector pin seizure. High viscosities can be tolerated if known and consistent, but large variations cannot be accommodated at a single injector setting.

Sulphur content is controlled for a number of reasons. Firstly, in many developed countries, the overall level of sulphur oxides emissions from all sources is controlled; and, since diesel exhaust is emitted virtually at ground level, this comes under close scrutiny. Secondly, in areas where sulphur oxides emissions may be less closely controlled, it is still necessary to control the sulphur content in order to prevent contamination of the lubricant and to give reasonable oil drain intervals. Thirdly, sulphur oxides in the exhaust gases can lead to severe corrosion of exhaust systems, or even of engine cylinders, if prolonged operation at low operating temperatures is carried out.

Domestic and Industrial Gas Oils

When gas oil is used as a domestic or industrial furnace fuel, the requirements are somewhat different from those for diesel engines. Ignition quality as such no longer has any relevance because the fuel is lit with an igniter and burns continuously, although some of the compositional considerations that lead to good ignition quality also benefit clean combustion in a domestic boiler.

Although domestic vaporising burners running on gas oil were widespread in Western Europe until well into the 1970s, their use since then has rapidly declined in favour of the pressure-jet burner. Low consumption pressure-jet burners do of course themselves impose special quality requirements, because of the very small orifice size of the fuel jet.

The major consideration in the operation of small domestic boilers is that continuous trouble-free operation takes place during the period of most severe ambient conditions. Since the fuel is normally stored for long periods in outside tanks which may be exposed above ground, and the offtake rate is relatively low compared to the tank capacity, low temperature characteristics and fuel storage stability are of prime importance.

Although the filtration in domestic installations is not so severe as that in normal automotive diesel applications, the fuel is likely to have been stored in bulk at low temperatures for significantly longer periods. In the ideal installation design, only the pour point is critical, since provided fuel can be moved from the tank there is sufficient heat available over the indoor pipe run length to ensure that the fuel is filterable before the pump. However, many installations are not ideal, and although the cloud point is a very conservative measure, and CFPP is conservative in most cases, the latter is chosen both to give some margin of safety and to be in line with diesel fuel practice.

As domestic gas oil is free from the ignition quality constraints of automotive gas oil, it is likely to contain a higher proportion of the aromatic and olefinic components from the various cracking operations in the refinery. These components, particularly if not hydrotreated, have lower storage stability than the straight run components from distillation. As domestic fuels may well be stored for up to one year, or even longer in some cases, it is necessary to ensure that the storage stability is adequate in order to prevent the formation of sludge and lacquer processors. Accelerated aging tests are used for this purpose. The storage stability can also be improved by the use of additives.

The fuel property that dominates the performance of a small pressure-jet burner is the viscosity at the operating temperature. The efficiency of combustion and heat transfer is very sensitive to changes in droplet size and spray pattern, and excessive viscosity leads to either failure to ignite, or incomplete (smoky) combustion and heavy deposit formation. The maximum viscosity tolerable by these burners is about 12 centistokes, which is equivalent to a specification limit of 6 centistokes at $20°C$, or 3.8 centistokes at $40°C$, assuming a minimum operating temperature of $0°C$. The recent development of small fuel pre-heaters, which can be located just before the fuel pump, will eventually largely remove the viscosity constraint.

The other cause of deposit formation in small boilers is the thermal degradation of the fuel either just before or during combustion. This again is a property relating to the fuel composition, and the presence of cracked components in the fuel has effects similar to those on storage stability.

Sulphur content does not have any serious performance effects, but, as mentioned earlier, the overall sulphur oxides burden in the atmosphere of many developed countries may well be controlled by limiting the sulphur content of all fuels.

Industrial Gas Turbine Fuel

Derivatives of the aviation gas turbine have distinct advantages in applications where almost instantaneous full power availability is required, and fuel economy is a secondary consideration. Such applications include the power units of naval vessels and the "peak shaving" of base load electric power generation. These units require a distillate fuel of extremely low ash content, although the security of operation required is not so critical as in aviation usage, and so aviation quality kerosine is not justified. However, since the efficiency of gas turbine operation is dependent upon the turbine inlet temperature, there will always be economic pressure to raise this temperature to the maximum compatible with reasonable maintenance intervals. The corrosivity of complex metallic deposits is dramatically increased at their melting point, and these temperatures are often exceeded on the blades of the turbine. Thus, in order to minimise corrosive attack, the

Figure 6.15 **Avon industrial gas turbine (Rolls-Royce).**

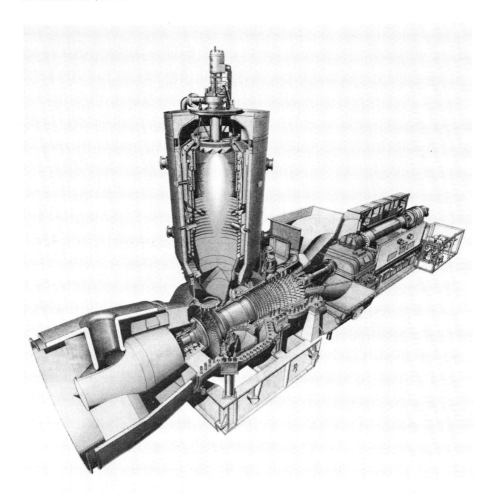

Figure 6.16 **Open cycle industrial gas turbine (Brown Boveri).**

physical quantity of ash-forming constituents must be limited. As already mentioned in the discussion on aviation turbine fuels, the gas turbine is extremely catholic in its tastes, running on most fuels from natural gas to the heaviest distillate, or in some cases on residual fuels. In practice, however, for logistic reasons, gas oil streams of commercial quality are generally used, with the normal fuel being the domestic and industrial grade. Special requirements limiting the total oil-soluble metals content to something of the order of 2 mg/kg, with individual metals (vanadium, lead, calcium, potassium) limited to 0.5 mg/kg or below, are added to the contract specifications for these fuels. Fuel clean-up facilities, usually consisting of water-washing and centrifuging, are virtually always applied to remove adventitious matter and water-soluble impurities.

Residual Fuel Oils

Residual fuel oils are by definition the products remaining from the various refinery processes after all the distillate or lighter fractions have been removed. These residues are complex mixtures of aliphatic and aromatic hydrocarbons in varying proportions, depending upon the source of crude oils processed and the complexity of the refinery. Many of the residues are highly viscous — some which contain high molecular weight hydrocarbons are solid at ambient temperatures and may be used in the preparation of roadmaking materials and other bitumen products. Some of the heavy residues are used directly as refinery fuels to provide process heat and electricity. The balance of these residues is diluted with distillate products to produce the many standard grades of residual fuel oils of varying viscosities available on the market for inland and marine use.

The increasing demand for automotive fuels such as diesel oils, kerosine and gasolines, coupled with the need to conserve energy and utilize natural resources more efficiently, has encouraged greater conversion of other fractions to these products, resulting in a smaller volume of residual fuels which then amount to 25%–30% of the barrel. The result is that residual fuel oils have become heavier, the mean molecular weight is higher together with an increased carbon to hydrogen ratio. In addition, in order to conserve distillate products the various grades of residual fuel oils are now produced close to specification maxima. Thus, whereas in the past residual fuels were often below the upper limits of viscosity, this now occurs infrequently while the density of fuels from many sources is now in excess of 1.000 kg/litre at 15°C. In practical terms this means that greater care and attention has to be given to the storage and handling of these fuels, and that interceptors have to be designed or modified to deal with fuels of varying density.

During the combustion process the carbon particles formed from these fuels take longer to burn, thus larger furnace sizes are needed to permit an increased oil droplet residence time. By limiting the combustion intensity of the furnace of a smoke tube type boiler to about 1.8 MW/m^3, with suitable burner equipment, residual fuels can be completely burned to meet the particular emission levels embodied in national legislation.

The predicted post-1973 decline in demand for residual fuel oils, other than that attributable to the current world economic depression, has not materialised. The main reasons are the success of many energy conservation schemes encouraged by the rising cost of all energy sources, the additional reserves of crude oils found in various parts of the world thus extending the availability of all types of liquid fuels well into the next century, and the high capital and running costs of conversion to coal firing. Additionally, a not inconsiderable factor is the demand for security of supply of energy, resulting in an increase in multi-fuel installations, particularly those using oil and gas. Such installations give freedom to purchase fuel selectively from differing sources, thus improving control of the cost of energy used in a particular factory or process. Figure 6.17 shows the trends from 1966 to 1980 of the relative use of various energy sources, and it could be well into the next century before the petroleum fuel–solid fuel curves cross over.

Some larger fuel users are considering the use of very high viscosity residual fuels to reduce their energy costs. Such fuels, if not diluted by expensive distillates, could cost less than the standard range of residual fuel oils blended with distillates to specific viscosity maxima. However, the use of such very heavy residues would be restricted to larger plants having the facilities to handle such materials.

Environmental legislation can place constraints upon the use of residual fuel oils due to the presence of elements such as sulphur and the inherent metals, vanadium, nickel and iron being concentrated from crude into the residues. The amount of sulphur and the metals present is entirely dependent upon the crude oil source from which the fuels are made and the degree of dilution required to

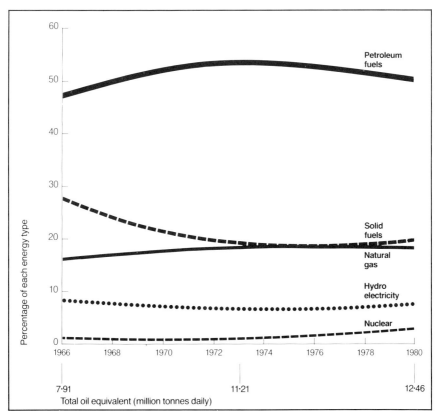

Figure 6.17 **Percentage consumption of commercial energy in world outside Communist areas**

meet any particular viscosity specification. The present commercial grades of residual fuel oils have sulphur contents that may vary from 0.7% to 4.0%, and the metals, of which vanadium is the major constituent, can be anything from 30 mg/kg to 1000 mg/kg. Environmentally, sulphur oxides (SO_2 and SO_3) in flue gases from the combustion of all fuels are causing concern in many countries to the extent that more restrictive sulphur limits may be placed upon fuels or flue gas desulphurisation made mandatory. However, it is suspected that little thought has been given to the ultimate disposal of the sulphur or sulphur compounds removed from the flue gases. Whatever course of action is taken can only increase the cost to the user of producing heat from fuels containing sulphur. Irrespective of concentration, the inherent metals do not normally cause problems except where contamination with sodium compounds from salt water occurs. Sodium reacts with some of the metals in the fuel to form comparatively low temperature melting point eutectics, which may foul heat transfer surfaces or attack refractory linings. Such problems can easily be controlled by preventing sodium contamination, or, in exceptional cases, by the use of an additive such as magnesium oxide.

On the more positive side it has been demonstrated by many workers that the presence of metals such as vanadium and iron in the fuel catalyses the burn-out of carbon particulates and thus reduces particulate emission levels. To achieve such an effect it is essential that the metals, particularly if added separately, are widely dispersed throughout the bulk of the fuel. It is pointless adding a bucketful of additives to the fuel tank during delivery of the fuel and hoping it will be properly dispersed. This is unlikely to occur and the additive will be wasted.

Residual fuel oil grades are generally specified by viscosity maxima, although in some cases brand names may also be used. The standard unit for viscosity measurement is the centistoke (cSt) at either 80°C or 100°C. Other units such as Saybolt and Engler are still used in some parts of the world. The use of Redwood Seconds is now obsolete and is of no value for design purposes, being a purely comparative test. Marine residual fuels traditionally have the viscosity quoted at 50°C, but it has been shown for the heavier grades of marine bunkers that the figures are unreliable at this temperature; consideration is being given to standardising all viscosity measurement temperatures at 80°C–100°C for both marine and inland grades.

Generally, there have been three grades of residual fuel oil available: light, medium and heavy, with some sub-divisions. The demand for light fuel oil in the inland market is falling and in some areas it is no longer available; however, it continues to be blended for the marine diesel engine market. It was previously remarked that fuel density will tend to increase in the future to above 1.000 kg/litre at 15°C, as a result of which some users will have to take steps to ensure that water separates from the fuel. Depending upon the density of the fuel as delivered and the actual storage temperature of that particular grade, water from, for example, leaking steam coils, ingress of rainwater or tank sweating, will settle at the bottom of the tank or rise to the surface of the fuel oil. Oil spillage is another matter that cannot be ignored as the fuel oil will then be at ambient temperature. Under these conditions, fuels with a density in excess of 1.000 kg/litre will settle to the bottom of the interceptor and not float on the surface as is the case with densities of less than 1.000 kg/litre at 15°C. With high-density fuels it is essential that all interceptors on site are modified to cope with such fuels; otherwise, contamination of watercourses and drainage systems will occur.

In some countries undiluted heavy residues are being supplied to specific installations. Such fuels could vary in quality and would not be subject to the stringent specifications applied to the current grades of residual fuel oils. At present these fuels have to be stored, handled and atomised at much higher temperatures than those used for the existing grades of residual fuel oil. The additional costs for storage, heating and handling would be offset by the reduced cost of the fuel compared with present grades. It should be emphasised that new technology is not required to burn these residues, as they are currently used as refinery fuels and can be successfully combusted whilst meeting the required environmental standards.

In spite of the many pressures to convert plant from burning residual fuels to solid fuels, liquid fuels will continue to provide a valuable source of energy for the foreseeable future for all types of plant.

Liquefied Petroleum Gas

Introduction

Liquefied petroleum gas (LPG) is the generic expression for propane, butane and mixtures of propane and butane. These are gases at atmospheric temperature and pressure, but they can be easily liquefied by compression, by cooling or a combination of the two. When in liquid form, their volumes are reduced considerably (about 240 times for butane and 270 times for propane) so that a small cylinder or tank of LPG can provide a useful energy reservoir, equivalent to a very large volume of gas.

Gaseous fuels have many advantages. They can be piped to the point of application, complete combustion is more readily achieved without smoke or deposits, and LPG in particular contains virtually no sulphur, which ensures that the products of combustion are non-corrosive, harmless to materials being heated and cause little pollution to the atmosphere. LPG evaporates readily when the pressure is reduced, and the resultant gas is piped away from the reservoir and is used for a variety of purposes. Gas emerges because the liquid is boiling, which causes a drop in temperature and if the off-take is substantial the temperature of the boiling liquid drops to such an extent that there is a considerable reduction or even a cessation of gas formation. For large offtakes the LPG can be withdrawn from the bulk tank as a liquid and fed to a vaporiser.

Where the final consumer is either purchasing bottle-size quantities of 5-50 kg of product in cylinders or receives small bulk deliveries, there is a heavy investment in both cylinders and small bulk pressure vessels which the supply company installs on customers' premises. There is also the need for depots and cylinder-filling plants, which are strategically located in relation to the market outlets. The high costs of storage and handling normally limit packed LPG to those applications where it can command a premium price.

It is difficult to store large quantities (exceeding a few thousand tonnes) in a pressurised LPG tank, because the required thickness of the tank wall increases with the diameter of the tank until an economically or technically feasible maximum is reached. Thus when very large volumes of LPG are to be handled, the product is often cooled down below its atmospheric boiling point, so that it can be stored at atmospheric pressure in refrigerated, insulated tanks.

An alternative, and environmentally attractive, method of bulk storage is the use of underground caverns. This, however, requires that suitable rock structures or salt strata are in the locality for the construction of the cavern.

Sources of LPG

LPG is produced from two distinct sources. First, it is produced from the processing of crude oil in refineries and in some countries as a by-product from chemical plants as well. Such LPG is normally only available in pressurised form and is mainly marketed locally in cylinders and small bulk tanks. This has been the traditional basis of the LPG business in the majority of consuming countries outside the United States. Second, it is produced as a result of the extraction of LPG from natural gas streams, and, in some instances, from crude oil, at or close to the point of production.

Natural gas from a well consists largely of methane, smaller quantities of heavier hydrocarbons — collectively known as natural gas liquids (NGL) — and varying amounts of water, carbon dioxide, nitrogen and other non-hydrocarbons.

NGL extracted from natural gas streams can be separated at a fractionation plant into a series of hydrocarbons increasing in density and complexity, comprising:
- Ethane (C_2H_6). This is gaseous at normal temperature and pressure.
- LPG-propane (C_3H_8) and the different isomers of butane (C_4H_{10}).

These gases, either individually or as mixtures, become available as refrigerated liquids at around atmospheric pressure, and not normally in pressurised form as has traditionally been the case with LPG from refineries. This means they will

APPLICATION, SPECIFICATION AND TESTING 425

require insulated storage at the loading and reception terminals and specially designed cryogenic ships for transport, as the markets are often at great distances from the sources and transport by pipeline may not be possible, in which case it can only be economically transported by sea.

- Condensate (also known variously as natural gasoline, C_5 plus and pentanes plus) is a liquid at normal temperature and pressure consisting of varying mixtures of pentanes (C_5H_{12}) and heavier hydrocarbons, extracted by compression or absorption from natural gas.

The Uses of LPG

LPG provides a clean fuel for cooking and heating in the domestic and commercial markets and it is suitable for many uses where a high quality fuel is required in the industrial market.

When the total LPG market was small and the uses were very specialised, and mainly served by cylinders, most of the LPG was filled into containers at central refinery filling plants or at a major special LPG installation. As the market has grown and diversified into both large and small bulk deliveries to many different types of customer and the cylinder business has expanded, in some countries decentralised filling depots have been established, sometimes operated by resellers. Linked to the filling depot there is sometimes an LPG appliance sales and servicing operation and a road vehicle delivery service for gas cylinders. Since the containers are pressure vessels they are manufactured and maintained to high standards. Consequently, they are expensive and their distribution, inspection, maintenance and collection require very careful supervision. Even when resellers are involved, the capital, or a substantial part of it, may have to be provided by the LPG supply company, and this is often reimbursed by means of a deposit or rental system.

Apart from these traditional markets, other markets are being established. Several countries, for example, are developing use of LPG as an automotive fuel. Algeria has decided that almost half a million vehicles should be converted to LPG; in the Netherlands some 9 per cent of all gasoline-engined vehicles have been converted, while in Tokyo over 95 per cent of all taxis run on LPG. Other countries of Europe in which there are automotive LPG markets in various stages of development are France, Austria, Belgium, Denmark, Eire, Italy, Spain and the United Kingdom.

Automotive LPG has an appeal to national governments in that its use helps to reduce the overall level of exhaust emission in urban areas. To a country with indigenous LPG production it has the added advantage of helping to reduce crude oil imports, and there may be instances in which the importation of LPG to provide part of the demand for gasoline will be less expensive than either increased crude imports or conversion of heavier fractions into higher value automotive products.

The decision of an LPG marketing company to enter this sector requires considerable investment at each service station selling LPG, and the economics of automotive LPG are such that these markets can develop only if the gas is given preferential tax treatment against gasoline, and the success of automotive LPG markets is proportional to the tax differential between the two fuels. There is, however, increasing interest among international bodies and national governments in developing the use of this fuel. LPG is used in the petrochemical industry as a feedstock to ethylene production and as feedstock to alkylation plants.

LPG can also be used as an enrichment agent in the pipeline gas industry and as a feedstock to synthetic natural gas (SNG) manufacture.

Major LPG Producing and Exporting Areas

Apart from the United States and the Communist areas, the existing or prospective major producing areas of LPG include North Africa, the Middle East and the North Sea. In addition, a number of other countries including Canada, Mexico, Venezuela and Indonesia, are already, or are expected to become, significant producers of NGL for export.

Algeria is currently expected to become one of the world's largest exporters of LPG. During the 1980s the quantities available for export may reach 7 million tonnes per annum.

Countries in the Middle East collectively have the facilities to export up to 20 million tonnes per annum of LPG. Actual production levels vary depending on associated gas feedstocks from crude oil production.

Saudi Arabia is currently the world's largest LPG producer with annual exports of some 10 million tonnes of LPG per annum. The completion of Petromin's Master Gas Scheme has enabled large amounts of associated gas, which was previously flared, to be stripped of NGL for export.

Other countries in the Middle East also have facilities to export significant volumes of LPG, although none of them has a capacity similar to Saudi Arabia's.

The volume of LPG from the North Sea will be much smaller than from the Middle East or North Africa. The United Kingdom sector is expected to be the largest producing area, with a capacity sufficient to allow for exports.

There are several other countries producing LPG and some of these have existing or potential export capability. Venezuela, one of the first countries to export LPG, is expected to export up to one million tonnes per annum in the mid-1980s. Estimates of Mexico's export potential range from one million tonnes to 5 million tonnes per annum of LPG by the late 1980s. Canada exports some 4 million tonnes per annum of LPG primarily to the United States.

Transporting LPG

As indicated earlier, propane and butane can be liquefied at normal ambient temperature by increased pressure or at atmospheric pressure by reducing the temperature (propane to minus 45°C and butane to between minus 1°C and minus 12°C, depending on the proportions of normal and iso-butane).

Traditionally, with the exception of bulk refrigerated LPG, mainly to Japan, transport has been in pressurised trucks, rail tank wagons, ships and barges while pipelines have been used for inland transport of large volumes, especially in the United States. For onward transport to consumers LPG is delivered in cylinders and bulk vehicles of various sizes.

The maximum capacity of a pressurised LPG ship is limited to a few thousand tonnes by the high manufacturing costs involved. It is therefore economically preferable to transport large quantities of LPG at atmospheric pressure in refrigerated tanks. The LPG can be unloaded into refrigerated storage tanks or, after heating, into pressurised tanks which supply conventional distribution facilities; alternatively, transshipment into coastal tankers may be used to deliver refrigerated products to smaller terminals that cannot receive large ocean-going ships.

In some instances LPG may also be transported by "spiking" — that is, enriching crude oil with small quantities of LPG.

The Major International LPG Markets

In the past, the volumes of LPG traded internationally have been relatively small for a number of reasons. Availability from the refinery has only been in general sufficient for local uses, and the difficulty and expense of transportation and handling, together with the relatively low prices paid for energy products in general, did little to encourage potential exporters to establish new projects. As a result of the reduced flaring of natural gas streams, the export availability of LPG in 1985 could be nearly 40 million tonnes compared with nearly 20 million tonnes at present.

WOCANA (the world outside Communist areas and North America) demand for LPG has been growing steadily in recent years for three main reasons:
- Increased recovery of associated gas in crude oil production fields.
- Refrigerated LPG ships allowing a significant increase in size of vessels and corresponding decrease in freight costs.
- Parallel with these developments which increased supply potential, increasing standards of living, combined with higher prices of firewood (scarcity) and of kerosine (competing with aviation demands), led to increasing demand for

LPG as a domestic fuel in the developing countries, while in the developed countries LPG was seen to provide additional security in the automotive fuel portfolio, with the added advantages of being high-octane and pollution-free.

These developments are expected to continue in the medium term but, with continuing declines in refinery crude runs, the move from refinery LPGs to field-produced LPGs will accelerate. European demands are expected to be met from the North Sea and North Africa, U.S. import requirements from Canada and Mexico, while requirements of other areas should come from the Middle East.

Japan is currently the largest importer of LPG, mainly for the traditional residential, commercial and industrial markets. These markets are large, owing primarily to the absence of significant volumes of indigenous natural gas. LPG has traditionally been used, beyond the reach of town gas supplies, for cooking and water heating. LPG may increase its role as an enrichment agent for natural gas or develop as a feedstock for SNG manufacture. In the chemical sector, LPG already constitutes 25 per cent of feedstock for methanol, but ethylene production is based on naphtha and few new plants are planned.

In 1980, consumption in Western Europe was some 19.4 million tonnes, where traditionally LPG markets have been based on refinery production. Most Western European countries are now net importers of LPG, which is a fundamental change compared with the situation 5 years ago. This is owing to the development of new markets and a consequent increase in demand coupled with low refinery production resulting from lower crude runs. In the next decade any major expansion of the use of LPG in Western Europe would thus be based on imported product, except in the UK and Norway.

The automotive market for LPG in Western Europe has been growing steadily over recent years, from 900 thousand tonnes in 1977 to 2200 thousand tonnes in 1981, largely as a result of supportive fiscal policy by the respective Governments, encouraged by the EEC. Growth in this market is likely to continue and there will be additional requirements for butane as feedstock to new alkylation plants.

There may be opportunities in the petrochemical industry where condensate and butane could supplement naphtha as a feedstock to ethylene production, but this will be dependent on the relative prices of naphtha and LPG.

In 1980, LPG consumption in the United States was estimated to be nearly 53 million tonnes. This market is expected to continue to increase over the coming years, but only slowly. However, the production of LPG from United States gas plants will decline in the future, as an increased proportion of this natural gas will be drawn from new and deeper gas wells, which tend to contain less NGL. Imported LPG could therefore become increasingly important as a petrochemical feedstock, in gasoline manufacture and eventually as a partial substitute for natural gas.

Energy Efficiency

An account of the use of oil fuels would be incomplete without a reference to "energy efficiency" and "fuel conservation"; terms which are met with very frequently these days. Although the efficiency with which oil and other fuels are used has increased slowly for many decades, it was not until the oil crisis of 1973/74 that a general realisation that fossil fuel reserves are finite led to a concerted drive to conserve them by using fuels more efficiently.

Before describing how increased efficiency is reducing fuel use, it is convenient to define terms:

Energy is "the capacity for doing work". It exists in many forms: the chemical energy of a fuel, which is released when it is burnt and may then be used to do the work of moving a car; the potential energy of water stored behind a dam, which does the work of driving a turbine when it flows to a lower level; the electrical energy from a power station, which can do the work of driving an electric motor. There are many other forms of energy: wind; solar radiation; heat; pressure, etc. Energy is measured in many units such as kilowatt hours, barrels of oil equivalent and calories, but the international standard unit is the joule (see section on Units of Measurement, p. 666).

Conservation of Energy. This term frequently appears in a context that implies that something has been saved. However, it is a rather meaningless phrase because energy is always conserved. The first law of thermodynamics states that "Energy can be neither created nor destroyed; the amount in the universe is constant". This means that whatever we do we cannot help but conserve energy — but we may not always conserve fuel. For this reason the term "energy efficiency" is more useful for discussing fuel saving.

Energy Efficiency is a measure of how good a use is made of energy, as it is converted from a valuable form (e.g. oil) to another less valuable form. This can be demonstrated by an example: the use of a litre of oil to heat a house.

When the oil is burnt, some of the energy it contained goes up the chimney as heat in the flue gas, which then slightly warms the atmosphere and is wasted. The remainder passes into the house via boiler and radiators in the form of heat. However, this also eventually passes through walls and windows to the atmosphere. The final result of burning one litre of oil is that the energy it contained has passed to the atmosphere, making it very slightly warmer. The energy still exists but it is no longer of any use for house heating.

To make the best use of the one litre of oil, the aim must be to use its energy to maintain the house at the desired temperature for as long as possible. This can be done by reducing the amount of heat lost to the chimney (by designing a more efficient boiler) and by slowing down the rate at which heat leaks away through walls, windows, etc. (by improved insulation).

Other examples of using energy efficiently are getting the most tonne-kilometres of transport or the most manufactured goods from the energy released by each tonne of fuel burned.

Fuel Conservation aims to reduce consumption of fossil fuels and is achieved either by improved energy efficiency or by going without.

Improved energy efficiency has already had a noticeable effect since 1973. At that time most developed countries were increasing their fuel consumption by about one per cent for each one per cent increase in GDP (gross domestic product), whereas by 1980 very much smaller increases in fuel use were needed to increase GDP. Past improvements in energy efficiency differed greatly between countries so that although Sweden and the U.S.A. have a similar GDP per capita, the latter country uses some 50 per cent more energy per capita.

Although much improvement in energy efficiency has already been made, there is still great scope for more. It has been estimated that over the next 40 years there is prospect of an average improvement of 46 per cent. In countries with a high fuel use per unit of GDP, such as Canada, the U.S.A., the U.K. and the Netherlands, the potential for improvement is much greater.

Such improvements might be brought about in various ways.

In transport, the number of litres of fuel per 100 kilometres driven in the private car may be halved over the next 10 or 20 years as improvements are made to engines, gear boxes, electronic driving controls, tyres, body shape and weight.

In industry, there are many ways of improving the energy efficiency of the production of goods. For example, about two-thirds of fuel used by industry goes to making materials (e.g. steel) rather than finished goods. This means that there is great scope for fuel saving: by improving materials processing; by recycling; by design for lightweight and low-energy-content materials (e.g. plastics rather than aluminium) and by design for easy repair and maintenance.

In buildings of all kinds, the fuel used for heating, cooling and hot water production accounts for nearly half of all energy consumed. Many existing buildings are badly designed and poorly insulated. Their fuel consumption can often be halved by improving insulation alone. Further savings can be achieved when heat-pumps are used or when the heating is provided from a combined heat and power scheme or a total energy system. As the real cost of energy in all its forms increases in the years to come, the attractiveness of energy efficiency (the "fifth fuel") will increase even more.

General Characteristics of Lubricants

Tribology

Lubrication depends upon many factors besides the nature of the lubricant. The shape and roughness of rubbing surfaces and the materials from which they are made are important. Whether or not there is a rolling as well as rubbing motion, and the relative speeds of moving parts, also need to be considered. The nature of the environment in which lubrication is required is critically important in many applications. In many modern machines lubrication is often required at extremely high or extremely low temperatures, or in the presence of high humidity or artificial atmospheres or nuclear radiation. When lubricants are used for metal-cutting and similar operations, where there is inevitably some contact between human beings and the lubricant, the possibility of a hazard to health is an important consideration.

Consequently, the sciences and technology associated with lubrication have become increasingly complex and it was for this reason that a new term "tribology" was introduced to embrace all the many aspects and phenomena. As industry becomes increasingly capital-intensive the cost of interrupted operation through the breakdown of only one component of a complex production system increases in proportion. Similarly, the real cost of unnecessary waste of energy rises with the real cost of energy. Sound tribological practices are therefore of increasing significance.

Lubrication, Lubricating Oils and Greases

The main object of lubrication is to reduce friction and wear in the moving parts of a machine. This reduces frictional heat and the consequent waste of energy, and it also avoids premature failure of the machine through excessive wear or seizure. Although there are some engineering plastics and resins that do not require the use of a lubricant, their use is limited to relatively light duties. Most of the moving parts of machinery are metallic and a lubricant is necessary. Lubrication can be achieved with gases and solids as well as with liquids, and there are special circumstances where either gas or solid lubrication is preferred. However, such cases are rare and attention here will be confined to lubricating oils and greases.

A lubricant, as well as performing its main function of reducing friction and wear, is often called upon to perform other roles. In many applications it is useful as a heat-transfer medium, removing heat from "hot spots" in a machine, thereby preventing damage by overheating. Cooling of engine pistons, of power-transmission systems and of metal-working operations are examples. Protection of machine parts against corrosion, particularly when shut down, is another important function in most applications. In some applications the lubricant acts as a carrier for solid contaminants, which can either be removed by oil filters, or be kept in suspension until the oil is changed. This is a particularly important function in piston engine lubrication, in which oil is contaminated by combustion products. In all applications, no matter how good the lubrication may be, some wear of mating surfaces occurs and minute metallic particles are detached from their parent body. These particles can cause serious abrasive wear if they are not carried away from moving parts and filtered out.

Friction, Wear and Viscosity

A very big reduction in friction is achieved by interposing a film of liquid between two metallic surfaces which are in relative motion. If the two surfaces are pressed hard together under heavy loading, the liquid film will be squeezed. If the film does not have sufficient strength and thickness, it will rupture and the metallic surfaces will make contact and scuff. The property of the lubricant that is important in determining film strength and thickness is the viscosity. Separation of the moving parts by a liquid film is generally referred to as fluid lubrication (or hydrodynamic lubrication). A very important example is a journal bearing in which the rotation of the shaft draws in oil and maintains a film on which the shaft "floats" (Fig. 6.18, p. 446). The greatest reduction in friction is achieved with an oil that is just sufficiently viscous for a film of adequate thickness to be maintained under the prevailing temperature and load conditions that are applied to the bearing.

In certain circumstances, lubrication can occur (that is, friction and wear are reduced) when a liquid film does not exist between the two surfaces. For such a condition to exist the lubricant must have the capacity to adhere tightly (i.e. to be adsorbed) on to the metal surfaces. This particular property of a lubricant depends upon the chemical nature of some of its constituents, which may be deliberately added in order to enhance the capacity of the lubricant to be absorbed on to the metal surfaces. Whenever the speed of the moving parts is insufficient or the load on them is too great for an oil film to be maintained, some lubrication still occurs provided that the lubricant possesses this capacity. This phenomenon is called "boundary lubrication" and is an undesirable condition. Friction with boundary lubrication is lower than without lubrication but nevertheless much higher than exists with fluid lubrication. Not only is energy wasted, but excess heat may be generated and surface damage ensue. There are many circumstances in which boundary lubrication cannot be avoided, and the quality of the lubricant is then critically important. Oils containing "EP" (extreme pressure) additives have been formulated for this problem. The problem sometimes occurs only during the initial hours of use of a machine, before the wearing surfaces have been "run-in". In such cases the initial fill may be a special "running-in oil", but once the running-in process is complete such an oil may not be required any longer.

Viscosity. The viscosity of a liquid is a measure of its resistance to shear or the rate at which it will flow through a hole. High viscosity oils are thick, low viscosity oil thin. In a lubricant, viscosity is strongly dependent on temperature, decreasing as temperature increases; the temperature must therefore always be specified in stating the viscosity of a lubricant.

Viscosity Index. The amount by which the viscosity of an oil decreases with rise in temperature depends both on the type of crude from which is is derived and the refining treatment to which it has been subjected. The relationship between viscosity and temperature is of significance for lubricating oils since most oils have to operate over a range of temperature. There are many ways of expressing the relationship but the one firmly established in the petroleum industry is the viscosity index (VI) system.

Oils are generally classed as high, medium and low (HVI, MVI, LVI) viscosity index oils. There are no precise definitions or strict lines of demarcation between them. However, oils with a VI of more than about 85 are usually classed as HVI, whilst those below 30 are generally considered as LVI. MVI oils fall roughly between these two. However, as will be indicated in the section on additives, VI can readily be increased beyond the levels normally available by refining alone. Hydrocracking techniques, however, can produce base oils having a VI higher than 140. These are classified as VHVI or XHVI oils.

Viscosity Classifications. The most widely used system of viscosity classification is that established by the Society of Automotive Engineers (SAE) in the United States of America. In this system, a certain SAE number defines a viscosity range at either a sub-zero temperature or 100°C. Numbers with the suffix W refer to viscosity at the low temperature, and those without a suffix are viscosities at 100°C.

In this system the W grades should not exceed the specified viscosity at defined low operating temperatures with 5°C intervals. Thus a 10W oil will not exceed the specified viscosity at $-20°C$, whilst a 5W oil will be limited at $-25°C$ and a 15W product will be limited at $-15°C$.

The SAE viscosity grade system applies to engine lubricants (0W, 5W, 10W, 15W, 20W, 25W and 20, 30, 40, 50 and combinations of these such as 10W/30, 15W/40 and 10W/40). There is another series of SAE viscosity grades that apply to gear and axle lubricants (e.g. 75W, 80W, 85W, 90, 140 and 250).

There is a third viscosity classification system which is widely used for lubricating oils used in industry. The selection of the correct viscosity grade for each of the many different uses that exist in a factory is important. It is highly desirable that an easily recognisable viscosity identification should be marked on every oil container. The system best suited to this need is the ISO (International Standards Organisation) Viscosity Grades (ISO VG). For each viscosity grade range a minimum, maximum and midpoint kinematic viscosity is specified at 40°C. For example, the ISO VG 10 range has a midpoint of 10 centistokes, a minimum of 9 and a maximum of 11.

Mineral Oil Lubricants

Originally, the only lubricants used were animal or vegetable fats and oils, such as palm oil. For a small minority of special applications these oils are still preferred. In some applications synthetic lubricants are essential. However, mineral oils derived from crude petroleum are dominant. They may be manufactured from either a vacuum distillate or a residual fraction of the vacuum distillation of a lubricating oil feedstock, both have a boiling range above that of gas oil. The residual fraction after further refining, provides a very viscous oil called a "bright stock", which is used as a blending component in the heavier (more viscous) grades of lubricant.

Vacuum distillates are the main source of lubricant base oils. LVI oils are made from naphthenic lubricating oil distillates by acid treatment or in recent years by hydrorefining, and these oils usually have such low wax contents that no costly dewaxing operation is required to obtain low pour points. LVI oils are used whenever viscosity index and oxidation stability are not important marketing characteristics.

MVI oils are produced from both naphthenic and paraffinic distillates and are distinguished as MVIN and MVIP oils; the MVIP oils have to be dewaxed. MVI oils are used as general purpose lubricants for applications where the low viscosity index of LVI oils is a disadvantage.

HVI oils are usually prepared by solvent extraction and dewaxing of paraffinic distillates. Solvent extraction improves not only the viscosity index but also the oxidation and colour stabilities. HVI oils are used wherever an oil of high viscosity index and/or good oxidation stability is needed, as in motor oils and turbine oils.

An increasingly-used process is hydrorefining, i.e. treatment with hydrogen over a catalyst at elevated temperatures and pressures. This process can be used to partly or completely replace solvent extraction. Additionally, this process is used to convert waxy feedstocks by isomerisation to oils with a very high viscosity index (up to 150 VI).

"Compound oils" are blends of mineral oil and one or more "fatty oils" derived from vegetable or animal oils or fats. Fatty oils possess the valuable property of being very strongly "adsorbed" onto metal surfaces. Consequently, they are particularly useful in some applications where "boundary lubrication" conditions exist. Fatty oils are unsuitable for high temperature use because they are thermally unstable and are easily oxidised. Rapeseed oil and various fish oils are the ones most commonly used.

Lubricating Greases

Lubricating greases are made by thickening lubricating oils with soaps, clays, silica gel or other thickening agents. Greases are used when the parts to be lubricated are not easily reached or are inadequately sealed, or where there is a danger of oil contaminating a product being processed. They are commonly used in anti-friction bearings. Greases range from soft semi-fluids to hard solids, the hardness increasing as the content of the thickener increases.

Greases are classified according to the type of thickener, e.g. calcium, lithium, organic, etc., and their consistency (consistency is measured in terms of "penetration", the distance a plunger penetrates into the grease under standard conditions), and are classed according to the penetration classification of the National Lubricating Grease Institute (NLGI) in the USA, 000 being the softest and 6 the hardest.

The consistency of a grease depends on the structure built up by the thickener in the oil, and this structure is affected by mechanical disturbance. A grease has a different consistency immediately after being stirred or churned from that after it has been undisturbed for a long time. Since greases in use are apt to be churned by the motion of the parts they lubricate, they are usually tested after being "worked" by a standard amount, in a container fitted with a perforated plate that can be forced up and down through the grease.

Lubricating Oil Additives

Straight mineral oils, together with "compound oils", were once able to meet all normal lubrication requirements of automotive and industrial practice. As these requirements became more severe with the progressive development of engines and general machinery, it was necessary to improve the quality of lubricating oils by new methods of refining and eventually to use "additives" either to reinforce existing qualities or to confer additional properties.

Additives are frequently used to enhance a particular function, but many additives are multi-functional. The chief functions are:
- to improve the viscosity index (VI improvers);
- to increase oxidation stability (anti-oxidants or oxidation inhibitors);

- to keep contaminants in suspension (dispersants);
- to prevent lacquers and deposits forming (detergents);
- to prevent wear (anti-wear agents) through the neutralisation of acids derived from fuel combustion;
- to reduce friction (friction modifiers);
- to prevent scuffing (extreme pressure or EP agents);
- to depress the pour point (pour point depressants);
- to prevent rusting (anti-rust agents);
- to prevent foaming (anti-foam agents).

Oxidation Characteristics

All petroleum products under normal conditions of storage and use come into contact with air and hence with oxygen, often at high temperatures. This contact sometimes occurs when the product is in a finely divided state, for example as a mist in an engine. Such conditions promote oxidation. Hydrocarbons vary in their susceptibility to oxidation; paraffins, i.e. saturated chain compounds, are generally more resistant than side chains in aromatic ring compounds, and unsaturated compounds are particularly readily oxidised.

Although there are many different oxidation tests, they mostly follow a procedure in which the product being tested is exposed to excess air or oxygen at the specified temperature, and the amount of oxygen absorbed in a given time is determined and the quantity and quality of the oxidation products are assessed.

A major difficulty of such tests is the choice of conditions. If these conditions (especially the temperature) are similar to those in service, the oxidation may be so slow that the test is too long to be serviceable for control or specification purposes. However, if the conditions are made more severe by increasing the temperature, or by using oxygen instead of air, in order to shorten the test to a serviceable time, the nature of the reaction may be altered and the results of the test may be irrelevant to performance in service.

Steam turbine oils used in electrical power generation are required to remain in use without appreciable oxidation for very long periods. In the Turbine Oil Stability Test (TOST), the oil is in contact with water at 95°C while air is blown through the oil in the presence of iron and copper catalysts. Portions of the oil are withdrawn at intervals and tested for acidity. The acidity rises very slowly for a period and then rises rapidly. The time taken to reach an acidity equivalent to 2 mg of KOH per gram of oil is called the induction period, which is regarded as a measure of the resistance of the oil to oxidation. The test primarily determines the life of the oxidation inhibitor and may not indicate the stability of the base oil once the inhibitor has been exhausted. Results show some correlation with practice in that the service life of the oil increases with the length of the induction period.

Load-Bearing Capacity

Under hydrodynamic conditions the ability of an oil to bear a load depends on its viscosity, but when metal-to-metal contact is imminent its ability depends on the effectiveness of the EP agent to prevent catastrophic wear.

Various laboratory machines have been devised to assess this property by causing test pieces to be rubbed together under load in the presence of the lubricant and measuring the extent of wear, the maximum permissible load without undue wear, the load to cause seizure, or some other property dependent on the ability of the lubricant to facilitate lubrication under the conditions imposed. The best known machines of this type are the Almen (or Almen-Wieland), Falex, SAE Shell Four Ball and Timken machines. Although these machines undoubtedly give some indication of EP activity, their results are often mutually contradictory and show no precise correlation with service performance. They are, however, useful for comparative testing of oils and additives.

Load-bearing capacity is of particular interest in gear lubrication, and as an alternative to the above-mentioned machines laboratory rigs in which actual gears are run are used. The best known are the IAE Gear Machine designed by the Institute of Automotive Engineers (UK), the Ryder Machine (USA) and the FZG Machine (West Germany). These also do not always agree amongst themselves in the rating of lubricants or correlate with service performance. Moreover, all these rigs use spur gears, whereas the major problems in gear lubrication arise with hypoid gears.

Hence tests in actual vehicle hypoid axles have also been developed to assess the lubricant performance and some are now specified in official specifications (e.g. US military MIL Specifications) or in motor or gear manufacturers' specifications. Particularly under conditions of high torque/low speed/shock loads, these tests appear to correlate better with practice than do any laboratory machine or rig tests.

Corrosion and Protection

It is important that lubricants should not corrode the metal surfaces with which they come into contact, such as iron, copper and silver. The lubricants should therefore not only be non-corrosive in themselves but should not become corrosive during service. They are also expected to provide positive protection against corrosive attack by moisture or other agents.

Oxidation of the lubricants with formation of acids is another potential cause of corrosion and this is one reason why oxidation stability is of such great importance.

In the case of engine oils, engine tests are run under conditions conducive to the formation of corrosive acids to assess the protective properties of the oil.

Health and Safety with Lubricants

Lubricants, and related products such as metal-working oils, heat-treatment oils, etc, present little or no hazard to users provided that they are properly used. This means taking reasonable care to keep them away from the eyes and to avoid breathing their vapours and mists. Undue contact with the skin is also to be avoided because frequent and prolonged exposure can in some cases cause irritation, or in exceptional circumstances, more serious conditions such as skin cancer.

Leaded oils should never be used in oil mist distribution systems as they present a potential hazard in liquid contact with the skin leading to absorption.

Hygiene and Safety Precautions. In order to minimise risks, workers in industry need to practice good standards of industrial and personal hygiene. They must ensure that machine splash-guards are correctly adjusted. They should wear protective clothing (including oil-impermeable gloves) if possible or else use oil-repellent barrier cream on exposed skin. Oily rags (and tools) should not be put into clothing pockets nor used for wiping hands. Cuts and abrasions should receive immediate First Aid treatment, and oil-soaked clothing should be removed and cleaned. In addition to normal hygienic practice, hands should always be washed before using the toilet and after work.

Product Precautions. Lubricating base oils which have not had polycyclic aromatics reduced in refining to the same extent as most highly refined oils may, under conditions of repeated gross contact, cause keratosis or even, in extreme cases, skin cancer. These base oils are not recommended for use where such conditions can occur, neither should they be used in oil mist application or where mists are likely to form.

When an oil product contains solvent, vapours may cause irritation to eyes and to mucous membranes. Some such products may incorporate aromatic hydrocarbons and particular care should therefore be taken to avoid inhaling vapours.

Some synthetic lubricants contain tri-aryl phosphate esters, which are potentially hazardous to health if ingested or absorbed through the skin. No eating or drinking should be permitted where they are used, and protective clothing and gloves must be worn. Any liquid that accidentally gets on to the skin must be quickly washed off.

Disposal of Products and Packages. Lubricants should be disposed of so that they do not contaminate drainage systems, rivers, waterways or groundwater. Packages should preferably be handled by recognised drum dealers and renovators.

When spillage occurs it should be absorbed with sand, earth or other mineral absorbent, which should then be disposed of in accordance with local industrial waste disposal regulations.

Lubricants Business

Although sales of lubricants amount to only just over one per cent by volume of sales of petroleum fuels, they are of higher unit value. A growing proportion of lubricants is sold by specialist companies who purchase base oils from major refiners and operate their own blending and packaging plants. Other specialist companies, including vehicle manufacturers, purchase pre-formulated and pre-packed oils from blenders to sell under their own brand name. Nevertheless, a high proportion of lubricants is marketed by major oil companies, with about a third being supplied by major international oil companies combined. Only the biggest companies attempt to carry a comprehensive stock, because of the very large number of special performance grades that are required to meet market needs. Many of the smaller lubricant marketing companies specialise in a narrow sector of the total business, e.g. metal-working oils.

Although there are several thousand formulated products the number of grades of base oils produced by a refinery is much smaller, usually less than 20. Hence the number of base oil tanks required for the supply of raw materials in a lubricant blending plant is not excessive. It must be remembered that additives account for an increasing proportion of the total output of finished products. A major blending plant may carry as many as 200 different additives in stock, of which a few are delivered in tank car loads but the majority in drums. The purchase of additives has become an important part of the lubricants business, since additives are mostly high value chemicals which account for a significant proportion of the cost of high quality lubricants.

Since the volume of lubricants is only a very small fraction of fuel production, it is not surprising that there are relatively few lubricating base oil refineries. The small volume problem is also complicated by the fact that only certain crude oils are suitable for lubricant production. Only one-fifth of the Shell refineries produce high viscosity index oils from selected paraffinic crudes, whilst medium viscosity index oils from essentially naphthenic crudes are manufactured at even fewer refineries. There is a primary distribution of base oils from refineries to marketing company blending plants, with the output of many being dedicated to their domestic market whilst others supply neighbouring countries. In an international oil company, as well as in independent lubricant companies, there is therefore a substantial operation for the purchase, supply and distribution of lubricants, which has to be organised by many marketing companies. With much of the new lubricating base oil refinery capacity being planned in oil-producing areas (particularly in the Middle East), the supply pattern of marketing companies will probably undergo considerable changes during the 1980s.

Blending plants vary in size between about 4,000 and 100,000 tonnes output per annum. Although some of their production is distributed in bulk, most of it is in a wide range of branded packages. Alongside the blending operation is a can-filling and drum-filling factory and a finished packed product warehouse. The introduction of automation in the most modern blending plants and warehouses contributes to the reduction of operating costs, the maintenance of high standards of quality assurance and a marked reduction in the production of off-specification product.

Many blending installations are supplied with base oils from refineries by inland transport, but there is a substantial traffic in ocean-going tankers. These vessels are usually in the range of 18,000 to 30,000 dwt and each year as much as 3.5 million tonnes of base oils crosses the seas, in ships which may each load and discharge at many ports. The capital cost of such special vessels was more than US$1,000 per dwt in 1980. Their scheduling and operation is markedly different from VLCCs (very large crude carriers). Although much of the finished packed product distribution is by road or rail, some also is moved by barge or by sea-going carrier.

A comprehensive lubricants marketing operation requires several thousand different speciality grades. More than half go to engine applications and are increasingly high performance, premium price products. A third go to industrial use either as lubricating oils, greases or protectives. The remainder are process oils and white oils. Prices have to be determined for each grade over a wide range of bulk and packaged quantities, and stock levels in depots must be closely monitored to ensure positive delivery on demand, but there must be no working capital tied up unnecessarily in excessive stocks. Maintenance of this fine balance is greatly facilitated by electronic data processing techniques. There is a worldwide trade in vehicles, ships and machinery and so customers' needs can only satisfactorily be met if there are recognised and approved branded products available everywhere. The development and maintenance of internationally recognised specifications, test methods and quality standards is a very important part of the lubricants business, and requires very careful central planning on a worldwide scale. A highly qualified technical sales staff, supported by an efficient technical service organisation, is an essential part of an effective lubricants marketing activity.

Engine Lubricants

Whether in gasoline or in diesel engines, the lubricant has the same basic function, lubricating the moving parts to enable them to operate without undue friction or wear. It is also necessary to prevent the formation of deposits derived from the combustion products of fuel.

Before World War II, engines were lubricated by straight mineral oils. In the post-war years, however, additive oils became the norm and extremely high standards of performance have been achieved. Not only is engine wear reduced considerably and the engines retained in a much cleaner condition than was previously the case, but the engines themselves are operating under much more severe conditions. On top of these demands, manufacturers require increased oil drain periods. We have now reached a situation where the products of reputable manufacturers achieve a very high standard of performance indeed. As the demands of engines, for example temperatures and pressures, become more severe, further improvements will be necessary, however.

In both gasoline and diesel engines a circulating system is used whereby a relatively large supply of oil is kept in circulation and is distributed under pressure to the various bearings and other mechanisms. The oil also provides piston cooling.

Some small two-stroke gasoline engines in light weight motor cycles, motor-assisted cycles, scooters, lawn mowers, outboard motors, and chain-saws, etc. use a different system in which the lubricant is mixed in suitable proportions with the gasoline to make a "petrol" mixture. In operation, this oil effectively lubricates the various components of the engine.

The American Petroleum Institute (API) has defined several types of service requirements for gasoline and diesel engines on the basis of engine type, operating conditions, and fuel type, and provided a simple code for oils meeting these requirements. There are three definitions of service for gasoline engines, MS, MM, and ML and three for diesel engines, DS, DM, DL covering, respectively, severe, moderate and light conditions of operation. A wide range of authorities, international and national, have issued test specifications which require minimum performance levels for gasoline and diesel lubricants. Additionally, major motor manufacturers often demand lubricating oils that meet engine test requirements to suit their own particular purposes.

The essential properties of a lubricant to meet these various requirements are that it should be of appropriate viscosity, should not deteriorate rapidly under the specified engine or other test conditions and should protect the engine from deposits, corrosion and wear.

Viscosity

Viscosity is an important property of an engine oil. It controls the speed of flow and ease of getting to the points to be lubricated. When the engine is cold, it is desirable for the oil not to be too viscous in order that rapid lubrication is achieved on starting. On the other hand, when the engine is hot, viscosity should still be sufficient to provide an adequate lubrication film. Clearly, it is desirable for the viscosity of a particular oil to change as little as possible with temperature. This end is achieved by using HVI (High Viscosity Index) oils for most formulations. The viscosity/temperature performance of an oil can also be improved by suitable VI improver additives. More recently, very high VI oils (VHVI or XHVI) have been developed based on hydro-treating of cracked wax materials. These oils permit the formulation of lubricants with outstanding viscosity/temperature performance without the use of VI improvers.

Deterioration of Oil in Use

Lubricating oil in the engine deteriorates in two ways: it undergoes chemical and physical changes due mainly to oxidation and it becomes contaminated by material from the combustion chamber, its own degradation products and airborne dust. Although physical contaminants can be removed by a filter which is fitted to all engines, acidic products of combustion are normally countered by the presence of alkalinity in the oil. Contamination of the oil from unburnt fuel and water of combustion, although a problem in the past, is of less importance nowadays due to improved thermostating of the engine, always provided that the engine is properly maintained.

Soot is formed as a result of imperfect combustion such as when idling or running under rich mixture conditions. Most of this is blown out by the exhaust, but some contaminates the oil on the pistons and cylinders and then drains down to the crank case. In gasoline engines, lead compounds are produced from the fuel additives and pass into the oil with the soot. These solid contaminants, in conjunction with oil and water, may form a sludge that is essentially an emulsion of water in oil stabilised by the solids. This sludge may settle out in the crankcase or be circulated with the oil and block oil ways. Oil filters are incorporated in the oil circulation system to remove these contaminants as far as possible, as well as any airborne dust that escapes the air filter and the air intake system. There is a limit to what the oil filter can remove, and it too may be blocked if sludge is excessive. The conditions under which oil is used in an engine are very conducive to oxidation. Oil mist is in intimate contact with air at a fairly high temperature in the crankcase and at even higher temperatures in the combustion chamber. Oil films are subjected to high temperatures on the cylinder walls, the pistons, and piston rings. Oxidation products are formed, their nature depending largely on the temperature. In the crankcase, acidic material and complex carbonaceous products known as asphaltenes are produced and, in association with the fuel contaminants already mentioned, help to form a stable sludge that is characteristic of low temperature operation. In the combustion chamber and around the piston rings, the oil forms deposits of "carbon" by a combination of oxidation and thermal degradation. Part of this deposit is washed back to the sump and part remains in the combustion space on the piston head and in the ring grooves. All these effects of oxidation aggravate the contamination of the oil beyond that resulting from the fuel. However, the presence of anti-oxidant additives retards the occurrence of oxidation and also prevents the oil from becoming acidic, when it would corrode bearing materials such as copper/lead.

Engine Oil Additives

Reference has already been made to the presence of additives in modern engine oils. Frequently, multifunctional additives are employed, each having a variety of properties. The net result, however, must be an oil that is resistant to oxidation, able to suspend contaminants and prevent wear of the engine. The viscosity index (VI) improvers already referred to are often incorporated, while wear on cams and tappets is prevented by the presence of extreme pressure additives. Corrosive wear in cylinders is prevented either by alkaline additives which neutralise the acids responsible for corrosion or by additives that absorb them and prevent them reaching metal surfaces. Oil-soluble contaminants are kept in suspension by dispersant additives and thereby prevented from settling out from oil or adhering to metal surfaces. In this way, the engine is kept clean and, although the oil may look darker than a corresponding non-additive oil, the suspended contaminants are in such a fine state of division that they are relatively harmless and do not block oilways and filters. Anti-rust additives may also be used; they are absorbed on metal surfaces and prevent the access of water and oxygen. It should be emphasised that not only should the new oil exhibit all these desirable characteristics but they must be maintained throughout the life of the oil.

Automotive Gasoline Engine Lubricants (Motor Oils)

Both single-grade and multi-grade motor oils are invariably based on HVI or XHVI oils in which the appropriate additives are incorporated. The tendency to form deposits in the piston and ring belt area is less in the gasoline engine than in the diesel engine and the lubricant does not need such marked anti-ring sticking qualities. The emphasis is rather on the prevention of low temperature sludge, the maintenance of adequate dispersancy to keep the engine clean, together with stability from the effects of high temperature and prevention of rusting and corrosive wear. Increasingly, oils have been treated as an engineering component in the overall design of equipment. Thus, whereas in the past oils were formulated to prevent wear of cams and tappets in certain engines, nowadays it is the combination of metallurgy and oil that achieves satisfactory performance of the engine. A major trend in recent years has been a tendency towards longer and longer oil drain periods. Coupled with the low rates of oil consumption in most modern engines, this results in extremely high demands on the oil, since the replenishment of additives by top-up oil is at a minimum.

Two-stroke engines demand somewhat different lubricants because part of the lubricant is burnt with the fuel; an undue amount of carbon deposits from the oil must, therefore, be avoided.

Automotive Diesel Engine Lubricants

Diesel fuels contain more sulphur than do gasolines and therefore produce more corrosive acid gases when burnt, so the lubricant needs to be more alkaline. Diesel engines usually run at higher loadings than do gasoline engines and are therefore more likely to form high temperature deposits and are more subject to ring sticking and ring packing. Low temperature sludge is not, however, unknown. Diesel lubricants are therefore formulated principally with a view to preventing interference with ring action by deposits, and the emphasis primarily is on anti-acid but also on anti-oxidation and dispersancy, together with anti-wear properties. Viscosity/temperature characteristics are relatively less important in commercial equipment, although passenger car diesels increasingly require the use of multi-grade diesel lubricants.

Lubrication of Marine Diesel Propulsion Engines

Marine diesel propulsion engines operate basically on the same principles as automotive internal combustion engines, although there are certain differences in design and in fuel and lubricant requirements. They are generally large, slow-speed engines (with speeds up to 130 rev/min) coupled direct to the propeller, but medium-speed engines (with speeds from about 400 to 750 rev/min), with geared or electric drives, are also used in large vessels. Practically all large marine diesels are of the direct injection type, the fuel being sprayed direct into the combustion space. These large engines require positive cooling of the pistons, which must be made of some metal that expands as little as possible on heating.

There are basically four types of diesel engine used in ships; the four-stroke single-acting engine working on a conventional four-stroke cycle; the two-stroke single-acting engine; double-acting engines and the opposed-piston two-stroke engine.

The increasing emphasis on reducing costs of operation has meant that the majority of large diesel engines in ships operate on residual or boiler fuels. These fuels have a high viscosity and must be heated to reduce the viscosity sufficiently for efficient injection and atomisation. However, the presence of considerable quantities of sulphur in many of these fuels places a very high demand on the lubricant. Since the large-bore engines have cylinders and pistons that are lubricated by a separate system from the crankcase, oil is sent by a mechanical lubricator driven from the engine to a number of injection points or quills dispersed around the cylinder. Used oil drains from the cylinder into a catch tank and is either burnt as fuel or disposed of as waste. In order to minimise wear in engines running on these fuels, cylinder lubricants containing high concentrations of basic additives to neutralise the corrosive acids are used. The first of these oils was an emulsion of water containing basic additives in an SAE/30 lubricating oil. Subsequently, single-phase lubricants containing highly basic additives have been used for this purpose.

Gas Turbine Lubricants

The vast majority of gas turbines are used in aircraft or are in fact aircraft-derived engines used for some other purpose such as propulsion of naval vessels or for standby generation of electricity or peaking generation of electricity.

In principle, the gas turbine engine is far simpler to lubricate than is the piston engine. No moving parts are involved in the combustion process and lubrication is required solely for the turbine and compressor bearings, the reduction gear and propeller of turbo-prop engines and the various auxiliary equipment drives. However, although the machinery makes little demand on the oil, the wide range of temperatures encountered in aircraft operations introduces problems. If an engine stops and has to be re-started at altitude, very low temperature may be encountered, perhaps down to minus 80°C and an aviation gas turbine lubricant must, therefore, be able to function down to very low temperatures. On the other hand, many of the bearings run at temperatures above 250°C and an oil that will withstand such conditions without undue oxidation or deposit formation is required. Highest temperatures may be reached immediately after an engine has been shut down on the ground when "heat soak" occurs. For these reasons, synthetic lubricants are used almost invariably for aviation gas turbines. The lubricants have to pass certain oxidation, corrosion, load-carrying and shear stability tests in addition to having the required viscosity/temperature characteristics.

Industrial gas turbines, other than those derived from aircraft, are robust and present few lubrication problems and normal steam turbine oil grades are used. These are relatively low viscosity HVI oils containing anti-oxidant and anti-rust additives.

Other Lubricants for Industry

All of the lubricants described under the heading of Engine Lubricants (p. 444) are used in industry as well as in other sectors of the economy. However, there are many other lubricants that have little use outside industry, and it is this diverse mixture of "industrial lubricants" that are dealt with in this section. The products that have most widespread use are those used to lubricate bearings and gears.

Bearings

The most common application of lubricants is to bearings, which are included in an endless variety of mechanical equipment. A bearing is a support provided to hold a moving member of a machine in its correct position, and there are two main types: plain bearings and anti-friction bearings.

Plain Bearings. In its simplest form the plain bearing (Fig. 6.18) consists of a hollow cylinder in which a shaft rotates. The portion of the shaft within the bearing is called the journal, and the hollow cylinder is called the journal bearing. If the revolving shaft is subjected to an end load, a thrust bearing must be provided and may consist of a collar on the shaft rotating against a flange on the support. A plain bearing may also be used to support a shaft or spindle that does not rotate but moves backwards and forwards longitudinally; it is then called a guide bearing.

Plain bearings are usually oil-lubricated, although grease may be used in some instances, as when sealing arrangements are inadequate for oil. When a journal rotates in a bearing, a film of oil is built up as a result of the rotation and prevents metallic contact between journal and bearing. The viscosity of the oil should be as low as possible in order to reduce friction, but not so low that the film will break down under the load imposed and permit metal-to-metal contact. Bearing oils have viscosities ranging from that of light spindle oil to that of heavy

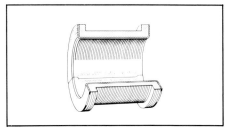

Figure 6.18 **Plain bearing**

cylinder oil, the choice depending on speed, load and operating temperature. Under mild operating conditions, LVI oils are adequate, but the more stable HVI oils are required for more severe conditions.

A special type of plain bearing is the porous bearing, made of sintered metal powder, usually bronze, and impregnated with oil during manufacture. Porous bearings need no additional lubrication in service, and have been particularly successful in lightly loaded apparatus such as domestic vacuum cleaners.

Anti-friction Bearings. In the plain bearing, the moving surfaces slide one over the other. A considerable reduction in friction can be achieved by replacing this sliding motion by a rolling motion, and this is done by attaching hardened steel rings, called races, to the moving and stationary members and inserting between them a row of steel balls or rollers. A cage or separator ensures that the rolling elements are adequately spaced and do not rub together (Fig. 6.19).

Theoretically, rolling contact needs no lubricant, but in practice some sliding occurs in anti-friction bearings, particularly in some types of ball bearings, due to distortion under load. There is also some sliding between the rolling elements and the cages, and lubricant is therefore necessary to control sliding friction, to act as a seal against ingress of moisture or other contaminant and to protect against corrosion.

Most anti-friction bearings use grease, which stays in place without elaborate

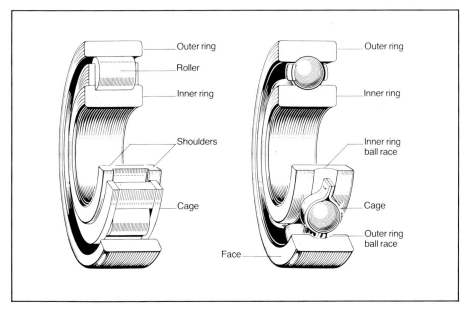

Figure 6.19 **Anti-friction bearings: roller (left) and ball bearings**

sealing, does not need constant replenishment, and is an effective seal against ingress of contaminants. Many bearings are filled with grease on assembly and then sealed, the original grease charge being expected to last the life of the bearing. The choice of grease depends on the speed, load and running temperature of the bearing, but has been considerably simplified by the advent of multi-purpose lithium greases. Performance requirements for some industrial and aviation greases are severe and cover a wider range of conditions than for any other application. In particular, good performance at very high as well as very low temperature is required, which necessitates good oxidation resistance at high temperatures without excessive hardness at low temperatures. Both automotive and aviation greases require high resistance to water, and little tendency to oil separation.

Oil is sometimes used in anti-friction bearings, but requires very effective sealing unless the bearing is in a totally enclosed, oil-lubricated system.

Gears

Gears are incorporated in many machines as a means of power transmission, for increasing or decreasing the speeds of shafts or for changing the direction of the drive. There are many types of gears to suit varying conditions of operation (Fig. 6.20).

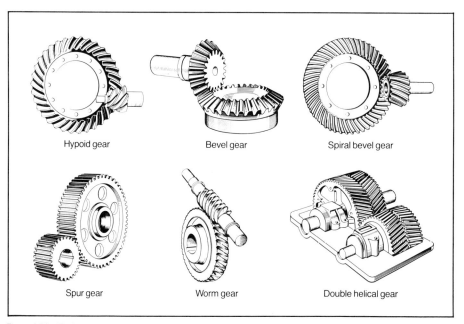

Figure 6.20 **Typical gears**

Spur and helical gears connect parallel shafts either with or without alteration in speed. Where speeds and loads are high, straight tooth spur gears tend to be noisy, and the helical gear with its smoother transfer of load from one tooth to the next is quieter.

Worm, bevel or crossed helical gears are used to change the direction of drive, with or without alteration in speed. A common example is the worm or bevel gear in the rear axle of many modern vehicles. The hypoid gear, a special form of bevel gear, is used in the back axle of passenger cars and operates under very severe conditions in that sliding speeds are great and tooth loading high.

The main functions of a gear lubricant are to reduce friction and wear by providing a film of oil between the working surfaces and to carry away the heat generated during tooth contact. If the load conditions are not severe, oils without additives are suitable. They must be viscous enough to maintain the film yet sufficiently free-flowing to dissipate the heat. HVI oils are used, with the viscosity grade depending on the speed of the gear. For the more highly loaded spur, helical and bevel gears, oils containing mild EP (extreme pressure) additives, such as lead naphthenate and sulphurised fatty oils, are often used. Worm gears are lubricated with high viscosity HVI oils, compound oils or mild EP oils.

Hypoid gears require special EP oils to prevent scuffing. Such oils incorporate additives containing sulphur, chlorine or phosphorus, which react with the gear surface when local high temperatures are developed and form a protective film that prevents metal-to-metal contact. Hypoid lubricants for the back axles of modern passenger cars have to meet stringent specifications imposed by car manufacturers and government authorities, which specifications include severe performance tests.

Whereas in the past the load-bearing capacities of hypoid gear oils were customarily assessed by various laboratory test machines or laboratory rigs involving actual gears, there is a trend nowadays to full-scale axle tests for both development and control purposes.

Compressors

A compressor is a widely used machine for compressing various gases. Compressed air can be used for such purposes as to provide power to operate tools, to spray liquids and to grease-lubricate vehicles. Gases of all sorts are compressed in order to store large quantities in small cylinders either as compressed gas or as liquid. Large compressors are used in natural gas transmission, steel works and chemical plants.

When a gas is compressed its temperature rises in proportion to the degree of compression. High compression may cause very high temperatures and so is usually achieved in stages, with the gas being cooled between the successive stages.

There are three main types of compressor, namely reciprocating, rotary positive displacement, centrifugal and axial flow. The reciprocating type is the simplest and uses a piston and cylinder with valves. Rotary positive displacement or sliding-vane compressors operate by trapping gas in a succession of cells. The gas is then compressed by the mechanical contraction of the cell enforced by the contour of the casing as the rotor revolves. Centrifugal and axial flow compressors operate continuously by the velocity of the gas.

The method of lubricating a compressor varies with the type and design of the unit.

Except where no internal lubrication is required, compressors bring lubricating oil and gas into intimate contact under the best possible conditions to promote reaction, i.e. high pressure, high temperature and fine state of subdivision or large area of exposed surface. This must be taken into account in the choice of lubricant. When compressing air, oils with good oxidation resistance are required; when compressing other gases, the possibility of chemical reaction with the lubricant must also be borne in mind.

Depending on the severity of the conditions, either high quality oils without additives, or oils with anti-oxidant and anti-rust additives (as for steam turbines), or oils with anti-oxidant and dispersant additives (as for diesel engines) are used. The latter are particularly preferred for high-temperature operations where severe oxidation and deposit formation is to be expected. For grease-lubricated anti-friction bearings, multi-purpose lithium greases are preferred.

Refrigerators

Most refrigerators have compressors, either reciprocating or rotary, to compress the refrigerant, the subsequent rapid expansion of which produces the cold. Although the main functions of a refrigerator lubricant are the same as those of compressor lubricants, conditions in the refrigerator impose additional demands such as low viscosity and good low-temperature properties. Moreover, since the refrigerant and the lubricant are brought together, they must not interfere with each other in their respective functions. Straight MVIN oils are generally used, with appropriate viscosities for the particular application.

Steam Turbines

Whilst most small and medium size electricity generating plant is driven by either diesel engines or gas turbines, all large-scale generating plant is driven by steam turbines. In the steam turbine, high-pressure steam impinges on blades set round the circumference of a rotor, causing it to rotate at high speed. Most turbines have several rows of blades on the rotor alternating with rows of fixed blades on the casing, which direct steam on to the rotor blades.

Oil is supplied by a circulating system to the bearings on which the rotor revolves, to the governor mechanism and to the reduction gears. It also serves as the hydraulic medium to operate the governor controls, acts as a coolant and protects metal parts against rusting. Turbines have become increasingly more compact and steam temperatures higher so that the coolant function of the oil has become increasingly important.

Turbine oils are expected to operate for years without replacement except for occasional topping up, and must therefore be very resistant both to oxidation at operating temperatures and to the catalytic effect of copper, brass and steel. Moreover, since water is always present in the system due to steam leakage past bearing seals, it is essential that the oil and water should separate cleanly so that the water can be periodically drained off. In addition, it is customary to centrifuge or filter the oil in the course of circulation. Turbine oils are therefore usually solvent-refined oils with good demulsification properties and containing oxidation inhibitors and anti-rust additives.

When turbines are used as marine power units, reduction gears have to be used and are lubricated by the same oil as for the turbine. Normal turbine oils are satisfactory for this purpose for moderately loaded gears, but for the more heavily loaded gears in modern ships a specially formulated EP turbine oil is necessary to prevent gear scuffing.

Electrical Equipment

Oil is used in several types of electrical equipment, both as a coolant and as a dielectric i.e. having the properties of an electrical insulator. Oils for this latter purpose are known as "insulating oils".

Transformers. Transformers are used to raise or lower the voltage of electric current to the desired level, and consist essentially of insulated copper windings around an iron core. All but the smallest transformers are immersed in oil, which assists in removing heat generated in the coils and core and provides insulation between core parts.

The oil may attain temperatures around 90°C and is exposed to some extent to air and also to copper, a well-known oxidation catalyst. Under these conditions it will slowly oxidise, with the formation of acids and sludge, one or other of which may predominate, depending on the type of oil and the operating conditions.

The insulating ability of a transformer oil depends largely on its freedom from contaminants, such as dust, fibres, moisture and its own oxidation products, all of which can seriously reduce the electrical resistance of the oil. For this reason particular care is taken in the packaging, distribution and general handling of transformer oils, and special tests have been devised to assess their performance, e.g. measurements of resistance to oxidation and of electrical resistance.

Although oxidation-inhibited transformer oils are available and show much longer life in laboratory oxidation tests, the service life of uninhibited transformer oils is generally so long that it is usually regarded as adequate, and uninhibited oils are in general use. Their quality is rigorously controlled by various national specifications all of which have long been established and each of which enjoys almost exclusive merit in its own territory.

Switchgear. Oil-filled switchgear is widely used over an extensive range of voltage. The function of the oil is to quench the arc formed between the contacts when the switch is operated and to prevent it from re-striking. Transformer oils are generally used for this purpose.

Capacitors. A capacitor stores electrical energy in an insulating material located between conducting surfaces. A common type consists of impregnated paper in between sheets of metal foil. Petroleum jelly and mineral oils traditionally were used as impregnants, although a variety of organic liquids is also used nowadays, while plastic films are of increasing importance.

Since capacitor oils are not exposed to air to any great extent, and capacitors operate at moderate temperatures, there is comparatively little risk of oxidation. However, even a slight degree of oxidation would have serious consequences and oils with high oxidation stability are always used.

The electrical stresses in the insulating material are higher in capacitors than in other electrical equipment, and the impregnant must be able to withstand them, so that leakage currents are very small and little heat is generated. Capacitor oils must therefore have high electrical resistance and resist chemical decomposition under electrical stress.

Metal-working

Metal-working operations may be divided into two categories, cutting and forming. Both require rather specialised lubricants.

Metal-cutting. Cutting operations include all those in which a portion of the metal is parted from the bulk, as in turning, screw-cutting, tapping, boring, planing, honing and grinding. A lubricant is required primarily to cool the tool and the work-piece, but also to provide lubrication between tool and work-piece, to prevent chips from being welded to the tool, to wash away the chips and to protect tool and work-piece against corrosion. In doing all this, the lubricant ensures longer tool life, higher precision and better surface finish, as well as reducing power and tool costs and increasing capacity.

APPLICATION, SPECIFICATION AND TESTING 453

Water is the best common coolant known and was indeed the first cutting fluid to be used, either as such or as aqueous solutions of soap or soda. However, such liquids are poor lubricants and promote severe rusting of ferrous metals. Fatty oils, mineral oils or mixtures of the two proved to be excellent lubricants and protectives but not such good coolants, and eventually "soluble oils" were developed. Soluble oils are mixtures of oil and emulsifier that form an emulsion of oil in water when diluted with water, thus combining good cooling with sufficient lubrication for many operations.

Three types of cutting fluid are widely used, soluble oils, water-base fluids and straight cutting oils, the choice depending on the severity of the cutting operations.

Soluble oils are used where cooling is more important than lubrication, as in turning and grinding. They consist of low viscosity mineral oils containing emulsifiers which on dilution with water form stable oil-in-water emulsions. They are normally used at concentrations of one to twenty per cent by volume of oil circulating over the seat of the operation. Sodium naphtha sulphonate (a by-product of the refining of transformer and technical white oils) is commonly used as the emulsifier. Soluble oils can be adapted to more severe cutting operations by the inclusion of EP additives to produce EP soluble oils. The emulsions formed by normal soluble oils are opaque, milky fluids. "Clear soluble oils", which contain more emulsifiers, produce transparent emulsions which enable the work-piece to be seen more clearly, at least for a limited time until the emulsion becomes opaque with use.

Water-base cutting fluids are used where greater and more permanent clarity is required. They consist of concentrated aqueous solutions of chemicals (such as sodium nitrite, triethanolamine sebacate and soaps of polyglycols) and are diluted with water before use. Although they are aqueous liquids, they do not promote rusting and have adequate lubricating ability.

Straight cutting oils are used in more severe operations where the lubrication provided by soluble oils is insufficient. Various grades are used according to the severity of the operation. For machining non-ferrous metals, mineral oils without additives or blends of mineral oil with 10–15 per cent by weight of fatty oils are used. For somewhat heavier operations on low alloy steels and certain non-ferrous metals, mineral oils are used that contain mild EP agents that do not stain copper such as sulphurised fatty oils. For even more severe operations on steels, more active EP oils are used that may contain free sulphur or compounds containing sulphur, chlorine or phosphorus individually or mixed.

The cutting performance of cutting oils is impossible to assess by means of ordinary laboratory tests, and actual cutting tests carried out on a statistical basis under rigorously controlled conditions are essential.

Metal-forming. Metal-forming operations include all those in which the metal is deformed to produce the required shape and dimensions, as in rolling, drawing, extruding and forging. Lubricants are used to reduce friction and wear, to remove heat generated in the process and to achieve good surface finish. Two of these operations, rolling and drawing, are of particular industrial importance.

Rolling. In the manufacture of metal plates, sheets and strip, heated billets of metal are rolled down to plate thickness of 3–10 mm (hot rolling) and then further reduced in thickness if required by cold rolling, which gives a better surface finish than hot rolling. Occasionally (as with some brasses) the cast ingot is cold rolled.

Hot rolling is done in the presence of water (steel) or dilute soluble oil emulsions (aluminium). For the cold rolling of steel to car body thickness, and for the production of mild steel strip in general, special soluble oils are used to give sufficient lubrication to permit great reduction in thickness, but not enough lubrication to cause slipping. The emulsion must also protect against rusting and leave no harmful deposits after annealing. For very thin steel plate (tin plate), palm oil or a palm oil/water emulsion is generally used. For cold rolling of aluminium, copper and copper alloys, straight oils, fatty oils and compounded oils are all used. For aluminium foil, kerosine or spindle oil/kerosine mixtures containing oleic acid, fatty oil or esters are used. They must give no staining on annealing.

Drawing. Many metal articles are manufactured by pressing metal sheets or plates into the required shape by means of suitable punches and dies — "pressing" and "deep drawing". Wire, rod and tubes are produced by cold drawing through dies.

These processes, often carried out in several stages, require lubricants to reduce friction and wear of both the die and the work-piece, and to produce a good surface finish. A variety of lubricants is used, from viscous mineral oils and compounded oils to aqueous emulsions of fats with or without mineral oils, according to the severity of the operation. Dry powdered soap is used for most steel wire-drawing because it permits long die life in spite of the high pressures involved, but for non-ferrous wire-drawing, and for some fine steel wire, oils or emulsions are used. "Drawing compounds", consisting of emulsions of partly saponified fats, are used for deep drawing and tube-drawing, and EP agents or fillers such as talc or zinc oxide are sometimes incorporated in them.

Heat Treatment. Heat treatment refers to any thermal operation to which a metal is subjected in the solid state with the object of modifying its physical properties. Two particularly important treatments, hardening and tempering of steel, make considerable use of oils.

APPLICATION, SPECIFICATION AND TESTING   455

Hardening involves heating to a certain minimum temperature and then cooling rapidly (quenching) in some suitable fluid. Tempering consists of re-heating the hardened steel, maintaining it at an elevated temperature for a period, and then cooling again, usually in air. The re-heating is sometimes done in oil-baths.

Heat Transmission

Oil is often used as a medium for supplying heat for industrial purposes. It is usually circulated through pipes from a heater to the required site and back again. The oil must be thermally stable so that deposits do not form in lines and heater tubes. In a well designed system there is little contact with air so that oxidation is not severe. To avoid difficulty in starting the circulation from cold, the oil should have as low a viscosity as is compatible with low volatility at operating temperatures. At one time it was customary to use rather viscous cylinder oils, but medium viscosity distillates are now generally preferred.

Hydraulic Equipment

Hydraulic power-transmission serves a wide range of purposes where multiplication of force is required or where accurate and dependable control gear must be provided. Various developments such as automation have greatly extended the use of hydraulic equipment.

The principal requirements of a hydraulic fluid are that it should be relatively incompressible and sufficiently fluid to permit efficient transmission of power. It must also be a good lubricant for the pumps and bearings in the system and form a good seal between the moving parts. It should not foam, should be stable against oxidation and provide good protection against corrosion. If water enters the system, it must separate quickly from the oil.

In older and larger hydraulic machinery, water is still used, but it is a poor lubricant and causes rusting. Soluble oil emulsions are used in some closed-circuit systems to minimise these defects, but their lubricating properties are not good enough to allow the use of high-speed pumps, and large slow-speed pumps, driven through reduction gears, have to be used. Rusting is not entirely avoided and rust particles can cause damage to packings and valves.

Mineral oils eliminate the danger of rusting and permit the use of high-speed, high-pressure pumps directly driven by an electric motor, with the hydraulic fluid serving as the lubricant for all working parts. HVI oils with a range of viscosities are used, containing anti-oxidant, anti-rust, anti-foaming and anti-scuffing agents.

If the fire risk is large and/or its consequences very serious as in aircraft or in some industries, fire-resistant fluids are used. These can be water-base fluids, containing glycol or similar materials, water-in-oil emulsions or synthetic materials such as chlorinated or fluorinated hydrocarbons or phosphate esters.

Textile Machinery

In all machinery a certain amount of power is used in overcoming friction within the oil film itself. This loss becomes excessive in the high-speed machinery used in the textile industry, and the lubricant must therefore be as thin as possible consistent with good lubrication. Textile machinery usually operates in humid atmospheres, and the oil must give protection against corrosion. Moreover, care must be taken not to soil the fabric being processed and so the oil is made non-splashing by incorporating small amounts of soap or polymer. Where there is still a risk of contamination, the oil must be easily removed by scouring, or where scouring is inadmissible, the oil must be highly stable and colourless. Suitable additives are used to promote scouring.

The modern high-speed oil-bath spindle rotates at 11,000 to 15,000 rev/min or even faster, and may run for 6000 hours or more on a single charge. Highly refined oils that will not thicken or otherwise deteriorate in use are therefore necessary.

Anti-friction bearings are used extensively in textile machinery under a wide range of temperatures and often in very moist situations. Multi-purpose lithium greases can be used for most of these operations instead of several types of grease, each for its own specific conditions.

A wide range of textile machinery oils is required to cover the many different conditions of operation; they range from oxidation-inhibited HVI oils to similar oils containing non-drip, anti-rust, emulsifying or other additives. Technical white oils are used when light coloured oils are required to avoid discoloration of textiles.

Textile Manufacture

Oils are also used to lubricate textile fibres (particularly wool) during the various manufacturing processes. Such oils must be removable by scouring at a later stage and must leave no residues or otherwise affect the fibre in any way that will interfere with subsequent processing. Emulsions of oleic acid or of a mineral oil are generally used. In the latter case the mineral oil, with suitable emulsifiers dissolved in it, is supplied for dilution with water before use.

Synthetic fibres also require lubrication to prevent breaking during processing. Emulsified oils made with low viscosity white mineral oils are generally used.

Mineral oil without additives is used as "batching oil" to soften the fibres in preparing jute for further processing.

Corrosion Protection

The protection of metals against corrosion is important because corrosion can cause very considerable losses in both materials and time. Relatively permanent protection is provided by painting, enamelling, etc. Such durable and expensive coatings are neither justifiable nor desirable in many instances where some measure of protection is nevertheless required to keep the metal surface in good order pending further processing or use. The need arises most frequently during transport or storage, and may be for periods ranging from a few days to a year or more. Products serving this purpose are known as "temporary protectives", temporary not only in duration but also in case of removal, as opposed to paint and similar "permanent protectives".

Protection from corrosion depends on preventing moist air from coming into contact with the metal, and the application of an impervious coating is a convenient and effective means. Temporary protectives provide coatings ranging from thin oily or hard films to thick grease-like films, and are commonly classified according to the type of film (oily, hard or soft) and the method of application (hot dip, solvent-deposited). The duration of protection depends not only on the type of film but also on the severity of the conditions to which the article is exposed.

Oily films are suitable only for short-term protection, but are easily applied and removed, and in some instances need not be removed at all before further use or treatment. They are used for example on precision tools, strip and sheet after rolling, enclosed gears on idle equipment or for indoor storage of spare parts. They are provided by mineral oil with or without rust inhibitors or other protectives such as petrolatum or lanolin.

Hard films provide protection for comparatively longer periods under severe conditions and are not readily disrupted by handling, as are oily or soft films. Hard films pick up contaminants less easily, but they are correspondingly more difficult to remove. They are usually deposited from solutions of rosin, waxes, lanolin and similar materials in a volatile solvent.

Soft films give more lasting protection than oily films, the duration of which depends on their thickness and consistency. They are used on metal parts during storage and on the exterior of machinery and on packaged parts and assemblies during transport and storage. Thick soft films can be used for storage or shipment under severe outdoor conditions. Soft films are provided by petrolatum or lanolin/petrolatum mixtures, lubricating greases or solutions of such substances in a volatile solvent such as white spirit. The solvent types are applied cold, the others either hot or cold.

Special water-displacing grades of solvent-deposited soft-film protectives are used on wet surfaces, the water being displaced from the metal and replaced immediately by a protective film.

Process Oils

A considerable quantity of mineral oil is used in a wide variety of industrial processes, and such oils are referred to as "process oils".

Typical applications are as ingredients in the manufacture of printing inks, rubber articles (rubber extenders), horticultural sprays, cosmetics, hair creams, etc, and as treating materials in leather-dressing, coal-briquetting, dust-laying, egg-preserving, etc.

The type of oil used naturally depends on the application in question; for many applications low-viscosity naphthenic distillates are suitable, but for others, such as cosmetics, hair creams and other toiletries, where it is important that the oil should be free from odour and colour and absolutely harmless to the skin, only the more highly refined white oils can be used.

White oils are produced by more drastic refining of MVI distillates to remove unsaturated compounds and other constituents that impart colour, odour and taste. They are usually solvent-extracted and then repeatedly treated with strong sulphuric acid or oleum and alkali. Two types of white oils are made, medicinal oils and technical white oils, differing in the extent of refining, the latter being rather less severely treated.

Several grades of medicinal oil are manufactured to comply with the various national pharmacopoeia specifications. These specifications differ in minor details but all require a water-white oil, free from all harmful components, odour and taste. In addition to their medical use as "liquid paraffin", medicinal oils are used as lubricants for food-handling machinery, where oil may come into contact with food, and as ingredients in cosmetics and other pharmaceutical preparations.

Technical white oils are either water-white or semi-pale (half-white oils) and are made in several viscosity grades. They are used as lubricants where light coloured oils are required, for example in textile machinery. They are the only legal lubricants in some countries for the lubrication of spinning mules, and their main use is in the cosmetic industry for the manufacture of hair oils and creams and other preparations. The non-toxic nature of these products is of course a necessary property in these applications. They are also used in horticultural sprays and for the impregnation of some types of wrapping papers.

Petroleum Waxes

About two million tonnes of petroleum waxes are produced each year, of which most is paraffin wax and the remainder micro-crystalline wax.

For many years paraffin wax was used almost entirely for the manufacture of candles, first supplementing and eventually supplanting the use of beeswax, stearine (from animal fats) and tallow. Today it has many and varied end uses, of which candle-making is but one. The end use pattern varies markedly in different areas according to the degree of industrialisation and to local custom. For instance, the use of pre-packed foods is far greater in the USA than in Western Europe, and the offtake of wax for frozen wrappers, bread wrappers, milk and fruit-juice cartons and the like is the major proportion of the total demand, whilst in Europe it is much less.

The principal applications of petroleum waxes are described in the following pages.

Wax as a Combustible

The proportion of the world production of wax used for candle-making has been declining for many years, but has now levelled out at about 20 per cent. However, the actual quantity used for candles is increasing because, although the demand for household candles has diminished, there has been an increasing demand for decorative and art candles, church candles, nightlights and hot-plate candles.

Candles were once made by repeated dipping in molten wax and draining, starting with the wick and finishing with a candle of the required dimensions. Some art and church candles are still made in this way, but most candles are now made by moulding paraffin wax around a wick in machines. Semi-refined paraffin wax with a melting point of 50°C and upwards is normally used. Paraffin wax candles tend to stick in the mould, to bend when warm and often to exhibit mottling. Small proportions of substances such as stearine, microcrystalline wax or polyethylene are added to overcome these difficulties, and they also produce an opaque white candle that is generally preferred to the translucent article produced by straight paraffin wax. Nightlights are also made by the same process, except that the wicks are inserted after moulding; they usually consist wholly of paraffin wax. An alternative method of manufacturing candles is by the cold extrusion of powdered wax through a die.

Paraffin wax is used in the manufacture of matches for impregnating the wooden splint so that it readily catches fire and continues to burn. The cheaper, lower melting point grades (match waxes) are used, the splints being dipped in the molten wax at a temperature that will give good penetration. Book matches, using cardboard instead of wood, are similarly treated.

Paraffin wax is also used in the manufacture of explosives and fireworks, and slack wax in the manufacture of fire-lighters from waste paper and wood fragments.

Wax as a Waterproofing Material

Wax is used for waterproofing a great variety of materials, of which paper and textiles are the most important. The treating of paper and board for packaging purposes has superseded candle-making as the major application for petroleum waxes — at least half of all petroleum wax produced being used in this way. Wax is ideal for the purpose; it is clean, odourless, tasteless, harmless and attractive in appearance; it is chemically inert, resisting attack by strong acids, alkalis and oxygen at normal ambient temperatures; it is insoluble in water, practically impervious to air and moisture, even in thin films, and produces a hydrophobic surface that repels water from the treated material.

Although little affected by oxygen in the air at ordinary temperatures, petroleum wax is oxidised more readily if kept molten for prolonged periods, as in dipping baths, and can discolour and develop an unpleasant odour. Therefore, when materials are waterproofed by dipping in molten wax, working temperatures are kept as low as possible, small baths are used so that make-up with fresh wax is reasonably frequent, and waxes containing oxidation inhibitors may be used.

Waxing of Paper. Paper can be either coated or impregnated with wax. In coating, wax is applied to one or both sides of the paper by passing it between heated rollers either or both of which carry molten wax picked up from a heated trough. A continuous film of wax is formed on cooling. Alternatively, the paper may be passed through a bath of molten wax and immediately chilled in cold water or on cold rollers so that the wax forms a continuous glossy film. The weight of wax applied varies considerably according to the application but may equal or even exceed that of the paper.

Straight paraffin wax is not very suitable for this application; it is too brittle, and the film is liable to crack when the paper is folded or creased, especially at low temperatures. Blends of paraffin wax with about 25 per cent by weight of microcrystalline wax are used, being much more flexible and less liable to crack. Coated papers are practically impervious to water vapour, and are accordingly used for wrapping foodstuffs, such as bread, that are affected by changes in atmospheric humidity or deteriorate by loss of moisture.

Impregnated papers are made by immersion in a bath of molten wax followed by passage between heated rollers, so that most of the wax is absorbed and little, if any, is left on the surface. Impregnated paper is not impervious to water vapour, but it does prevent the passage of liquid water and it has a hydrophobic surface. It can therefore be used for paper cups and bags. Since cracking or creasing is not a problem with such articles, straight paraffin wax can be used.

Waxing of Board. Board is waxed either by coating or by impregnating according to the end use. An important application is in the manufacture of cartons for milk, fruit-juice and other food stuffs. Such cartons are coated with wax, either by dipping a preformed carton in molten wax or by assembling the cartons from blanks that have been coated by passing between heated rollers carrying molten wax, as described for paper coating.

Drinking cups are impregnated with wax by dipping or spraying after assembly and then passing through a hot oven to allow the wax to soak into them.

Blends of paraffin wax with microcrystalline wax or polymers or both are used for coating cartons so as to give a tough, flexible, non-cracking, non-flaking film; similar blends are used for impregnation.

Waxing of Textiles. Wax is used to make textile materials water-repellent or "shower-proof". The wax is generally applied as an emulsion, either by spraying or dipping, so that the textile fibres are wax-coated but the interstices are not sealed. The fabric remains porous to air (thus giving adequate ventilation) but is water-repellent. The grade of wax used is not critical.

Wax is also used to treat yarn to facilitate further processing and generally to improve the "handle" of the finished fabric.

Waxing of Foodstuffs. Besides the use of waxed wrappings for the preservation of foodstuffs, wax may be applied direct to the surface of some foods. The wax may be applied as a thick overall covering, surrounding the article with an airtight package, as in the waxing of cheese, or as a thin layer which slows down evaporation of water but still allows breathing to take place, as in the waxing of fruit and vegetables, e.g. oranges and cucumbers.

Thick films are produced by dipping in molten wax at about 66°C, and thin films by dipping in a wax emulsion or by spraying with an emulsion or a solution of wax in a volatile solvent.

Wax films not only hinder or prevent the foodstuff drying out, but also inhibit the growth of moulds and bacteria (for example on cheese).

Polishes

Natural waxes and petroleum wax have long been used in the manufacture of polishes, and about 10 per cent by weight of the total production of petroleum wax is so used. Natural waxes are undoubtedly superior to all others in their durability and gloss when rubbed to a polish, but many of them are too hard to be easily worked and they are expensive. On the other hand, paraffin wax has little polishing value, giving a dull, easily smeared film; microcrystalline wax is not much better, giving somewhat sticky films. However, blends of natural, paraffin and microcrystalline waxes make excellent polishes, the petroleum waxes serving to modify the consistency and texture of the polish, assist in solvent retention and eke out the more expensive natural waxes. Such blends are used in a great variety of polishes, e.g. for boots and shoes, furniture, motor vehicles and leather articles. Beeswax, carnauba and candelilla waxes are the chief natural waxes that are blended with paraffin wax or microcrystalline wax or both to give the required consistency and texture.

Wax polishes are prepared in three main types, pastes, creams and liquids, although powdered or flaked waxes, usually blends, are used for treating dance floors.

Paste polishes are mixtures of wax and solvent, usually white spirit, made by dissolving the wax in the hot solvent and allowing to cool. The wax separates out as a network of wax crystals in which the solvent is enmeshed, the whole forming a fine-structured paste, the consistency of which depends largely on the solvent content and the way in which it is dispersed, which in turn is influenced by the nature of the wax blend and its "solvent retention" properties. When the paste is spread and rubbed out, the solvent evaporates and a continuous film of wax is formed.

Creams may be either dispersions of water in wax–solvent mixtures or dispersions of the wax–solvent mixture in water, stabilised by various dispersing agents. On rubbing it, the water and the solvent evaporate, leaving a wax film.

Liquid polishes are usually wax–solvent mixtures, such as are used in paste polishes, but with sufficient solvent to give a liquid product, i.e. the solvent is not enmeshed in a crystalline structure, but the wax is either in solution or dispersed in the solvent. Some liquid polishes are dispersions of wax in water with insufficient wax to form a cream.

Miscellaneous Uses

Among the many other uses of petroleum wax, the following are some of the most interesting.

Wax is used in a metal-casting process known as the "lost-wax process". A replica of the article to be cast is shaped in wax and is then coated with a suitable refractory material. The mould is then heated and the liquid wax removed, thus leaving an accurate replica into which the metal is cast.

Wax is blended with rubber to protect it from deterioration or to modify its properties.

Microcrystalline wax is a good insulator and resistant to water and is therefore used in the manufacture of cables as an impregnant for covering fabric or as an external coating.

Wax is a feedstock in the chemical industry for various processes. It may be cracked to olefins for the production of raw materials for detergents, oxidised to acids which are then esterified to form wax-like compounds used in the manufacture of polishes and crayons, or chlorinated to make "chlorinated waxes" used as plasticisers and for electrical insulation.

Wax baths are used as a heat transfer medium in the physiotherapeutical treatment of limbs affected by arthritis, rheumatism or post-operational stiffness.

Testing of Wax and Waxed Products

Properties of petroleum waxes such as oil content, melting point, hardness, viscosity, density, colour, etc., serve to characterise the various grades of wax, and can be determined by standard tests.

These tests do not measure the performance of wax in many of its applications, and functional tests in which the conditions of the application are simulated are in common use, although few of them have yet been standardised. Some of the more important are listed below.

Tensile strength is defined as the longitudinal stress required to break a test specimen of specified dimensions in a standard ASTM test (D 1320).

The rigidity modulus is the ratio of stress to strain in a material subjected to a shearing force. It gives an indication of the force required to shear the material, and in the case of paraffin wax provides a better correlation with product quality for certain applications than do other less precisely defined properties such as hardness or penetration or even the above-mentioned tensile strength.

Some food wrappers are sealed by pressing the portion to be sealed successively between heated and cooled plates. The force required to pull apart two strips of standard paper, coated with wax to be tested, and sealed together in this way, is a measure of the suitability of the wax for this application.

Waxed papers tend to stick together, which is undesirable. This tendency is measured by a "blocking test" in which sheets of waxed paper are laid on one another, pressed together by a weight, and kept successively at a range of temperatures; the temperature at which sticking first occurs is the blocking point. This test has been standardised by the ASTM (D 1465).

When wax cartons are bent or deformed, as in closing and opening, the wax coating should not flake and allow particles of wax to drop into the contents. This quality is assessed either by flexing tests or by dropping a weight on to wax-coated board.

When waxed paper is passed between rollers in the course of waterproofing, the wax should not "scuff" off the paper and build up on the roller. This quality is assessed by passing a waxed paper strip between two rollers, one of which is loaded, and determining the loss in weight of the strip.

The glossiness of a waxed paper and its retention of gloss on storage are important characteristics in relation to customer reaction. Visual inspection is highly subjective and is liable to give variable ratings, and a more objective test is used in which the reflected light is measured by special instruments.

Permeability to water vapour is measured by sealing a moisture-absorbing salt such as calcium chloride in a waxed container, storing the container in a humid atmosphere under standard conditions, and measuring the increase in weight of the salt.

Bitumen

In the UK the term "bitumen" is applied to the solid or semi-solid residues from the distillation of suitable crude oils. The same products are known as "asphalts" in the USA, a term reserved in the UK for natural or mechanical mixtures of bitumen and mineral matter.

Bitumen also occurs naturally alone or mixed with mineral matter, and was first known to man as seepages exuding from the ground. Substantially pure bitumen deposits occur in various parts of the world, the best known being gilsonite in Utah and Colorado, USA. Natural rock asphalts, impregnated with up to 14 per cent by weight of bitumen, are quarried in several parts of Western Europe, and the names Val de Travers, Neuchatel and Ragusa are well known. The most renowned natural asphalt deposit is the Trinidad Lake containing a mixture of about 39 per cent by weight of bitumen, 32 per cent by weight of mineral matter and 29 per cent by weight of water and gas.

Bitumen obtained by distillation of crude oil is generally a black or dark brown material ranging from a highly viscous to an almost solid substance at normal ambient temperatures, depending on the amount of light fractions removed. Light brown varieties known as "albino" bitumens are produced from a few crudes. On heating, bitumen softens gradually and eventually becomes fluid; the temperature at which it reaches a certain consistency under arbitrary conditions is called the softening point. Commercial grades have softening points ranging from 25 to about 135°C. The hardness or consistency of a bitumen is determined by measuring the distance in tenths of a millimetre to which a needle penetrates into the bitumen under standard conditions of load, temperature and time, and ranges from zero for very hard to 500 for very soft bitumens as measured at 25°C.

Bitumen can be oxidised, or rather dehydrogenated, by blowing air through it at high temperatures. "Blown" grades have a somewhat rubbery consistency and are less temperature-sensitive than are the straight distillation grades.

Both straight and blown bitumens are made in a number of grades with fairly narrow ranges of penetration and softening point. The straight grades are usually designated solely by their softening points for hard grades or their penetrations for medium and soft grades, while the blown grades are described by the combination softening point/penetration. They all have to be heated to well above their softening points in order to apply them. There are, however, alternative methods of using bitumen at lower temperatures, that is as a solution in a suitable solvent or as an emulsion in water.

Bitumen Emulsions

Bitumen emulsions are made by emulsifying 50–70 per cent by weight of bitumen in water in the presence of 0.5–1.0 per cent by weight of an emulsifying agent, usually a soap. They are generally used cold for both road-making and industrial purposes.

Road emulsions are generally made with 200–300 penetration bitumen; the emulsions must remain stable in storage and during transport but "break" soon after application to the road, and be of suitable viscosity for spraying. Modern road emulsions contain approximately 70 per cent of binder and are sprayed at about 80°C.

Industrial emulsions generally use somewhat harder bitumens and usually contain various admixtures such as clay. They are used for roofing and flooring and in the paper industry, in all of which applications they are required to mix with other materials without breaking and to break only slowly after application. 40–50 penetration grades are used in roofing and flooring emulsions and very hard, high melting point grades in emulsions for paper.

The bitumen particles in these emulsions are negatively charged (anionic emulsions). Many aggregates such as quartzite are also negatively charged, and difficulty is sometimes experienced in coating them, especially when damp, due to the failure of the bitumen to adhere. Special emulsions are made in which the bitumen particles are positively charged (cationic emulsions) and adhere readily to the aggregate; modern road emulsions are mainly of this type.

The annual production of bitumen from petroleum in the world outside the Communist areas is about 17 million tonnes, most of which is used in the construction and repair of roads and airfields, and the remainder in industrial applications and hydraulic construction.

Road Construction

There are two classes of modern roads, flexible and rigid. The flexible road consists of a thick layer of crushed stone or similar material, perhaps coated with a bituminous product, with a wearing surface bound with bitumen or tar. The rigid road is made of cement concrete slabs, which carry the load and also serve as a running surface.

The flexible road has many advantages over the rigid road. It permits some settlement of the foundations without cracking or otherwise failing, and is less liable to crack when exposed to wide variations in temperature. It is jointless, gives a better riding surface and is easier to repair. With rigid roads there is no alternative to heavy construction, regardless of the amount of traffic to be carried, whereas with the flexible road there is considerable scope for varying the thickness of the foundation and the type of wearing surface. With flexible construction it is therefore possible to build up the road in stages as traffic requirements increase, a very great advantage in developing countries.

The components of a flexible road are the wearing surface, the base, and the subgrade. The function of the wearing surface is to provide a waterproof, non-skid cover to the road and to withstand the shear and abrasive action of traffic. It may consist of only a very thin layer of bituminous material for a lightly trafficked road or several inches of high quality bituminous mixture for heavy traffic. The function of the base is to carry the traffic load. It is frequently made up of a number of layers of materials of different strength, and in modern road construction it is becoming increasingly common to build a large part of the base with bitumen-bound material.

In designing a flexible road, the object is to provide sufficient thickness and strength of base and surfacing so that traffic stresses will not deform the subgrade. Until recently, flexible pavements were designed by empirical methods which had proved satisfactory with uncoated granular base materials. The advent of bitumen-coated bases with their far superior load-spreading properties has focused the need for a more scientific approach to road design, and roads are now designed more on engineering principles, taking into consideration the stresses involved to calculate the constructional requirements. With the more rational design methods, uncoated base material can now be used to better advantage than hitherto, although the use of bitumen-bound base is preferred.

The simplest form of bitumen application to roads is surface dressing, the bitumen being sprayed over the surface, which is then covered with stone chippings. This method may also be used to provide a wearing surface on a previously untreated road so as to make it better able to carry traffic, in which case, as traffic increases and the road has to be strengthened, a bituminous carpet of suitable material can be superimposed to whatever thickness is warranted by the traffic.

For road surfacing of a higher quality there are a considerable number of mixes that can be used, the choice depending on the traffic conditions and on the materials and plant available. The most important of these surfacings, bitumen macadam, cold asphalt, sand mixes, asphaltic concrete, hot rolled asphalt and mastic asphalt, are discussed in the following pages.

Aggregates used in making road mixes include broken stone of various sizes, crushed slag, gravel (either natural or crushed), and sand. For certain mixes "filler" is also sometimes added. This is finely ground material most of which passes through a 200 mesh sieve, and examples are limestone dust and Portland cement. The grading of aggregate is of great importance in the designing of road mixes. The larger stone, in general, forms the main structure, the interstices being filled with the smaller stone, sand and filler, the whole being bound together with bitumen to give a compact, durable construction.

Each particle of the aggregate is coated with bitumen, which provides an adhesive, ductile film to bind the particles together and form a resilient, flexible and waterproof structure. The bitumen can be applied as such or in the form of a "cutback" or of an emulsion, the choice depending on the nature of the construction.

Bitumen Macadam

The expression "macadam" perpetuates the memory of John Loudon McAdam who did much to revive the science of road making in the nineteenth century. His principle was to apply to an existing surface a layer of broken stone of suitable quality and size, which when compacted would give inherent strength by interlocking. To provide a denser construction it became the practice to add water and fine aggregate, and this gave rise to the term "water-bound macadam". When modern traffic conditions made such surfaces unsuitable, the water slurry was replaced by hot tar or bitumen poured into the interstices of the stone layer, a process known as "penetration macadam" or "grouting". Penetration macadam has, however, been replaced in many countries by precoated material known as tar or bitumen macadam. The name "tar macadam" or "tarmac" is often loosely applied to any coated aggregate of this type, irrespective of whether tar or bitumen has in fact been used. Bitumen macadam consists of a mixture of coarse aggregate and a relatively small quantity of fine aggregate and filler, coated with a bitumen or cutback. The stability or load-bearing capacity of the mix depends to a great extent on the mechanical interlocking of the aggregate particles. Mixes can vary from very densely graded mixtures with low void content to very open graded mixtures of high void content.

Bitumen macadam can be laid in one or two courses, the thickness varying from half to two inches for the former and from two to four inches for the latter. Single-course mixes are used for general resurfacing work and two-course mixes on new roads or on roads that need strengthening. The macadam may be spread manually or mechanically.

Cold Asphalt

Cold asphalt provides a clean even surface with a "sandpaper finish". It is normally used as a thin carpet material for the resurfacing of existing roads, but it is also very suitable for surfacing footpaths, railway platforms, playgrounds, and for general maintenance and patching work. Two varieties are recognised: fine cold asphalt all passing a 1/4 in. BS sieve, and coarse cold asphalt substantially all of which passes a 3/8 in. BS sieve.

It consists of a finely graded aggregate, the most suitable being crushed slag, coated with a small quantity of soft bituminous binder. Cold asphalt, in contradition to its name, is always mixed hot and is frequently laid hot. The principle of the design of the mix is the provision of mechanical stability by the interlocking of the aggregate particles on consolidation; the bitumen initially assists compaction by its lubricating effect and then functions as a waterproofing binder, giving the necessary cohesion to withstand the combined action of traffic and weather. By virtue of the relatively low binder content the mix is easily workable and, with a suitable choice of binder, it is possible to produce a mix that can be stored and then laid cold.

Sand Mixes

In many parts of the world, stone aggregate is not readily available, and the only cheap aggregate available locally is sand. Sand can be used in road surfacing work in a variety of ways — hot sand mix, cutback sand mix and wet sand mix, and each of these will be considered briefly.

Hot Sand Mix. These mixes are made using naturally occurring sand either as such or blended with quarry fines. They are not so carefully graded as sand sheet asphalt and do not, therefore, have the same durability or stability, but they provide a satisfactory surfacing for roads carrying fairly heavy traffic.

Their stability depends largely on the hardness of the bitumen and therefore as hard a bitumen as convenient is used compatible with the climatic conditions. They can be made in a fairly simple hot mix plant and are frequently laid by grader.

Cutback Sand Mix. "Cutbacks" consist of bitumen, usually 100 penetration grade, "cut back" with a solvent, usually kerosine or creosote but sometimes white spirit or gas oil. Cutbacks become more viscous and bind more firmly as they lose solvent, and the rate of "setting" or "curing" depends on the nature and content of solvent used and the climate. The particular type and grade to be used depends entirely on the application. Cutbacks are seldom used cold but generally at 90–100°C, considerably below the temperature necessary for straight bitumens. Where dry sand is readily available, sand mixes using a cutback binder will provide a low-cost method of road surfacing. They are suitable only for light and medium pneumatic-tyred traffic, but after a period of service will provide a suitable base on which to lay a better quality surface able to carry heavy traffic. Cutback sand mixes have been used with considerable success for roads and airfields in a number of Middle Eastern and African countries and in the USA.

The sand should be clean and preferably even graded, as this will increase the stability of the mix. A wide range of cutbacks can be used, the choice depending upon the climate and the type of sand and mixing equipment available. Cutback sand mixes can be spread by hand or mechanically. Initially, the mix remains somewhat plastic but eventually sets up sufficiently to carry traffic, at which stage a final surface dressing is necessary to prevent surface fretting.

Wet Sand Mix. The wet sand mix was developed in Western Europe during World War II as a rapid means of airfield construction where only wet sand was available. In order to coat the wet sand, cutbacks containing a wetting agent were developed which worked in conjunction with hydrated lime or cement as an activator. Wet sand mixes proved to be superior to dry sand mixes in that they gained stability or bearing capacity much more rapidly.

Hot rolled asphalt can be laid as a resurfacing material or in new construction, where it is normal practice to lay a two-course construction to a total thickness of 3–4 in. It provides a very durable surfacing with a life of 15 to 20 years and is widely used in the UK for surfacing heavy duty roads.

Mastic Asphalt

Mastic asphalt consists of a mixture of bitumen, fine limestone aggregate and filler in proportions that give a voidless impermeable material.

The binder may be petroleum bitumen, natural asphalt or a mixture of the two. The fine aggregate is added in increments to the molten bitumen and cooked in a special boiler which keeps the contents stirred for up to 5 hours. Small chippings are added and the mastic asphalt is then usually cast into blocks, which are remelted at the laying site in a special boiler. Mastic asphalt is spread by hand and provides an impermeable and highly durable surface. In order to improve its non-skid properties, precoated chippings are often rolled into the surface or a rough finish given by an indenting roller.

In some West European countries a slightly different form of mastic asphalt is made consisting of coarse aggregates with a very high filler content. The material is first mixed in an asphalt plant and then further mixed in a mastic boiler while being taken to the site of work. The asphalt is spread mechanically as a thin wearing surface on an asphaltic concrete base. To provide a non-skid finish the surface is rolled with an indenting roller, or small chippings are rolled in.

Bitumen-Bound Bases

As mentioned previously when discussing the design of flexible roads, the modern tendency is to use a bitumen-bound base in place of uncoated or cement-bound material. A bitumen-bound base consists of a well compacted, high stone content mix with a low void content. The bitumen should be as hard as is compatible with ease of mixing and laying and its quantity kept to a minimum.

In designing a bitumen-bound base, use can be made of any cheap aggregates that are locally available. Quarry wastes and pit or river gravel and sand are widely used. A bitumen-bound base can be built up in layers to any required thickness. Traffic is usually allowed to use the bitumen-bound base for up to a year, as this allows any settlement to take place before the expensive final surfacing is placed.

Figure 6.21 **Asphaltic concrete (section showing three courses).**

APPLICATION, SPECIFICATION AND TESTING

Since the war, this process has developed in a number of countries where wet sand is the only available aggregate. It has been widely used for minor roads in Belgium, France and the Netherlands, and for roads and airfields in Africa and the Far East.

Very lean wet sand mixes (and other types of lean sand mix) are proving a satisfactory and economical means of stabilising sand as a road base material, and this form of construction is likely to be used on an increasing scale in developing countries where low-cost roads are urgently needed.

Asphaltic Concrete

Asphaltic concrete consists of a very carefully proportioned mix of coarse aggregate, fine aggregate and mineral coated with bitumen. It provides a surfacing of exceptional durability and is widely used for heavily trafficked roads, motorways and airfield runways under all types of climatic conditions.

An asphaltic concrete mix is designed with a continuous grading from the

Figure 6.22a **Hot rolled asphalt.**

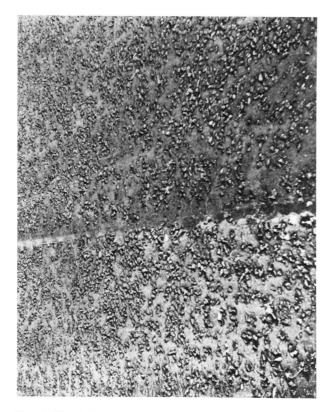

Figure 6.22b **Surface texture of hot rolled asphalt with precoated chips.**

maximum size down to the filler with the object of obtaining a controlled void content. The amount of bitumen added is related to this void content and to the traffic and climatic conditions. The binder used is normally a penetration bitumen.

For making asphaltic concrete an elaborate hot mix plant is required where the aggregates of various sizes, filler and bitumen can be correctly proportioned. Mixing temperature depends on the grade of bitumen but is around 150°C. The mix must be laid and compacted while still hot. Asphaltic concrete can be laid in two or three layers where a thick surfacing is required or in a single course as a resurfacing material. Careful control must be maintained during manufacture and laying, but on a suitable base, asphaltic concrete should have a life of 15 to 20 years without requiring any maintenance (Figs. 6.21 and 6.22).

Hot Rolled Asphalt

This material evolved from sand sheet asphalt, which consisted of a mixture of carefully graded sand, mineral filler and bitumen. The inclusion of stone chippings not only improves the stability but decreases the cost. The material is made in a hot mix plant and is laid by a mechanical finisher. Precoated chippings are rolled into the surface to increase its non-skid properties.

Bitumen-bound bases are also being increasingly used for airfields, where the saving in design thickness provides considerable advantages both from the constructional and the economic point of view.

Road Maintenance

The maintenance of a safe and sound road surface becomes increasingly important as traffic densities, loads, and speeds increase. Also, with high cost modern road construction, failure to correct faults quickly can result in extremely costly repair projects being required. It is particularly important that there should be a waterproof surface in order to prevent penetration of water.

Surface Dressing. A surface dressing is the least expensive method of maintenance. It consists of a thin layer of bitumen or other binder sprayed over an existing road surface, which is then covered with stone chippings and rolled. Cutbacks or emulsions are usually used for this purpose, but in warm climates hot bitumen may be employed, provided it is possible to apply the chippings while the surface is still sticky.

Although the process may appear relatively simple, workmanship of a high standard is essential for satisfactory results. The binder must be suited to local conditions and applied at the correct rate, whilst the chippings should be clean, dry and cubical in shape. If the weather is wet, either an adhesion agent is used in the bitumen and/or the chippings are precoated with bitumen in order to ensure that they are not stripped off by traffic.

Whilst surface dressing is an acceptable maintenance technique for many types of road, it is not used on motorways and similar high-speed roads because of the risk of damage to vehicle windscreens from flying chippings. If there is serious deformation of the road surface a surface dressing would not be appropriate. When such a fault occurs, or when the wearing course becomes detached from the base, it is usual to remove the surface layers of asphalt by "planing" and to re-lay the surface with fresh material. This is a more costly process and alternatives have been developed which offer cost advantages without sacrificing quality.

There are three basic options available, all of which are a form of "road recycling". The longest established process is applicable to macadam roads that have sound foundations but have cracked or crazed or become deformed. The road surface is lifted and broken into lumps to a depth of about 80 mm with tines fixed to a roller or with a road milling machine. The pavement material is then scarified and treated with reciprocating harrows. After profiling with a blade grader, anionic bitumen emulsion is applied with a sprayer before the road is compacted with a rubber-tyred roller. During rolling, chippings are spread to fill voids, and a finishing surface dressing can be applied if required.

A somewhat similar treatment can be applied to hot rolled asphalt roads, but in this case before the road surface is broken up and scarified it is necessary to preheat the road surface in order to soften it. In both processes the existing wearing course materials are re-used in situ. Where circumstances do not permit in-situ recycling, a surface layer of asphalt can be planed off and removed to a hot mix recycling plant which is capable of producing hot mix products, containing a proportion of reclaimed material, which are of quality equivalent to a conventional hot mix plant producing virgin asphalt products. The rising cost of aggregate and bitumen has given a stimulus to recycling process development, especially in the USA.

Special Paving Materials. The availability of synthetic resins from the petrochemical industry (such as epoxy resins) has provided an opportunity for the development of paving materials with properties that cannot be achieved when bitumen is used alone. By extending visco-elastic bitumen with epoxy resins, the binder is transformed into a high-strength elastomer. This type of pavement binder was used to create the first purpose-designed anti-skid road surfacing material (Shellgrip). The application of this material at accident black spots results in a marked reduction in the rate of accidents (Fig. 6.23). The high strength binder prevents the embedment of the special calcined bauxite aggregate by the weight of traffic and so the aggregate is not polished. As a result, this surface treatment is able to retain its anti-skid properties for more than 10 years. Other thermosetting binder compositions have been developed (e.g. Erophalt) that are intermediate between a conventional bitumen surface dressing and Shellgrip. They have a wider potential application than Shellgrip and can be used where conventional surface dressing is not applicable.

Premium quality wearing courses were also introduced in the mid-1960s and are based on epoxy asphalt. These mixes have remarkable properties, including high fatigue resistance, high solvent resistance and no flow or deformation at high temperatures or under heavy axle loads. Originally, application was concentrated on special situations such as bridge decks, container terminals and airfields. However, the combination of increased truck axle loads, channelled traffic on motorways and high ambient temperatures causes excessive plastic deformation in conventional bitumen asphalt wearing courses. In extreme cases, resurfacing is necessary after only three years. Under these circumstances, the superior properties of epoxy–asphalt mixes can provide an alternative to frequent resurfacing.

Increased availability of sulphur in many countries (as a result of oil or natural gas desulphurisation) has stimulated the development of a new family of

Figure 6.23 Hammersmith flyover. Westbound down-ramp and eastbound up-ramp. Both treated with Shellgrip.

sulphur–bitumen asphalts. There are two basic types of mix. The first, sulphur-extended asphalt (SEA) incorporates a small percentage of sulphur as a diluent for bitumen in a conventional asphaltic concrete mix. In conditions where an asphalt mix is required that is less susceptible to temperature variations from summer to winter, SEA mixes are markedly superior. Since sulphur is substituted for bitumen, increasingly scarce bitumen is conserved; and in some instances where the price of sulphur is low there can be a cost advantage.

The second sulphur–bitumen system uses a higher proportion of sulphur. In this case the excess sulphur acts as a filler and permits the achievement of high quality paving materials based on inexpensive sand. The sulphur to bitumen ratio is greater than one and the hot mix fluidity is increased sufficiently for pavement construction to be completed without roller compaction. These high sulphur–bitumen mixes (such as Shell Thermopave) are structurally equivalent to conventional bitumen asphaltic concrete.

Industrial Applications

The uses of bitumen in industry are many and diverse. It is used because of its waterproofness, durability, flexibility and resistance to chemical action, properties in which it excels. Most grades of bitumen are employed, but the harder grades, blown grades and special filled bitumens find particular outlets in this field. In the following paragraphs some of these industrial applications are described.

Bitumen is widely used in construction work of all kinds, and the manufacture of bituminous products for use in this field often forms an industry in itself, for example mastic asphalt and roofing felt manufacture.

Mastic asphalt, which has been referred to previously in connection with roads, is also employed for heavy duty industrial flooring, as a base for other types of floor coverings, as a damp-proof course, for waterproofing below-ground structures, and for waterproofing and protecting insulation and other materials. In addition to the normal mastic that is mixed and laid hot, clay-filled bitumen emulsions provide a type of cold-worked mastic.

Bitumen is used extensively in the manufacture of roofing felts which consist of felt impregnated with a penetration grade of bitumen and then coated on one or both sides with a blown grade. Blown bitumen is used for the coating because it resists flow at high temperatures. The coated felt is normally dusted with a fine mineral powder to prevent sticking in the roll or is sometimes finished with relatively coarse granular material.

Roofing shingles are manufactured in a similar way, but a harder bitumen is used for impregnation, and special coloured granules are often employed for the surface finish. The treated felt is then cut into the required shapes and sizes.

Bituminised felts are also used for damp-proof courses and for tanking applications of all kinds. For this purpose, impregnated, but uncoated, felts are sometimes used.

Linoleum substitutes are a specialised form of bituminised felt provided with a decorative finish. The bitumen used for impregnation must be hard enough to resist indentation by furniture but sufficiently flexible to resist cracking during rolling and unrolling.

Though blown bitumens are excellent materials for most roofing applications, they do not possess sufficient elasticity, flexibility and resistance to flow and fatigue to satisfy the most extreme roofing requirements. Development of thermoplastic rubbers (TR) has enabled these requirements to be met through blends of selected bitumens and TR. The largest application is the coating of roofing felts on flat roofs, and the felts themselves can also be improved by manufacturing them with a bitumen–TR blend. The range of bitumen–TR products also covers applications such as joint sealants, pressure-sensitive adhesives and mopping adhesives.

Bitumen is used extensively for waterproofing and protecting all types of surfaces from attack by corrosive atmospheres, aggressive soils, and chemicals. Depending on the degree of protection required, various thicknesses and types of coating can be provided, and the grade of bitumen and technique employed are selected according to the application and exposure conditions.

Thin coatings can be provided by the use of solutions of bitumen in volatile solvents, sprayed or brushed on to the prepared surface to be protected. Such applications provide a relatively cheap, simple protection for a limited time. Somewhat more expensive but more durable thin coats are provided by more specialised bituminous paints in which drying oils, resins and other additives are incorporated. Such paints may be coloured by the addition of pigments or dyes, but colours are confined to the darker shades.

Heavier coatings for more lasting protection and for more critical conditions are provided by spraying with or dipping in hot bitumen, normally a blown grade, possibly containing filler. Alternatively, the filled bitumen may be applied cold either as an emulsion or mixed with a volatile solvent.

Bitumen is extensively used for the internal protection of pipes carrying water and other liquids (except of course liquids such as petroleum products which would attack the coating). External protection of all types of pipeline is effected by coating with bitumen. A filled, blown grade is frequently used and a coating thickness of up to about 4 mm adopted. If the pipeline is to be buried the external coating often incorporates a membrane reinforcement, and an outer wrapping in particularly severe conditions.

A rather unusual application for bitumen is in the construction of shafts for coal mines. Modern shafts have to be sunk to increasingly greater depths to exploit deep coal seams, and this accentuates the difficulties due to stresses on the shaft lining resulting from movements of the surrounding soil. Application of a layer of bitumen has provided a solution to the problem, and this method has been adopted for a number of mine shafts in Western Europe, using bitumen compounds specially designed to meet the particular conditions obtaining at each shaft.

Effective tunnel linings must be impermeable, durable and crack-resistant. In the Netherlands and West Germany, bitumen and bituminised felt have been employed to waterproof new road tunnels under canals and waterways.

Bitumen is widely used as an adhesive in the building and other industries. Particular applications include the laying of roofing felts, floor tiles, and insulating materials. The type of bitumen employed depends on the temperature conditions and the slope of the surface on which it is to be applied. Blown and penetration grades are used as such or as solutions or emulsions.

Bitumen is used in the manufacture of various types of waterproof paper. The paper is impregnated or coated with bitumen or prepared in the form of duplex paper, which consists of leaves of plain paper laminated with a continuous film of bitumen. More complex systems of paper, reinforcing membrane and bitumen are employed for paper sacks for fertilisers, etc.

The grade of bitumen employed depends on the type of paper being produced; a medium or hard grade is used for impregnated paper and blown grades for coated and duplex papers. Paraffin wax is often added to coating bitumen to reduce stickiness.

Bitumen board is made either by impregnating the board with bitumen or by incorporating the bitumen during the course of the board manufacture.

Coal briquettes have been manufactured in many countries for many years. Bitumen is a suitable binder because of its non-toxic nature, resulting in better working conditions during manufacture, and lower level of smoking during burning. Briquettes are manufactured either by the "dry process", in which the binder is mixed in a pulverised state with the coal fines, or by the "wet process", in which the bitumen binder is sprayed hot on the coal fines before pressing. The grade of bitumen employed depends on the type of coal and on the briquetting process. Special grades with high temperature susceptibility are sometimes used.

More knowledge of bitumen as an engineering material has facilitated its use in major hydraulic works such as the lining of irrigation canals, navigation canals and reservoirs, the construction of dams and the protection of coasts and harbours.

Bitumen is particularly suitable for these applications because of its impermeability, flexibility and binding properties which govern its behaviour under stress and its durability during the working life of the structure of which it is part. It is used either as such or in combination with fillers and aggregates such as limestone dust, sand and crushed stone. The type of mix used depends on the nature of the application, availability of plant, labour and materials, the size of job, and the speed with which the work has to be done (such as between tides).

Bituminous layers used in hydraulic construction have two distinct functions; firstly as an impermeable layer to prevent the seepage of water and the possibility of resultant structural failure in the underlying material, and secondly as a protective layer to prevent erosion of underlying material by waves and currents.

Chapter 7

TRANSPORTATION — MARINE AND PIPELINES

MARINE

History and Development

When the economic history of the twentieth century comes to be written, the amazing developments in the transport of oil by sea will form an interesting chapter. The first seventy years saw spectacular technological development and almost equally spectacular growth, followed by sudden and sharp decline. A brief survey of developments up to the present day will give some indication of the revolutionary changes which have challenged the tanker industry.

When oil first entered into international trade over one hundred years ago its transport by sea was, like that of most other cargoes, effected in specially made containers. At first wooden barrels were used, but these were subsequently replaced by large iron tanks fitted into the hull of the ship. As the economies of bulk transport became evident, the idea was conceived of using the hull of the vessel itself as the oil container. This necessitated the use of iron ships, instead of the wooden vessels previously employed, and constituted the main principle in the development of the tanker as we know it today. Probably the first ocean-going vessel constructed on these lines was the s.s. Glückauf built in 1885, with a gross tonnage * of 2,307 tons.

---

* References are made in this chapter to gross and deadweight tonnages. Gross tonnage, broadly speaking, represents the total capacity of all enclosed spaces in the ship, measured in "tons" of 100 ft^3. Deadweight tonnage (dwt.) represents the weight of the cargo, stores, bunkers and water which the ship can lift, expressed in metric tonnes. It is customary when referring to merchant shipping generally to express tonnage figures in terms of gross tons, but statistics relating only to tankers are more often quoted in deadweight tonnes, which is the measurement used in this chapter except where otherwise stated. With a normal tanker of average size, the gross tonnage is usually about two-thirds of the deadweight tonnage. Reference is also made to the knot, which is a speed of 1 nautical mile per hour (or 1.1515 miles per hour or 1.8532 kilometres per hour).

The use of steam machinery and coal-fired boilers in a vessel engaged in oil transport was still in its infancy and was attended by grave risks in view of the highly inflammable nature of the cargo; but such risks are the lot of the pioneer, and subsequent experience showed that running them was well worth-while.

The next landmark was the passage of the Suez Canal by a fully laden tanker, the first Shell tanker, the s.s. Murex of 5,010 dwt., built in 1892 at West Hartlepool. After lengthy negotiations with the Canal authorities she undertook her maiden voyage from Batum on the Black Sea to the Far East. The passage through the Suez Canal was completed without incident, despite fears that she might prove a danger to other shipping using the Canal. A sister-ship, the s.s. Conch, is shown in Figure 7.1.

From this small beginning some ninety years ago, the average quantity of oil moving through the Canal continually increased and reached 15 million tonnes per month in early 1967. It is interesting to note how one single geographical feature, the availability of the Suez Canal for the transit of Middle East crudes to North West Europe, had a remarkable impact on the development of the tanker industry in the 1950s and the first half of the 1960s.

In the early years of the present century the pattern of the oil trade underwent an important change with the opening up of the Sumatra and Borneo oilfields. Whereas previously oil was carried only from west to east and dry cargo on the return voyage, the new requirement for oil movements from the East Indies to

Figure 7.1 s.s. *Conch;* 5,010 dwt; built 1892. A sister-ship of the first tanker through the Suez Canal.

Europe made it possible to employ vessels exclusively for the carriage of petroleum. This resulted in a change of tanker design and encouraged some increase in size. At the beginning of the century the total tanker tonnage was half a million tonnes, with an average of 5,000 dwt. per tanker; by 1914 the total tonnage had increased to 2 million dwt., with an average of something over 6,000 tonnes per tanker.

During this period preceding World War I a further development, in which the Royal Dutch/Shell Group played an important part, was the construction of the first ocean-going motor vessel of any kind, the Shell tanker Vulcanus, built in 1910. Although only a small vessel of 1,215 dwt., she amply proved the advantages of motor propulsion and was the forerunner of a large number of motor tankers which today represent 64 per cent of the vessels in the world tanker fleet (39 per cent of the dwt.).

The years between the wars were characterised by a policy of consolidation and gradual development in size, speed and technical improvements. The size of the standard tanker rose to 12,000 dwt. with a speed of 11 knots. Half-hearted attempts were made to push the maximum size up to more than 20,000 dwt. but few vessels of this size were built. By 1939 the world tanker fleet had grown to more than 1,500 vessels totalling $16\frac{1}{2}$ million dwt., of which over $1\frac{1}{2}$ million dwt. were owned by Shell Group companies.

The vital need for oil during World War II gave a tremendous impetus to tanker building. The lead was taken by the USA, which, having developed a standard tanker of 16,600 dwt. with a speed of $14\frac{1}{2}$ knots, proceeded to turn out such vessels in large numbers. During the years 1942 to 1945 nearly five hundred of these ships, known as T2s, were built. Thus, despite heavy war losses, the world tanker fleet by the end of the war had risen to a total of 24 million dwt..

Thereafter began one of the most spectacular advances of shipping history. A number of influences were at work. The growth of the Middle East as a producing centre, the new policy of building refineries in the consuming areas instead of near the oilfields, the enormously increased demand for oil in the industrialised regions and the growing realisation of the economies to be secured from the use of larger vessels, all combined to establish a revolution in outlook. Gradually at first, and then at a quickened tempo, tanker owners began to build bigger ships. From 24,000 dwt. claims were made for the world's largest tanker by successive stages until 132,000 dwt. was reached by 1966, and the world tanker fleet totalled 97 million dwt..

The next seven years can perhaps now be regarded as the greatest period of expansion within the oil industry and consequently in the tanker trades. Output of the world's shipyards increased by leaps and bounds, from some 10 million dwt. in 1966 to over 28 million dwt. by 1973. At the end of that year the largest tanker in service was 476,000 dwt. and total tanker tonnage amounted to some

Figure 7.2 m.s. *Auricula*; 12,248 dwt; built 1946. A typical tanker of the late 1940s.

Figure 7.3 s.s. *Leda*; 276,860 dwt; built 1973. An example of a VLCC, developed in response to the closure of the Suez Canal.

219 million dwt.. This figure did not include a new sector of the shipping world, namely the combination carriers, i.e. vessels capable of carrying either oil or dry bulk cargoes, of which some 37 million dwt. existed in 1973.

This period of expansion was not free from difficulties. The underlying situation of confrontation throughout the period between Israel and the Arab States and more specifically the physical closure of the Suez Canal from October 1956 to March 1957, and subsequently from June 1967 to mid-1975, had a profound effect on tanker design, planning and operations. Compelled to sail both laden and ballast via the long route around the Cape of Good Hope, the shipping and shipbuilding industries reacted swiftly to the need for additional capacity. Being free from the draft limitations of the Suez Canal the Very Large Crude Carrier (VLCC) was born, initially of about 200,000 dwt. but subsequently much larger.

The speed of the reaction to the challenge is illustrated by the fact that in mid-1966 there was one tanker of over 150,000 dwt.; by mid-1973 there were 351. The growing importance of the larger ships is shown in Table 7.1.

Table 7.1 **World Tanker Fleet (2,000 dwt. and over)**. Percentage of Total Carrying Capacity

	Up to 25,000 dwt.	25,000 to 45,000 dwt.	45,000 to 80,000 dwt.	80,000 to 160,000 dwt.	Over 160,000 dwt.
1951	93	7	–	–	–
1961	50	37	11	2	–
1971	15	15	22	22	26
1981	4	8	11	19	58

In economic terms, the importance of the larger vessel is reflected in lower transportation costs as illustrated in Figure 7.4. This graph is based simply on newbuilding capital costs and owners' and charterers' operating costs; it does not take into account the effects of market conditions, which may yield the owner a return not commensurate with his capital investment, or of scheduling problems, which may result in inefficient trading patterns such as the need to call at many ports on a voyage in order to handle the volume of cargo carried. Problems such as the latter tend to provide a natural inhibition to progressing to even larger vessels.

By 1973 the largest size category already included a few ships over 350,000 dwt. and plans were being drawn up for ships of 500,000 dwt.. The era of the mammoth tanker had indeed arrived. This dramatic increase in ship size was not, however, matched in the earlier years by equivalent expansion of the shore facilities to accept these ships, particularly in the West European discharging

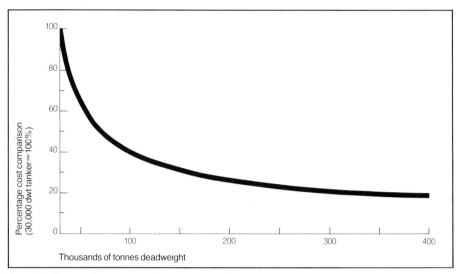

Figure 7.4 **Comparative transportation costs by size of oil tanker**
(Basis: newbuilding capital charge plus owner's and charterer's operating costs – per tonne of cargo)

areas, and lightening at sea, in port, or multi-port discharges became the order of the day. Shell pioneered the expertise required to lighten these very large vessels in open waters.

By this time the pattern of use of various sizes of ships had become more clearly defined. Vessels up to 40,000 dwt. were almost exclusively reserved for the carriage of refined products; the next group up to 100,000 dwt. was used for crude oil supplies to draft-restricted ports; and the VLCCs handled long-haul crude oil movements.

1973 began on a surge of optimism. Continuing substantial growth was foreseen, shipyard order-books were full, and freight rates reached levels hitherto undreamed of, or indeed seen since. The bubble burst in the latter half of 1973/early 1974, when the OPEC countries quadrupled the price of crude oil and progressively nationalised the oil concessions and facilities previously owned by the oil companies. The immediate result of this and later actions by the OPEC producers was to restrict the availability of oil to the consumer countries and to reduce oil demand. This occurred during a period when tanker availability was rising and shipyard order-books were full. Although some orders were cancelled and others rescheduled, the overall size of the tanker fleet continued to increase until 1979 as these new ships were delivered, and had only fallen by about 4 per cent from this peak by the end of 1981.

The period since 1973 has been characterised by rising oil prices. These were given a further sharp boost by the Iranian Revolution in 1979 and the Iran/Iraq war in 1980, and led to a significant drop in the demand for oil and a substantial tanker surplus. Efforts have been made to reduce the tanker surplus by scrapping older vessels (75 million dwt. scrapped in the period 1973/81), the laying-up of idle vessels, and general slow steaming. Demand for tankers, however, continues to fall, current availability of tankers exceeds demand by nearly 50 per cent, and equilibrium between supply and demand still (in the early 1980s) seems a long way off.

The Suez Canal reopened in the middle of 1975 but, because of the advent of the VLCC which was too large to transit the Canal, its significance was much reduced. The Suez Canal Authority then undertook an expansion programme, deepening the Canal so that from mid-1981 vessels up to 150,000 dwt. could safely transit fully laden, and ships up to 350,000 dwt. in ballast. However, the amount of oil moving through the Canal in 1981, about 2.5 million tonnes per month, is still only a fraction of the peak volumes of early 1967. Plans are in hand for further expansion of the Canal, but this may be delayed by current economic conditions and the existence of so much surplus tonnage.

Organisation of the World's Tanker Fleets

Nearly two-thirds of the world's tanker tonnage operates under the flags of five countries. Before World War II the British and Commonwealth flag fleets were larger than those registered under any other flag, but the severe losses sustained by British owners during the war, coupled with the large quantity of wartime tanker tonnage constructed in the USA, brought the latter country into top place by 1945. Subsequently, there were large transfers of US tonnage to other flags, with the result that by 1965 the USA had dropped to fourth place with 11 per cent of the world total. Norway and the UK held second and third places respectively, with 15 and 14 per cent. Japan, with virtually no tanker tonnage after World War II, had risen to fifth place with 5 million tonnes, or 6 per cent, and was still rapidly increasing her tanker fleet. The most striking change, however, has been the emergence of the Liberian flag, from nothing at all in 1945 to first place with $15\frac{1}{2}$ million dwt. in 1965 and 111 million dwt. in 1981, representing 30 per cent of the total. It is interesting to note that tonnage owned by the OPEC countries reached 5 per cent by 1981.

The practice of registering ships in the "open registers" (often called the flags of convenience), of countries such as Liberia, has developed because of the advantages enjoyed by the owners of ships so registered in freedom of choice on crew nationalities/manning scales/financing arrangements and in tax benefits. This has been the subject of considerable controversy at international political level for many years. Certain governments consider that there should be stronger economic links between vessels and the states in which they are registered; clearer definitions as to ownership, manning scales, standards of pay; greater control by the flag countries in enforcing standards of safety. In certain quarters it is further considered that the existence of the open registers acts as a barrier to the ambitions of the developing countries in building up their own national fleets. Other countries, while favouring enforcement of safety standards and greater transparency of ownership, disagree that abolition of the open registers is the best means of achieving these goals.

Apart from considerations of flag, the three main classes of owners are the oil companies, who run their ships as part of an integrated industry; independent owners without any other stake in the oil industry; and governments which in some cases desire, for reasons of internal policy, to control the commercial transport of the oil which they import/export, or to engage in international trade with national flag ships.

Before World War II, more than 50 per cent of total world tanker tonnage was owned by the oil companies, but the rapid post-war expansion in the demand for oil, with its consequent heavy burden of capital expenditure in all segments of the industry, made a reduction in this proportion almost inevitable. Many independent shipowners were quick to foresee the new situation and took steps to increase their share of the tanker trade. Table 7.2 shows the extent of the changes in ownership in the post-war period.

Table 7.2 **Ownership of World Tanker and Combination Carrier Fleet** (Percentage Share)

	Oil companies	Independents	Governments
1957	38	56	6
1961	38	57	5
1971	27	67	6
1981	25	65	10

The role of the independent owner is a vital one. In making their long-term plans for the provision of tonnage, the oil companies, after taking account of their own ships, proceed to charter from the independents the balance of their requirements by an amalgam of long-, medium-, and short-term contracts. In this way the companies are able to obtain the necessary flexibility to cope with fluctuating demand as well as changes in supply patterns.

For all that, the system has its demerits, since the law of supply and demand falls upon a comparatively small amount of marginal tonnage with the result that tanker freight rates for these ships are subject to violent changes. In 1951, 1956, and 1973, when world requirements tended to outstrip tanker availability, rates rose to very high levels; on each occasion substantial newbuilding programmes, which ultimately proved larger than necessary, were put in hand, and these resulted in a surplus of tanker tonnage in the following years and a consequent drop in rates. With the decline in tanker demand which followed the 1973/74 oil price increases, and which shows every sign of continuing in the 1980s, freight rates have remained at depressed levels. The bulk of the surplus tonnage exists in the VLCC/ULCC (Ultra Large Crude Carrier) size ranges, i.e. large crude oil carriers between 160,000 and 550,000 dwt.. Rates for these ships have been so low that they have for long periods not even covered owners' operating costs.

Obsolescence has naturally been a key feature in the struggle to reduce the surplus position. Some years after World War II, tankers were scrapped, on average, after a life of about thirty years. Gradually this was reduced to twenty years for most sizes of tanker, and to considerably less for VLCCs. The average age of the 39 VLCCs scrapped in 1981 was $11\frac{1}{2}$ years.

Class of Tanker

The Modern Crude Oil Carrier

In the early 1960s a typical large crude oil carrier was a ship of 70,000 dwt., optimised for the carriage of a single grade of crude oil from the Middle East to a West European refinery terminal. By the mid-1960s three factors were influencing the thinking of tanker designers:

(i) The major shipbuilders were confident that they had the resources to handle a further jump in size.
(ii) To design on the basis of the much longer voyage to Western Europe via the Cape would require a substantial jump in size if the essential economy was to be achieved.
(iii) The success of the single buoy mooring terminals connected to shore by submarine pipeline made it possible to think in terms of "taking the port out to the ship" instead of having to bring the ship into the port. This freed the design to a significant extent from the constraints imposed by the relatively shallow water access to many ports.

The first VLCCs entered service in 1968. They were mostly of around 200,000 dwt. and powered by steam turbines giving a loaded speed of about 16 knots. The power was delivered by a single screw. Steam was supplied by a single large efficient boiler often in combination with a much smaller "get you home" boiler for use in emergencies. The main engine boiler was also used as the power source for cargo handling equipment which was designed for even more rapid loading and discharge of homogeneous cargoes of crude oil, with a small number of very large cargo compartments and large diameter pipelines. The sheer size of these ships and their equipment as compared with their predecessors, made it necessary (as well as economically desirable) to introduce many features designed to replace manual effort, e.g.:

(i) The mooring ropes and wires could no longer be manhandled, and mechanical systems were introduced.

(ii) The pipeline valves could no longer be turned manually — which led to development of centralised, "push-button" cargo control rooms.

(iii) The engine rooms were too vast for the traditional watch keeping and engine control duties to be undertaken from within. The engine controls and monitors were therefore extensively automated, which made it possible to introduce control of all engine functions from the bridge of the ship.

(iv) The deck and overside areas of the ship were much too large for the traditional maintenance of painting surfaces. This provided the stimulus for the development of new paints capable of surviving much longer in a hostile marine environment.

All these and many more similar developments made it possible for the VLCC to be operated with greater reliability and safety, but with a smaller total complement than the much smaller ships built in earlier years.

Subsequent VLCCs and ULCCs have not changed in basic design in a fundamental sense. The trend towards greater automation was maintained, but the number of separate cargo compartments was increased. This facilitated the use of these ships for carrying more than one grade of cargo.

The increase in size continued steadily for a few more years. However, by the mid-1970s the effects of the oil price increases and the recession in the major oil-consuming nations led not only to the surplus of tanker tonnage already referred to, but also introduced a need for flexibility in shipping that the very largest ULCCs were not well suited to provide. The increase in oil price has also had a profound effect on the thinking of tanker designers on the subject of bunker fuel consumption. A few owners have re-engined VLCCs, replacing steam turbines with diesel engine propulsion in view of the significant reduction in fuel consumption, but it will only be in the future when new VLCCs come to be needed again that the full results will be seen of the designers' efforts to build ships with maximum fuel efficiency.

The Modern Products Carrier

Although the total demand for tankers to carry refined products diminished steadily in the 1970s, a number of newbuildings has been called for each year to replace some of the immense number of such ships built twenty years earlier and thus coming to the end of their economic lives. This has given the tanker designers the opportunity to revise regularly their basic design concepts. Indeed, looking ahead, with the prospect of additional supplies of refined products becoming available from refineries at or near the source of crude oil instead of

MARINE 489

being sited within the consuming countries, there is the prospect of an increasing need once again for the type of tanker most suitable for the transportation of these products to markets worldwide. There seems little doubt, therefore, that this is a part of the tanker tonnage picture to which designers will continue to pay close attention.

A typical modern products carrier is of approximately 30,000 dwt., and is powered by a slow-speed diesel engine capable of burning heavy fuel oil efficiently. The compartmentation of the cargo tank is such that 6 grades of cargo can be carried with complete segregation, but up to 12 grades if a minimum degree of admixture in the pipelines between certain grades is acceptable. These ships can safely carry "black" and "white" oils in a single cargo, a facility generally lacking in earlier product carriers. Tanks are fitted with steam heating coils so that the heavier grades that require heating to make them readily pumpable can be carried. Vessels of this type are frequently fitted with a waste heat recovery plant so that heat in the exhaust gases can be used to drive generators, thus making it possible to reduce bunker consumption still further.

Perhaps the most marked change from earlier product carriers, however, is in

Figure 7.5 m.s. *Eburna;* 31,374 dwt. A typical 1979-built products carrier, capable of carrying different grades on the same voyage. Mt. Fuji in the background.

regard to the manning of these ships. Total complements are in the range of 20 to 25 men, with possibilities of some further reduction. Most routine operations are automated and the engines are capable of operating for long periods unattended.

Luboil Carriers

Lubricating oil, of which there are many grades, is a high-quality product and requires the greatest care in handling if contamination is to be avoided. Sufficient luboil is transported by sea to the world markets to classify this as a bulk movement. Individual parcel sizes are generally small, requiring vessels with a variety of tank sizes with a highly sophisticated pipeline/pumping system so that the many grades can be carried completely free from risk of any contamination between the separate grades of cargo. Thus the trade is ideally suited to be carried by the vessels commonly known as Parcels Tankers. These are very sophisticated vessels of 20,000 to 40,000 dwt. having a high degree of segregation, pipeline and pumping flexibility. Some luboil has for many years been carried in conventional product carriers but the older vessels are fast disappearing. Product carriers of the new generation are nearly twice the size and their tank sizes tend to be too large to accommodate the individual parcels of luboils economically. Thus it is probable that increasing quantities will be transported by Parcels Tankers.

Chemical Carriers

The transportation of chemicals similarly demands a high degree of purity and segregation, calling either for specially protective coatings on ships' tanks, or for stainless steel tanks and pipelines. Because of the large number and diversity of chemical products, special equipment is provided for the cleansing of tanks and lines after each cargo to avoid any possibility of contaminating subsequent cargoes. Vessels are constructed to recognised industry standards.

Bitumen Carriers

Bitumen is solid or nearly solid at ambient temperatures and so in order to carry and easily discharge this product in bulk, vessels have to be equipped to maintain product temperatures in the region of 120°C, depending on grade. Because of the stresses caused by the wide range of temperatures experienced on laden and ballast voyages, vessels are normally strengthened to counter the buckling effect on steel plates which otherwise would occur. The number of large bitumen vessels of 18,000 to 25,000 dwt. has remained fairly stable for several years. There have been a few additions to the number of vessels in the 4,000 to 10,000 dwt. sizes, which handle movements from small production plants, or to markets with low demand, or requiring special grades.

LPG Carriers

In the early days, LPG was compressed into a liquid state to reduce the volume to manageable proportions for transportation by sea. "Pressurised vessels" (i.e. vessels carrying LPG under pressure in tanks) reach their optimum size at around 5,000 cubic metres. Until about ten years ago the major proportion of LPG was moved in this type of ship. There are still some four hundred ships, ranging in capacity from 100 to 5,000 cubic metres, distributing pressurised LPG on coastal and short-distance voyages throughout the world.

With the development of the extraction processes by certain large-volume crude oil producers, e.g. Saudi Arabia, Kuwait, Abu Dhabi and the UK, the availability of LPG for distribution into world markets has gradually increased from about fourteen million tonnes in 1977 to about twenty-five million tonnes in 1982, and is expected to reach around thirty-five million tonnes by 1985. As the major consumers of the increasing supplies of LPG were situated long distances from production areas, cheaper methods of transportation were sought. Reduced costs were achieved by liquefaction by cooling, i.e. by reducing the product

Figure 7.6 m.s. *Isomeria*; 58,950 m^3; built 1982 by Harland and Wolff. An LPG carrier.

temperature to between minus 42°C and minus 50°C for propane and to between 0°C and minus 6°C for butane at atmospheric pressure. Both shore and ships' tanks need to be fabricated from special low-temperature steels and insulated. There are various designs of ships' tanks. Due to heat transfer through the insulation, liquid cargo vaporises or "boils-off". To avoid this loss of product, vessels are equipped with compressors and coolers to reliquefy this gas which is returned to the tanks. More importantly, as LPG is heavier than air and can form flammable mixtures when diluted with air, venting of gas to atmosphere is normally prohibited. To give some idea of the growth in the transportation of refrigerated LPG, in the 5 years 1966 to 1970 31 vessels were built of 10,000 to 100,000 cubic metres capacity, whereas in the eleven years 1971 to 1981 some 120 vessels were built or odered, of which 19 are between 50,000 and 60,000 cubic metres and over 50 are larger.

Liquefied Natural Gas (LNG) Carriers

LNG is a very different cargo from LPG. It consists predominantly of methane, but contains other substances such as ethane, propane and butane to an extent which varies according to the quality of the natural gas from which it is made, the demands of the market, and the method of manufacture. Once re-gasified, it is either burned in power stations to generate electricity, or it is supplied to domestic and industrial users as gas.

The natural gas is liquefied by cooling to below its boiling point of about minus 161°C at atmospheric pressure. This means that the ship and shore storage tanks, as for refrigerated LPG, have to be fabricated from special materials and heavily insulated. Even so, some LNG still vaporises on voyage. This "boil-off" gas is piped to the engine room for burning in the ship's boilers. It is possible that future designs of LNG carrier may have diesel engines and reliquefaction plants on board to reliquefy the boil-off gas and return it to the cargo tanks.

The technology associated with the carriage of a liquid at such an extremely low temperature is complex. It is no surprise therefore that these ships are among the world's most expensive types of commercial shipping; for instance, a large new LNG carrier for delivery in 1985 could cost something like $200 million.

Another feature which distinguishes LNG shipping is that, more than in any other trade, the shipping is dedicated to the project producing the LNG and is an integral part of a chain of operations from production through liquefaction and transport to re-gasification and end-use. Hence ships tend to be built for specific projects, and to remain on the same trade route for most or all of their working life. This requires a high standard of performance in every aspect of the operation, from scheduling and time-keeping to operational consistency and safety. In fact, LNG ships have an excellent safety record.

Figure 7.7   s.s. *Gastrana*; 75,000 m³; built 1974. An LNG carrier dedicated to the "closed loop" Brunei–Japan.

There are currently 64 LNG carriers with over 25,000 cubic metres cargo capacity, and the majority of them (39) have over 120,000 cubic metres capacity. The largest built to date has 133,000 cubic metres capacity. Since the density of LNG is about half that of oil, and furthermore because of the need to have completely separate tanks for ballast, the dimensions of these larger ships are comparable with those of a VLCC of 200,000 tons dwt. capacity.

While the growth of the world's LNG carrier fleet has been very marked since the first commercially carried LNG cargo was loaded in 1964, LNG still makes up less than 3 per cent of the world's current gas consumption. Nonetheless, if the remotely situated reserves of natural gas are to play their full part in future energy consumption, there will need to be a further significant expansion in the LNG trade.

Offshore Production / Offtake Tankers

No listing of the types of tankers in use today would be complete without a brief reference to the special tasks, often carried on in the most hostile marine environment, associated with offshore oil production. An overall picture of the development of offshore oil fields is described in Chapter 3 (Exploration and Production) and reference is made there to the many different types of mooring and loading arrangements in use, in locations where it is either physically or economically impracticable to construct a pipeline to carry the oil to shore. In such circumstances, existing tankers are often specially modified, or new ships built with the features judged necessary to provide reliable storage and shipping, so essential for optimum production from the field. These mooring and loading systems are frequently based on the concept of the single point mooring, in order to reduce the combined forces of wind, wave, tide and current to a minimum. Because of the inability of support craft to operate in bad weather, systems have been developed whereby offtake tankers can "self-moor". Such tankers are often dedicated to a particular system and have special mooring and bow loading equipment fitted, for example the s.s. Medora on the Fulmar Field in the North Sea (Fig. 7.8).

Figure 7.8 **Floating storage unit on the Fulmar Field (North Sea), showing the converted tanker s.s. *Medora* attached to a single anchor leg mooring. Fulmar A production platform and jack-up rig *Cicero* in the background.**

Systems are now being developed and implemented which, in addition to the storage and offtake functions, have production facilities on the storage unit. Tazerka in Tunisia is an example.

Tankers and the Environment

Safety for both ship and cargo has always been a prime consideration of responsible shipowners, operators and crews. Nevertheless, accidents small and large have occurred ever since men took to sea. With the increased size of tankers these accidents have a far greater impact than ever before on the environment and consequently on the public as a whole. Notably, accidents close to the densely populated areas such as North Western Europe and the seaboards of the United States have attracted much public interest.

It was the grounding of the s.s. Torrey Canyon on 18th March, 1967 on the Seven Stones Reef off Land's End (UK) which drew the attention of a worldwide public to the consequences of a major tanker disaster. However, the biggest oil pollution involving a tanker that the world has ever witnessed occurred on 16th March, 1978 when the Liberian tanker m.s. Amoco Cadiz ran aground off Portsall on the Brittany coast of France.

In the 11 years between these two tanker disasters there were unfortunately several other marine casualties involving tankers that resulted in loss of life and/or substantial pollution of coastlines. The ensuing investigations into the causes of these accidents revealed that, whereas in some cases a major contributing cause has been a failure of equipment, more frequently the underlying error could be traced back to human failure — a lack of competence in navigation, in basic seamanship in the operation of the equipment available on the ship, or in the appreciation of a situation of potential hazard until it was too late for corrective action to be effective.

The years since s.s. Torrey Canyon have been a period not only of intensive study and research, but also of action by governments and by the oil and shipping industries aimed at achieving substantial improvement in regard to safety and pollution avoidance. Action has been taken along several different lines, including:

(i) Changes in the design requirements for new ships to segregate completely the carriage of oil cargo from the carriage of water ballast, and to limit the size of individual tank compartments.

(ii) Changes in operational procedures and equipment on all tankers, existing as well as newbuildings, aimed at ensuring that any ballast water pumped into the sea could not contain residual oil particles, and that the navigational equipment available on board was substantially improved.

(iii) The installation of inert gas systems on all large tankers, which make it

virtually impossible for a potentially explosive mixture of air and hydrocarbon gas to exist anywhere in the ship.
(iv) A uniform approach internationally to the standards of training and competence of sea-going personnel.
(v) The improvement and dissemination of knowledge worldwide on the techniques appropriate for the clean-up work after an oil spill and the provision of equipment in suitable locations.
(vi) The establishment of funds which can be made available quickly following an oil spill, so as to ensure that clean-up efforts are nowhere frustrated through lack of funds and to ensure that genuine third party damage is effectively compensated.

No-one connected with the industry would claim that the possibility of serious marine accidents has been eliminated. The sea remains an often difficult and sometimes hostile environment and accidents will happen. It is, however, generally believed that significant progress has been made with tanker operations in making the sea a safer and a cleaner place. Good operating practice in the tanker industry can thus be seen to incorporate two main objectives:
- The provision of an efficient oil transportation service, flexible enough to adjust to changing patterns of trade, at reasonable cost.
- The operation of this service in accordance with the international community's rising expectations for safety and environmental standards.

There is no conflict between these two objectives.

PIPELINES

Most industrialised countries have long had large networks of pipes for the distribution of water and gas, whilst pipelines to move commodities over long distances originated in the oil industry well over a century ago. The first successful crude oil pipeline was built in 1865 in Pennsylvania, a screwed cast-iron pipeline of 2 inch (5 centimetres) diameter and six miles (9.7 kilometres) length. Its life was short for it was torn up by the infuriated Teamsters it had put out of work, but it demonstrated the feasibility of the method.

The three basic functions of pipelines in the oil and natural gas industry are:
(i) To transport crude oil, from oil fields on land or offshore to terminals for export, and from import terminals and oil fields on land to refineries.
(ii) To carry refined products from refineries or tanker terminals to consumers or local distribution depots.
(iii) To transport natural gas from the fields to local distribution centres, or direct to large consumers.

Pipelines are designed and constructed in such a manner that the transport of hydrocarbons through them is virtually unaffected by climatic conditions and other natural hazards, such as floods, fog and frost. They help to avoid congestion on inland waterways, railways and highways.

The total length of main pipelines in the world (including the USSR) amounts to some 280,000 kilometres (150,000 kilometres for crude oil, 100,000 kilometres for natural gas and 30,000 kilometres for products), but if smaller or less significant pipelines are included, considerably greater lengths are involved. The largest and most recent main pipeline is the 48 inch (1.22 metre) diameter TransAlaska crude oil pipeline covering 1,250 kilometres from Prudhoe Bay to Valdez. It will be overshadowed by the huge Soviet natural gas pipeline under construction during the early 1980s.

Within the confines of this chapter it is not possible to review the development of crude oil and product pipelines throughout the world. In order to illustrate some of the problems presented by pipeline systems, the situation in Continental Western Europe is described. When reading this section it should be noted that whereas further significant construction of crude oil and product trunk pipelines is unlikely to occur in this particular area during the foreseeable future, elsewhere in the world major activity continues.

Main Crude Oil Pipelines in Continental Western Europe

In Western Europe the development of crude oil pipelines has been relatively recent. Up to the late 1950s, refineries were built and expanded at or near coastal ports. The relatively long coast line, with relatively short distances to the main inland consumption areas and the existence of excellent road and rail networks, of several large navigable rivers and of a developed system of canals, allowed a fairly easy and economic access from coastal refineries to the interior of Western Europe.

Oil consumption grew dramatically in the 1950s and reached a level in the late 1950s such that the oil industry built inland refineries close to the main consumption areas of Western Europe and simultaneously constructed joint venture crude oil pipelines for their supply. This alternative was more attractive than further expanding the existing coastal refineries and building product pipelines to serve inland markets because, due to the relative homogeneity of crude oil, fewer pipelines were needed. The first joint venture crude oil pipeline was the 28 inch (71 centimetre) diameter pipeline "Nord–West Ölleitung" (389 kilometres in length) between Wilhelmshaven and Wesseling (Köln) which was opened in 1958. It was followed in 1960 by the 24 inch (61 centimetre) diameter Rotterdam–Rhine Pipeline, feeding other refineries in the Ruhr area of Western Germany.

Throughout the 1960s, and until the energy crisis in 1973, the demand for crude oil in Western Europe continued to grow rapidly, requiring new main crude oil pipelines to feed newly built and expanded inland refineries at rapidly expanding centres of industry (Bavaria, Upper Rhine, Northern France). In the period 1960 to 1970 the other main crude oil pipeline systems went into operation (SEPL, RDO, CEL, TAL, AWP) while just before the energy crisis in 1973 two systems were considerably expanded (NWO, SEPL). The development is shown in Table 7.3.

Since 1973, the peak-year for throughput in most of the above mentioned pipelines, crude oil throughput has been falling, reflecting a decreasing demand for products in the market place with a steep decline since 1979. Today (early 1983) the crude oil pipeline capacity, as installed, is greatly under-utilised in most of the above mentioned pipelines which have, in fact, an even greater potential capacity which could be achieved by adding pumping capacity. In the second half of 1982, the 40 inch (102 centimetre) NWO pipeline was closed because at today's level of throughput the operation of the 28 inch (71 centimetre) pipeline alone is more economical. In the same period the 24 inch (61 centimetre) diameter section Wesseling–Raunheim of the RRP was taken out of operation because of the closure of the refinery which it served in Raunheim. Installed and potential

Table 7.3 **Main crude oil pipeline development Western Europe**

Year	Pipeline	Year	Pipeline
1958	North-west oil pipeline (NWO) 389 kilometre, 28 inch diameter, Wilhelmshaven-Wesseling	1968	Rotterdam–Rhine pipeline (RRP) 172 kilometre, 36 inch diameter, Europoort–Venlo replacing the 153 kilometre, 24 inch diameter, Pernis–Venlo which has been converted to products service
1960	Rotterdam–Rhine pipeline (RRP) 299 kilometre, 24 inch diameter, Pernis-Wesel/Wesseling		
1962	South European pipeline (SEPL) 769 kilometre, 34 inch diameter, Fos–Karlsruhe	1969	North-west oil pipeline (NWO) 87 kilometre, 28 inch diameter parallel sections, Wilhelmshaven–Wesseling
1963	Rhine–Danube oil pipeline (RDO) 286 kilometre, 26 inch diameter, Karlsruhe–Neustadt	1970	Adria–Wien pipeline (AWP) 415 kilometre, 18 inch diameter, Würmlach–Schwechat
1963	Central European pipeline (CEL) 340 kilometre, 26/18/16/12 inch diameter, Genoa–Ferrera–Collombey	1971	South European pipeline (SEPL) 260 kilometre, 24 inch diameter, Fos–Lyon
1963	Rotterdam–Rhine pipeline (RRP) 156 kilometre, 24 inch diameter, Wesseling–Raunheim	1972	South European pipeline (SEPL) 714 kilometre, 40 inch diameter, Fos–Strasbourg
1966	Central European pipeline (CEL) 568 kilometre, 26/24/22/18 inch diameter, Ferrera–Ingolstadt	1972	Rhine–Danube oil pipeline (RDO) Company fully merged with Transalpine pipeline (TAL)
1967	Transalpine pipeline (TAL) 464 kilometre, 40 inch diameter, Trieste–Ingolstadt	1973	North-west oil pipeline (NWO) 244 kilometre, 40 inch diameter, Wilhelmshaven–Gelsenkirchen
1967	Rhine–Danube oil pipeline (RDO) Flow of oil reversed to Ingolstadt–Karlsruhe	1973	Central European pipeline (CEL) 82 kilometre, 32 inch diameter, Genoa–Ferrera

PIPELINES

Table 7.4 **Main crude oil pipelines in Continental Western Europe –** actual throughput versus available capacity

Million tonnes per annum (mta)		Potential capacity[1]	Installed capacity end of 1982	Throughput			
				1968	1973	1979	1982
NWO	North-west oil pipeline	80	45[2]	20.7	25.1	24.2	10.4
RRP	Rotterdam-Rhine pipeline	36	23	16.6	18.4	15.1	14.8
SEPL	South European pipeline	90	65	23.7	42.3	39.9	28.8
CEL	Central European pipeline[3]	8	8	7.0	7.1	7.9	7.1
TAL	Transalpine pipeline[4]	44	32	14.6	23.7	23.1	13.9
AWP	Adria-Wien pipeline	10	10	–	6.0	9.0	6.2
Grand total		268	183	82.6	122.6	119.2	81.2

(1) After adding pumping capacity
(2) Of which 20 mta closed in second half of 1982
(3) Capacities and throughputs to Germany only
(4) Excluding capacities and throughputs to Austria (AWP)

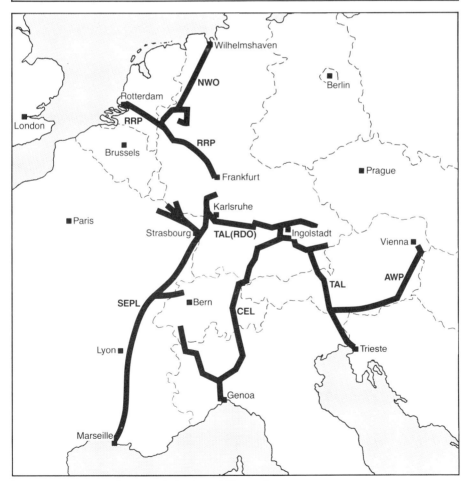

Figure 7.9 **Major crude oil pipelines in Continental Western Europe**

capacities are compared with actual throughputs in Table 7.4 which clearly shows the under-utilisation. The layout of the pipelines is further shown in Figure 7.9.

Oil Products Pipelines

Products pipelines are subject to criteria different from crude oil pipelines. Product volumes are usually smaller, which results in pipelines of smaller diameter and higher capital cost per unit of capacity. Also, for the same pipe size, products pipelines require more investment because of more elaborate installations and usually more intake and more delivery points. Product pipelines are more complex in operation, require more instrumentation and, combined with the smaller volumes, incur higher operating costs per unit of throughput.

Products pipelines can rarely compete in economy and flexibility with alternative means of transportation, such as express block trains, push barges and coastal tankers for quantities below 2–3 million tonnes per annum. However, for larger quantities multi-products pipelines are economic; they are almost always under joint-venture arrangements to secure the required volumes.

The operation of a multi-products pipeline is similar to that of a crude oil pipeline, but more complex because quality control is of great importance. Since a whole range of products may be carried, ranging from aviation gasoline to light fuel oils, proper sequence monitoring of products batches is necessary to reduce contamination to a minimum. Separation tools, in the form of spheres or special product buffers, may be inserted between batches to minimise interfacial mixing. A great degree of automation and computer-assisted control has been integrated into the pipeline systems, to achieve reliable and flexible product scheduling, quality control, pipeline operation and integrity monitoring.

Main Oil Products Pipelines in Western Europe

As an example, in Western Europe there are three main multi-product pipelines, which like the crude oil pipelines were mainly developed in the 1960s.
(i) The TRAPIL system (Société des Transports Pétroliers par Pipeline). It consists mainly of three parallel pipelines between the Le Havre and Paris areas, which became operational respectively in 1953 (10 inch, 27 centimetre diameter), 1961 (12 inch, 32 centimetre diameter) and in 1964 (20 inch, 51 centimetre diameter). It was continuously expanded and extended until 1974 and today it comprises a pipeline network connecting depots at Le Havre and various refineries along the River Seine with the products distribution depots in the areas of Paris, Rouen, Caen, Orléans and Tours and with underground storage facilities at May sur Orne (near Caen).

(ii) The RMR system (Rhein-Main-Rohrleitungstransport GmbH). Its main line of 20 inch (51 centimetre) diameter between Dinslaken and Ludwigshafen and the branch line of 18 inch (46 centimetre) diameter to Raunheim went into operation in 1967 while the 24 inch (61 centimetre) diameter Rotterdam–Dinslaken connection became operational in 1968. Today the system connects Rotterdam, Dinslaken and Godorf refineries with products distribution depots and also with the chemical plants along the rivers Rhine and Main.

(iii) The SPMR system (Société du Pipeline Mediterranée-Rhône). A system which went into operation in 1968 and which connects depots at Marseille and refineries in the Marseille and Lyon areas with products distribution depots alongside the River Rhône up to Lyon and depots between Lyon and the Swiss border. At the Swiss border the system extends via a 9 kilometre 12 inch (32 centimetre) diameter pipeline to distribution depots at Geneva, which pipeline is called SAPPRO (Société Anonyme du Pipeline à Produits Pétroliers sur territoire Genèvois) and became operational in 1972.

Natural Gas Pipelines

Pipelines are the only practical method of transportation of large volumes of gas, both overland and for relatively short distances across seabeds. The most advanced gas pipeline at present is the 437 kilometre 36 inch (91 centimetre) FLAGS pipeline from the Brent field in the North Sea to St. Fergus in Scotland, which was commissioned in May 1982, but even more ambitious projects in other areas around the world are under review. Pipeline diameters are constantly increasing due to growing gas transport demand; 42 inch (107 centimetre) high-pressure gas pipelines already exist in the Netherlands and larger pipelines of up to 56 inch (142 centimetre) diameter are being used in the Soviet Union.

Construction methods and problems are similar for oil and gas pipelines but there are significant differences in the economics of the two types for a number of reasons. Gas pipelines are not generally in competition with other forms of transport, and the problems of seasonal peak demand for gas are usually more pronounced and more costly to overcome than for oil. Also, gas pipeline compressor stations tend to be more costly than oil pipeline pumping stations. As transporters of energy, gas pipelines are less efficient than oil pipelines since the energy content transported in a gas line is only about one-fifth to one-quarter of that of an oil pipeline of the same size.

Ways of increasing the "specific energy" of a gas stream are being constantly sought and theoretically there are two likely possibilities. One is to increase pipeline operating pressure; the other is to transport natural gas in liquefied form (LNG).

The first approach, however, is limited by the physical limitations for pipe and supplementary equipment (valves, fittings, compressors), and by inherent logistics (particularly weight), economic and safety considerations.

For the second approach, one cubic metre of LNG at the cryogenic temperature of minus 161°C will expand to approximately 600 cubic metres at ambient temperature and pressure. However, this massive change in volume efficiency is obtained only by using expensive refrigeration plant and insulated pipelines. This explains why LNG pipelines are used only between storage tanks and jetties, in schemes where LNG is transported across the sea by special LNG carriers.

An example of the use of cryogenic loading lines exists in Brunei. It consists of two 18 inch (46 centimetre) insulated lines on a trestle extending 5 kilometres out to sea. The feasibility of submarine LNG pipelines is the subject of further research.

The Economics of Oil Pipelines

Generally, before initiating a pipeline venture it must be shown that for the economic life of the pipeline the users will ship sufficient volumes at the established tariff to enable the owners of the pipeline to repay their loan, meet the operating and maintenance costs and return a profit to compensate for the risk element and warrant the investment.

The comparative costs of different forms of oil transport to a large extent depend on the physical characteristics of the forms. Pipeline economics are identified with large initial investments, relatively low operating costs and decreasing unit costs for larger volumes (since capital cost varies directly with the pipe diameter, whereas capacity varies with the square of the pipe diameter).

As energy costs are much higher now than twenty years ago, it has become more economic to install pipelines of larger diameters than in the past in order to reduce the amount of energy required for pumping.

In order to take advantage of the lower transport cost, competing oil companies often join together in the building of pipelines. A separate company is usually formed to design, construct and operate the pipeline on behalf of all the participants, with each owning shares in the company in proportion to throughput requirements. A further advantage of a joint company is that a large part of the investment is usually borrowed direct from the money market.

The tariff charged by a pipeline company for transport and ancillary services includes operating and maintenance costs, depreciation, interest on borrowed capital, taxes and profit. The ancillary services may include marine and storage facilities, reception of tankers, inspection and handling of cargoes, quantity measurement and quality control.

Pipeline Legislation

Although in many countries the construction and operation of pipelines is still subject to general legislation, the increasing use of pipelines has led to a growing tendency for specific pipeline legislation dealing with such matters as rights-of-way, technical aspects of construction and operation, transport for third parties and government control or assistance.

Planning and Preparation

Specialised pipeline engineering contractors are frequently employed for the design and construction of pipelines. Basic data, from which the choice of pipeline diameter and location of pumping stations is made, include the quantity and characteristics of the fluid, the average and extreme temperatures to be encountered, the profile of the terrain to be crossed and the length of the pipeline.

Selecting the route is a vital part of planning. As soon as adequate surveys have been made, permits are obtained from the authorities, and rights-of-way are negotiated. Special permits are often required for the crossing of railways, roads and rivers, and this usually involves the submission of detailed maps, drawings and specifications.

Finally, the necessary materials and equipment have to be selected, specified and ordered, after which the construction contract can be awarded.

Materials and Equipment

Materials involved in a pipeline project represent a considerable proportion of the total cost; they range from about 20 per cent for offshore pipelines to about 40 per cent for land pipelines. Pipe-laying apparatus as well as welding and non-destructive testing have improved greatly in recent years, and the tensile strength of the steel has been significantly increased without sacrifice of bending or welding qualities. Today, steel with a yield strength of up to 70,000 psi (4,800 bars) is available, allowing the use of large diameters combined with relatively thin walls. This has provided a technologically and economically attractive solution to the requirement for increasing pipeline diameters and operating pressures. The improvement is partly due to improved metallurgy (such as the addition of small quantities of such metals as niobium, titanium and vanadium) and partly to more sophisticated manufacturing techniques, combined with rigid quality assurance.

Steel pipes for trunk lines may be seamless or welded, welding being particularly applicable to large diameter pipes. There are two methods employed in pipe welding, namely SAW (submerged arc welding) and ERW (electric resistance welding). The latter method has undergone significant improvements, and is finding more and more acceptance in the industry. Advances have also been made in the employment of spiral weld pipe, formed from steel strip spirally wound with the edges joined by submerged arc welding.

Changing economic circumstances have contributed towards the increased production of sour crude oils and gases, which in turn has enhanced the development of materials which are resistant to HIC (hydrogen induced cracking). This phenomenon, a form of corrosion which only occurs in the presence of hydrogen sulphide and water together, has just recently been recognised. Advanced technology is being devised to meet this challenge.

Technological progress has also been made in the area of coating materials. Bonding, abrasion and corrosion resistivity of polyethylene and epoxy coatings have been improved, and epoxy coatings can now withstand temperatures of some 80°C.

Plastic pipes have specific advantages such as low weight, resistance to corrosion and chemical attack, and ease of handling, but they are still inferior to metal pipes in their ability to withstand extremes of temperature and pressure. The use of glass reinforced epoxy (GRE) pipe offers an improvement. Such material is available for pressures and temperatures up to 363 psi (25 bars) and 110°C, respectively.

Valves are installed at intervals along the pipeline so that sections of the line can be isolated when necessary. "Full opening" valves such as ball valves are used, to permit the use of scrapers to clean the line periodically and of "intelligent pigs" to monitor the condition of the line and detect possible internal damage.

Normally pumps or compressors are required to create the pressures needed to move fluids through a pipeline. For low viscosity oils, centrifugal pumps are widely used as these pumps readily handle variations in throughput. For high viscosity oils, a positive displacement type of pump is usually preferred, and multi-stage, high-speed reciprocating pumps are generally used. For gases, centrifugal or reciprocal compressors are used. Depending on the circumstances, almost any type of prime mover may be used, such as electric motors, gas turbines, diesel engines and gas engines.

Following a demand for increased operational flexibility, turbine prime movers are nowadays usually equipped with dual fuel (gas/fuel oil) systems. Spacing of the pumping or compressor stations is part of the technical and economic evaluation in the design phase of the project, which considers criteria such as the installation cost of different sizes of pipeline compared with the horse power of the pumping or compressor stations, as well as operating and maintenance costs.

Construction

Construction of land pipelines is generally carried out by a number of self-contained groups or "spreads" each working on a separate section of the pipeline. The size of a spread is governed mainly by the diameter of the pipe and the type of country being traversed. Under reasonably good conditions a spread for laying "big-inch" pipe, i.e. 16 inch (41 centimetres) or larger, averages 1.5 kilometres per day and may achieve as much as 2–3 kilometres.

The construction phases consist of clearing and grading the right-of-way, hauling and "stringing" the pipe, ditching, bending, lining-up and welding, inspecting welds, cleaning, priming, coating and wrapping of joints, lowering in, backfilling and cleaning up, followed by pressure testing. Extra phases of construction are entailed whenever roads, railways or rivers have to be crossed. For all these activities a great deal of specialised equipment and machinery has been developed. Since it is costly to immobilise all this equipment and the crew, especially the welders, for even a day, rights-of-way and permits must be obtained well in advance.

Although pipelines are sometimes laid on the surface in desolate country, practically all modern pipelines are buried. Buried pipelines offer physical protection from interference, especially in congested areas, while stresses due to temperature fluctuations are much reduced.

The construction of offshore pipelines is carried out by specially developed vessels, which are pulled forward by anchor winches step-by-step along the route, while gradually depositing the pipeline through an S-shaped curve down to the seabed. The welding together of the lengths of pipe, the application of protective coating to the welds and inspection are carried out at a number of fixed positions along the length of the ship. With the development of offshore oil and gas production into deeper waters and extremely severe weather and sea conditions, sophisticated pipelaying barges of the semi-submersible type have had to be developed. Such vessels require a large crew and a supporting fleet of tug boats, supply boats and survey vessels. As a result, the cost of laying a pipeline offshore is generally much higher than that of laying a comparable pipeline on land.

Practically all pipelines have welded joints. The electric arc hand welding process has been used almost exclusively in the past. However, automatic welding is being used more and more, especially offshore where the additional time and effort saved aids the efficiency of the pipelaying barge "production line". Today, both methods are used equally. Radiographic inspection of welds, especially in vulnerable areas such as river crossings or offshore, is normal practice.

In special cases, such as offshore "J-laying" of pipe and for some flow lines (lines from a wellhead to a production platform), alternative jointing methods (e.g. mechanical connectors or threaded joints) could be potentially advantageous. Hence work is in progress to determine the feasibility of such systems.

Without special protection, buried steel pipelines would be subject to corrosion, which is essentially electro-chemical in nature. Buried pipelines in the past have therefore been coated with layers of bitumen or coal tar, asbestos felt and/or glass fibre, or with synthetic tapes consisting of one or several layers of polyethylene and butyl rubber. The quality of these "over-the-ditch" applied systems depends, however, to a great extent on the environmental situation during installation and may be influenced by factors such as adverse climate (e.g. dust in desert conditions) and the quality of site equipment and/or personnel.

Factory-applied coatings, i.e. polyethylene or epoxy coating systems, on the other hand, have attained such high quality that this method has generally replaced the "over-the-ditch" coating. Only coatings of joints are still applied at site, often in the form of "shrink sleeves" composed of heat-shrinkable cross-linked polyethylene sheet mixed with additives and lined with an adhesive layer.

In recent years the increased use of cathodic protection in combination with insulation has reduced external corrosion of steel pipelines to negligible proportions. In this method an electric current is made to flow towards the surface to be protected; that is to say, the whole pipeline is rendered cathodic.

Operation and Maintenance

The requirements regarding quantity, flow rate, timing and sequencing of batches to be transported, and the monitoring of pumping stations and storage tanks have to be coordinated.

Whilst different crude oils may be either mixed or kept separate, refined products should always be kept separate, because only a small percentage of one product can be blended with another without affecting its specifications. As long as the fluid in a pipeline moves fast enough, mixing between two adjacent product batches amounts to only a fraction of one per cent. Elastomer spheroids have come into use in product pipelines for separating batches even more effectively by reducing the interfacial mixing.

The operation of a pipeline system, especially one with several points of origin and destination, requires careful planning and control, for which a good system of communication is essential. Modern pipeline systems make extensive use of microwave transmission, telephone and teletype for communication. Computer-controlled systems are widely used, including automatic remote control of pumping and take-off stations with the aid of telemetering systems. Computers are also used to assist in planning and programming for the most economic operation.

Maintenance includes control and monitoring of both external and internal corrosion of the pipeline. External corrosion is monitored by measurement of the cathodic protection potential and of the electricity consumption at the corrosion protection stations. Monitoring of the internal as well as of the external corrosion is carried out with "intelligent pigs". These are sent along the pipeline and measure the intensity of an electromagnetic field induced in the pipewall, in order to locate corrosion by measuring wall thickness. Based upon such measurements, corrective action can be taken which may include corrosion inhibition and/or repair.

Safety Measures

The oil industry's need for safe, reliable and efficient pipelines to serve terminals, refineries and depots coincides with the need for safety of the public. Pipeline safety is achieved chiefly by building a well-designed pipeline and operating it to the best engineering practices, thus avoiding possible failures.

To ensure that all materials and equipment incorporated in a pipeline system are suitable and safe for the conditions under which they are used, they must comply with strict specifications and standards.

It is a standard rule for pipe manufacturers to inspect and test the tubes continuously throughout the process of manufacture. Auxiliary equipment, such as valves, fittings, and pumps, are subjected to similar exacting testing and inspection procedures which are often carried out by independent inspection agencies. Construction, too, is carried out with the most careful inspection, supervision and testing procedures, and the adoption of automated control systems for pumping stations and storage systems largely reduces the possibility of human error.

In addition to factory testing and inspection of the completed pipe and auxiliary equipment, it is standard practice to test the pipeline coatings electrically for breaks or pin-holes, known as "holidays", before burying. Finally, the pipeline is hydrostatically pressure-tested to determine the overall integrity of the pipeline. These precautions prior to commissioning, greatly reduce the possibility of a leak during operation. However, failures still occur as a result of third party activity (such as farming and road-repairing), natural hazards and mechanical failure. To monitor the degree of spillage and to reduce risks still further, the CONCAWE committee (Conservation of Clean Air/Water, Europe) was set up by Shell and other oil companies. This has helped to identify and analyse the causes and frequency of these failures, and to establish the basis for further improvement of the already high standard of pipeline safety.

The Future of Pipelines

The development of oil and gas reserves offshore led to a requirement for large submarine crude oil and gas pipelines. Basically the procedure to determine the optimum economic diameter is the same as that for land pipelines, as is the actual design. However, submarine pipelines are more difficult to construct (with the risk of buckling during laying operations), inspect and repair; they are thus much more expensive than land pipelines. In general terms, submarine pipelines are three to five times as expensive to build as comparable land pipelines.

Much of the primary work for the modern submarine pipeline has taken place in the North Sea where deeper water, harsh and unpredictable weather, peculiar seabed conditions, the effects of currents and tides and of shipping/fishing operations make pipelaying a difficult and potentially hazardous operation. Various methods of laying are available depending on the type of pipeline and diameter.

As exploration reaches into deeper water, pipelines will eventually have to follow. At present, pipelines for depths down to 300 metres are being designed. However, before greater depths can be attempted, improvements in laying methods, materials, welding quality and inspection, and environmental data of greater reliability and accuracy, are required, not to mention parallel developments in the field of the supporting services such as diving. However, not all future developments are offshore-related. The present level of high technology and efficiency of today's pipelines has been made possible as a result of faster and more economical pipe installation due to improved field equipment, X-ray inspection devices and higher tensile-strength steel (which reduces the weight/strength ratio). Operationally, improvements in the reliability of pumps, prime movers and motorised valves and improved telemetry and remote controls permit an entire pipeline system to be operated from a central control room.

Research and further development work is in progress on the transportation of Liquefied Petroleum Gas (LPG) and Liquefied Natural Gas (LNG) at low temperatures, special offshore pipelaying methods for deep water, underwater pipeline welding, improvement of seabed pipe-burying methods, and transport of slurry. Efforts are also being made to increase pipeline capacities by reducing the flow resistance arising from pipe-wall friction. Methods used include the development of friction reducers (added in minute quantities to the transported medium) and buffers of low viscosity fluid between the oil and the pipe wall. The further development of the "intelligent pig" which can be passed periodically through a pipeline (without interruption of oil flow) to detect and locate any minute pipe damage is one facet of pipeline research into means of preventing pollution and increasing safety.

Chapter 8

NATURAL GAS AND GAS LIQUIDS

WHAT IS NATURAL GAS?

Its Composition

The term "natural gas" is applied to gas produced from underground accumulations, the composition of which will vary from field to field. Most natural gases consist largely of methane and other light hydrocarbons; nitrogen, carbon dioxide, hydrogen sulphide, water and other materials may be present in varying proportions.

The principal hydrocarbon is methane, the member of the paraffinic series of hydrocarbons with the lowest boiling point (see Table 8.1). Other paraffinic compounds with higher boiling points, namely ethane, propane, butanes, pentanes and heavier hydrocarbons, are usually present in decreasing proportions. Collectively, they are generally known as natural gas liquids (NGL). While small

Table 8.1 **Paraffinic hydrocarbons in natural gas**

Name	Chemical formula	Boiling point (°C) at atmospheric pressure	
methane	CH_4	−161.5	
ethane	C_2H_6	−88.6	gaseous at normal atmospheric temperature and pressure
propane	C_3H_8	−42.1	
isobutane	C_4H_{10}	−11.7	
normal butane	C_4H_{10}	−0.5	
isopentane	C_5H_{12}	27.9	
normal pentane	C_5H_{12}	36.1	liquid at normal atmospheric temperature and pressure
normal hexane	C_6H_{14}	68.7	
normal heptane	C_7H_{16}	98.4	
normal octane	C_8H_{18}	125.7	

quantities of the lighter hydrocarbons such as ethane and propane may be left in the gas as marketed, heavier hydrocarbons are extracted to avoid technical problems in the transport of gas to end-users, and/or to bring the gas to the client's specification.

When separated from other components, methane, ethane, propane and the butanes are gases at ordinary atmospheric temperature and pressure, while pentanes and heavier hydrocarbons are liquids. These heavier hydrocarbons may remain dissolved in the natural gas to a certain extent (like water in air), but when there are changes in temperature or pressure, a liquid phase may be formed (just as a drop in temperature can condense out water in air as fog or rain).

Its Origin

The geological factors that result in the creation of various fossil fuels, i.e. oil, coal and natural gas, are discussed in Chapter 3 (Exploration and Production).

It is, however, important to indicate two basic processes that can give rise to natural gas accumulations. First is the decomposition of coal. This results in a gas very rich in methane, but containing only small quantities of the other hydrocarbons listed in Table 8.1. The second main form of natural gas is created when large molecules of oil break up as a result of heat and pressure over millions of years. Gas found separate from crude oil is known as "non-associated gas", whereas gas found with crude oil is known as "associated gas", as the name implies. Associated gas may be found as a "solution gas" dissolved within the remaining crude oil, or as "gas-cap gas" adjacent to the main layer of crude oil. Associated gas is usually much richer in the larger hydrocarbon molecules (ethane, propane, butane) than non-associated gas.

EXPLORATION AND PRODUCTION

Exploration

Whatever its origin, natural gas must have accumulated in a suitable reservoir rock if it is to have been trapped in sufficient quantity to be discovered and produced on a commercial scale. In many parts of the world the reservoir rock is limestone or sandstone. These rocks in turn must be covered by a layer of impermeable cap rock in order to have prevented the gas from finding its way to the surface and dispersing in the atmosphere.

A further requirement is that the rocks must have been twisted or faulted by the earth's internal convulsions into a formation which trapped the gas and thus prevented it from escaping laterally. Anticlines are one of the commonest structures bearing gas or indeed oil. These traps may be a considerable distance from the source from which the gas was originally formed. In order to find such traps, it is necessary to identify those parts of the world where the right geological conditions existed and where favourable structures were formed.

As with oil, these theories are today most commonly tested by seismic methods, whereby a recording is made of shock waves bouncing back from rock layers below the surface. Such techniques, and the methods used for evaluating the data, have been improved over the years, to the extent that more accurate predictions can now be made of the possible location of a hydrocarbon deposit. Thus a good idea can often be gained of whether a structure may contain oil or gas, but nevertheless drilling still provides the only certain method of ascertaining if hydrocarbons are present, and if so, whether these are oil, gas or both.

Production of Associated and Non-Associated Natural Gas

As described above, reserves of natural gas can be classified as either associated or non-associated, and this distinction is extremely important in deciding how and when such gas should be produced.

In general, the production of associated gas is determined by the rate and manner of production of the accompanying crude oil. Usually only solution gas is produced in the initial stage, as gas-cap gas may be used to maximise the recovery of crude oil (which normally commands a higher value at the wellhead). In some instances it is the impulsion provided by solution gas that enables the oil to be recovered, in which case the aim will be to minimise the amount of solution gas produced.

Where it is advantageous to reinject associated gas to prolong the productive life of the oilfield, the gas is separated from the oil at the wellhead and then pumped back into the field at another point to re-pressurise the field and enable a greater proportion of the remaining oil to be recovered. In this way, gas may be recycled a number of times without being wasted, since it may still be recovered and used towards the end of the active life of a field. Whether this is desirable will depend not only on the geological structure of each individual field, because in many cases reinjection would actually hamper the flow of oil, but also on related economic considerations.

Where reinjection is not practicable for any reason, the rate of oil production may result in gas production substantially in excess of oilfield requirements. Unless there are markets that can be economically supplied with such gas, there may be no alternative to "flaring" (venting and burning off) the surplus gas. In 1980, worldwide flaring of associated gas probably amounted to at least the equivalent of some three million barrels a day of oil, or as much as the total amount of natural gas usefully consumed in Western Europe in that year.

However, with the general rise in energy prices and increased awareness that a valuable energy resource is being wasted by flaring, an increasing number of schemes are being developed to gather and utilise such gas.

For non-associated gas such problems do not apply and the gas need only be produced as and when a suitable local and/or export market is available.

WORLD RESERVES

Historically, natural gas has generally been regarded as a much less valuable commodity than oil, with the result that drilling efforts were often directed with the sole aim of finding oil, to the point where in some locations, reservoirs of non-associated gas were abandoned as being "worthless" and not even recorded. Consequently, knowledge of the extent of the world's recoverable reserves of natural gas is far less precise than is the case with crude oil. Nevertheless, some experts have suggested that future reserve additions will eventually turn out to be at least as large as existing proven gas reserves, and possibly up to three or four times larger.

Moreover, proven reserve figures only include what is known as "conventional" natural gas, i.e. gas which can be produced with existing techniques at an economic price in today's conditions. In several parts of the world, in particular the United States, there are known to exist large quantities of natural gas which are difficult to produce for geological, locational or other reasons. Examples of these non-conventional deposits of natural gas are geopressurised gas, gas in tight formations and gas hydrates. These known resources of "unconventional" natural gas could eventually provide a useful supplement to conventional reserves when

Table 8.2 **Estimated world proven recoverable reserves of natural gas** * (trillion (10^{12}) cubic metres)

	Associated	Non-associated	Total
USSR **	2.2	23.8	26.0
Middle East	10.0	11.3	21.3
North America (incl. Mexico)	1.9	7.8	9.7
Africa	1.3	4.6	5.9
Western Europe	0.7	3.8	4.5
Asia/Far East/Australasia	0.5	4.0	4.5
Central and South America	1.5	1.2	2.7
Total	18.1	56.5	74.6

* Source: various published data. The split between associated and non-associated gas shown above only purports to indicate broad orders of magnitude.
** Includes small quantities in Eastern Europe and China.

Table 8.3 **Comparison of oil and gas reserves in 1980** (milliard (10^9) boe)

	Oil	Gas
Total proven recoverable reserves	600	500
Current world consumption	23	10
Approximate ratio of proven reserves to consumption	26:1	50:1
Potential undiscovered reserves *	2200	1650
Approximate ratio of total reserves to consumption	120:1	215:1

* World Energy Conference estimate 1980.

the necessary technological problems have been resolved and when economic circumstances favour such developments.

This section has concentrated on proven recoverable reserves of conventional natural gas, which give a more realistic indication of the gas available for economic exploitation with proven techniques. Most current estimates of world proven reserves are about 75,000 milliard (10^9) m³, which is equivalent to almost 500 billion (10^9) barrels of oil (more than three-quarters of world proven reserves

Table 8.4 **Natural gas: reserves/production ratios in 1980**

Representative countries / regions	Proven reserves milliard m³*	Gross production (less re-injection) milliard m³*	Reserves to production ratios (rounded)
Net importers			
USA	5,400	510	10
Western Europe	4,500	185	24
Japan	20	5	4
Net exporters			
USSR	26,000	420	62
Iran	13,700	20	685
Algeria	3,700	45	82
Canada	2,500	75	33
Mexico	1,800	40	45
Indonesia	700	30	23
Total world	74,600	1,600	47

Source: various published data
* Cubic metres in original quality as reported

of conventional crude oil). These reserves would be sufficient to maintain present rates of world gas consumption for about another 50 years.

In spite of the high degree of uncertainty for some of the figures given in Tables 8.2 and 8.3, a number of conclusions can be drawn:

- About three-quarters of world proven reserves of gas are non-associated and require planned development on their own account rather than as a by-product of oil. Indeed it is generally expected that a very high proportion of reserves yet to be discovered will also be non-associated gas. So although the efficient utilisation of associated gas is an important issue, the development of non-associated gas to the mutual benefit of both producing and consuming nations is of greater long-term significance for the world's energy supply.
- Because a substantial proportion of the world's remaining gas reserves is located in areas remote from the main markets, increasing quantities of gas will have to be transported from those areas with abundant reserves to the resource-short consuming countries if natural gas is to make its full potential contribution to world economic growth.

As shown in Table 8.4, the United States, Western Europe and Japan, on the basis of proven reserves, are likely to become increasingly dependent on imported gas. At the same time the USSR and a number of countries in the Middle East and Africa have the potential to supply a substantial part of their needs.

CONSUMPTION OF NATURAL GAS

World Perspective

Table 8.5 shows the proportion of primary energy supply (domestic consumption) provided by natural gas in comparison with other fuels in various parts of the world. Natural gas has now grown to become the third largest source of primary

Table 8.5 **Percentage breakdown of 1980 primary energy supply by areas**

	North America	Communist areas	Western Europe	Japan	Rest of world	World
Oil	46	32	53	68	57	46
Coal	18	42	19	16	19	26
Natural gas	26	20	15	6	11	19
Nuclear	3	1	4	4	1	2
Hydro/other	7	5	9	6	12	7
Total	100	100	100	100	100	100

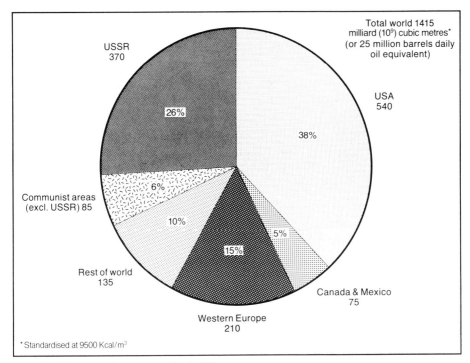

Figure 8.1 **World natural gas consumption in 1980**

energy, lagging only in those areas with limited indigeneous gas reserves, or in those less developed parts of the world which, although possessing natural gas, have yet to develop the necessary infrastructure to make use of it.

It is apparent from the above variations in energy patterns that growth of gas consumption in each of the major consuming markets merits individual examination. The following section discusses the major differences between consuming areas in terms of the structure of the gas industry and the relative importance of the main market sectors. The countries or regions that consume the largest quantities of natural gas are highlighted in Figure 8.1.

The United States

Market Development

Pride of place must go to the United States as the earliest natural gas market of any significance and still the largest.

The natural gas industry in the United States began in the middle of the nineteenth century, when gas was used in small quantities for lighting and

Table 8.6 **1980 Primary energy supply in the United States**

	Million b/doe	Percentage of total
Oil	16.0	45
Coal	7.3	20
Natural gas	9.6	27
Nuclear	1.3	4
Hydro/other	1.6	4
Total	35.8	100

cooking. During the next fifty years, growth was slow and usage was limited to those states where it was produced, primarily in association with crude oil. Even so, much of the gas was flared.

This situation changed with the development of seamless welded pipe made from high-strength steel, mechanical ditch-digging equipment and other technological improvements and techniques permitting the transmission by pipeline of natural gas at higher pressures to more distant markets. Because of the parallel development of gas manufactured from coal in areas where natural gas was not available, local distribution grids already existed in most major cities. It was not until the 1940s and 1950s that further advances in pipeline technology facilitated the movement of natural gas on a major scale from production areas in the south to the industrial markets in the more northerly states. By 1980 there were more than one million miles of natural gas pipelines in the United States serving almost 50 million customers, manufactured gas having been replaced to all intents and purposes by natural gas some years beforehand.

Growth in natural gas use matched that of the pipelines and between 1950 and its peak in 1972, consumption increased about fourfold. At its peak, natural gas supplied one-third of total primary energy consumption in the United States; it is still the second largest contributor (Table 8.6).

In 1980, natural gas was the leading supplier of energy to the domestic/commercial sector (46 per cent), where the main uses are space heating, cooking and

Table 8.7 **1980 Gas consumption in the United States** (milliard m^3)

Domestic/commercial	205
Industrial	230
Power generation	105
Other * (mainly pipeline use)	55
Total	595

* Excluded from Figure 8.1.

water heating. It is also the major energy source (36 per cent) in the industrial sector, and supplies 15 per cent of the power generation market. The breakdown of gas use is shown in Table 8.7.

Supply

In 1980, some 95 per cent of the natural gas consumption in the United States was derived from indigenous production, with the remainder being imported, mainly by pipeline from Canada and to a lesser extent Mexico, supplemented by small quantities of gas delivered in liquefied form from Algeria.

Legislation and Pricing

As a result of the administrative practice under which individual states control business within their own boundaries, but only federal authorities regulate trade between states, two main segments of the gas business developed:
- Intra-state trade, where the producer or transmission company sells gas within the producing state.
- Inter-state trade, where a transmission company moves gas from a producing area in one state to a consuming area in another.

While intra-state trade was subject only to local state legislation, inter-state gas transmission has for many years been subject to federal legislation. Strict controls on wellhead prices and transmission charges on inter-state gas resulted over the years in gas prices being held down below their free market level as the price of alternative energies, notably oil, increased. At the same time, regulation of wellhead prices has resulted since the late 1960s in additions to proven reserves failing to match levels of consumption. The inevitable consequence has been that the reserve base was eroded and supply began to be constrained. By 1975, proven indigenous reserves of natural gas had fallen to less than 230 trillion cubic feet (6,200 milliard m^3), or less than 11 years' production, and it was clear that these resources would be insufficient to maintain production levels for any length of time.

In an attempt to remedy this situation, the extremely complex Natural Gas Policy Act (NGPA) was passed in 1978 after a great deal of political debate. This attempted on the one hand to regulate demand by preventing gas from being supplied for new under-boiler uses, and on the other hand to increase supply. Wellhead prices for gas discovered after 1977 would move towards their market value on a predetermined schedule, with price controls being removed from all such gas in 1985, affecting probably some 50 per cent of indigenous gas, regardless of whether it was sold intra-state or inter-state. However, because of the unanticipated level of increases in oil prices since the NGPA was approved, wellhead prices for natural gas, with certain exceptions, were in the early 1980s still regulated well below those of competing energy forms, resulting in a continuing debate as to how and over what time scale the provisions of the NGPA should be changed.

The USSR

The Development of the Soviet Gas Industry

The USSR remains the world's second largest producer and consumer of natural gas and will probably overtake the United States during the 1980s if official production plans are achieved.

As indicated earlier, the USSR, with over one-third of the world's proven reserves of natural gas, is in a strong position to become an even more significant producer. Development of the USSR's gas resources began in the late 1940s, but it was not until the mid-1950s that significant use was made of natural gas. Previously, most of the reserves of natural gas had been found as a consequence of exploring for crude oil, rather than as a deliberate policy to find gas. But around this time, planners began to appreciate the magnitude of Soviet reserves and the important contribution that could be made by natural gas to the industrialisation of the country. Table 8.8 shows how prominent that contribution to internal consumption has now become.

Although a natural gas network was built up covering most main centres of population, household consumption rates remain low, as space heating is often provided by district heating systems operated by municipal undertakings or combined heat and power plants. The main priority was to develop the industrial sector and to use gas as a fuel for power stations, and it is in these markets that most gas is still used. The virtual doubling of gas consumption during the 1970s was met almost entirely by indigenous gas, although for logistical reasons

Table 8.8 **1980 Primary energy supply in the USSR**

	Million b/doe	Percentage of total
Oil	8.9	38
Coal	6.3	27
Natural gas	6.6	28
Nuclear	0.3	1
Hydro/other	1.4	6
Total	23.5	100

relatively small quantities were imported from Afghanistan and Iran to meet demands in southern regions.

By far the largest proportion of the unexploited gas reserves lie in the West Siberian Tyumen region some 3,000 kilometres from Moscow and its surrounding industrial areas. Indeed, Siberia is expected to supply virtually all the additional production planned over the next few years. While in 1965 only 15 per cent of production was from east of the Urals, this figure has already risen to more than 50 per cent. This changing supply pattern has required major pipeline developments, and in twenty years the gas pipeline transmission system has expanded more than fivefold to a total of well over 100,000 kilometres. The first high-capacity line from Siberia was commissioned in 1974; new challenges of logistics and technology had been overcome in its construction, in particular in the crossing of large distances of permafrost country.

The USSR is also the world's largest exporter of natural gas, with exports of about 55 milliard m^3 in 1980 valued at more than $5,000 million. About half is supplied to a number of East European (COMECON) countries, and these exports provide an important element of integration within that organisation, as well as offering some degree of diversification of energy supply to economies which in most cases have relied largely on coal. There are several pipeline routes, including two major trunklines, of which the most recent is the Orenburg pipeline, an ambitious cooperative venture completed in 1979.

The remaining exports have been supplied to Western Europe. Since the initial agreement with Austria in 1968, export contracts have been concluded with West Germany, Italy, France, Finland and Yugoslavia, and major pipelines bring the gas 4,500 kilometres from West Siberia. In 1980 the USSR supplied about a quarter of Italy's gas, almost 20 per cent of West Germany's and nearly 15 per cent of total Continental European gas consumption. The conclusion in the early 1980s of another series of major gas import deals between the USSR and various West European countries could result in some 30 per cent of Continental Europe's gas consumption being met by Soviet gas in 1990.

Further information about energy in the USSR is given in Chapter 2 (Oil and Gas in the Centrally Planned Economies).

Western Europe

Until the later 1950s natural gas in Western Europe was a very localised and small industry based on relatively modest reserves in Austria, France, Italy, West Germany and the Netherlands. This picture was changed by the discovery of the huge Groningen non-associated gas field in the Netherlands in 1959, and by the subsequent discovery of significant reserves under the North Sea.

Figure 8.2 illustrates the extent of this change by the beginning of the 1980s.

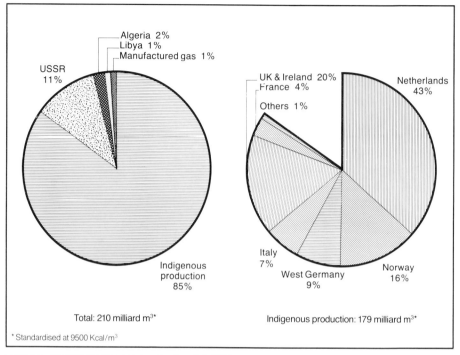

Figure 8.2 **Western Europe: gas supplies by origin 1980**

Only 1 per cent of total gas use is now manufactured from coal or oil, and a further 14 per cent is imported natural gas. The remaining 85 per cent is produced in a number of West European countries, with the Netherlands still dominant, but followed ever more closely by the North Sea producers (the UK and Norway) whose production has built up rapidly during the 1970s.

Natural gas has now attained an important position in the primary energy supply of Western Europe, although coal still retains second place due to its

Table 8.9 **1980 Primary energy supply in Western Europe**

	Million b/doe	Percentage of total
Oil	13.2	53
Coal	4.8	19
Natural gas	3.6	15
Nuclear	1.0	4
Hydro/other	2.2	9
Total	24.8	100

CONSUMPTION OF NATURAL GAS

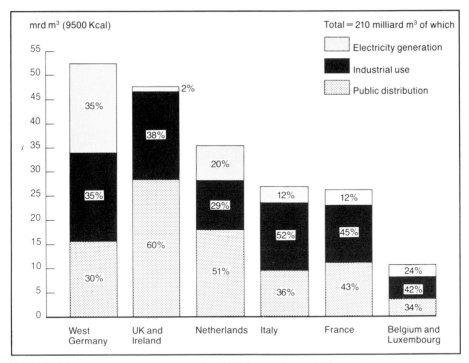

Figure 8.3 **Western Europe: gas consumption by principal country/sector 1980**

availability in some countries in large quantities and at competitive prices (Table 8.9).

However, there are considerable differences between countries, with natural gas supplying some 40 per cent of primary energy in the Netherlands in 1980, but being insignificant in parts of Southern Europe (Spain/Portugal) and much of Scandinavia. It is equally clear from Figure 8.3 that there are a number of important country-by-country variations in the way in which gas is used in Western Europe. For this reason, developments in several of the major consuming countries are discussed separately.

1. The Netherlands

The discovery of the Groningen field led to the establishment of a marketing organisation, Gasunie, in which the producers and the government (directly or indirectly) had equal shareholdings. This organisation increased natural gas sales at a rapid rate, offering supplies of indigenous gas at a price that was below that of competing oil products. As a result, gas penetrated deeply into all market sectors in the Netherlands and by the mid-1970s provided over half the national primary energy consumption and supplied heating to some 85 per cent of households. Export contracts had also been signed with West Germany, Belgium, France, Italy and Switzerland, usually on contracts of about twenty years' duration, with the result that export sales were approximately as large as inland sales in the Netherlands.

Although further small discoveries of natural gas were made, mostly offshore, it became apparent that Dutch reserves were being depleted more rapidly than was regarded by the government as desirable in the interests of national supply security. As a result, four major decisions were made in the mid-1970s:
- to maintain export contracts at their existing levels with no increase in the volumes committed;
- to withdraw natural gas from under-boiler uses in power stations where it could be substituted by residual oil and/or coal;
- to seek to import gas from Norway and, if possible, from other sources;
- to develop as a priority the smaller offshore reserves, leaving as much as possible of the remaining Groningen reserves (some 1,300 milliard m^3 in 1980) as a strategic reserve to supply internal needs into the next century.

By the mid-1970s natural gas production had become the largest single source of revenue to the Dutch government, with the government take being regulated through a "Maatschap" or financial partnership. In the late 1970s, while negotiating for increased gas prices, Gasunie relaxed to some extent the conditions of offtake imposed on its export customers. The result was that Dutch gas increasingly became the balancing producer for Continental Europe, producing at a low rate during the summer (when demands in other countries of Continental Europe are mainly met from indigenous sources or imports from sources other than the Netherlands), and at a higher rate during the winter or at times of supply disruption. In this way the Netherlands continues to play a major role in the West European gas industry.

2. West Germany

West Germany is currently the largest natural gas market in Western Europe. It had the initial advantage of having a large number of local distribution companies distributing "town gas" (gas manufactured from coal or oil) and several transmission companies using coke oven gas from steelworks, but it was not until the discovery of indigenous natural gas and the opportunity to import from the Netherlands and elsewhere that gas became significant, making up by 1980 some 17 per cent of the national supply of primary energy.

In contrast to several other West European countries, the major gas transmission companies remain in private ownership, although the majority of the local distribution companies are controlled by municipalities. In 1980, 30 per cent of the natural gas consumed in West Germany was from indigenous reserves, with the remainder being imported from the Netherlands (36 per cent), Norway (16 per cent) and the USSR (18 per cent). This substantial reliance on imports meant that the major West German companies have been extremely active internationally in the search for further imports and have been among the leaders in discussions with the USSR, with the intention of securing increased pipeline supplies in the mid-1980s.

In the early 1980s, a large proportion of West German gas consumption was in the electricity generation sector and the non-premium industrial market where prices were relatively low. In 1980 only some 25 per cent of households used natural gas for heating. If, as seems likely, West Germany is forced to rely increasingly throughout the rest of the century on more costly imported gas, the gas companies may need to develop more premium markets if they wish to maintain their share of the energy market. Another potential concern was supply security, as up to 30 per cent of supplies may be derived from the USSR alone by the end of the 1980s. Nevertheless, as illustrated later, the growing complexity of the international supply network offers some reassurance on that score.

3. The United Kingdom

Like West Germany, the UK had an extensive town gas distribution system, but because this coal-based gas was expensive to produce and distribute, sales were largely limited to meeting demand for cooking and water heating in urban areas. Nevertheless, the switch to cheap naphtha feedstock in the early 1960s brought about the first significant inroads into the residential market. Natural gas was then discovered in the Southern Sector of the North Sea in the West Sole field in 1965, followed shortly afterwards by a larger find in the Leman Bank, and thereafter by a number of other fields. Also in 1964, natural gas began to arrive via the world's first commercial trade in liquefied natural gas (LNG) from Algeria (see later section of this chapter), although supplies from this source provided only a small proportion of total requirements.

All transmission and distribution within the UK is undertaken by the nationalised British Gas Corporation and its area boards. Conversion of the whole system to natural gas, including the appliances of some 14 million customers, was successfully completed in the decade prior to 1978, and during this period a further substantial increase in market penetration was achieved in the residential sector. In 1980 over 80 per cent of households were able to receive gas, with some 35 per cent using gas central heating, and a further 25 per cent using other gas-fired appliances as their main source of household heating. Industrial sales played an important part in the rapid build-up of natural gas consumption but are now increasingly being restricted to those customers who have a particular need for high-quality fuels, or where interruptible sales are required to balance demand with supply. Sales to power stations have likewise almost ceased. In total, natural gas now supplies some 20 per cent of primary energy in the UK.

British Gas Corporation, having sole purchasing rights for natural gas produced in UK waters, was able to buy UK gas at lower prices than those prevailing in Continental Europe, though rather higher prices were paid for imports from the Norwegian Frigg field. Concentration on the heating of homes and commercial buildings and the ensuing restriction on growth in base load industrial demand has resulted in a very seasonal market with particularly high winter demand. British Gas Corporation has therefore sought to develop economic methods of gas storage, and has shown particular interest in sources of non-associated gas (which can be produced in the winter only to meet peak demands).

Increasing competition between the UK and Continental European buyers for the acquisition of uncommitted North Sea gas is likely in the coming years. Changes are also expected in the role and position of British Gas Corporation.

4. Norway

Norway had no internal consumption of natural gas in 1980 but has become an important element in the West European picture because of the reserves found in its waters. It supplies the UK from the Frigg field straddling the UK/Norway median line, and gas from Ekofisk, Albuskjell and other neighbouring fields is landed at Emden in West Germany for a West European consortium of buyers. Contracts have been signed for the delivery in the mid-1980s of additional supplies to Western Europe from Statfjord, the 34/10 Block and Heimdal via the Statpipe line, which after landfall at Kaarstoe in Norway for the extraction of NGL will also be piped to Emden.

The real importance of Norway for the future of Western Europe almost certainly lies in its more northerly gas fields. A major gas discovery has been made in the Block 31 area, and drilling north of the 62° parallel has indicated further commercial finds. However, some of these fields pose new production problems and are unlikely to become a major factor in Western Europe's gas supply until well into the 1990s. By the end of the century, however, Norway may have largely replaced the Netherlands as the main supplier of indigenous gas for Continental Europe.

5. France

Natural gas in France was originally based on indigenous supplies from the Lacq area, although the quantity produced has always been relatively small. As in the case of West Germany, imports have been secured over the last decade from the

Netherlands, Norway and the USSR, together with LNG from Algeria. Although, at 11 per cent in 1980, natural gas still plays a smaller role in energy supply than in the large industrialised countries of North West Europe, this contribution is expected to increase steadily in the coming years.

Nearly all gas transmission and distribution is handled by a nationalised entity, Gaz de France, which has in recent years shown interest in additional imports from Norway, the USSR, and a wide variety of potential sources outside Europe. France already has three LNG terminals at Fos (near Marseilles), Le Havre and Montoir (near St. Nazaire) in order to receive Algerian LNG, and is therefore well placed to receive further supplies of LNG from other sources as they become available.

6. Italy

Italy has relatively modest quantities of indigenous production, but has received pipeline imports from a number of sources, principally the Netherlands and the USSR, supplemented by LNG supplies from Libya through a terminal at La Spezia. As a result, natural gas provides 16 per cent of primary energy, close to the current West European average.

The major development in the 1980s is expected to be the start in 1982 of natural gas deliveries from Algeria via the Trans-Mediterranean Pipeline, the world's first long-distance, deep-water pipeline. The construction of the pipeline was completed in 1981, and if successful it could be the precursor of similar pipeline gas supplies from North Africa to Western Europe.

7. Other West European Countries

Elsewhere, Belgium, though having no indigenous reserves, has developed its gas industry throughout the 1970s on the basis of imports from the Netherlands by Distrigaz, the national transmission company with a 50 per cent government holding. Distribution is carried out by companies jointly owned by municipalities and private companies. More recently, Dutch supplies have been supplemented by purchases from Norway, and natural gas provided 19 per cent of primary energy in 1980. It is also planned to import LNG from Algeria, with initial deliveries being made in 1982–83 via France, pending completion of the reception terminal at Zeebrugge. Interest has also been shown by Distrigaz in pipeline gas supplies from the USSR and LNG imports from various sources.

In 1980, Austria met about 30 per cent of its gas requirements from indigenous sources, the remainder being supplied from the USSR and West Germany. Spain purchased small quantities of LNG from Libya and Algeria delivered to its terminal near Barcelona; in the late 1970s and early 1980s some indigenous reserves were discovered and their development is being planned during the 1980s. In Switzerland, natural gas contributed 4 per cent of primary energy supply in 1980 and was imported from the Netherlands and West Germany.

At the time of writing, Finland is the only user of natural gas in Scandinavia, taking modest quantities from the USSR. However, Denmark is planning to

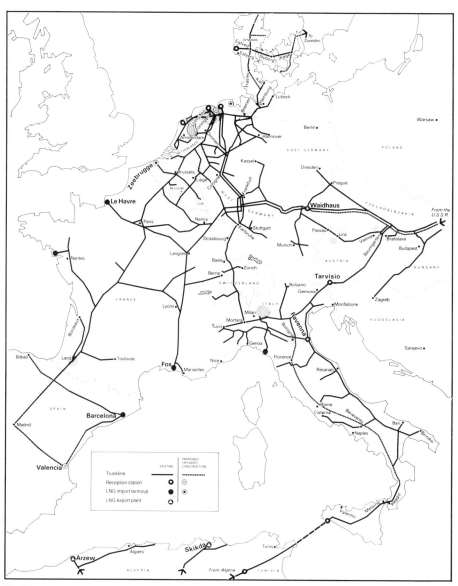

Map 8.1 **Major pipeline systems in Continental Europe**

develop reserves in the Danish offshore by the mid-1980s, and is in the process of constructing a pipeline grid linking into the European system.

8. The Pipeline Network

Throughout this section the degree of interconnection between the various gas pipeline systems in Western Europe has become obvious. Map 8.1 illustrates the resulting network, which in Continental Europe alone now comprises more than 80,000 kilometres of transmission pipelines and a distribution grid of more than 500,000 kilometres. This and the supply flexibility that it provides is likely to become more important as Western Europe comes to depend increasingly on imported natural gas.

Japan

Japan has only very small reserves of indigenous gas and relies very largely on imports of LNG for its supplies of natural gas, which in 1980 provided some 6 per cent of primary energy consumption. Although some 15 million residential customers (nearly half of all households) are connected to a gas supply, total usage of natural gas is relatively small. In 1980, consumption of natural gas was less than 30 milliard m^3, or under 2 per cent of world consumption. One of the reasons for this is that gas is not yet utilised on any scale for residential space heating, partly because of the absence of competitive tariffs and partly because of the availability of kerosine at controlled prices.

The first LNG import contract was for supplies from Alaska to Tokyo Electric and Tokyo Gas, with the major part being purchased by Tokyo Electric, the supplier of electrical power to the Tokyo area. Because of the pollution problems in many major Japanese cities, severe regulations have been imposed which have in effect forced major fuel users to utilise so-called "zero sulphur fuels" such as low-sulphur crude oil, LPG and naphtha. In comparison with these, LNG offered an attractive and economic alternative, with the result that about three-quarters of all natural gas currently imported and consumed in Japan is used for the generation of electric power.

Following this initial supply of LNG from Alaska in 1969, larger quantities were contracted from Brunei (first deliveries 1972), Abu Dhabi (1977) and Indonesia (1977). The buyers are mainly groupings of power companies and gas utilities. Among the latter, the three largest (Tokyo Gas, Osaka Gas and Toho Gas) resolved to convert all their customers to natural gas. Conversion began in Tokyo in 1972, and was expected to be completed by the mid-1980s; several other major cities were in various stages of conversion to natural gas.

Although natural gas does not dominate any market sector, nevertheless it had

Table 8.10 **1980 Penetration of natural gas into various market sectors in Japan**

Market Sector	Natural gas as per cent of energy used in market sector
Electricity generation	15
Domestic/commercial	12
Industry	3

made substantial progress during the last ten years, as indicated in Table 8.10.

Since Japan was keen to expand its use of LNG and to reduce its degree of reliance on imported oil, natural gas's share of these market sectors is expected to continue to increase as new import schemes are realised from Indonesia, Malaysia and other potential exporters.

Other Markets for Natural Gas

In a survey of this kind, it is only possible to touch on some of the developments that have been taking place elsewhere in the world up to the early 1980s.

In the Pacific Basin, various countries such as Bangladesh, Pakistan, Thailand, Malaysia, Australia, New Zealand, etc., have been increasing the level of natural gas use within their domestic economies, one of the major objectives being to reduce their level of dependence on imported oil. In several cases, plans included the separation of natural gas liquids for local use and/or export. Within the same region, other countries such as South Korea and Taiwan, which lack substantial indigenous reserves, also considered the import of LNG, in order to diversify their sources of energy. As a result, natural gas consumption in this region is likely to grow steadily over the rest of this century.

Another area with considerable potential for growth is the Middle East. The development of networks to collect and make use of associated natural gas in Saudi Arabia, Abu Dhabi, Kuwait, Qatar and other neighbouring countries is already well underway, together with the separation and export of natural gas liquids. One such system in Abu Dhabi (which began operation in 1981) is shown in Figures 8.4 and 8.5. Increasing quantities of natural gas will be used within the local economies to fuel power stations, refineries and export industries. In certain countries, such as Iran, Saudi Arabia and Kuwait, plans have been made to exploit non-associated gas for the same purposes, and in others, such as Egypt, schemes were in progress to develop pipeline networks to serve domestic, industrial and power generation customers. These patterns of development are also

Figure 8.4 NGL field extraction plant under construction at Asab, Abu Dhabi, UAE, showing in background associated gas flares which would be extinguished with commencement of plant operation.

spreading to those African countries with indigenous gas reserves, substantial populations and growing energy demand.

Latin America is another region where the use of natural gas has been slow to become established. However, natural gas finds in Chile, Argentina, Bolivia, Trinidad and Venezuela among others, have quickened interest in its exploitation by pipeline for use locally and for export, where appropriate, to adjacent countries. In certain cases, exports of LNG to the United States, Western Europe or other potential markets have been contemplated. One country in this region with significant gas production potential is Mexico, which has substantial proven reserves, both associated and non-associated. Mexico is seeking to increase local use of the gas produced in association with crude oil, as well as to maintain and probably increase exports by pipeline to the United States.

Further to the north, Canada's home market had grown to some 45 millard m^3 by 1980, based principally on substantial proven reserves in Alberta and to a lesser extent British Columbia. In addition, there was a well established export trade via a number of pipelines to the United States. It was also planned to

Figure 8.5 **NGL fractionation plant and storage facilities at Ruwais, Abu Dhabi, UAE, where the NGL stream is separated into propane, butane and pentanes plus.**

extend the coverage of the gas transmission system beyond Quebec in the early 1980s to include the states of the eastern seaboard. The ultimate level of reserves in the Canadian Arctic regions and elsewhere was thought by many to be sufficient not only to supply home market demand but also to provide a basis for the development of LNG exports to the Far East or Western Europe.

In summary, in most areas of the world where there are proven reserves of natural gas within reach of potential markets, production is already taking place or else there were plans for its exploitation in the 1980s. As a result, gas is likely to become a much more significant source of energy in many countries where its use has so far been limited or non-existent.

TRANSPORT OF NATURAL GAS

The natural gas industry comprises three main phases of activity illustrated in Figure 8.6. First is production. Second is transmission (the transport of large

TRANSPORT OF NATURAL GAS

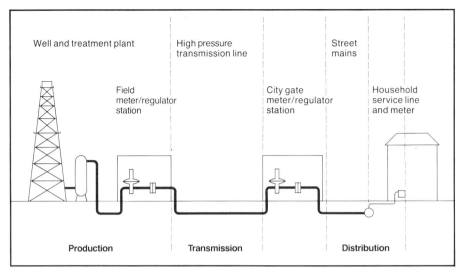

Figure 8.6 **Natural gas supply: the three phases**

quantities of gas from producing to consuming areas), which will be dealt with in this section. Finally, there is local distribution to end-consumers, which will be discussed later.

The transmission phase is usually thought of in terms of long-distance deliveries of substantial quantities of natural gas by large-diameter high-pressure pipelines, either within a single country or across international borders. However, where large expanses of water or other barriers intervene between source and market, an alternative form of transportation which may be employed is shipment in liquefied form (LNG); this method is also described.

Economics of Gas Transport

In order to appreciate the inherent differences between transport of gas and oil by whatever means, typical heating values for identical volumes of fuel oil and natural gas are compared in Table 8.11.

These wide differences in heating value explain why a gas transport or storage system must be physically much larger than its oil equivalent in order to handle a similar quantity of energy. A typical oil pipeline can transport about five times as much energy per day as a high-pressure gas pipeline of the same diameter, in spite of the fact that the velocities of the natural gas are normally higher than those of oil.

Figure 8.7 shows the approximate relationship for the cost of transport of gas and oil over long distances. It will be noted that the greater the distance involved,

Table 8.11 **Volumetric heating value of fuel oil and natural gas**

	MJ/m^3 *	Btu/ft^3 *
Fuel oil	41,000	1,100,000
Natural gas		
as gas at atmospheric pressure	37.26	1,000
as gas at 70 bars (about 1,000 psi)	3,000	80,000
as gas at 140 bars (about 2,000 psi)	6.700	180,000
as LNG at minus 160°C	25,200	675,000

* At atmospheric temperature, except where otherwise indicated. Natural gas, whether as gas or LNG, has a much lower density than fuel oil, and thus a lower energy content per unit volume, but the energy content per tonne is higher than for fuel oil.

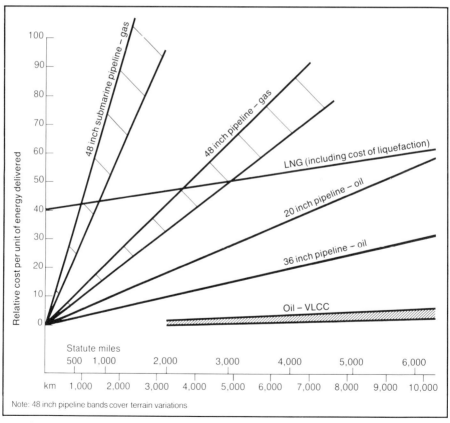

Figure 8.7 **Comparison of oil and gas transportation costs**

the greater the disparity between oil and gas transport costs. Even for 48-inch-diameter pipelines, the cost of transport for gas is much higher. If a gas pipeline is to match the carrying capacity of an oil pipeline, its diameter must be around two-and-a-half times as large, while the wall of the gas pipeline must be considerably thicker to withstand the higher operating pressure, both of which contribute significantly to its capital cost.

Figure 8.7 also illustrates that when long-distance movement of natural gas is required, liquefaction can become an increasingly attractive option on economic considerations.

Transport by Pipeline

Pipelines have been used for centuries, the Chinese being recorded as the first to transport natural gas through bamboo pipes. Wooden pipes were used near New York in the early part of the nineteenth century but were soon followed by lead and cast-iron pipes. None of these early systems, however, could be used at high pressure or for transporting large volumes over substantial distances.

The development of the modern high-pressure transmission line made with steel took place largely in the United States in the early part of the twentieth century as knowledge of metallurgy increased and techniques for welding, pipe-laying and corrosion prevention improved. Fifty years ago, pipelines with a 20-inch diameter were operating at a pressure of 20 bar. Today, gas transmission lines as large as 56-inch diameter are in service, with operating pressures up to 70 bars (approx. 1,000 psi) or even higher.

When gas flows in a pipeline there is a drop in pressure along the line. In order to avoid excessive pipe diameters and very high input pressures, it is usually necessary to install compressor stations at suitable intervals along the pipe route. For heavily loaded lines, they may be needed every 100 to 150 kilometres. Each usually comprises two or more compressors driven by gas-fuelled engines, gas turbines or electric motors. On many modern lines, they are operated by remote control from a central control point.

Even when compressor stations are installed, the operating costs of a pipeline are relatively low so that the recovery of capital costs usually accounts for 75–85 per cent of the costs of transmission by pipeline.

Gas transmission networks now exist in all major gas-consuming markets except Japan. In recent times, moreover, deep-water pipe-laying techniques have brought a new dimension to the transport of gas. One outstanding example is the existing Trans-Mediterranean Pipeline, which is expected to convey up to 12 milliard m^3 per annum of gas from Algeria to Italy. This line includes a 160-kilometre stretch of three 20-inch pipelines from Tunisia to Italy at water depths up to 600 metres operating at a working pressure of 150 bars. All over the world, an increasing proportion of natural gas is being produced offshore and brought to land for processing and distribution, in areas as diverse as the North Sea, the Gulf of Mexico and South East Asia.

The importance of pipelines to the gas industry is underlined by the fact that over 95 per cent of all the natural gas consumed in the world in 1980 was delivered from wellhead to the final consumer entirely by pipeline.

Shipment of Liquefied Natural Gas (LNG)

As shown in Figure 8.7, the cost of natural gas transmission by submarine pipeline is much higher than by a comparable pipeline on land (due almost entirely to the higher capital costs), while the cost of transporting gas by either method is much higher than that of shipping oil by very large crude carriers (VLCC). This disparity has proved one of the major deterrents to the international trade of gas. As was demonstrated in the section on reserves, a substantial proportion of the world's proven natural gas reserves is separated from potential customers either by many thousands of kilometres of land or by oceans through which it would not be feasible, even with modern technology, to lay pipelines. In such circumstances, a different approach is needed. The answer has been to liquefy natural gas by cooling it to about minus 160°C, and then to transport the resultant liquefied natural gas (LNG) in purpose-built insulated ships.

Because of the large capital costs involved in the liquefaction plant and for storage/re-gasification facilities at the receiving end, as well as the need for insulation and special materials resistant to low temperatures for the construction of LNG pipework and tanks, liquefaction is uncompetitive with pipelines for moving natural gas over short distances. Since the cost of producing, liquefying and re-gasifying a given quantity of gas in a given location will not vary, however far that gas is to be carried, the total supply cost of natural gas by this method rises less rapidly with distance than for pipelines. As a result, it can become an attractive option even in cases where long-distance pipelines could physically be built. As illustrated in Figure 8.7, if the distance between producer and market exceeds about 5,000 to 6,000 kilometres, it would usually be cheaper, if there is a corresponding sea route, to transport gas as LNG, rather than by overland pipeline; if the alternative is a submarine pipeline, the break-even distance would be significantly less, possibly less than 1,500 kilometres, depending on the depth of the sea.

Early History

Based on experimental work by Faraday and his colleagues, natural gas was first liquefied using a combination of compression and cooling by Cailletet, a French ironmaster, in 1877. It was not until 1917, forty years later, that natural gas was liquefied on a practical scale by the United States Bureau of Mines. Shortly afterwards, the world's first commercial-scale liquefaction plant was built in West Virginia in the United States. It was not, however, until 1959 that the first trial cargo of LNG was shipped from Lake Charles, Louisiana, in the United States to Canvey Island in the UK in "Methane Pioneer".

The success of this experiment led to the first commercial LNG ocean transport scheme from Arzew, Algeria, to Canvey Island, using two carriers, "Methane Princess' and "Methane Progress", each capable of carrying some 12,000 tonnes of LNG on each voyage. First deliveries were in 1964, and since that time a number of larger projects has been undertaken (the forerunners of a trade that is rapidly growing in importance).

LNG Plant

LNG is produced in a liquefaction plant by cooling the feed gas to produce a low-viscosity liquid boiling at about minus 160°C at atmospheric pressure. Up to twenty per cent of the original input may be used as fuel in the process of cooling the gas to such a low temperature. Most liquefaction plants are built in a series of parallel modules, or "trains" as they are usually called. Apart from technological limitations, the train concept enables part of a plant to be taken out of service for periodic maintenance and repair while the remaining trains continue to operate.

The facilities required at any particular LNG plant will depend on a number of factors, especially the composition of the input gas, and the sales specification of the LNG. The main elements of a liquefaction system are shown in Figure 8.8 and described below.

These may be divided into three categories:

(1) Primary Components of the Plant

Feed Inlet Station (for Associated and/or Non-associated Gas). To catch any liquid slugs, to adjust the gas pressure, to take any fine liquid droplets out and to measure the quantity of gas received.

Condensate Removal. To remove heavy hydrocarbon vapours from the natural gas, because these vapours could freeze at low temperatures and block the liquefaction unit.

Acid Gas Removal. To remove corrosive sulphur compounds and also carbon dioxide, which would otherwise freeze in the liquefaction plant.

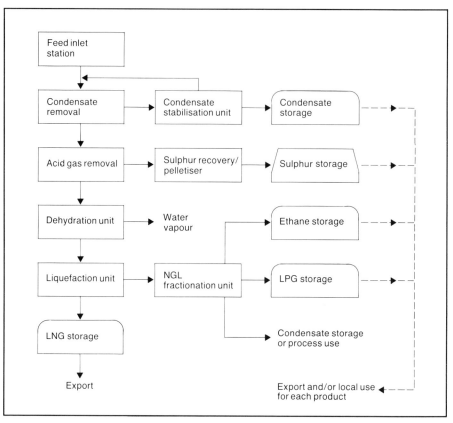

Figure 8.8 **Schematic view of typical liquefaction system**

Dehydration Unit. To remove water from the gas to be liquefied, because water vapour would freeze in the liquefaction unit.

Liquefaction Unit. This employs the same principles as a domestic refrigerator, although the equipment is much more complicated. Gaseous refrigerants are first compressed and liquefied. When the liquid refrigerants have evaporated, they serve to cool the natural gas. The gases used as refrigerants are normally propane, ethane, methane and nitrogen; of these, the first three may be extracted from the natural gas, while the nitrogen is obtained from the air.

Natural gas liquids can be produced as a by-product at this stage, depending on the composition of the feed gas.

(2) Essential Ancillary Units

Condensate Stabilisation Unit. A conventional stripping process, removing light components from the condensate so that it will satisfy the vapour pressure specification.

Fractionation Unit. A conventional distillation process, handling raw NGL removed in the liquefaction unit and producing refrigerant make-up and marketable NGL products.

(3) Additional Facilities (where Appropriate)

If the feed gas contains a high proportion of sulphur, the following facilities may also be required:
- Sulphur Recovery Unit.
- Sulphur Forming Unit (Pelletiser).

Product Storage and Handling Facilities

Storage Tanks and Loading. A storage tank for LNG must have an inner tank surrounded by insulation, which in turn will need an outer covering. LNG tanks are now usually designed on the double integrity principle, i.e. the outer barrier as well as the inner can withstand low temperatures and can therefore contain any leak from the inner tank. The whole system may consist of an open-topped inner tank, surrounded by an outer dome-roofed tank with a concrete outer wall surrounded by an earthen bank.

LNG is usually loaded through two identical parallel pipelines. When loading is not taking place, LNG is circulated through the loading lines to keep them at working temperature.

Storage tanks for refrigerated LPG and ethane are designed on the same principles as those for LNG.

Sulphur (where appropriate). Sulphur pellets will normally be transferred from the sulphur pelletiser by a conveyor belt system to a stockpile in a sheltered building. The loading facilities consist of a reclaimer, conveyor belt and shiploader system.

Marine Facilities. To load LNG and any other products that the plant may produce, ships must be berthed as close as possible to the plant storage area, so that an artificial harbour or special loading system may be required if a natural deep water harbour is not conveniently available.

Utility Systems. Most LNG plants generate their own electricity. The refrigerant compressors and the generators are most commonly driven by steam turbines.

Steam for mechanical drives, electricity generation and heating purposes is generated in a bank of boilers, sufficient capacity being provided for the plant to operate if some of the boilers are shut down for inspection or repairs. Normally, all available boilers are kept in operation.

Each LNG train, the process services area and the utilities area will need to be provided with fresh or salt water for cooling; in addition, fresh water is required for the boilers.

General Facilities. These include plant, buildings, safety and telecommunication systems.

Photographs of a liquefaction plant under construction at Bintulu in Sarawak,

Figure 8.9 **A liquefaction train under construction at Bintulu, Sarawak, East Malaysia.**

Figure 8.10 A general view of one of five liquefaction trains at Lumut, Brunei.

East Malaysia (Figure 8.9), part of the liquefaction plant at Lumut in Brunei (Figure 8.10), and the loading of an LNG carrier (Figure 8.11) are shown here.

LNG SHIPPING AND TERMINALLING

Ocean transport of liquefied natural gas (LNG) has now been developed to the point that the technology involved can be regarded as fully proven. As an example, in 1980 the 1000th cargo of LNG was shipped from Brunei to Japan, following an eight-year period of incident-free operation, with annual deliveries reaching over 5 million tonnes a year.

Since LNG has a calorific value per unit volume little more than half that of crude oil, an LNG tank can hold only about half as many heat units as an oil tank of the same size. The capacity of the largest modern LNG carriers is around 135,000 m^3 of LNG, equivalent to around 70,000 tonnes of crude oil, compared with VLCC cargoes of several times that quantity. Technological and operational

Figure 8.11 **LNG carrier SS *Gastrana* loading at Lumut, Brunei for Japan. The loading crane is approximately 45 metres above sea level and 4.5 kilometres offshore.**

matters in respect of the shipping of LNG are discussed in Chapter 7 (Transportation — Marine and Pipelines).

On arrival at the receiving terminal, the carrier discharges its LNG cargo into insulated shore tankage. Subsequently, the LNG is re-gasified and fed into the transmission pipeline in accordance with market needs.

The Closed-Loop System

In its practical operation, an LNG chain (comprising gas production, treatment, liquefaction, storage, loading, ocean transport, unloading, storage, re-gasification and transmission/distribution to end-customers) is akin to a pipeline gas project. To all intents and purposes it is a "closed-loop" system with dedicated facilities supplying LNG (gas) at a high load factor to specified customers. In this sense the LNG shipping phase can be regarded as a flexible pipeline (see Figure 8.12).

As more liquefaction plants and receiving terminals are built, a somewhat

DISTRIBUTION AND MARKETING

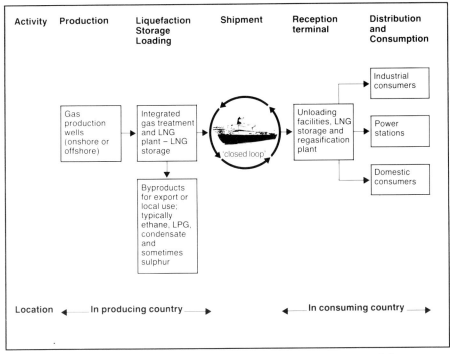

Figure 8.12 Schematic view of an integrated LNG project showing interdependency of phases

greater degree of trading flexibility could evolve in the future, but the bulk of LNG trade is always likely to be of a dedicated nature.

DISTRIBUTION AND MARKETING

Having discussed the methods employed to bring natural gas to the area where it is used, this section examines both the problems and challenges facing the local distribution company and the main markets for gas.

Local Distribution of Natural Gas

While high-volume users such as large factories and power stations may be supplied directly from a gas transmission pipeline, the majority of consumers receive their gas through the mains of a local gas distribution and marketing company.

Pipeline systems have the advantage of being extremely reliable because they are rarely affected by factors such as bad weather, strikes and traffic congestion.

However, they must be operated with maximum efficiency because:
- the customer has no storage;
- if supply fails even for a short period and is then re-established, unburnt gas could escape from any appliance that had not been turned off;
- if the supply fails, air may enter the system and a combustible mixture of air and gas could build up in the pipes.

A piped system is, however, inflexible, because:
- it cannot be moved once installed;
- it cannot exceed a predetermined capacity without considerable expense;
- it cannot be withdrawn from service for maintenance or repair unless there is a duplicate or alternative system.

The essence of distribution is to receive gas at high pressure from transmission pipelines and then to deliver it at medium to low pressure through a network of pipes of decreasing diameter to a large number of end-users. The unit costs of distribution and marketing are normally several times as high as those of transmission, though in any individual case the ratio will be affected by factors such as:
- average consumption rates per end-user;
- the geographic distribution of customers, i.e. whether in close proximity or widely dispersed;
- the relative lengths of the transmission and distribution systems.

In many cities and towns around the world, distributors set up business in the nineteenth century as producers and sellers of so-called "town gas" manufactured from coal. This gas had a calorific value of about 500 Btu/ft^3 (4,500 kcal/m^3) or less. Localised distribution networks were developed to distribute this low-pressure, medium-calorific-value gas from small works to compact areas of supply. Large "trunk" mains followed the main roads from the manufacturing point, and complex networks of smaller "street" mains supplied the areas in between, to form a distribution grid.

These early mains used cast iron pipes with hemp and lead joints and were limited to pressures measured in inches of water column. Mechanically jointed spun-cast iron pipes have been used in more recent years, and such grids can be operated at significantly higher pressures. Modern distribution systems make better use of the high delivery pressure of natural gas by using all-welded steel pipe up to the house regulator at pressures of up to 120 psi (8 bars), or polyethylene pipes up to 60 psi (4 bars).

When natural gas became available, the old distribution networks and the appliances were converted to accept natural gas with its higher calorific value (around 1,000 Btu/ft^3) and different characteristics. This has meant at least a doubling or trebling of the thermal carrying capacity of the system, due to natural gas's higher calorific value and the higher pressure required by the consuming appliances. Nevertheless, demand has often increased to the point where a grid of medium-pressure lines has had to be added to inject gas into the main system at areas of low pressure. Fortunately, the cost per household of these systems is low, compared with the total connection cost.

The vast majority of town-gas appliances required significant modification or conversion to burn natural gas efficiently and safely. Some appliances were unsuitable for conversion and had to be replaced. In the early 1980s, Japan remains the last major gas market still being converted to natural gas distribution, but there are a few smaller manufactured gas systems in various parts of the world which will need conversion if and when local supplies of natural gas become available.

In recent years, another problem has arisen where a distribution system has been converted or designed to cope with a particular quality of natural gas. As indigenous reserves begin to decline, there may be the need to import supplementary supplies of gas, which can be of a different calorific value and composition. In such cases there are a number of options, such as:

- using imported gas in specified areas only;
- blending a gas of a higher or lower heating value with the imported natural gas;
- changing the gas flow metering orifices in each appliance. This is a similar and cheaper operation than conversion from town gas to natural gas.

Whatever method may be used, variations in gas quality must be within a limited range if the consumers' appliances are to operate satisfactorily.

Load Balancing

Since pipelines involve relatively high capital expenditure but low operating costs, the tariffs imposed by the supplying transmission company encourage a distribution company to operate as near as possible to maximum loading throughout the year. Due attention must therefore be given to meeting peak demands of a seasonal or similar nature in planning a distribution system.

Because demand for gas is influenced by the weather, domestic habits, industrial working hours and many other factors, the winter peak offtake in a northern hemisphere market may be several times greater than the summer trough. A typical daily offtake pattern is illustrated in Figure 8.13. Furthermore, hourly peaks of even greater disparity occur between day and night and at various times during any 24-hour period. Gas supply systems have to deal with both types of peaks and troughs, in order to avoid both unacceptable fluctuations in the delivery pressure of gas to customers' appliances and any interruption to supplies.

Hourly peaks can be met in a variety of ways, for example by using high-pressure gas holders, or buried lengths of large-diameter piping as storage, or "line

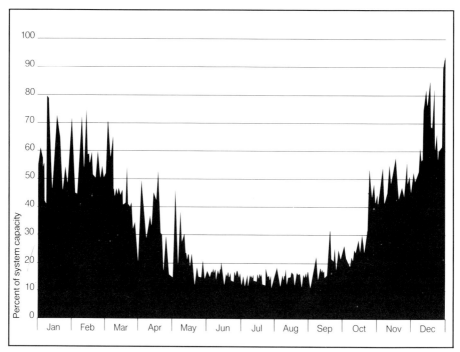

Figure 8.13 **Typical daily gas loads for a year**

pack". The last involves allowing the gas pressure in the high-pressure pipelines within the system to rise towards maximum during hours of low offtake, and to drop during peak hours, with the pressure to each customer's premises being kept constant by a pressure regulator.

Seasonal peaks can be more difficult to deal with, and more costly measures may have to be employed. The top of the seasonal peak can be "shaved" by feeding into the system a mixture of liquefied petroleum gases and air, of similar combustion characteristics to natural gas. This method has a low capital cost but a high energy cost. A lower energy cost but higher capital element method is suitable for the next slice of the peak, such as a small-capacity liquefaction plant that feeds LNG into storage at atmospheric pressure for most of the year. When gas is needed for peak-shaving purposes, the liquid can be pumped through a vaporiser for re-gasification and fed into the pipeline system. For the lower and larger slices of seasonal demand, underground storage or interruptible sales are used. SNG (substitute natural gas) plants may also be attractive, depending among other things on the size of the demand and the cost of feedstock.

Where available, depleted oil or gas fields may be used for underground storage, since it is probable that gas pumped into such formations will be

Table 8.12 **Storage of natural gas**

Storage method	Maximum volume	
	($\times 10^6$ m^3)	($\times 10^6$ ft^3)
Above Ground		
Low-pressure gas holders (per holder)	0.6	22
High-pressure tanks (per tank)	0.03	1
Below Ground		
High-pressure pipes (per installation)	3	100
Underground caverns, aquifers or depleted fields	up to 1,000	up to 3,500
LNG, per 100,000 m^3 tank *	60	2,200

* Some companies or authorities limit individual tanks to much smaller sizes.

recovered without major losses. Similarly, gas can be pumped into underground water pockets (aquifers), salt domes or mined caverns where they exist, though a cushion of gas must be first built up and then retained within the structure.

Typical maximum quantities of natural gas that can be stored by each method are given in Table 8.12.

The Public Nature of Gas Distribution Companies

In order to gain from economies of scale, and because of the obvious undesirability of installing duplicate gas mains in urban streets, gas distribution to end-users tends to fall within the same public utility type of service as water, telephones and electricity. In most countries each operator is awarded a franchise area by the authorities and accepts some degree of regulation. In other words, gas distribution is usually a monopoly within the franchise area, although still subject to competition from other fuel suppliers, not least from electricity. Such monopolies are almost invariably subject to government regulation at central or local level, whether in respect of pricing matters or more general policies. In some countries, a single government-owned distributor may have responsibility for all gas distribution, or at least the sole legal right to supply gas to certain geographic regions or classes of consumers. In other countries, gas distribution is carried out either by private companies or municipalities and is subject to complex controls and regulations governing conditions of service, prices, rates of return, and technical and safety standards.

Because of its monopoly status, and its large share of the residential energy market, gas distribution is usually subject to more direct control or influence from local and central governments than, for example, oil or coal distribution. Government policies therefore play an important role in shaping the development of the gas business; the desire to protect domestic (residential) customers from abrupt price changes is one of the more frequent and significant distortions that can and do arise from such controls.

Markets for Gas

Natural gas is a very attractive fuel for many purposes. It provides a clean-burning flame, relatively unpolluted exhaust gases, easily controlled rates of heating and, where required, high heat intensity. Moreover, because it is piped direct to the point of application, the customer needs no storage, pumping or processing equipment on his premises and supplies should be almost invariably reliable and regular. It thus has many advantages over and above its inherent thermal efficiency so that potential markets are extensive. Indeed, natural gas can be used in virtually all stationary fuel applications (which typically make up around two-thirds of the primary energy consumption in an industrialised country in the northern hemisphere). In particular circumstances it may also be used as an automotive fuel in compressed (CNG) or liquefied form (LNG). In addition, serious consideration is now being given in some parts of the world to the conversion of natural gas into methanol or gasoline for automotive use. This subject is discussed briefly near the end of this chapter.

Natural gas, because of its inherent properties, can, in the absence of price controls, command a higher price from the customer for many applications than strict thermal parity with alternative fuels such as gas oil, fuel oil and coal. Many of these so-called "premium" or high-value uses are to be found in the residential, commercial and industrial sectors of the market. In certain situations, even power stations can be regarded as premium users, particularly in areas with problems of atmospheric pollution.

(i) The Domestic Market

Coal gas distribution systems were used mainly to supply cooking and water heating requirements, for which the daily demand is fairly constant throughout the year. Lighting was also an important outlet for manufactured gas before it was displaced by electricity. The advent of natural gas has, by increasing the calorific value of the gas distributed, widened the options available. Natural gas has even made a limited recovery in the lighting market, mainly for outdoor lighting in gardens, patios and such like.

The versatility of natural gas enables it to be used safely and efficiently for a wide range of domestic space heating systems, ranging from individual room heaters to complete central heating. It particularly lends itself to automatic control, which improves the efficiency and convenience of utilisation, while new gas technology is expected to raise the heating efficiency of domestic gas-fired appliances by twenty per cent or more above that of conventional appliances.

In most industrialised countries, electricity is the main competitor for cooking and water heating in urban areas, while in less developed countries the range of alternative competing fuels may be rather greater. Kerosine and gas oil are most commonly the alternative fuels for space heating, although the electrically driven heat pump could become an important competitor in years to come, with the added attraction of its ability to provide air-conditioning in the summer months. Gas-driven heat pumps for the domestic market are also being developed, and if energy prices continue to rise, the greater capital cost of such heat pumps could be justified by the lower energy costs required to maintain a given standard of heating.

The ability of gas distributors, after the introduction of natural gas, to penetrate the space-heating market has been a major factor in their success in several countries. In this way, additional volumes can be sold to existing residential customers with little or no extra investment in distribution facilities.

(ii) The Commercial and Small Industrial Market

In this market, which includes public buildings, shops, offices, hospitals, schools, hotels, small industrial users and the like, gas applications are, in general, similar to those in the residential market, but on a larger scale. Gas, however, has an important advantage in city centres over liquid and solid fuels, which require space for storage and adequate access for delivery vehicles. The reduction in air pollution also favours gas in such locations. In addition, recent improvements in the technology of using gas to produce hot water or steam (the major use of energy in this sector) have increased the potential utilisation efficiencies considerably. In certain types of building, the introduction of gas-fired infra-red radiant heating has achieved very large fuel savings over conventional methods of space heating.

(iii) The Industrial Market

Individual industrial customers may often have a large enough demand to justify separately negotiated gas prices or special tariffs. In certain industries, such as metal-working, textiles, glass and food, natural gas can command a high price because its inherent advantages of cleanliness and controllability offer a positive benefit in the processes employed by those industries (two examples are shown in Figures 8.14 and 8.15). These applications frequently combine premium value with a high year-round load factor, with the result that they are very important to

Figure 8.14 **High temperature industrial radiant burners, in various operating phases, fired with natural gas. With modern gas technology it is possible to operate at furnace temperatures up to 1400°C.**

the gas supplier, and efforts are continually made to expand this category of uses by further research and development work.

In other processes such as the production of steel, cement-making, etc., little or no technical advantage over other fuels is gained from the use of natural gas. The same usually applies in those industries where fuel is primarily used for steam-raising in conventional boilers. Because the competing fuels in such cases tend to be lower priced, e.g. heavy residual fuel oil and coal, the prices attainable for gas will be generally lower than in the premium sectors. As a result, the sale of large quantities of natural gas for these non-premium applications usually occurs only when one or more of the following conditions applies:

- Natural gas can be drawn from plentiful indigenous sources and is available at a competitive price.
- There are regulations (e.g. in Japan and parts of the United States) which limit the permitted emission of pollutants. In such cases, industrial consumers may only be able to use fuel oil or coal if expensive capital investment is made in desulphurisation and other equipment, so that a clean fuel such as natural gas

Figure 8.15 High temperature radiant burners in use for the direct firing of sintered metal products. Only because natural gas burns so cleanly is it possible to use direct firing for materials which are highly sensitive to air quality.

may be a more economic alternative. In such instances, the use of gas may be regarded as falling within the premium category.
- Gas is sold to industry on an "interruptible" contract, whereby under specified terms of notice the gas distributor has the right to stop supplying gas when demand from customers on "firm supply" contracts is at its peak. This is a very useful way of levelling out gas demand over the year. However, in order to provide sufficient incentive for the industry in question to install facilities both for gas and an alternative fuel, the price of gas will usually have to be lower than that of competing fuels, unless the user gains particular benefit from the use of gas. This is usually the lowest-value gas market of all, but may none the less be extremely important to the distributor, as it may be the lowest-cost method of correcting the load factor of the higher-priced firm supply.
- Total gas requirements for the preferred high-value applications are initially insufficient to justify building a pipeline, or demand cannot be built up rapidly

enough to match the availability of gas. In such cases the economics of the system can be improved by selling gas to large consumers for a limited number of years at a relatively low price, in order to provide additional revenue while the higher-value markets are building up.

As a general principle, a high-quality resource like natural gas should be directed preferentially to the highest-value uses. Nevertheless, non-premium industrial applications provide a fairly steady base load, with limited seasonal and hourly variations. As a result, significant quantities of natural gas are used in such markets in the majority of countries.

(iv) Electricity Generation

As in the case of non-premium industrial use, natural gas does not normally offer any significant economic advantage compared with liquid or solid fuels. However, a significant consumption of natural gas for power generation has developed in Japan, where air-pollution legislation is a major consideration. Companies operating power stations in densely populated cities like Tokyo and Osaka can more easily meet regulations by burning natural gas or a similarly clean fuel than by adopting the complex and expensive measures needed if cheaper fuels are used. In such circumstances, electricity generation has become a high-value market for natural gas.

In several other countries, the power generation market has developed as an important user of natural gas at a price similar to that paid by non-premium industrial users. Consequently, as gas prices have risen relative to other fuels, consumption in this sector has tended to decline in recent years. Nevertheless, in oil-exporting countries, where natural gas is plentiful and where power stations and/or desalination plants represent a substantial proportion of energy consumption, natural gas is likely to be the main fuel for those purposes, thus freeing more crude oil for export. Since natural gas is inherently more expensive to transport than oil, the export of crude oil tends to give a higher netback value to the producer than natural gas.

(v) Chemicals

Methane is relatively stable chemically, which limits the use of natural gas as a petrochemical feedstock. Because of its high hydrogen to carbon ratio, it is used for the production of ammonia, in terms of tonnage the most significant chemical derived from oil and gas. Ammonia is the raw material for nitrogeneous fertilisers and a range of other chemicals such as synthetic fibres, thermosetting resins and explosives. Methane is also used for the production of methanol as an intermediate product. However, in general terms, methane does not have the same degree of versatility as some other hydrocarbons for conversion into a wide range of chemical products.

DEVELOPMENT OF THE INTERNATIONAL GAS TRADE

The Economics

Pipelines and the liquefaction route are as yet the only proven methods for transporting large quantities of natural gas economically over long distances, although there is interest in converting smaller quantities of gas into other liquid fuels or feedstocks in order to reduce transport costs. There are a number of important factors that distinguish the long-distance movement of natural gas from that of oil. They arise from the capital-intensive nature of both the pipeline gas and LNG businesses, where a single modern-day, major project can cost several thousand million dollars.

From the date when the decision is made to undertake an LNG or major pipeline export scheme until the receipt of the first revenues may be anything up to ten years. It will then be several more years before the capital can be repaid even on an undiscounted basis, and still further years before the project can be said to show a true return. Not surprisingly therefore, the level of risk combined with capital exposure is usually too large to be undertaken by a single entity, even though it may be possible to finance a substantial proportion of the cost of the project by borrowings from the international capital markets. As a result, most major gas projects are undertaken on a partnership/joint venture basis, with the involvement in many instances of the host government in one form or another. The need to reach a consensus between all parties concerned is likely to increase the lead time in establishing such projects.

In order to raise sufficient finance for such schemes, it is necessary for the promoters to have reliable outlets, backed by "take-or-pay" or similar contractual obligations for the duration of the project, which is usually twenty years or more. In the case of a pipeline gas project, the inflexibility of supply is obvious. For LNG, the physical reasons may be less obvious; nevertheless, the economic reasons have proved sufficiently real to necessitate what has become known as the "closed-loop" system (see Figure 8.12), with sellers using dedicated facilities and ships to supply one buyer or group of buyers. Unlike oil, there were by 1980 virtually no opportunities for the spot trading of LNG, although this situation may gradually change over the long term if and when there are a larger number of liquefaction plants and reception terminals around the world. To summarise, the international gas trade (whether pipeline gas or LNG) is much less flexible and less subject to rapid change than the oil industry.

A further point results from the capital-intensive nature of the business. For economic reasons, it is desirable for the supplier to utilise his liquefaction or pipeline facilities at the highest possible level (i.e. load factor) all the year round in order to reduce unit costs. However, as indicated earlier, many natural gas markets are not of this nature and require a higher rate of supply at peak periods of demand. If gas were to be supplied at a rate matching the fluctuation of demand, the supplier would need to charge very high prices in order to spread the costs of meeting peak demand. This fact has major implications not only as described for the gas distribution and marketing business, but also for the nature of the international trade.

Finally, it is difficult to establish an international "price" for gas in the same way as is done for internationally traded oil. In the first place, around 85 per cent of all natural gas is consumed in the country of origin, often at prices that are regulated directly or indirectly by the state. As a result, widely different market prices have evolved, depending on the energy policies of individual consuming governments. Furthermore, where gas is traded internationally, the cost of transporting gas is typically between five and ten times higher than that for oil (depending on the distance, size of cargo, and other factors) and, as such, forms a substantial part of the delivered cost of LNG or pipeline gas. If there were a standard producer (fob) selling price, this would mean a very wide range of delivered gas prices in the consuming country, depending on the distance and cost of transport. For some remotely located gas, this could well result in a delivered cost either much higher than the market could bear or uncompetitive with more favourably situated suppliers.

Integration

Any international gas project necessitates a major commitment from each and every participant, such as national government, oil company, and contractors. There is a need to handle at one and the same time such major issues as production, transport, construction, marketing and finance to ensure that all phases are welded together into a coherent package, in which no one element either dominates or is weaker than the others, to the ultimate detriment of the whole project. In addition to meeting any conditions imposed by governments, it will be necessary to reconcile the interests of sellers, buyers, financiers and all other parties who may be involved. Continuous planning, sophisticated computer programing, coordination and optimisation of all activities from inception are vital to successful project implementation. There are as yet relatively few entities who have demonstrated in a positive way that they have the resources, experience and ability to undertake such a massive and complicated task.

The Growth of International Gas Trade

Looking first at pipeline trade, there are four outstanding examples which collectively account for the greater part of the world's international gas trade today (see Figure 8.16):

- Dutch gas, based primarily on the Groningen field, has played a vital role during the 1960s and 1970s in supplying a significant proportion of the gas consumed in West Germany, Belgium, France, Italy and Switzerland.
- Supplies from the Soviet Union to both COMECON countries in Eastern Europe (from 1967) and to Western Europe (from 1968).
- Supplies from Canada to the United States.
- Supplies from Norway to various West European countries.

All of these will continue to be major elements of the world's international pipeline gas trade for many years to come.

Natural gas is also exported by pipeline from Mexico to the United States, between various South American countries, and from Iran and Afghanistan to the USSR. Some of these trades will increase, and new schemes, such as the piping of gas from North Africa to Southern Europe, are in various stages of planning or implementation.

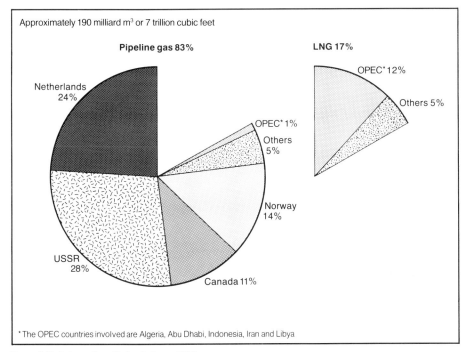

Figure 8.16 **Internationally traded gas 1980**

Turning to LNG, this accounted for only around 2 per cent of total world gas consumption in 1980 and 17 per cent of total internationally traded gas. However, its rise has been quite dramatic during the 1970s, as illustrated in Figure 8.17, with Japan becoming the world's largest market for imported LNG.

In 1980, imports into the USA from Algeria were severely restricted by a pricing dispute, which accounted for the decline in total trade in that year compared with 1979. In spite of such setbacks, the outlook for international LNG trade remains encouraging, with many new projects around the world in various stages of construction, negotiation and planning.

It will be seen from Table 8.13 that the number of exporters and importers as yet remains limited. However, it is expected that deliveries of LNG will continue to increase over the rest of the century.

New projects or expansions of existing projects in Malaysia, Algeria and Indonesia are currently under construction and will be added to the above list of operational international LNG schemes during the early 1980s. Other sources in Africa, the Middle East, South East Asia, Australia, Canada, South America and

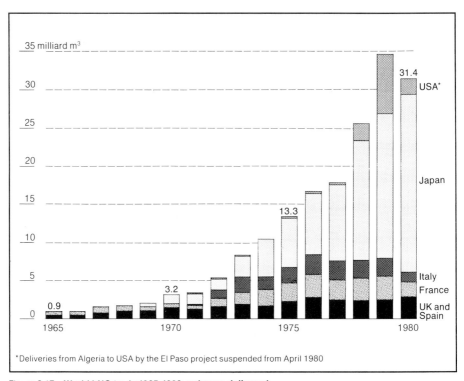

Figure 8.17 **World LNG trade 1965-1980: volumes delivered**

Table 8.13 **1980 Operational LNG projects**

From	To	Start-up year	Plateau volumes fob (milliard m³ p.a.)
Algeria	UK	1964	1.1 *
Algeria	France	1965	0.6 **
Alaska	Japan	1969	1.6
Libya	Italy	1970	2.4
Libya	Spain	1970	1.1
Brunei	Japan	1972	7.5
Algeria	France	1972	3.7 **
Algeria	Spain	1976	4.8
Abu Dhabi	Japan	1977	3.0
Indonesia (East Kalimantan)	Japan	1977	4.2
Indonesia (North Sumatra)	Japan	1978	6.3
Algeria	USA	1978	10.9 ***
Algeria	USA	1978	1.3

* This particular 15-year contract had been fulfilled at the time of writing and may or may not have been extended subsequently.
** These two projects to Gaz de France, together with the third contract which was not operational in 1980, were renegotiated into one contract during 1981.
*** Supplies under this contract to El Paso were suspended in April 1980.

elsewhere are expected to be developed, and will be added to the world's existing portfolio of LNG projects during the second half of the 1980s or early 1990s.

NATURAL GAS LIQUIDS AND GAS-DERIVED LIQUID FUELS

So far, this chapter has concentrated on natural gas, and its major uses. To complete the gas story, however, it is necessary to examine two related activities which have been mentioned in passing but remain relatively modest contributors to world energy supplies in 1980. They are:
(a) the marketing of natural gas liquids extracted from natural gas; and,
(b) the chemical conversion of natural gas into methanol or gasoline or ammonia.

Natural Gas Liquids

A natural gas stream at wellhead may contain varying quantities of heavier hydrocarbons, collectively known as "natural gas liquids". Where it is economi-

Figure 8.18 **Extraction plant, compressor and demethaniser tower at St. Fergus, UK.**

cally justified, these may be removed from such natural gases at a separation plant, such as the one at St. Fergus in the UK (Figure 8.18).

The natural gas liquids may then be further processed at a fractionation plant (see Figure 8.5) into three main product groups for marketing purposes, viz.:
- ethane (C_2);
- liquefied petroleum gases — LPG (C_3 and C_4);
- condensates (C_5 and heavier fractions).

Each has distinct properties and is used in different markets. A detailed description of LPG and its uses will be found in Chapter 6 (Marketing of Oil products).

Ethane

Ethane is gaseous at normal atmospheric temperature and pressure, and in some respects is not a "natural gas liquid" in its full meaning, since ethane is seldom liquefied in isolation. Outside North America, it has in the past usually been left in the natural gas stream to enhance or maintain the calorific value of marketed natural gas.

Ethane does, however, have one major use, namely as a feedstock for the production of ethylene, a major building block of the modern petrochemical industry. Compared with other potential ethylene feedstocks (usually liquid oil products such as naphtha or gas oil), a high proportion of ethane can be converted to ethylene, whereas the heavier feedstocks produce a significant quantity of co-products such as propylene, butadiene and butylenes. The capital costs of an ethane cracker are lower than those of a liquids cracker for the same ethylene capacity, but the extent of the economic advantage gained by using ethane will also depend on the relative values of the various chemical products in any particular market.

In the United States, ethane extracted from natural gas has become the main feedstock for ethylene, providing the source of more than half the ethylene produced in that country. This was a logical and economic development because ethane was worth more to the ethylene producer than naphtha or gas oil, but had a much lower value in natural gas sold as such. Elsewhere in the world, the same combination of circumstances has not generally applied in the past. Either quantities of natural gas produced have been too small to provide enough ethane to support a world-scale ethylene plant, or (as in the case of Western Europe and Japan) the petrochemical industry became established on the basis of relatively cheap naphtha or gas oil before natural gas became available in substantial quantities. As a result, the petrochemical industry outside North America produces far more co-products, which in the United States are produced at oil refineries.

Nevertheless, the future is likely to see an increase in the extraction of ethane under the following conditions:

(i) where gas production is relatively plentiful and markets are limited so that ethane has a higher value as a chemicals feedstock than in alternative fuel uses; examples include Saudi Arabia and Norway;
(ii) where the use of ethane as a chemicals feedstock can reduce imports of crude oil or oil products, e.g. New Zealand;
(iii) where the removal of ethane may make the leaner natural gas more compatible/acceptable to the receiving market, e.g. countries like Canada and Mexico which export natural gas to the United States;
(iv) where the extraction of ethane offers added value prospects without reducing

the calorific value of the resultant natural gas below acceptable limits, e.g. the UK.

In general, however, it is unlikely that special provisions will be made to strip out ethane in isolation. It is therefore only likely to become available as a feedstock where either the liquefaction of natural gas or the fractionation of natural gas liquids is being undertaken.

Condensates (or Pentanes plus)

Unlike the other products mentioned, these heavier hydrocarbons are liquid at ambient temperature and normal pressure. Dependent on the composition of the natural gas, these condensates appear in various forms, ranging from a light product similar to naphtha to what is virtually a light crude oil. As such, condensate can be transported in conventional oil carriers and may be used (dependent on composition) either as a refinery or chemicals feedstock, or directly as a fuel.

Methanol, Gasoline and Ammonia

A very high level of capital expenditure is required to liquefy, ship, store and re-gasify natural gas, or to transport it by pipeline, and so alternative ways of bringing the energy content of natural gas to distant markets have been studied.

One of the more interesting is the possibility of converting natural gas into methanol, a liquid used in the chemical industry, but which could also be used as a supplement or as a replacement for gasoline in automotive engines, or yet again as a low-sulphur fuel in its own right. The advantage of this would be that methanol could be shipped in conventional oil tankers, with certain modifications, and there would be no need for expensive insulated ships or storage tanks.

However, there are also disadvantages. Not only is the capital cost of a methanol plant much higher than for a liquefaction plant per unit of energy converted, but chemical conversion of this kind is less energy-efficient than the physical conversion of natural gas to LNG. Typically, the energy loss through methanol conversion and transport may be 35–40 per cent compared with 15–20 per cent for LNG. Current economic studies suggest that, other things being equal (e.g. tax rates, cost of gas, product value in the market, desired rates of return), savings in shipping, storage and handling costs will only match the increased cost of chemical conversion when the supply point and the market are very widely separated (by at least 10,000 kilometres). Under such conditions, conversion to methanol has not been competitive with LNG as an economic means of transporting natural gas where both are practical options. Nevertheless, opportunities for methanol could still be attractive under circumstances where:

(i) Methanol has a substantially higher market value than natural gas. This may be true either in the chemical industry or as an automotive fuel. However, the chemical industry demand is relatively small (at around 11 million tonnes per annum in the world outside the Communist areas in 1980) and so is not likely to form the basis of a major international trade, while the automotive use of methanol and its derivative MTBE (methyl tertiary-butyl ether) has not yet been developed to a significant extent.

(ii) The reserve of natural gas in question is insufficient to support LNG or pipeline gas exports, but sufficient to supply an economically sized methanol plant.

At present, the latter situation, i.e. the development of smaller reserves, appears to offer the best opportunity for developing an international methanol industry on a project-by-project basis, unless major consuming countries decide to promote the use of methanol on a wide scale as a diversification measure.

Another technique (which is expected to be used on a commercial scale in the mid-1980s in New Zealand) is the further conversion of methanol into gasoline. This has yet to be proven on a commercial scale, but may become an interesting alternative in areas which have plentiful natural gas but little or no indigenous crude oil. Conversion to gasoline has the advantage of avoiding the need for major redesign of engines, or changes in the existing infrastructure of service stations which would be necessary if methanol were used directly as an automotive fuel. The disadvantage, however, would be the higher level of capital expenditure and the lower efficiency of energy utilisation.

Alternatively, countries with a surplus of natural gas may seek to convert it into ammonia-based fertilisers or chemical products. However, the existing production of ammonia (a little under 50 million tonnes in 1980 in the world outside the Communist areas) is already based almost entirely on natural gas where this is available. There are no signs in the early 1980s of any new uses for ammonia being developed that could be of major significance in the context of the quantities of natural gas available, in contrast to the so far unrealised potential of fuel methanol.

THE FUTURE

Almost all geologists are agreed that substantial quantities of natural gas remain to be discovered. The ultimate size of the world's gas reserves is open to speculation, but, all things being equal, proven reserves would be sufficient in theory to support increasing gas production until well into the next century. Given that natural gas is an extremely acceptable fuel for a very wide range of uses, an expansion of the gas business would appear almost inevitable. Nevertheless, reality may not be quite so simple as this.

Cost, Price and Value

In 1980, more than 85 per cent of the world's natural gas consumption occurred in the country of origin but, as has already been made clear, this cannot continue much longer in those main gas markets of the world where the ratio of indigenous reserves to production is declining. If natural gas is to retain or increase its share of energy supplies, imports will become increasingly necessary. However, a large proportion of the world's uncommitted reserves of natural gas is situated in areas remote from the large existing gas markets. This represents a major challenge for the future well-being of the gas industry.

Major potential import markets such as the USA and Continental Europe are still production areas and hence tend to think of natural gas as a substitute for coal and fuel oil. They therefore place a value upon natural gas that is similar to that of fuel oil. Due to the much higher cost of most imported natural gas, such a market value will net back a much lower return to potential exporters than would crude oil. While most producers have accepted the fact that netbacks for gas must be somewhat lower than for oil, they would still argue that a better market value can be developed by bringing the price of gas to the end-consumer into line with competitive energies and by promoting high-value uses for gas. In most cases the return to the producer would still be lower than that for oil but a better balance of benefits between consuming and producing countries could be achieved.

The manner of resolving the problems of sharing out the benefits of producing and using natural gas is likely to be one of the biggest single influences on the future size and shape of the natural gas business, and hence on the growth of the international gas trade.

The Energy Picture

Another dilemma is the inevitable uncertainty surrounding the supply and demand for other energy forms and the price at which they will be available. In a world of national protectionism and self-reliance within tightly grouped trading alliances, it is difficult to forecast precisely how and within what time scale the huge natural gas reserves situated in the Middle East and elsewhere may successfully be brought to market. Likewise, if energy conservation and the substitution of alternative fuels mean that major oil shortages are postponed well into the twenty-first century, then the incentive to exploit natural gas to the full may become less pressing and be postponed. However, the need for an increased diversification of energy resources may counterbalance this to a certain extent.

Nevertheless, it remains likely that at some time in the not too distant future the momentum of economic and population growth in many areas of the world will lead to more effective utilisation of these proven reserves of high-quality energy.

THE FUTURE

Possible Trends

However the uncertainties of the energy picture may be resolved, a number of developments can already be perceived which seem likely to be major features of the gas industry for the rest of this century:

(i) Increased exploration for natural gas in its own right, rather than as a co-product of oil, particularly in areas close to existing major markets. With developing technology, areas such as the deeper waters off Western Europe and North America continue to become increasingly attractive, since natural gas found in such locations seems likely to obtain a high value under almost any set of circumstances.

(ii) Producers of associated gas are becoming increasingly concerned to reduce levels of flaring and to utilise the gas in various ways. These efforts are likely to be intensified, although the total elimination of flaring may never be achieved for economic and other reasons.

(iii) New gas markets will be established. Such developments may be justified not only on grounds of reduced foreign exchange outgoings for imported energy supplies in countries struggling to meet this growing burden, but also to improve security of energy supply. In some cases the most suitable way of achieving this aim may be to convert the gas into other fuels for use, for example, in the transport market or as a chemicals feedstock.

In summary, the extent of the world's gas reserves does not present a constraint *per se* to the development of the gas business, but there are locational, economic, technical and other obstacles to be overcome. Many existing markets will need to rely to an increasing extent on imported gas supplies. Market-use patterns will probably change, with greater emphasis on higher-value applications. New uses for gas can be expected, as well as the development of new markets, particularly in the developing countries.

Chapter 9

OIL SUPPLY AND TRADING

INTRODUCTION

The activity concerned with the procurement and movement of oil from producing areas to its distribution in centres of consumption is commonly known in the industry as "supply operations". It involves the coordination and organisation of many variables so that, at the end of the chain, markets are supplied reliably and competitively with the products they require.

In the early days of the oil industry, supply operations were largely carried out in an integrated manner by the major oil companies. These companies had operational control of the oil from the wellhead to the points of consumption. As the industry grew in scope and economic significance both producer and consumer governments became increasingly interested in participating in this activity, directly and indirectly. Producer governments, who in many cases were heavily dependent upon oil revenues, led the way and in the early 1970s OPEC member governments took increasing control of the production facilities in their countries.

In retrospect, 1973 can now be seen as a watershed year. Before 1973 supply operations were mainly managed by large integrated companies; a system that functioned well in a world of steadily expanding demand. After 1973 this integrated system became fragmented. Crude oil production came within the full control of the countries that produced it, and with this control over production levels came the power (subject eventually to market constraints) to determine the price of crude oil. Briefly, the significant events of 1973 comprised difficult and unresolved negotiations between the major oil companies and the principal members of OPEC, compounded by the outbreak in October of the Arab–Israeli War. The hostilities were accompanied by the imposition of embargoes on oil supplies to the USA and the Netherlands by Saudi Arabia and Kuwait, resulting in a widespread acute shortage of oil. In this turbulent situation, OPEC raised the price of crude oil from $2.80 per barrel, first by $2.00 and then by a further $6.00 from January 1974. The effect of this dramatic rise was to deepen the subsequent world economic recession and to halt the hitherto progressive growth of oil demand. This in turn created spectacular surpluses of capacity in oil tanker fleets and refineries, which were all the more difficult to remedy because they were industry-wide.

FACTORS AND CONSTRAINTS IN OIL SUPPLY

Apart from the changes wrought upon the supply scene by structural change within the industry, many variable factors have always affected day-to-day supply operations. The supply planner has traditionally been continuously concerned with a delicate balancing act between these factors, which are described below (Sections 1 to 8). Although the numerous supply options involved in day-to-day activities lend themselves nowadays to the aids to decision-making that computer technology offers, there is nevertheless no substitute for sound commercial judgment based on experience. This has particularly proved to be so in the times of emergencies and disruptions that have occurred in recent years.

1. The Geographical Factor

As oil has mainly been found far from the principal centres of consumption, the transportation of very large volumes of oil over great distances is an important factor in its supply. Table 9.1 shows the differences in production and consump-

Table 9.1 **Oil production and consumption by region** (million barrels per day)

	1950		1975		1980	
	Production	Consumption	Production	Consumption	Production	Consumption
North America (incl. Mexico)	6.2	6.6	12.6	18.6	14.0	19.6
Caribbean and South America	1.8	0.6	3.6	2.6	3.8	3.2
Western Europe	0.1	1.3	0.6	13.7	2.6	14.2
Africa	neg	0.3	5.1	1.2	6.1	1.4
Middle East	1.8	0.2	19.7	1.3	18.9	1.9
Japan	neg	0.1	neg	5.1	neg	5.1
Rest of Asia and Australasia	0.2	0.4	2.2	2.9	2.8	3.9
Communist areas	0.8	1.2	11.9	10.4	14.6	13.0
World total	10.9	10.7	55.7	55.8	62.8	62.3

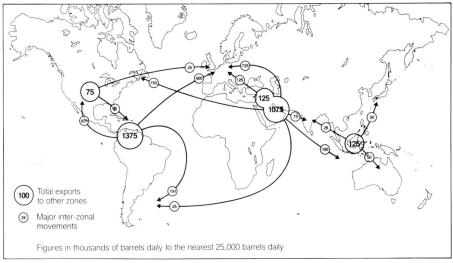

Figure 9.1 **Principal movements of crude oil and products between zones, 1950**

tion in the various geographical regions, and Figures 9.1, 9.2 and 9.3 show how supply patterns have changed during the last thirty years.

The spectacular growth of production in the Middle East between 1950 and 1975 can be clearly seen, as can the developments in the African region that took place during the 1960s. This growth in production was absorbed almost entirely by the growth in demand in the industrialised countries of Western Europe and in

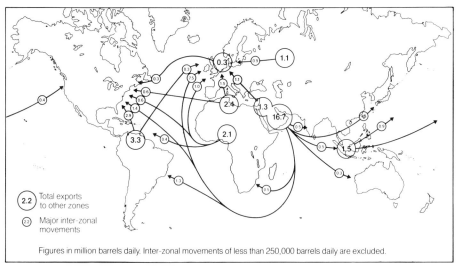

Figure 9.2 **Principal movements of crude oil and products between zones, 1975**

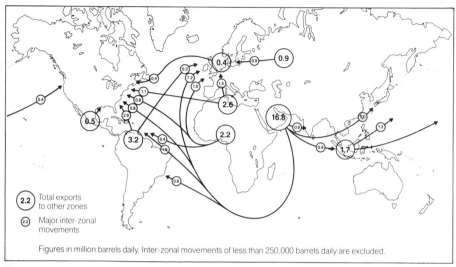

Figure 9.3 **Principal movements of crude oil and products between zones, 1980**

Japan. The United States, for long self-sufficient, also became a net importer of crude oil.

The 1970s saw the expansion of production in the North Sea and Mexico and a slight decline in the relative importance of the Middle East. However, it should be noted that the gap between production and consumption in the industrialised countries of Western Europe and of the Far East (mainly Japan) is still considerable.

2. Differences in Types of Crude Oil

Crude oils produced from different fields or reservoirs have different properties. Each crude oil has its own yield of refined products and even the quality of the products produced from different crude oils varies considerably. Table 9.2 illustrates these differences; it should be noted that the main products of primary distillation are often not marketable as such, but have to undergo further processing and blending, which may alter the overall yields shown.

3. Diversity of Product Demand in Consuming Countries

The demand for oil products varies considerably by type as well as by volume between countries. Developed countries with large passenger car populations consume a large proportion of their oil as gasoline. Developing countries with their lower standards of living use less electricity and consequently have a high

Table 9.2 **Yields of main products from primary distillation of various crude oils** (per cent by volume)

Country (crude oil name)	Gases	Gasoline	Kerosines	Gasoil/diesel	Residue (fuel oil component)	Remarks
North America						
USA (Alaska)	0.8	13.4	11.6	21.5	53.3	Low sulphur: 1.0% on crude, 1.5% on residue
Western Europe						
UK (Forties)	4.3	22.5	12.2	21.9	39.5	Very low sulphur: 0.3% on crude, 0.6% on residue
Norway (Ekofisk)	3.3	31.2	13.6	21.6	30.8	Very low sulphur: 0.1% on crude, 0.3% on residue
Middle East						
Saudi Arabia (Arabian Light)	1.7	20.5	12.0	21.1	45.1	Very high sulphur: 3% on crude, 4.5% on residue
Qatar (Qatar)	4.4	29.1	15.9	20.6	30.7	Medium sulphur: 1.1% on crude, 2.6% on residue
South & Central America						
Venezuela (Tia Juana Pesado)	—	1.4	3.6	14.7	80.8	High sulphur: 2.7% on crude, 3.0% on residue
Mexico (Isthmus)	1.8	22.9	13.1	22.0	40.4	High sulphur: 1.6% on crude, 3.0% on residue
Africa						
Nigeria (Nigerian Light)	2.9	25.8	14.4	27.7	29.4	Very low sulphur: 0.09% on crude, 0.028% on residue
Libya (Libyan Light)	2.8	21.6	12.9	22.1	40.9	Low sulphur: 0.5% on crude, Less than 1.0% on residue
Far East and Australasia						
Indonesia (Sumatran Light)	0.5	11.5	9.5	20.6	58.3	Very low sulphur: 0.08% on crude, 0.1% on residue
Malaysia (Miri Light)	1.9	28.1	16.7	32.1	21.3	Very low sulphur: 0.04% on crude, 0.1% on residue
Australia (Gippsland)	2.3	36.0	13.8	24.7	23.5	Very low sulphur: 0.08% on crude, 0.2% on residue
USSR, China						
Russia (Ural)	2.2	20.9	14.7	19.5	43.1	High sulphur: 1.5% on crude, 2.6% on residue
China (Daqing)	0.5	9.8	6.9	16.5	66.4	Very low sulphur: 0.1% on crude, 0.15% on residue

demand for kerosine for lighting, cooking and heating. Countries with little or no coal or natural gas require more fuel oil, particularly those like Japan which are industrialised.

Table 9.3 shows typical variations between countries in product demand patterns. It will also be noted that shifts in product patterns have occurred over the five-year period, as countries have undergone economic development and have also adjusted to higher oil prices.

Table 9.3 **Comparison of main product demand in selected countries** (per cent by volume)

	United States		West Germany		Japan		Brazil		India		Indonesia	
	1975	1980	1975	1980	1975	1980	1975	1980	1975	1980	1975	1980
Gases	8.8	9.0	4.2	2.8	6.7	9.2	7.2	8.8	2.1	2.0	0.2	0.2
Gasolines	46.0	45.5	27.2	33.2	21.6	22.6	34.3	30.0	17.1	14.9	18.4	16.5
Kerosines	6.1	6.2	1.7	2.1	8.3	9.6	4.5	4.7	18.1	18.2	39.8	36.0
Gasoil/ diesels	18.6	20.2	44.1	42.7	12.6	15.4	24.3	30.0	35.5	38.1	27.8	33.0
Fuel oil	15.7	14.6	17.1	14.4	47.8	39.8	26.3	23.4	20.4	21.2	10.7	11.7
Other residues	4.8	4.5	5.7	4.8	3.0	3.4	3.4	3.1	6.8	5.6	3.1	2.6
Total consumption (million b/d)	15.6	16.1	2.6	2.7	4.9	4.9	0.9	1.1	0.5	0.6	0.2	0.4

4. The Price of Crude Oil

The price factor is of fundamental importance in its effects on the supply/demand relationship.

While geographical factors have a major impact on most oil movements, commercial considerations play the decisive role. Traditionally, the relative prices of different crude oils are determined by the product yield and quality of each crude oil, and the distance to the area of consumption.

The range and yield of products that can be manufactured vary widely. Light crude oils yield more gasoline and other generally higher priced products. Heavy crude oils usually contain more of the heavier products such as fuel oil, which are less valuable. Some heavy products can be "up-graded" or converted into lighter products, but conversion is expensive. The lightness of a crude oil is therefore a rough guide to its value.

The quality of the crude oil affects ease of processing and the quality of finished products. For example, crude oil with a high sulphur content usually requires more expensive refining, and even then some of the refined products may still have a high sulphur content, and so cause unacceptable pollution when burnt. High-sulphur crude oils are accordingly generally less valuable than low-sulphur crude oils.

A third element is the distance of the area of production from that of consumption. Venezuelan, Mexican, North African and North Sea crudes are nearer than Middle East supplies to the main consuming areas of North America and Western Europe, and so the former command a "location premium".

Changes in demand can also affect crude oil values. For example, cold winters increase the demand for heating oils and so the relative prices of crude oils with higher yields of these products tend to rise. On the other hand, the higher demand for gasoline increases the relative prices for lighter crude oils during the summer.

It will be noted that the factors that determine the relative prices of crude oils are dynamic rather than static; they change continually. Even the factor of geographic distance changes in importance as transportation costs vary. When freight rates are low, the premium that short-haul crude oils attract is reduced.

In addition to the influence on price of market factors, other considerations may influence and on occasions determine the price of crude oil. There are times when government-to-government deals, barter schemes, and other forms of trading arrangements may influence the effective price of some oil.

5. Transportation Costs

Mention has already been made of the effect of transportation costs on the relative prices of crude oils. Transportation costs also affect another aspect of supply operations, the balancing of products.

In the selection of crude oils for a given market, care is taken to choose where possible those that have product yields that match demand. It is, however, seldom possible to do this matching exactly. Moreover, market requirements themselves vary, particularly in temperate countries where seasonal fluctuations can be considerable (e.g. seasonal fluctuations in demand worldwide can amount to as much as 8 million barrels per day). Product surpluses or deficits are therefore unavoidable and usually have to be adjusted by moving products from one area of consumption to another. To some extent these imbalances can be minimised by costly refining processes such as "cracking". However, the extent to which these additional refining processes are used will depend on their costs, relative to the transportation costs of disposing of the imbalances by moving them from one area to another. Thus transportation costs of products are an important factor in supply operations.

6. Abrupt Changes in Production and Demand

The availability of crude from different countries can vary dramatically as the result of political changes. The circumstances surrounding events in Iran that precipitated the drop in supplies in 1978 form a clear example. Early in the year Iran had been exporting around 5 million barrels per day, 18 per cent of OPEC exports, and second only to Saudi Arabia. By the end of 1978, exports had ceased.

The immediate shortfall was met by stocks and eventually by increased production from Saudi Arabia, Iraq, and Kuwait (so much so that by 1979 OPEC production was 1 million barrels per day more than in early 1978). The increase in world production was even bigger at over 2 million barrels per day, of which 700,000 were used to restore the stocks. At the global level there was no significant oil shortage. However, in the local areas of consumption there was a perception of shortage, fed by anxiety as to whether future supplies would be sufficient. This feeling of uncertainty is now much stronger than before 1973 because the link between consumers and producers through the integrated oil companies is no longer there. The lack of confidence in the future created strong pressure from buyers to purchase any oil that was available and so enabled the producers to increase their prices. The result was that prices nearly trebled from early 1978 to mid-1980.

A similar episode took place in September 1980 when the Iran–Iraq hostilities broke out and availabilities from these two countries dropped. The heavy buying that ensued gave producers the opportunity to charge very high premiums during the supply pinch. Yet again an over-supply emerged in a short while.

The sharp rise in prices in 1979 aggravated the economic downturn and

Table 9.4 **Total oil products market demand in selected countries** (thousand barrels per day)

	1979	1980	% change
United States	17,600	16,110	−8
Japan *	5,380	4,880	−9
West Germany	3,000	2,690	−10
France	2,370	2,200	−7
Italy	1,940	1,890	−3
United Kingdom	1,810	1,540	−15
Spain	930	950	+2
Netherlands	880	830	−6
Belgium/Luxemburg	570	500	−12
Sweden	560	490	−12

* Including crude for burning.

deepened the worldwide industrial recession. It also accelerated the substitution of oil by other fuels and promoted conservation. The result was a downturn in demand of around 5 per cent between 1979 and 1980 compared with an increase of about 1 per cent between 1978 and 1979. Table 9.4 shows the demand in some countries in 1979 and 1980.

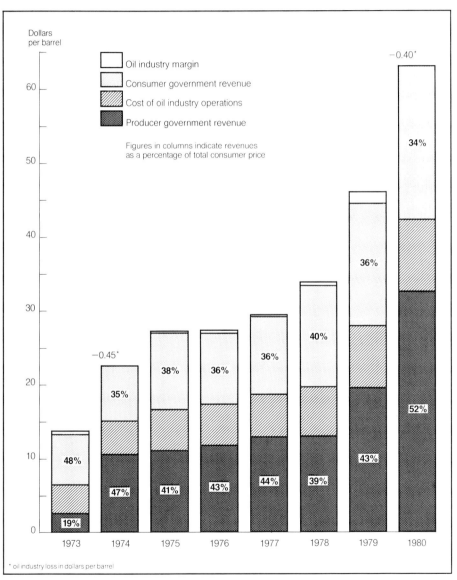

Figure 9.4 **Consumer cost breakdown: typical OPEC barrel sold in Western Europe**

7. Effects of Consumer Government Taxes

Not all of the reduction in demand has been caused by the increase in crude oil prices by the producers. A certain amount is attributable to consumer governments' direct and indirect taxation, which has its effects on consumption. Figure 9.4 shows the approximate cost breakdown of a typical OPEC barrel sold in Western Europe.

8. Non-technical Constraints

Soon after the take-over by governments of oil concessions in the producing areas, OPEC members pursued a policy of reducing their dependence on the larger oil companies and transferred part of their crude oil sales to other buyers. Of these, about 3 million barrels per day (1980 figures) were sold directly to governments. Most of this oil went to Brazil, Italy, France and Argentina, countries that had long decided that a large part of their internal oil market should be served by state companies. Japan, which has previously depended on third party purchases from oil majors, shifted to direct deals via trading houses when the oil companies were unable to supply. In the case of some smaller developing countries whose economies are strained to the limit to meet oil payments, the expectation of secure supplies at no more than government selling prices, with hopes of credit facilities, at times make government-to-government deals appear genuinely attractive.

These moves tend to remove some supplies from the more flexible commercial stream and tie them up into rigid tranches. If the trend continues, the loss in flexibility in the global movement of crude oil will be considerable.

Another constraint that has crept into supply operations is "destination restriction" for political reasons. Other restrictions, such as tightening of loading schedules, insistence on the use of producer country ships, and prohibition on exchanging and re-selling the oil, have also been imposed. In times of crude oil surplus, however, these constraints tend to be lifted.

THE OIL SUPPLY SCENE POST-1973: FRAGMENTED WITH DIMINISHING FLEXIBILITY

With the demise of the integrated system, the pre-eminent role of the major oil companies has been replaced by several entities who now control the various elements of the supply chain. The first of these are the producing governments, represented by OPEC. The production of crude oil is largely in their hands and the setting of its price is basically their prerogative. This development affecting the ownership of oil and control of production levels is the most significant change in the supply system. It is also the area where the change has been most nearly complete. The raw material production end has been almost totally uncoupled from the rest of the supply chain. From a situation where the production of crude oil was in the hands of the companies concerned with the continuous supply of crude oil to their customers, the production levels are now controlled by countries whose objectives and priorities may be different from that of meeting the needs of oil users. Therein lies the most far-reaching consequence of this change on the supply scene.

The spiralling of crude oil prices and the uncertainty over supply has stimulated consumer governments to take a more active interest in oil supply to their countries. Their actions vary with their individual situations. Some, like India, have taken over the entire responsibility for supply. Others have taken over the purchase of crude oil. Still others have adopted supervisory roles such as the control of product prices, the licensing of crude oil importation and the maintenance of stock levels. Their actions, whatever they have been, are understandably aimed at safeguarding their own individual positions, but they often lead, albeit unintentionally, to rigidities in the supply system.

The effort of the oil producers to switch crude oil away from the major oil companies has given new importance also to another link in the supply chain, namely the independent traders and their operations on the spot market.

Even in the days of the integrated system there was need for a spot market because of the inherent problems of exactly matching refinery output with market requirements. Through spot market deals product surpluses could be traded and deficits made good. Historically, only a small percentage of total oil traded has been dealt with in this market and in consequence prices have tended to be volatile. For a marketer faced with a sudden exceptional product requirement (for example, for winter heating oil during a sudden cold spell) purchase of marginal amounts on the spot market, even at more than the going price, can be attractive financially. In such circumstances the spot price of heating oil may jump while at the same time the price in the mainstream market, where perhaps 95 per cent or more of the oil is sold, remains unchanged.

The advent of independent traders has given new importance to spot trading in crude oil and products. Nevertheless it continues to be a marginal market with rapidly fluctuating prices that cannot be taken as a true index of real market value but can often provide a leading indicator to the trend that prices in general may follow.

For all the growth of the independent traders, it is still the case that the major oil companies retain a large role in the supply of oil products to end-users. The facilities to carry out this role are in their hands, such as the tankers, the refineries, and the products' distribution network. They are therefore still the largest buyers of crude oil. Thus they retain the ability to accommodate substantial fluctuations in demand levels in consuming areas, to the benefit of consumers. This flexibility is of utmost importance to supply operations, for the reason that all the factors discussed above are continually varying, and continuity can only be achieved by flexibility. Security of supply may, however, diminish, as the volume handled by the Majors shrinks and more complex but also more rigid supply arrangements prevail.

THE SUPPLY SYSTEM IN THE EARLY 1980s

The fragmentation of the industry and the proliferation of participants with varying degrees of involvement mean that a number of supply channels are now in use. It is obvious that all of them are achieving the main objective of supply operations, which is to provide oil to the end-users. What is less apparent is the degree of efficiency (or inefficiency) with which this is done and the vulnerability of today's systems to disruption.

One consequence of fragmentation and the entrance of new participants is the decrease in the degree of optimisation of supply operations. New participants mean new facilities, more storage tanks, more ships, more stocks. It is, of course, the consumer that ultimately has to pay the extra costs involved.

On security of supply, the traditional system had two very important advantages over the present one. The first was a very close link between the crude oil producers, the oil companies' upstream end, and the product suppliers, the oil companies' marketing subsidiaries. To a large extent success was achieved in keeping difficult commercial negotiations separate from national political concerns, with consequent benefits. The second was the Majors' very diverse sources of supply coming from widely spread production operations, so that in the event of a political upheaval or an accident or some natural calamity in one country or region, oil could be made available from elsewhere.

This security arising from diverse supply and flexible distribution proved very effective during past emergencies such as the Suez Canal closure in 1956/57, and the oil embargo of 1973/74. Today with the Majors no longer able to accept responsibility for overall continuity of supply, consuming countries have attempted to tackle the problem by participating in the IEA (International Energy Agency), which was specifically set up to deal on an international basis with possible disruptions.

The above observations do not argue for the return to the integrated system. It will be to everybody's benefit, however, to give thought to the virtues of the integrated system, and where applicable to try and incorporate these virtues into present practice.

The ownership or control of crude oil production by oil companies is not a necessary condition for operational efficiency. What is relevant is an all-round appreciation of the worldwide benefits of maintaining continuous supply at stable prices. As mentioned previously, the main reason for the phenomenal growth of the industry in the years before 1973 was ease of supply and stability of prices. A reversal of this trend could, in time, diminish the importance of the contribution of oil to the world's energy sources.

Chapter 10

PETROCHEMICALS

THE ORIGIN OF PETROCHEMICALS

Leisure, shopping, building, driving — all these activities have an impact on our lives and all involve the use of materials, including those now known as petrochemicals. There is no universal agreement on the meaning of the word "petrochemicals", but in this chapter it is taken to cover the bulk organic chemicals including polymers, which are primarily derived from crude oil and natural gas, but not the equally large tonnage of lower-valued inorganic chemicals (such as ammonia) derived from the same sources.

The origins of the modern petrochemical industry go back to the mid-nineteenth century, long before petroleum became the major raw material. It is important to recognise that many petrochemicals (such as polyethylene) were originally produced from non-petroleum feedstocks (such as fermentation alcohol, coal and cellulose). The impact of petroleum upon the industry (in the 1930s in the USA and 1950s elsewhere) was to stimulate greatly its growth by the provision of a large and dependable supply of a relatively cheap raw material. It is therefore logical to consider petrochemicals in the broader historical context of bulk organic chemicals.

It is instructive to examine briefly the development of an early organic product, cellulose nitrate. Despite the apparent triviality of this example taken from more than a century ago, it is remarkably illustrative of future trends. Billiard balls were made from ivory and the shortage of elephants' tusks caused a US company to offer a prize for the development of a substitute. This challenge was taken up and the resultant winning product was cellulose nitrate. This product was not entirely satisfactory, however, as by itself it tended to be highly inflammable and could detonate on impact (distinct disadvantages in a billiards saloon!). The problem was overcome by the addition of camphor to the nitrate, and thus in 1862 celluloid was born. This new product could be moulded into billiard balls by heating and pressing and it was not long before a range of other moulded items were produced such as jewellery cases, buckles, and boxes. Thus the intention of a particular material for a specific outlet, namely celluloid for billiard balls, gave rise to a whole new range of opportunities unrelated to the original substitution. This phenomenon of substitution leading to innovation has been repeated many times in the history of materials in general and of petrochemicals in particular. This is a key characteristic of petrochemicals' development to which we shall return.

The advent of organic chemistry as a science resulted in many different products being synthesised or isolated in the laboratory in the last century, which remained solely of academic interest for decades but which are now produced commercially in hundreds of thousands of tonnes. Examples of such products are styrene and vinyl chloride, two precursors of some of the most widely used petrochemicals today. These were first identified in the 1830s but not commercially manufactured until over a century later. The successful development of products of this type arose from a combination of circumstances. Market need coincided with the availability of technology for the economic manufacture of both the base chemical and the product derived from it. Most important of all, there were business managers with the ability to spot the market opportunity and the determination and resources to develop the means to satisfy it.

The earliest organic chemical manufactured from petroleum was IPA (isopropyl alcohol). This was produced in the USA and resulted from the requirement in the First World War for large quantities of acetone (a derivative of IPA). The acetone was used in the manufacture of explosives and as a solvent for aeroplane "dopes" used to stabilise and seal fabrics. After IPA, the next significant landmark in the development of petrochemicals was the commercial production of ethylene glycol by Union Carbide in 1926. This was a landmark of probably even greater significance than the production of IPA, in that it was the first petrochemical derived from ethylene; the availability of ethylene in abundance has been the kernel of petrochemical development. Ethylene glycol was first produced as a non-evaporating antifreeze for cars. This use is also noteworthy in that the car industry has played an important historical role in petrochemical development, not only as a major consumer of petrochemicals but also because the production of gasoline from crude oil resulted in by-products that became major petrochemical feedstocks.

The range of petrochemical products produced commercially expanded considerably in the 1930s to include solvents, detergents, fibres, plastics, resins and rubbers, but it was not until after World War II that the petrochemical industry really boomed. By 1980, nearly 70 million tonnes of petrochemicals were produced in WOCA (world outside Communist areas), as shown in Table 10.1. This compares with about 2 million tonnes in 1950.

It will be seen that plastics and resins dominate petrochemicals and that these

Table 10.1 **1980 Petrochemical production in WOCA**

	Million tonnes
Plastics and resins	43.0
Synthetic fibres	9.3
Synthetic rubbers	6.0
Chemical solvents	7.5
Synthetic detergents	3.3
Total	69.1

Table 10.2 **1980 Materials production in WOCA**

	Million tonnes
Steel	455
Paper and board	150
Petrochemicals	69
Aluminium	12

together with the other polymers (fibres and synthetic rubbers), constitute about 85 per cent of the total. (The table does not separately identify the small-volume, high-added-value speciality products such as agrochemicals and fine chemicals, which are not generally regarded as petrochemicals.)

In order to indicate the scale of petrochemical production, it is compared with that of other materials in Table 10.2.

In money terms, the petrochemical industry in 1978 in Western Europe was comparable in size, on the basis of value added, to the iron and steel industry and rather more than half the size of the motor vehicle industry.

THE IMPORTANCE OF PETROCHEMICALS

How has this growth been achieved and what is the importance of petrochemicals to the world today?

The seminal ideas of the use of synthetic organic materials arose from the need to find substitutes for relatively unmodified natural products, such as ivory, which already had well defined outlets but for which raw material availability was limiting expansion. The development of petrochemicals was a particular case of this movement; other examples were synthetic organics produced from wood and coal. Although massive volume growth came only after World War II, the initial commercial development of some important petrochemicals preceded the war and in some instances received a massive push as part of the war effort. Thus it was in the 1930s that products such as nylon, synthetic rubber and polyethylene were initially produced, albeit not from petrochemical feedstocks in the initial stages. Nylon was the outcome of work by W. Carothers in the USA on the development of new types of synthetic fibres; it was in Germany that styrene–butadiene rubber (SBR) was developed as a result of efforts to find a product with better properties than natural rubber and which could be produced domestically, whilst polyethylene was discovered accidentally in the UK during some experiments on chemical reactions under high pressure and found to have excellent electrical insulation properties.

Commercial production of these three products would have remained at relatively low volume levels if their use had been solely as direct substitutes for cotton, natural rubber or gutta percha cable insulation. The essence of petrochemical growth, however, lay in the fact that the successful products were frequently not simply substitutes but also "enabling" materials having properties which

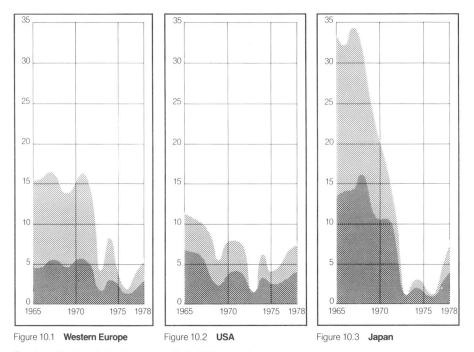

Figure 10.1 **Western Europe**   Figure 10.2 **USA**   Figure 10.3 **Japan**

Trend growth rates of petrochemical and industrial production

▓ Petrochemical production (5-year moving average in %)
▨ Industrial production (5-year moving average in %)

resulted in many other applications being developed. Thus petrochemical growth has been the result not just of substitution but of innovation, both acting over and above the carrier wave of general economic growth.

The high economic growth rates experienced in the developed countries in the period up to 1973 were reflected in the demand for all types of goods, and hence for the materials, including petrochemicals, from which these goods were made. The actual level of growth of petrochemicals was, however, considerably above that of the economy in general, as it was boosted by the substitution/innovation effect, as shown in Figures 10.1, 10.2, and 10.3. This resulted, for example, in petrochemical growth in Western Europe being about 15 per cent per annum at a time when industrial production grew at about 5 per cent per annum.

It is the inherent great versatility of petrochemicals in general and polymers in particular that has resulted in their being used in such large volumes and by virtually every industry, to the extent that life as it is known today would be impossible were petrochemicals not available. Figure 10.4 shows the end-uses of petrochemicals in Western Europe, whilst Table 10.3 lists a few more specific outlets of some petrochemicals.

The potential of petrochemicals was recognised and already being exploited

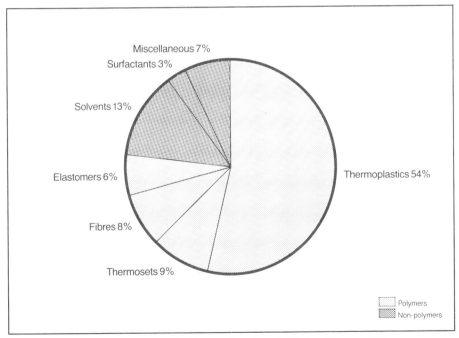

Figure 10.4 **Breakdown of petrochemical demand by volume in Western Europe – 1977**

Table 10.3 **Petrochemicals in use – a few examples**

Packaging	Building	Other uses
Washing-up liquid bottles	Shuttering and moulds for concrete	*In the home . . .*
Mineral and carbonated drink bottles	Damp course film	Moulded chairs
Caps for bottles and aerosols	Thermal insulation	Mattresses
Carboys for industrial chemicals	Window frames, guttering, drainage pipes	Decorative laminates
Coatings for tinplate cans	Gas, sewage and water pipes	Vinyl wall coverings
Bottle crates	Electrical conduit	Washing-up bowls
Tote boxes	Foaming agents for plasterboard	Canisters for food
Flexible packaging for frozen food	Binders for particleboard and plywood	Buckets
Yoghurt and cream containers	Solvents for paints	Bristles for brushes
Stretch and shrink wrap for pallets	Insulation and sheathing for power, telephone and TV wire and cables	Floor tiles
Industrial strapping	Electrical plugs, sockets and switches	Soles, heels and shoes
Refuse sacks	Vandal-proof and security glazing	Washing machine drums and agitators
Sterile packs for medical use	Building film for site cover during construction	Microchip toy casings
		. . . in the office . . .
		Telephone handsets
		Pocket calculator casings
		VDU housings
		. . . and all over the place
		Printed circuit laminates
		Audio and video tapes

Textiles

Fibres for:			
Coats	Skirts	Carpet pile fibres	Tarpaulins
Suits	Socks	Carpet backing fabrics	Soil stabilisation fabrics
Underwear	Stockings	Pillow fillings	Industrial clothing
Shirts	Curtains	Quilt fillings	Fireproof clothing
	Upholstery fabrics		

Cassette housings and drives
Adhesives
Telephone poles
Road markers and paints
Fishing nets, ropes, string
Fish and vegetable boxes
Horticultural film
Powder and liquid detergents
Dry cleaning fluids
Disposable syringes
Contact lens
Artificial hips and other replacement surgical items
Anti-skid road surfacing

Transport

Dashboards	Knobs	Antifreeze
Bumpers	Seat foam	Lorry cabs
Radiator grilles	Seat covers	Train bodies
Gasoline tanks	Interior linings	Bus and coach parts
Tyres and inner tubes	Anti-corrosion treatment	Aircraft parts
Laminate film for safety glass	Paints	Yacht parts
Battery cases	Brake fluids	Space shuttle elements
Wire insulation	Degreasing solvents	
	Binders for brake linings	

immediately after World War II. In the USA for example, the petrochemical polymers then available in the largest volumes were PVC (polyvinyl chloride) and polystyrene. In 1949, US applications for PVC included wall and floor coverings, upholstery fabrics for cars and furniture, margarine packaging, garden hose and gramophone records. At the same time, outlets for polystyrene included car dashboard mouldings, refrigerator boxes, radio cabinets and wall tiles, whilst trials were well underway on the use of sheet for refrigerator linings.

Technical versatility in production, processing and application would, however, have remained but a theoretical possibility if the products themselves had not been competitive. The high growth of the industry has resulted also from the fact that there was in the 1960s and early 1970s a very favourable trend in petrochemical material costs relative to those of competing products, arising from improved technology and increasing scale of manufacture as a feedback effect of increased demand. The basis for this movement was the fundamental change made in the 1950s when petroleum and natural gas feedstocks began to be used rather than coal byproducts or fermentation alcohol.

During the 1960s, commercial production began of the new types of plastics such as the now high-volume products polypropylene and high-density polyethylene and the more specialised lower-volume but higher-unit-cost engineering plastics such as polycarbonate. Polymers are now such an integral part of everyday items that paradoxically they tend to be noticed as such only when material-related problems arise.

A brief examination of petrochemicals in the following sectors will illustrate their role:
- Building and construction
- Cars
- Packaging
- Textiles
- Detergents

In building and construction, polymers often play a vital role as, for example, pipe for drainage, sewage, gas or fresh water supply, as electrical insulation and sheathing for both power cable and telecommunication wires, and as thermal insulation in old and new buildings, as well as in the foundations of roads to prevent frost damage. Opportunities for greater use of plastics in building continue to open up. One example is the PVC window frame, an outlet which has developed particularly strongly in West Germany, where consumption rose from

Table 10.4 **Plastics use in Renault cars**

Type	First year of production	Weight plastic (kg)	Total weight (kg)	Plastic content (%)
R4	1961	13	665	2.0
R8	1962	18	725	2.5
R6	1968	22	780	2.8
R12	1970	33	900	3.7
R5	1972	33.5	730	4.6
R30	1975	66	1320	5.0
R14	1976	53	865	6.1
R18	1978	63.5	920	6.9

Source: Plastiques Modernes et Elastomers, October 1978 and Regie Renault

a mere 9 kilotonnes of PVC in 1970 to 130 kilotonnes by the end of the decade.

In car production, the availability of polymers has enabled costs to be reduced by, for example, the facility with which a part may be moulded in one operation and as a single item in plastic, whereas it would require many fabrication and assembly operations if made in metal. From the user's viewpoint the widespread use of plastic items in car production has also resulted in overall weight reduction, and thus of vehicle fuel consumption, which is an aspect of great importance in times of high energy costs. An illustration of the trend towards increasing use of plastics in cars is shown in Table 10.4.

A major outlet for petrochemicals in the consumer field is in packaging, particularly, but by no means exclusively, of food. Developments in this sector have had a major impact on retailing with the development of self-service shops and supermarkets, with their wider selection of goods. This trend has been made possible to a large extent because food and other items can be packaged in unit packs on an automatic basis, thus reducing unit labour costs, improving hygiene through reduced handling and better protection, increasing the useful life of foods and thus reducing spoilage. Polymers are used not just on the supermarket shelf for the individual pack (whether of flexible film forming a bag of frozen peas, or the rigid container holding yoghurt), but also in the collection and distribution system. Thus plastic boxes and crates are used in place of wood or cardboard, and shrink wrap film is used to overwrap bulk quantities of containers in place of a corrugated board box. To give some idea of the volumes involved, one of the most widely used plastic packaging materials is LDPE (low-density polyethylene); of the total 1980 West European consumption of LDPE of 3.7 million tonnes, over 70 per cent was for packaging use alone. New outlets continue to be found for plastics in packaging and one of the more recent high growth outlets is the PET (polyethylene terephthalate, a polyester) bottle for gaseous drinks. In the USA in 1976, PET consumption for this purpose was 5 kilotonnes. By 1980 it had increased to 150 kilotonnes.

The advent of synthetic fibres such as nylon, polyester, acrylics and more recently polypropylene, has also had a profound effect on both the textile industry and its customers. Within the industry, the availability of synthetic fibres has enabled cost-reducing techniques to be developed for the spinning, fabric-making and assembly of textile products. As a result, textile consumption has increased in a wide range of applications. Thus, in many northern European countries, wall-to-wall carpeting in domestic housing is now commonplace in most rooms, whereas previously linoleum or polished boards would have been the norm. This evolution has not only improved comfort levels but eased the burden of housecleaning. Synthetic fabric garments have likewise reduced the effort involved in washing, as they are less easily soiled, more easily laundered and also often require no ironing.

Very considerable quantities of both liquid and powder synthetic detergents based on petrochemicals are now used, not simply as substitutes for soap but also because of other positive attributes. These include their relative indifference to the hardness of water, improved soil-lifting and dispersing characteristics and suitability for laundering at lower wash temperatures, thus reducing energy requirements. It is of interest to note that the first synthetic detergent plant in Europe was the Shell "Teepol" plant at Stanlow, building of which began late in 1939, based on work carried out on a pilot plant in Amsterdam built in 1935. This can be considered to be the first truly petrochemical plant in Western Europe, as the feedstock was derived from wax obtained from crude oil. Other and earlier West European plants producing what are now considered petrochemicals in fact used feedstock from other sources; thus the first UK polyethylene plants were based on ethylene derived from fermentation ethanol.

MANUFACTURE

The conspicuous growth of petrochemical usage during the 1950s and 1960s encouraged an increasing number of companies to invest in petrochemical manufacture, with the objective of participating in what seemed initially a profitable sector. Expectations are not, however, always fulfilled and the increasing number of participants resulted in some less than profitable ventures.

The major participants in the expansion of the petrochemical industry were the oil companies, seeking to benefit from the upgrading to chemical feedstock of their sometimes surplus refinery streams, and the established chemical companies anxious to widen their product base.

Base Chemicals

The vast majority of petrochemicals are manufactured from relatively few base chemicals i.e. the lower olefins (mainly, but not only, ethylene, propylene, butadiene and butylenes) and the aromatics (mainly benzene, toluene and xylenes). In Western Europe and Japan, these monomers are almost exclusively produced through the cracking of naphtha and gas oil obtained as a result of the distillation of crude oil. In the USA, the position is somewhat different in that, although oil distillates are used on an increasing scale, more than half of the major olefin (ethylene) is obtained from the cracking of natural gas liquids which were available in abundance. Whereas in Western Europe almost all propylene and butadiene is co-produced with ethylene, in the USA approximately two-thirds of chemical propylene is recovered from refinery gases, and only one-third co-produced with ethylene.

The cracking process is thus the cornerstone of petrochemical manufacture. As the word implies, the major feature of cracking is the reduction of the size of the molecules of the feedstock. A cracker is the very complex unit in which this process takes place. Cracking for olefins is carried out in tubular coils inside an oil-fired furnace. Naphtha or other feedstock is passed through the furnace tubes and is heated to a high temperature (700 to 850°C); residence time is under one second.

In order to avoid coke formation, steam is introduced and it is for this reason that the units are commonly referred to as steam crackers. The introduction of steam has another very important effect in that it reduces the partial pressure of the hydrocarbon vapour, thus maximising yield of gaseous products, particularly olefins.

After leaving the furnace, the hot product gas has to be cooled rapidly in order to avoid degradation of the olefins which would not only reduce ethylene yields but result in the formation of tarry polymer and coke. Heat recovered in heat exchangers is used to produce high-pressure steam.

The mixture of gases from the furnace is separated into the individual constituents in the cold section of the cracker where they are compressed up to about 500 psi (35 bars) and cooled to temperatures as low as minus 160°C in order to recover the ethylene and propylene.

It can be seen from this brief description that cracking to produce olefins is very energy-intensive, requiring about one tonne of fuel per tonne of ethylene, and thus it is important to optimise energy utilisation particularly in the hot/cold recovery circuits.

The tremendous increase in demand for ethylene which took place in the 1950/70 period has brought with it big changes in cracker technology and even bigger changes in the size of plants. The first ethylene plants in Western Europe were built in the early 1950s and in comparison with today's standards they were very small. For example, the first cracker built at Rheinische Olefinwerke, West Germany, was in 1954 and this had a capacity of 15 kilotonnes per annum, a level which falls within the range of error for a typical 1980 cracker with a design capacity of 500 kilotonnes per annum.

The modern flexible naphtha/gas oil/LPG cracker, capable of handling this wide range of feedstocks, can be optimised as feedstock supply and price fluctuate. The economics of ethylene production depends greatly on the earnings of the by-products as well as on the price of the feedstocks. The viability of older crackers can sometimes be improved, by modifying or replacing the furnaces, both to widen the choice of feedstocks and to increase the thermal efficiency. An ethane cracker is relatively simple compared with a naphtha cracker. The former produces virtually only ethylene, whereas the latter produces a host of useful by-products which require further processing equipment for their separation and purification.

MANUFACTURE

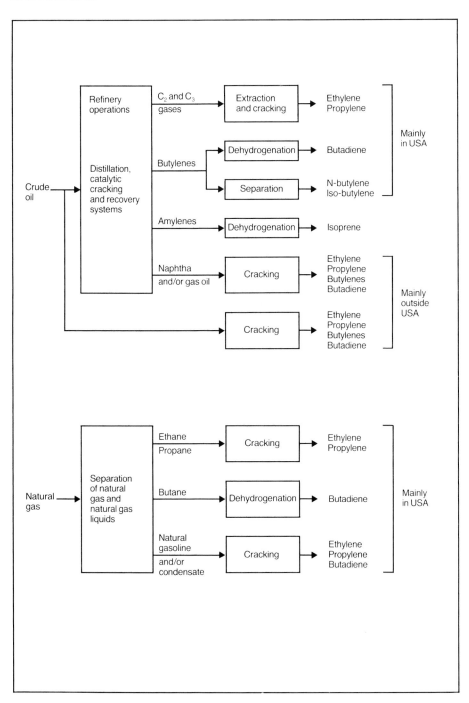

Figure 10.5 **Lower olefins from petroleum**

Another point to be noted is the variation in the amount of feed required per unit of ethylene. For one tonne of ethylene, the following tonnages are typically required of each of the major alternative feedstocks:

>   ethane   1.25
>   naphtha   3
>   gas oil   4

Thus, an ethylene plant of a given capacity based on liquid feedstocks is significantly bigger than one using ethane.

Figure 10.5 gives an overall view of the production processes for the range of lower olefins and an indication of the major routes used in the USA and elsewhere at the beginning of the 1980s. Most of the new crackers planned for the mid-1980s, in for example Canada and the Middle East, will be based on ethane, and as such their output will consist essentially of ethylene with effectively no propylene or other olefins.

Although the lower olefins are the most important base chemicals for petrochemicals, aromatics are also of major significance. The three most important products in the group are benzene, toluene and the xylenes. Figure 10.6 gives an outline of the process routes for the production of this group of base chemicals.

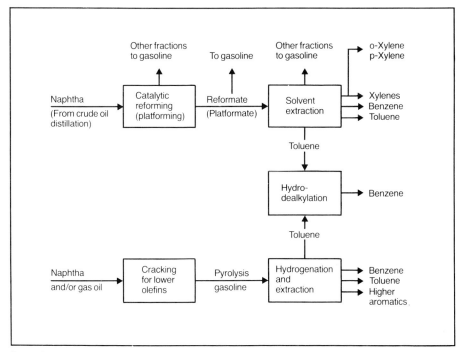

Figure 10.6 **Aromatics from petroleum**

Polyethylene

To give some idea of petrochemical production processes subsequent to base chemical manufacture, a brief outline will be given of the manufacture of polyethylene, for which the base chemical is ethylene. There are several different processes for the production of polyethylenes, some yielding relatively similar products, others producing distinctly different materials.

The original process for production of polyethylene is the high-pressure process discovered and developed by ICI; this is still the most important route for LDPE (low-density polyethylene) manufacture with 1980 WOCA capacity being about 13 million tonnes. As the name indicates, the process involves the reaction of ethylene at very high pressures. Highly purified ethylene (greater than 99 per cent) is compressed in two stages, firstly to 3,000/4,500 psi (200/300 bars), and then to 30,000/45,000 psi (2,000/3,000 bars).

Before starting the second stage compression, the ethylene is heated to about 200°C and a polymerisation initiator and chain-transfer agent are added. The reaction takes place in either tubular or stirred autoclave reactors. The LDPE produced is molten and removed through a let-down valve whilst unreacted ethylene is recycled to the secondary compressor. The molten LDPE is fed to an extruder for pelletising and subsequent packaging or transfer to bulk storage.

LDPE produced in this way has a density of about 0.92 kg/l and a molecular weight of between 50,000 and 300,000. LDPE consists of highly branched chain molecules. Side chains of between two and four carbon atoms occur at intervals of about 20–50 carbon atoms in the main chain, which is itself branched. This high degree of branching leads to a relatively low level of crystallinity of 65 per cent.

It is also possible to produce polyethylene which has hardly any branches. This is known as HDPE (high-density polyethylene) and has a high degree of crystallinity of 85–90 per cent. The first commercial process for the production of HDPE was that discovered by K. Ziegler in 1952. This uses organometallic compounds as catalysts for the conversion of ethylene to polyethylene. The reaction may be carried out in solution, suspension or in the gas phase and operates at low pressures ranging from 150 to 300 psi (10 to 20 bars) and 60°C to 100°C.

High-crystallinity HDPE is much harder and has a higher softening point than LDPE; it has a density between 0.94 kg/l and 0.96 kg/l. Whilst LDPE is mainly used for film and sheeting, HDPE has a more diverse end-use pattern with outlets which include film, injection-moulded items such as crates, and blow-moulded containers and bottles.

Towards the end of the 1970s a series of new processes for the production of LDPE became available. These had one major common characteristic namely, they were all low-pressure processes. Product from these processes is called LLDPE (linear low-density polyethylene). LLDPE is a copolymer of ethylene and small quantities (2–10 per cent) of alpha-olefins, and has a polymer structure with short side chains as in LDPE, but an unbranched (i.e. linear) main chain as in HDPE. LLDPE therefore has an intermediate degree of crystallinity and its properties tend to be intermediate between LDPE and HDPE. LLDPE has attracted a considerable amount of interest and investment, because it offers a favourable combination of production costs and performance characteristics. For example, LLDPE film has a higher strength than LDPE film, and so significantly thinner films can be used for equivalent performance. Savings up to 40 per cent have been reported when strength is the limiting factor.

The polyethylenes are relatively simple petrochemicals, being formed by the linkage of the basic ethylene unit. They also represent one of the major petrochemical groups i.e. thermoplastics. The following sections give a brief indication of the chemical processes used in the manufacture of a representative selection of some other major groups of petrochemicals, i.e. thermosetting resins, synthetic fibres, synthetic rubbers, solvents and detergents.

Thermosetting Resins

One of the earliest resins produced was phenol formaldehyde, more generally known as phenolic resin. This dark brown product found widespread use in the first half of the twentieth century, particularly for electrical and radio applications; it still has many applications but these may be less obvious, e.g. as a binder for friction materials used in car brake shoes.

Phenolic resin is an example of a thermosetting resin which is first produced as a liquid or pliable solid that can be moulded, but which on further processing sets to a hard material which, unlike thermoplastics, cannot be softened by heat.

Epoxy resins are another example of thermosetting resins, and these materials, first produced commercially in 1947, are mainly based on ECH (epichlorohydrin) and bisphenol A. However, the term epoxy resin applies more generally to any thermosetting resins that in uncured form contain one or more reactive epoxide or oxirane groups.

There are three steps in the manufacturing process for ECH:

CH_3—CH=CH_2 + Cl_2 ⟶ CH_2ClCH=CH_2 + HCl
propylene   chlorine   allyl chloride   hydrochloric acid

CH_2Cl—CH=CH_2 + $HOCl$ ⟶ $CH_2Cl\,CHOHCH_2Cl$
allyl chloride   hypochlorous acid   glycerol dichlorohydrin

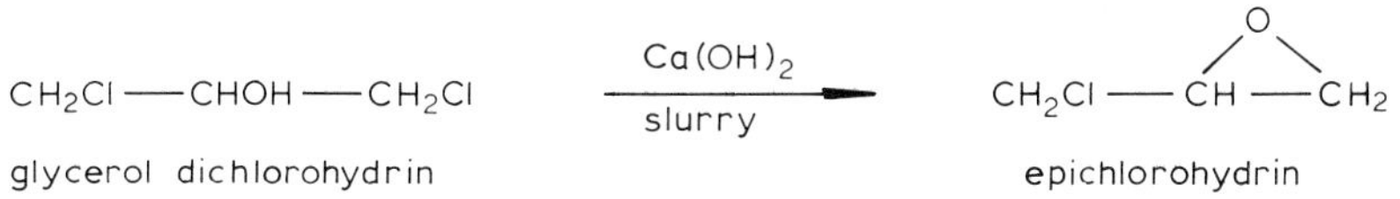

$$\text{CH}_2\text{Cl}-\text{CHOH}-\text{CH}_2\text{Cl} \xrightarrow{\text{Ca(OH)}_2 \text{ slurry}} \text{CH}_2\text{Cl}-\overset{\overset{\displaystyle O}{\diagup\diagdown}}{\text{CH}}-\text{CH}_2$$

glycerol dichlorohydrin → epichlorohydrin

In the second step above, 1,2-dichlorohydrin is also formed which in the third step is also converted into epichlorohydrin..

Bisphenol A, or diphenol propane, is manufactured by the reaction of acetone and phenol in the presence of an acid catalyst:

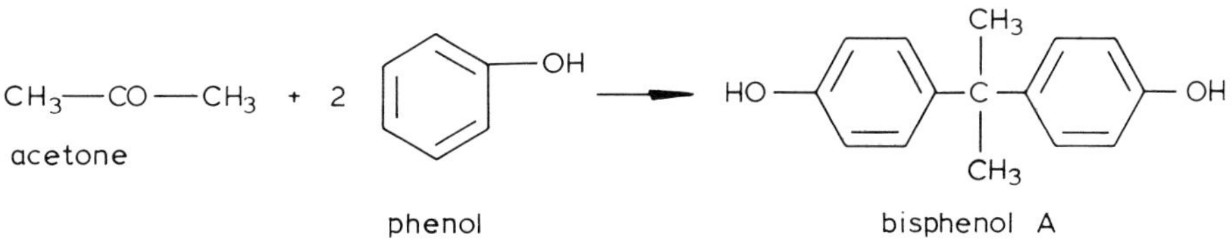

CH₃—CO—CH₃ (acetone) + 2 phenol → bisphenol A

Epoxy resins themselves are produced by the reaction of ECH and bisphenol A:

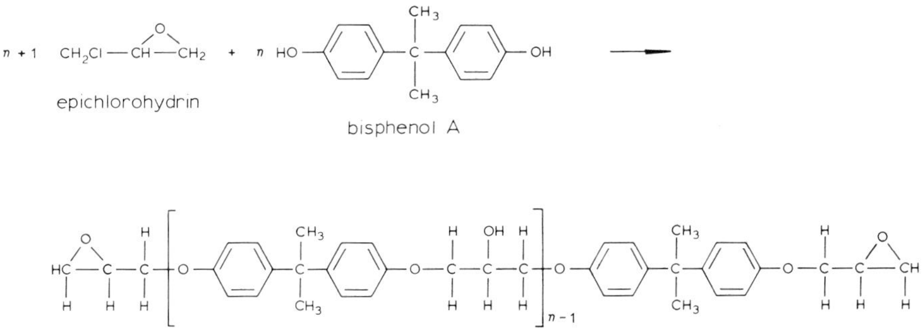

n + 1 epichlorohydrin + n bisphenol A →

[epoxy resin structure shown]

The ratio can be varied between the two basic components, as can reaction conditions, to yield resins of different molecular weights.

Epoxy resins are hardened or cured using a variety of curing agents to effect the cross-linking. The choice of curing agent or hardener depends on the type of resin, the application, and the speed and temperature of curing. Aliphatic amines and polyamides are among the curing agents most widely used.

Synthetic Fibres

Man-made fibres are normally divided into two classes, namely the cellulosics (derived from cellulose and thus of little relevance to this chapter) and the synthetics (nowadays mainly derived from petroleum). The four most important groups of synthetic fibres are:
- Polyesters
- Nylons (polyamides)

- Acrylics
- Polypropylene

Of the synthetic fibres, the most important group in volume terms is the polyesters. Polyesters may be produced in various ways but the main commercial route in the 1970s was the reaction of dimethyl terephthalate and ethylene glycol, with the tendency being to move towards an alternative process based on terephthalic acid (TPA) and ethylene glycol.

Terephthalic acid is mainly manufactured by the AMOCO process which involves the liquid-phase air oxidation of *p*-xylene in the presence of a catalyst:

$$\text{p-xylene} \longrightarrow \text{terephthalic acid}$$

Ethylene glycol is produced by the direct hydration of ethylene oxide, which is itself produced by the catalytic oxidation of ethylene:

$$2\ CH_2{=}CH_2 + O_2 \longrightarrow 2\ CH_2\text{—}CH_2\ (\text{ethylene oxide})$$

$$CH_2\text{—}CH_2\ (\text{ethylene oxide}) + H_2O \longrightarrow CH_2OH\text{—}CH_2OH\ (\text{ethylene glycol})$$

The production of the polymer is achieved by the reaction of ethylene glycol and highly purified TPA:

$$n\ CH_2OH\text{—}CH_2OH\ (\text{ethylene glycol}) + n\ \text{terephthalic acid} \longrightarrow [-CH_2CH_2OOC\text{—}C_6H_4\text{—}COO-]_n$$

Excess glycol is removed by vacuum distillation, whilst the polymer may be fed directly to the fibre spinning unit or may be cooled and converted into small granules or chips.

Synthetic Rubbers or Elastomers

The oldest and most widely used type of synthetic rubber or elastomer is SBR (styrene butadiene rubber), which is formed by the reaction of butadiene and styrene:

$$n\ CH_2=CH-CH=CH_2\ +\ n\ C_6H_5-CH=CH_2 \longrightarrow [-CH_2-CH=CH-CH_2-CH_2-CH(C_6H_5)-]_n$$

butadiene   styrene   SBR

Styrene is produced via ethylbenzene from ethylene and benzene:

benzene + ethylene $\longrightarrow$ ethylbenzene

ethylbenzene $\longrightarrow$ styrene + H_2

SBR does not signify one specific polymer but a range of solid and latex materials produced by a variety of processes having differing proportions of the two precursors. As butadiene can be polymerised to yield three isomers, the proportions of each in a particular polymer can also vary. Most SBR is produced by either hot or cold emulsion polymerisation, with "cold polymerisation" being by far the most dominant group of processes. The manufacturing process involves the mixing together of styrene, butadiene, water, emulsifiers, activators and other ingredients and keeping the reaction vessel cool.

Solution SBR is produced by reacting styrene and butadiene in organic solvents together with suitable additives, including a catalyst such as an organometallic complex.

Solvents

Solvents is a general term for a range of organic products which may have one of the following functions:
- Conventional solvent use in surface coatings, i.e. paints, varnishes and lacquers.

- Simple ingredient, e.g. alcohol in polishes.
- Processing agents, e.g. extraction of oils by paraffins from oilseeds.
- Intermediates, e.g. manufacture of acetone from isopropyl alcohol.

Solvents can be divided into two types, namely hydrocarbon solvents and chemical solvents. The former consist of hydrogen and carbon only; although originally direct by-products of oil refining, they are now produced in special plants. Examples of such solvents are benzene, toluene and white spirits and their main use is in surface coatings.

The most important chemical solvents are the oxygenated solvents (alcohols, ketones, esters and glycol ethers) and the chlorinated solvents.

A brief outline of the production of acetone will be given to illustrate manufacturing routes for this product group.

Acetone or dimethyl ketone is commercially produced from propylene by two main routes, via IPA (isopropyl alcohol) or cumene. IPA is itself a solvent.

In the IPA route, propylene is first hydrated to IPA either directly or by absorption in sulphuric acid and subsequent hydrolysis:

$$CH_3CH=CH_2 + H_2SO_4 \longrightarrow CH_3CH(OSO_3H)CH_3$$
$$\text{propylene} \quad \text{sulphuric acid}$$

$$CH_3CH(OSO_3H)CH_3 + H_2O \longrightarrow (CH_3)_2CHOH + H_2SO_4$$
$$\text{IPA}$$

Acetone is formed by the dehydrogenation of IPA:

$$(CH_3)_2CHOH \longrightarrow CH_3COCH_3 + H_2$$
$$\text{IPA} \quad\quad\quad \text{acetone}$$

This reaction takes place in either the vapour or liquid phase using various catalysts.

The cumene route for the production of acetone is more complex but yields another important petrochemical (phenol) as well as acetone. Benzene is alkylated with propylene in the reaction:

$$C_6H_6 + CH_2=CHCH_3 \longrightarrow C_6H_5CH(CH_3)_2$$
$$\text{benzene} \quad \text{propylene} \quad\quad \text{cumene}$$

After purification, the cumene is oxidised to cumene hydroperoxide by first emulsifying with dilute sodium carbonate solution and then heating in contact with air:

$$C_6H_5CH(CH_3)_2 + O_2 \longrightarrow C_6H_5C(CH_3)_2OOH$$
$$\text{cumene} \quad\quad\quad\quad \text{cumene hydroperoxide}$$

Cleavage of the hydroperoxide into phenol and acetone is effected by treatment with dilute sulphuric acid and then heating to initiate the strongly exothermic reaction:

$$C_6H_5C(CH_3)_2OOH \longrightarrow C_6H_5OH + (CH_3)_2CO$$

cumene hydroperoxide → phenol + acetone

The reaction products are then separated by distillation.

Detergents or Surfactants

The first synthetic detergents manufactured in Europe, as noted earlier, were secondary-alkyl sulphates based on higher olefins produced by cracking of slack wax obtained from the refining of lubricants. These detergents were used for industrial purposes. The alkyl sulphates were followed by alkyl benzene sulphonates or detergent alkylate in the 1950s and these rapidly replaced soap in washing powder for domestic use. The original alkylates were resistant to biological attack and were thus not fully degraded in sewage treatment plants. Small traces of detergent in the effluent from these plants caused foaming of rivers, which was obviously undesirable. Once the problem was identified, new more easily biodegradable products of various types were successfully developed. One such group of surfactants is based on linear alpha olefins.

Linear alpha olefins may be produced by the oligomerisation of ethylene using the Shell Higher Olefin Process or SHOP. This process enables linear olefins to be produced in the C_4–C_{40} range; those of specific interest for detergents are mainly in the C_{10}–C_{18} range. There are various processes for conversion of the linear olefins to detergent alcohols, including a Shell process which is a modified Oxo reaction. The general Oxo reaction involves reacting linear olefins with synthesis gas $(CO + H_2)$ to form intermediate aldehydes, which are subsequently reduced to produce detergent alcohols containing one more carbon atom than the initial olefin. The Shell process catalytically converts the olefins to alcohols in one step.

Alcohol ethoxylate-type detergents are produced from the alcohol by reaction with ethylene oxide in the presence of a base catalyst such as potassium hydroxide:

$$ROH + \underset{\text{ethylene oxide}}{CH_2\overset{O}{\overline{}}CH_2} \longrightarrow \underset{\text{alcohol ethoxylate}}{RO(CH_2CH_2O)_nH}$$

detergent alcohol

This type of detergent is used in detergent powder for washing machines (as part of the formulation) and in heavy-duty domestic liquids. It is also used industrially and in institutions (e.g. hotels and restaurants).

This section on manufacture has given a brief indication of the production processes for a few of the major petrochemicals; there are a multitude of other processes used in petrochemical production for which details can be obtained in various reference works. The aim of the section has partly been to give some flavour of the complexities of petrochemical production. Complexities, as such, may be interesting but their greater significance is that they arise from the continuous efforts to develop new products based on the multitude of petrochemical feedstocks, and to use these resources with increasing efficiency.

THE FUTURE

The petrochemical industry, as has been shown, is a major materials industry supplying products to virtually all material-using sectors of the economy. It is an industry which has reached a certain degree of maturity but in which innovation and substitution continue and give rise to growth above that of the economy in general. The focus of petrochemical manufacture is changing geographically, with an increasing share of production and consumption now outside Western Europe, Japan and the USA. It is not envisaged that this trend will in itself cause a major shift in the types of outlet for petrochemicals; technological advances in the user industries will, however, place new demands on the petrochemical producers which will in turn result in, and partly enable, new outlets to develop.

In the 1950s and 1960s when feedstock costs were relatively low, capital costs represented the major element of the total production costs of base chemicals. It was therefore important for producers to try to operate crackers as close to capacity as possible because the capital cost being a fixed charge, the greater the volume output, the lower the unit capital charge on each unit volume of cracker product, and thus the lower the total product cost. A similar argument also applied to the downstream plants on which the profitability of the cracker was dependent, and there was thus great pressure to load both the cracker and downstream plants to achieve maximum capacity utilisation. The market demand for each of the downstream products, such as low-density polyethylene or PVC, did not necessarily respond in a parallel fashion to capacity increases as they would in the ideal case and periods of oversupply were not uncommon. The complexities of the petrochemical manufacturing operation thus made, and continues to make, it extremely difficult to manage individual product sectors efficiently.

Despite the problems of the petrochemical industry in the developed world, prospects are good for the continued use of petrochemicals on a massive scale worldwide; growth will, however, vary significantly from country to country. In the developed countries, growth rates of consumption are expected to be modest

Table 10.5 **Plastics markets in the USA, analysed using systems approach**

Resin consumption in million (10^6) pounds
Energy consumption in trillion (10^{12}) Btu

Market	Resin	Resin consumption (10^6 lb)	Total energy consumption (10^{12} Btu)		
			Plastics system	Alternative system	Net difference
Agricultural film	LDPE	129.8	6.50	1.25	−5.25
Disposable diapers	LDPE	110.0	42.21	41.59	−0.62
Disposable film	LDPE	19.8	1.39	0.82	−0.57
Window units	PVC	57.2	2.22	4.90	2.68
Tumblers/cocktail glasses	PS	85.8	3.93	10.97	7.04
Home insulation	PS	123.5	5.96	10.40	4.44
Bottle basecups	HDPE	33.0	4.96	7.74	2.78
Disposable syringes	PP	50.0	2.52	1.49	−1.03
Automotive	PP	413.3	17.20	31.19	13.99
Automotive	ABS	163.0	8.06	13.19	5.13
Blow-moulded bottles	Polyester	189.2	18.89	32.93	14.04
Subtotal		1,374.6	113.84	156.47	42.63
All other markets		26,571.4	1,067.46	1,859.03	791.6
Total resin included in energy analysis		27,946.7*	1,181.3	2,015.5	834.2†

* An additional 2,940 million pounds of resin were examined in terms of end-use. However, no non-plastics alternatives were determined.
† With energy savings from wood-derived fuel omitted, this figure would be 733.1 trillion Btu.

Source: Franklin Associates Ltd, Kansas, USA

and show only a slight increase over general economic growth. However, there may be higher growth rates in many of the developing countries, where the scope for penetration is considerably greater, as they have low existing levels of petrochemical consumption.

The inherent usefulness and economic attractiveness of petrochemicals is an important characteristic that will help this growth to take place. It is paradoxical that numerous studies have shown that petrochemical products, despite being produced from oil, generally require less total energy input than similar items made from other materials; the results of one such study are shown in Table 10.5. The use of petrochemicals can thus contribute to the more efficient use of energy.

Concern has been expressed that, as supplies of crude oil and other petrochemical feedstocks are "limited", the production of petrochemicals could cease to be possible. However, as discussed in Chapter 1, crude oil supplies are likely to continue to be available for many years to come, even though the cost of obtaining them will increase. In addition, the production of petrochemicals can be regarded as a more noble use for oil than simply burning it and, as already indicated, petrochemical products generally represent an energy-efficient use of materials. It is also relevant to note that much of the chemistry of petrochemical production is not dependent on the origin of the base chemicals, and so other feedstocks can and will be used, if economic and other factors indicate that alternative feedstocks are more appropriate.

However, alternative feedstocks are already being investigated for various reasons. In some instances, the motivation may be strategic where, for example, a country does not have its own crude oil or natural gas supply but does possess coal deposits. In other instances, the target may be to identify new routes to certain products using, for example, biotechnology. It is likely that at least some of these routes will be adopted commercially and indeed coal is being used for "petrochemical" manufacture on a large scale in South Africa.

This very brief and simplified retrospective analysis of the development of the petrochemical industry has shown that the industry's foundations lie neither in petrochemical production in one particular country, nor in the success or otherwise of a single end-use, but rather in the fact that petrochemicals play a very important part in the efficient exploitation of the world's resources. As such their future is secured.

Chapter 11

UNCONVENTIONAL RAW MATERIALS AND SYNFUELS

INTRODUCTION

"Unconventional Raw Materials" (URM) may be defined as feedstocks other than crude oil. The rapidly increasing cost of crude oil and the decreasing security of supply are the main motives for the interest in such feedstocks. This chapter deals with the manufacture of synthetic fuels (synfuels) from unconventional raw materials.

The emphasis, as in crude oil refining, is on the manufacture of products for high-added-value markets, in particular transport fuels, and clean gaseous fuels.

In the early stages of synfuel developments (mid-seventies) there was also an interest in clean liquid fuels for heating/underboiler use. However, the high cost of synfuels will in general prevent their application in low-value markets, and emphasis is shifting to upgrading to higher-value products.

The synfuels scene at present is not only complex but also subject to continuous change and, as a consequence, not very transparent. It contains the following complicating factors:
- Many feedstocks are involved (coal, natural gas, tar sands, shale, biomass) and not all feedstocks are available to everybody.
- Many different conversion routes are available.
- Most routes have two options:
  (i) commercially proven, expensive processes,
  (ii) second-generation, more economic processes, but still under development.
  Both the timing of commercial availability and the relative economics are highly uncertain.
- Many synfuel options are unconventional (methanol, ethanol, medium-calorific-value gas, hydrogen).
- The strategic need for synfuels exists, but the economic conditions are not yet right and vary from country to country.

Given the above, it will be appreciated that a complete, detailed description of

all aspects of all synfuel options would lead to confusion rather than insight. This analysis will therefore be confined to:
- a description of the criteria that characterise a URM and its conversion;
- a review of liquid and gaseous synfuel options;
- a survey of the technology involved;
- some aspects of importance for ranking the options.

CHARACTERISTICS OF URMs

Any fuel consists of one or more compounds containing carbon and hydrogen, with or without oxygen. Other components are a nuisance or a contaminant. From an energy point of view pure hydrocarbons are to be preferred, since

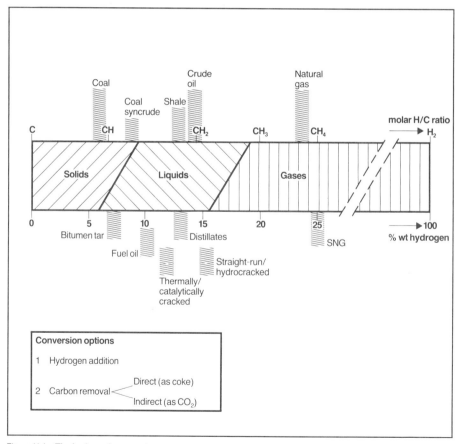

Figure 11.1 **The hydrocarbon spectrum**

CHARACTERISTICS OF URMs

oxygen does not add to the energy content of the fuel. Figure 11.1 shows that hydrocarbon liquids really exist only in a relatively narrow hydrogen/carbon (H/C) molar ratio of approximately 2 and that crude oil and its distillate products appear well positioned. Pure hydrocarbon gaseous fuels exist at higher H/C ratios.

Liquid fuels of interest also containing oxygen are alcohols, in particular methanol and ethanol. Gaseous fuels of interest are methane, hydrogen and carbon monoxide or mixtures thereof. Via the latter components a clean, non-solid fuel can be obtained with a very low H/C ratio.

In general, three factors determine the attractiveness of a URM: its availability, its H/C ratio and its content of contaminants.

The Availability Factor

A prime factor in this respect is the size of the resource base (Fig. 11.2). Coal and

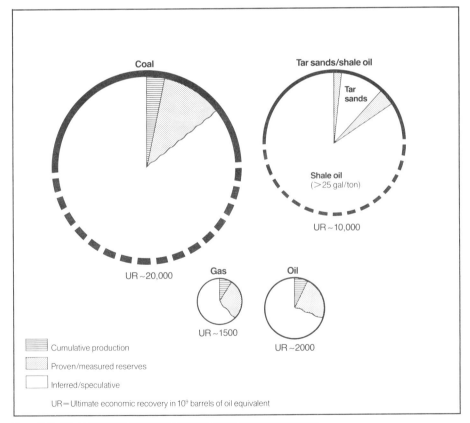

Figure 11.2 **World fossil fuel resources** Estimated ultimate recoverable resources

shale loom very large on the basis of their inferred reserves. Coal is dominant on the strength of proven reserves.

Another prime factor is geographical distribution. In the synfuel business, whether a resource is indigenous or not is often a decisive factor. Furthermore, for those countries without indigenous resources a wide distribution will in general improve the security of supply. The latter aspect is of importance only for coal and natural gas, since the other URMs by their nature cannot be traded internationally.

The Hydrogen Factor

The dominant factor in considering the economic value of alternative feedstocks is the degree of hydrogen deficiency. Hydrogen is a very expensive energy source and the cost of conversion and upgrading raw materials to distillate quality is largely determined, directly or indirectly, by the amount of hydrogen that has to be added.

Figure 11.1 indicates that shale, an aged biomass relatively rich in paraffins, is from this point of view an attractive feedstock, trailing crude oil. Coal is very hydrogen-deficient and, even when liquefied by the addition of hydrogen, requires considerable further upgrading (hydrogen addition). Tar sands, a mixture of bitumen and sand, are also hydrogen-deficient.

In the distillate products too the hydrogen factor remains important. Gasoline can contain a high percentage of hydrogen-deficient aromatic components and can therefore have a relatively low hydrogen content. Gas oils (diesel fuel, avtur) are more paraffinic and relatively rich in hydrogen. One could therefore argue the point that there is an advantage in producing gasoline from aromatic feedstocks and preferentially using hydrogen-rich raw materials for the production of gas oil.

Biomass and natural gas differ from the previous three feedstocks in the sense that they are hydrogen-rich. However, this advantage is reduced because liquid hydrocarbons cannot be produced directly from these feedstocks.

The exception to this rule is a relatively simple biomass option, at least from a technical point of view, through the use of certain vegetable oils (palm oil, soya bean oil) as a diesel fuel. This can be done in principle (transesterification), but requires not insignificant further development work.

The more natural option for conversion of biomass and natural gas into liquid fuels is the production of alcohols.

The Mineral Factor

In comparing coal, shale and tar sands, availability and the H/C ratio are not the only important factors. All three feedstocks have a high mineral content, much

Table 11.1 **Hydrogen-deficient synfuel feedstock**

	Shale	Tar sands	Coal	Syngas (ex coal)
"Formula"	$CH_{1.6}(SiO_2)_{3.2}$	$CH_{1.4}(SiO_2)_2$	$CH_{0.8}(SiO_2)_{0.05}$	$CH_{1.9}(CO_2)$
HC Content (% wt)	5–10	10–15	70–90	(25–30)
H/C Ratio "Syncrude"	$CH_{1.6}$	$CH_{1.4}$	$CH_{1.2}$	$CH_{1.9}$
Contaminants	N, O, As	S, N	S, N, O	–
Aromaticity	Low	High	Very high	"Optional"
"Best" product	Gas oil	Gasoline	Gasoline	Methanol/gas oil

higher than that of crude oil, and also the nature of the "chemical contaminants" built into the structure of the molecules is different (see Table 11.1).

"Oil Shale" is often a marlstone that is mostly clay which contains a brown to dark grey organic matter called kerogen. When heated to about 500°C in a retort the kerogen reacts to form shale oil plus gas that can be recycled to heat additional shale. In the process, the sensible heat contained in the "rock" may well exceed the energy content of the shale oil. Thus efficient heat recovery is of the utmost importance in shale conversion. The mineral residue, which is normally of greater volume than the original rock, is an aggressive material which represents a significant disposal problem.

Major oil shale deposits occur in the Western United States and Australia; there are also significant resources in Morocco and elsewhere. The quality of shale varies considerably both in hydrocarbon content (usually 5 to 10% wt) and in water content. Presence of excessive water can well make recovery of hydrocarbons uneconomic. Most recent technical and economic evaluation of oil shales has related to those in Colorado, USA, which are relatively speaking of high quality. It should not, therefore, be assumed that cost projections made on the basis of these particular shales can necessarily be applied to those occurring elsewhere.

Tar sands occur in a number of places, notably the Athabasca tar sands in Northern Canada, the Orinoco tar sands belt in Venezuela and also in Madagascar. Usually tar sands are richer in hydrocarbons than shale, and separation can occur under milder conditions (hot water extraction at approximately 100°C). Tar sands that occur at or close to the surface may be mined by conventional techniques and then retorted. Deeper deposits would have to be recovered by other methods, such as the injection of steam or hot water to drive liquids to the surface. Conceivably, underground combustion could be applied to tar sand recovery.

Compared with shale and tar sands, coal is a hydrocarbon-rich feedstock. This is a very significant factor, since it allows (hard) coals to be transported before conversion and it makes hard coal a raw material that can be internationally traded. Coal and ash can be separated before conversion (beneficiation), but it is more typical to separate the intermediate conversion products (liquids or gas) from the ash.

As regards the chemical contaminants, it is important to note that, besides sulphur, nitrogen is usually present to sometimes appreciable levels. Nitrogen is fairly hard to remove and requires severe (new) conversion/removal processing. Coal and shale oil also contain substantial amounts of oxygen.

Natural gas has the advantage of being a clean feedstock. Contaminants, if present, for instance hydrogen sulphide, carbon dioxide, carbonyl sulphide or nitrogen, are usually fairly simple to remove at source. Although biomass does not have a very high hydrocarbon content (20–30%), its contaminants are relatively harmless, mainly oxygen and water. Nitrogen and sulphur levels are typically low.

It will be obvious that a high level of physical and chemical contaminants has serious environmental implications.

CHARACTERISTICS OF SYNFUELS

Liquid Synfuels

Liquid fuels have a number of intrinsic advantages over solid or gaseous fuels. Compared with solid fuels, liquid fuels are much simpler to convert and handle, while compared with gaseous fuels or electricity, liquid fuels are much simpler to store. These advantages are particularly relevant if the fuel is used in a transport system. During transportation the energy storage and conversion systems have to

Table 11.2 **Characteristics of transport modes**

	Module size	Speed	Index*
Sea	(very) large	low	1
River	large	low	10
Road			
private	small	medium	10^5
goods	small	medium	10^4
Rail	large	medium	10^2
Air			
commercial	medium	high	10^4
military	medium	very high	10^5

* Index: Speed divided by module size with sea = 1.

be carried along simultaneously, and consequently these systems have to be light and simple. It is from these factors that transport fuels derive their high added value and are the prime end use of liquid synfuels. Liquid fuels cannot demand the same premium value in all transport applications. In general, the effect of the intrinsic advantages becomes smaller according as the transport module is larger and/or slower.

In comparison with other modes of transport, road transport has a combination of module size and speed in which the advantages of liquid fuels are felt most strongly. Air transport is also highly dependent; water and rail transport less so (Table 11.2).

Hydrocarbon liquids are most attractive transportation fuels, since they have a very high energy density and good combustion characteristics. Consequently, distillate products are the prime candidates in many synfuels options. However, they are difficult to manufacture from certain URMs (coal, natural gas, biomass), and alcohols may then provide an attractive alternative. Alcohols have excellent burning properties. Their main intrinsic disadvantages compared with pure hydrocarbons are a lower energy density (illustrated in Table 11.3) and complete miscibility with water.

A specific problem related to the use of alcohols is their introduction into the market, specifically the distribution and the development of suitable engines. But the problems might well be more general. The world has grown used to having one feedstock for all transportation fuels, all having similar broad quality standards. In a synfuel world this might be completely different.

Many feedstock options will only be available locally (shale, tar, biomass). Product quality standards of conventional distillate products may have to be adjusted to meet conversion limitations, and this could require changes in engine and/or performance characteristics. This holds particularly for diesel fuels, since high cetane number gas oils appear hard to manufacture from tar sands, shale or coal syncrude.

Alcohols can be introduced in two ways: neat and as a blend. With the latter approach, entering the market is relatively simple at low alcohol-in-blend concentrations. The introduction of neat alcohols requires a separate distribution

Table 11.3 **Alcohols and hydrocarbons compared**

	"Formula"	Energy density (Mcal/m^3)
Hydrocarbons	$(CH_2)_n$	gasoline 8,530
		gas oil 9,090
Methanol	$CH_2 \cdot H_2O$	3,800
Ethanol	$(CH_2)_2 \cdot H_2O$	5,600

system for a market that may remain relatively small for a considerable time. Special attention will have to be paid to the ingress of water into the distribution system, particularly in the case of a blend (phase separation) both from natural causes or man-made. In general, ethanol will be somewhat simpler to introduce than methanol.

It should be noted that alcohols are used to best advantage in Otto engines. Use of alcohols in ignition compression engines is problematic and would require significant further R & D effort for car/engine manufacturers (spark-assisted diesel, ignition improver, manifold injection).

There are several alternatives to liquid transport fuels under active study.

Compressed Natural Gas (CNG) is already applied on a small scale in a few countries. Restrictions are cost, safe handling and the weight of the CNG bottles.

Hydrogen, stored as a hydride, is attractive as a very clean fuel in the road transport market. The problems regarding safety can probably be solved to the satisfaction of technical people, but whether this will be acceptable to the public is an open question. The economics of hydrogen also appear not too attractive. In air transport the lightness of hydrogen offers a specific bonus and for special aircraft, which would use the cold stored in the liquid hydrogen to reduce drag, the efficiency advantage could offset the higher cost of the aircraft. This option is a very long-term one.

Electric vehicles are at present limited by batteries that are heavy and of relatively low energy content. Even if suitable advanced batteries are developed, their use is likely to be restricted to vehicles on well-defined duties such as taxis, delivery vehicles, and urban buses. It seems most unlikely that the sporadic use pattern of most cars or the high energy requirements of long distance vehicles could be satisfied by anything other than a liquid fuelled power unit. Thus, unless an economically viable fuel cell is developed, internal combustion engines are likely to remain the dominant source of motive power for road transport.

Gaseous Synfuels

Just as distillates are the reference for liquid synfuels, so natural gas (NG) is the reference for gaseous fuels. It is a clean, high-calorific-value gas with good burning characteristics, particularly suited for space heating and steam raising. Its high heat content allows transportation over large distances and distribution to small outlets. This makes it a premium fuel in the residential and commercial market. Many countries without an indigenous supply of NG or anticipated shortages are developing processes/plans to manufacture SNG (substitute natural gas) from coal, preferably indigenous. The critical issue is what value (S)NG can command, since the manufacture of SNG is expensive. In steam raising it competes with fuel oil, or coal, in domestic heating with NG, LPG or domestic heating oil. In the former markets its value is less than that of crude oil, in the latter market it could be claimed to be equal to or higher than that of crude oil. An alternative is the use of Industrial Fuel Gas (IFG) (a medium-calorific-value gas), which is a mixture of carbon monoxide and hydrogen, sometimes with small amounts of carbon dioxide and methane. IFG is much cheaper to produce than SNG but has problems with transport/distribution owing to its lower heat content and its content of carbon monoxide, which is poisonous. Use of syngas in industrial complexes near gasification complexes alleviates these problems.

THE STATUS OF THE TECHNOLOGY

Figure 11.3 surveys the various technologies. In a very generalised way, two main options can be distinguished.

(i) Conversion routes via a syncrude opt for hydrogen addition. This is an expensive route, but product make can be relatively high (60–70%). Product quality, however, may be problematic and the severe further upgrading required can reduce the efficiency by some 10–15%.

(ii) Conversion routes via syngas, the indirect route to liquids, use a carbon dioxide or hydrogen bleed (via the water shift reaction) to balance the H/C ratio. It is worth noting that for hydrogen-rich feedstocks a hydrogen bleed implies a loss; but for hydrogen-deficient feedstocks (coal) the shift reaction offers a relative advantage, since the correct H/C ratio is obtained via carbon removal (as carbon dioxide) rather than hydrogen addition. Product make is of the order of 50–55%, but product quality is satisfactory. Methanol manufacture from syngas is a commercially proven process.

The syngas route also offers an indirect route to SNG. In general, the efficiency of this option is slightly lower than the direct route (about 65 as against 70%), but the difference may be small, since direct processes too have to convert syngas into methane but do so internally rather than via the separate process. This improves the economics, but to what extent is still uncertain.

The choice between hydrogen addition and carbon removal in upgrading hydrogen-deficient feedstocks is a very general one. Although local/regional features tend to dominate synfuel economics, as a rule of thumb conventional economics would indicate a preference for the hydrogen addition route when the relative upgrading is below 50 % (i.e. hydrogen added/hydrogen present < 0.5) and for carbon removal if more than 50% hydrogen has to be added.

Hydrogen-Addition Technologies

Tar sands can be converted into transportation fuels by extraction and upgrading. The technology as such can be considered to be commercially proven. The

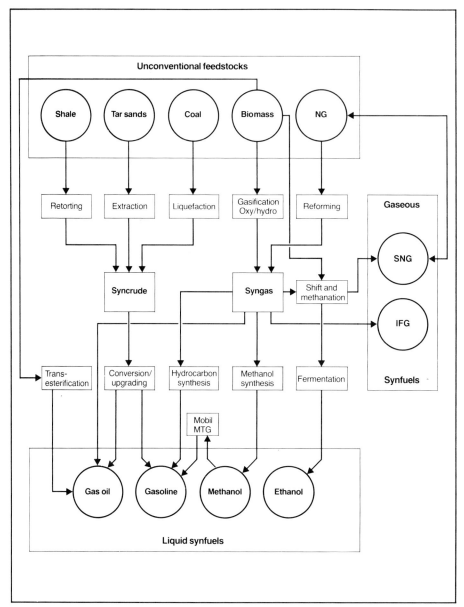

Figure 11.3

emphasis now is on improvement, i.e. higher energy efficiency, higher yields and better products.

Shale oil can be obtained by the retorting of shale. This has to be followed by upgrading to obtain products of acceptable quality. No commercial plants based on shale are now in operation; shale technology is not as far advanced as tar sands technology.

There are several retorting techniques under development: surface retorting (direct and indirect) and *in situ* (true and modified). In surface retorting, shale is mined, crushed and heated in large retorts by burning residual carbon in the shale. With direct retorting, this is done at the bottom of the retort; with indirect retorting, it is done separately, allowing better heat recovery. The modified *in-situ* technique creates large underground "retorts" by partially mining and blasting. With true *in-situ* processing no mining is done except for the drilling of injection and recovery shafts. Surface retorting is far more developed than *in-situ* retorting, and the first commercial projects will be based on the former techniques. *In-situ* techniques can be considerably cheaper but negative aspects are lower recovery and less control technically and environmentally.

When considering coal and particularly natural gas as feedstocks we have to realize that these raw materials, in contrast to tar sands, shale and biomass, can also be used as such. The conversion of coal into high-added-value products can be performed in several ways. The processes involved can be described as being in a development stage. During World War II the Germans were already making 4.5 million tons/year of synthetic liquid fuel from coal and the South African Sasol plants now in operation and under construction will eventually produce some 100,000–150,000 barrels per day of liquid products from coal. However, the processes involved are cumbersome and environmentally questionable, which is the reason for the industry-wide effort to improve these processes or develop "second-generation" processes.

One way to upgrade coal is direct liquefaction by the addition of hydrogen at high pressures and temperatures. The first process of this kind, the Bergius process, which was used in Germany during the World War II, is no longer used commercially. Several companies are developing improved versions. Initially, the emphasis was on the production of clean underboiler fuel, but developers now realize that this type of synthetic fuel cannot compete economically with coal as such in low-value markets, and they are concentrating on further upgrading by means of increased recycle, gasification of liquefaction bottoms and deep hydro-treatment/cracking of heavy fuels. The status of development of these processes can best be described as entering the demonstration phase, and if positive decisions on prototype projects are taken in the mid-eighties the first commercial units could operate in the nineties. However, considering the economics of coal liquids and the severe environmental problems during processing, this time schedule could well be considerably delayed. For a long time to come coal liquefaction projects seem viable only if there are strong strategic reasons for them.

Carbon Removal Technologies

An alternative way to upgrade coal is to gasify it with oxygen/steam to form synthesis gas (carbon monoxide plus hydrogen). This synthesis gas can be used as a clean medium-calorific-value fuel gas or as feedstock for the synthesis of hydrocarbons and chemicals. A number of coal gasification plants are being operated worldwide, mainly for the production of ammonia. The technology used is basically that developed before and during World War II, i.e. gasification by the Lurgi, Koppers–Totzek or Winkler process and synthesis to hydrocarbons by the Fischer–Tropsch process. This technology can be considered to be proven. Second-generation processes based on the above technologies have advanced to the point where operating experience is available from a large pilot plant and prototype commercial units are being planned or are under design, with an expected start of operation from 1985 onwards.

As mentioned above, the medium-calorific-value (300 Btu/cu ft) synthesis gas could be used as a fuel or as a feedstock for synthesis. Two processes are currently commercially available for producing liquid fuels:

(1) methanol synthesis according to ICI or Lurgi technologies;
(2) Fischer–Tropsch synthesis to form a mixture of hydrocarbons, mainly paraffinic but partly oxygenated (ARGE, Synthol).

Under development by Mobil is a process to convert methanol into gasoline over a synthetic 5 Å molecular sieve catalyst (ZSM 5). This process could be commercial by 1987 after the facilities under design for New Zealand have come on stream.

A further development, still only in the laboratory stage, is the synthesis of hydrocarbon liquids without methanol as an intermediate product. However, such a process will not become commercial before the nineties.

Like coal, natural gas has a dual role. In addition to being a high-quality product in its own right, natural gas can be a feedstock for synfuels. It can be converted by means of conventional steam reforming into synthesis gas, and from then on the same options are open as for synthesis gas from coal, although the difference in H_2/CO ratio, 0.5 from coal and up to 4 from natural gas, has to be taken into account. In effect, the major part of today's world methanol production is based on natural gas. Whether natural gas can be seen as a feedstock for synfuels, in competition with its use as such, is very much a question of the alternative netbacks to the producer as compared to sales as pipeline gas or LNG.

Biomass Technologies

Biomass covers a wide field of feedstocks varying from wood to sugar; the application may be direct burning, gasification or fermentation.

Biomasses rich in starch or sugar (sugar cane, sago, cassava) are best converted into ethanol (fermentation). Cellulose biomass (wood, straw) can also be fermented subsequent to a hydrolysis step or gasified to synthesis gas and further converted into methanol.

Most attention is being devoted to the traditional fermentation techniques to produce ethanol as a gasoline extender or pure motor fuel. The technology is well known but is being applied on only a small scale and is rather old-fashioned. Considerable improvements are possible and are being made. Restrictive factors might be the large variety of feedstocks and the importance of local circumstances, together with the cost of transport, which limits the application of economy of scale. The obvious disadvantages of biomass (poor economics; vast land use) are to a certain extent compensated for by more imponderable aspects, such as short lead times, simpler introduction into the market, scope for development, a favourable public image and the possibilities of small-scale manufacture.

RANKING THE OPTIONS

Only time will tell which synfuel options will become viable, and their timing will vary considerably from country to country. Any ranking of opportunities can thus only be highly tentative. A proper ranking would include availability and quality of resources, quality and marketing aspects of products, economics and technological risks of conversion processes, etc. But even if all these factors were known, a general ranking can be of only limited value, since national or even local circumstances often dominate. With security of supply often the driving force, the URM that is indigenous has a decisive advantage and for instance the question whether coal or shale is "better" can become irrelevant. However, when a country has more than one URM, or for internationally traded URMs, the issue remains important.

The previous sections have presented some generalised observations on the pros and cons of the various synfuel options. The important factor missing has been the economics of the various routes. And not without reason. It is extremely difficult to analyse economics of the various routes in a general way.

In the first place there is the complexity. For many options there are several competitive technologies available, often in a different stage of development, i.e. ranging from commercial processes, second-generation processes in the demonstration phase to processes being researched in laboratories. Consequently, the reliability and accuracy of cost figures vary enormously. The effect of cost escalation during the various stages of development can be very large; a factor of 2 or 3 is the rule rather than the exception. To obtain some sort of consistency is almost impossible and a cost picture may well be more misleading than instructive.

Secondly, as stated above, the synfuel business is to a large extent a regional or even local business. Global cost analyses therefore have less relevance, since national factors may be overriding. The most dominant factor could well be the tax/financing facilities offered for a certain project. Under current conditions most synfuel projects are not yet economic, and the way strategic interests (balance of payments, security of supply) are translated into financial incentives is often decisive. In other words, an economic/competitive analysis makes sense for specific opportunities but less so for a general survey.

Thirdly, there are many actors on the synfuel scene: mining, manufacturing and marketing companies, local and national governments and various "pressure groups" representing specific interests: labour unions, environmentalists, consumer organisations. Each actor may well have a different perspective of the relative merits of the various cost factors involved.

To obtain some insight into the relative merits of the various options, the energy efficiencies involved will be analysed. This has many advantages. First of all, the figures are relatively easily available and not subject to major changes in time. In the second place, there is often a fairly good relation between efficiency and cost for comparable processing options. A higher efficiency implies lower feedstock cost, lower environmental impact (lower heat/material losses) and often lower capital cost.

Finally, an efficiency analysis puts everything on a common basis and in this way it is possible to grasp the whole picture without the complications caused by

Table 11.4 **Ranking of synfuel options**

Resource	Product	Conversion efficiency (%)
New oil		
North Sea	Distillates	85–90
Tertiary recovery		60–85
Short residue		65–75
Shale/tar sands		50–70
Natural gas	LNG	80–85
	Methanol/distillates	55–65
Coal	Syngas	75–80
	Methanol	50–60
	Distillates	45–60
Biomass	Ethanol	35–45
	Methanol	45–50

RANKING THE OPTIONS 613

differences in cost of capital (i.e. rate of return) for different sections of the routes.

Table 11.4 illustrates this point. For a number of liquid synfuel options the conversion efficiency is given, based on technologies expected to be available in the nineties. The table shows that, compared with crude oil (efficiency of 90-95%), all other feedstocks produce synfuels with a lower efficiency (i.e. higher cost), more or less in the ranking order shale/tar sands, natural gas, coal, biomass.

A further general observation refers to the underlined high-efficiency options. These options are estimated to cost $30 per barrel oil equivalent (boe) (1981 US$) or less and are or could be applied now. The other options will cost some $50 per barrel or more, and commercialisation may take one or more decades. It should be noted that cost of conversion is not the only important parameter in an economic sense. This is illustrated in Tables 11.5 and 11.6, where the overall (resource to end use) efficiencies for several options are given, with space heating and private road transport as respective markets.

The fact that conditions for the nineties are analysed should be kept in mind, because it significantly affects the results. Table 11.7 illustrates this point for the efficiency of car engines.

The table clearly shows that the scope for development of the Otto engine running on gasoline is relatively large and that a new generation of engines designed for efficiency (high compression, lean burn) rather than performance and comfort may almost fully erode the large advantage that the diesel engine holds currently.

Table 11.6 also indicates that synthetic transport fuels will be considerably less efficient and thus more costly than even marginal, expensive transport fuels produced from the heaviest part of crude oil.

Table 11.5 **Tentative future energy conversion efficiencies for domestic space heating**

Feedstock	NG	Coal	
Product	LNG	SNG	Electricity
Mining/transport	0.94	0.96	0.96
Conversion	0.85	0.65	0.40
Distribution	0.98	0.98	0.89
End use			
gas boiler	0.85	0.85	
resistance heater			1.0
heat pump	1.5	1.5	2.5
Energy at home	0.67 1.17	0.52 0.92	0.34 0.85

An interesting comparison is presented by the various options for introducing coal into the transport market. Conversion of methanol into gasoline, technically possible, implies a double loss; the manufacture is less efficient/more expensive and the end-use is less efficient. The electric car is an interesting, efficient option, provided that the advanced battery with the performance assumed is available and that electricity is generated efficiently via coal gasification followed by a combined-cycle power station. However, acceptability to the user is questionable.

In space heating (Table 11.5) the dominant effect of the use of heat pumps is obvious. Insulation in combination with gas or electric heat pumps can reduce energy requirements by a factor of 2–4. The figures strongly support the view that gas is an excellent space heating fuel.

Table 11.6 confirms that biomass is an expensive option, mainly because of its low efficiency of conversion even with improved processes. This effect is further aggravated by the small scale that is intrinsic to biomass.

However, one point strongly in favour of biomass is not reflected in the table (nor in today's economics). In contrast to the other resources, biomass can be renewable. In today's economics the value of non-depletion is not accounted for. The relevance of this omission differs with viewpoint but is very evident to everybody when depletion actually occurs. The significance of this aspect can be illustrated by starting the energy chain not at the crop, but at the seeds. The cultivation process is a very effective user of solar energy and, if solar energy is taken as free, then the output of a plantation in terms of energy is a factor of 5–10 higher than the energy input (irrigation, fertilisers, harvesting, transport).

Table 11.6 **Tentative future energy conversion efficiencies for private road transport**

Feedstock	Short residue		Coal			NG (remote)	Biomass (sugar cane)
Product	Gasoline	Gas oil	Gasoline	Methanol	Battery	Methanol	Ethanol
Mining/transport	1.00	1.00	0.96	0.96	0.96	0.91	0.87
Conversion *	0.75	0.60	0.50	0.55	0.40	0.60	0.40
Distribution/charging	0.97	0.98	0.97	0.96	0.67	0.96	0.97
Engine **	0.22	0.23	0.22	0.22	0.75	0.22	0.22
Drive train	0.90	0.90	0.90	0.90	0.92	0.90	0.90
Energy at road	0.14	0.12	0.09	0.10	0.18	0.10	0.07
Weight correction factor (Gasoline = 1.00)	1.00	0.96	1.00	0.98	0.80	0.98	0.99
Efficiency index (Gasoline = 100)	100	81	64	68	99	71	46

* Typical rounded-off value; actual value depends on feedstock quality and process (configuration).
**Values are best estimates but are open to debate.

Table 11.7 **Possible engine efficiency development**

	Engine efficiency (%)	
	Current	Future
Gasoline/Otto engine	0.16	0.21–0.22
Gas oil/diesel engine	0.19	0.22–0.23
Alcohol/Otto engine	0.18	0.22–(0.25)

Thus the efficiency index in Table 11.4 would not be 48 for ethanol from sugar cane but 370, an order of magnitude higher than for other synfuels. A major problem with biomass is that of scale. Even in the tropics a large (10,000 hectare) plantation is unlikely to produce more than 2,000 barrels per day.

FUTURE OUTLOOK

Two observations may sum up the description of the synfuel scene:
- There are many options, but they are all expensive. Time will tell which options will be chosen/needed. Very often two crucial choices will have to be made:
  (i) start quickly with expensive, current technology or wait for more economic processes at the expense of prolonged insecurity of supply;
  (ii) manufacture synthetic fuels that imitate currently used fuels or produce unconventional fuels, generally less expensive but requiring a new infrastructure to be built up.
  There will be an advantage in aiming for rational combinations of feedstock–fuel–end-use appliance, i.e. certain feedstocks are more suitable for the manufacture of certain fuels, and certain fuels are used best in certain applications.
- It should be re-emphasised that there is no global optimum. Each country or region will have to do its best with the options available. And usually the options will be fewer than the limitations.

A final word of caution: very few things change so rapidly as the outlook on the attractiveness of synfuels. What is offered here is an attempt at a balanced perspective as viewed from the context of the early eighties.

Chapter 12

RESEARCH AND DEVELOPMENT

INTRODUCTION

The purpose of industrial research is to generate technology which a company and its employees can use to continue to earn their living in a competitive world. The research undertaken by companies ranges from basic research, such as the study of the mechanism of catalytic reactions, which can lay the foundation for novel processes and products, to the operation of pilot plants with throughputs of up to 100 tonnes per day to test new processes. Companies also undertake extensive research to ensure the safety of their operations and products vis-à-vis their own employees and society at large.

The size of the R&D (research and development) effort associated with the petroleum industry can be gauged from the fact that Shell companies spent almost $500 million on R&D in 1980 and employed some 7,500 people (approximately 5 per cent of total employees) in 17 laboratories throughout the world. Other major oil companies maintain similar R&D efforts.

Research is carried out in all major sectors of the business, viz. Exploration and Production, Manufacturing, Marketing, Chemicals, Natural Gas, and Transport, though some research, such as that into materials and engineering, is relevant to several business sectors.

Industrial research activities are not undertaken in isolation from other parts of the company, but form an integral part of the business. Thus research programmes are planned and carried out in close liaison with the business sectors that will use the results in their operations, and the forward business planning of these sectors takes account of the technologies that research is developing: industrial R&D is a highly interactive process.

EXPLORATION AND PRODUCTION

In a broad sense, research for exploration and production can be considered as falling into four main categories:

- Improving knowledge of natural phenomena, mainly of geological processes and the behaviour of fluids in subsurface formations.
- Developing new or better techniques for the identification and measurement of natural subsurface structures and features such as rock types, their properties and fluid content.
- Investigating artificial methods for increasing recovery from reservoirs by introducing additional energy or by favourably changing fluid flow processes.
- Developing and improving designs and materials for equipment and installations that recover, process, store and remove hydrocarbons near the point of production.

Many of the projects in these categories overlap or require the integrated efforts of several scientific disciplines.

Current research is very much geared to the discovery of more elusive accumu-

Figure 12.1 **Laser light-scattering equipment for research into enhanced oil recovery.**

lations, to developing difficult reservoirs, to increasing recovery from established fields and to working in more hostile environments.

Natural Phenomena

In the study of basic natural processes, new analytical methods and the creation and use of large data bases have widened the scope for this type of investigation. Information has become more accessible and more readily correlatable to support detailed theoretical concepts covering the whole habitat of hydrocarbons, from generation in the source rock to migration into the reservoirs.

Information of more significance can be extracted from thoroughly drilled areas, with known conditions of geology and hydrocarbon accumulation, to provide prediction models, at quite a detailed level, for new and sparsely explored provinces.

A typical subject of interest, to disciplines in both exploration and petroleum engineering, is the role played by faults and fractures in the subsurface formations as either pathways or barriers to fluid migration. It is important to be able to distinguish the different types, both on a regional scale and within an individual field. Similarly, new concepts on the behaviour of fluids in producing reservoirs are being evolved together with computer programs to simulate this behaviour under more complex and heterogeneous conditions, so it is incumbent upon the reservoir geologist to provide his part of the information in ever-increasing detail. This is particularly difficult in the case of carbonate rocks, which are susceptible to alterations in their pore-space geometry, subsequent to deposition caused by percolating mineral-rich solutions.

Thus, in spite of considerable strides in basic theory over the last decade or so, much remains to be done and now can be undertaken with the aid, not only of computers but also of advanced laboratory equipment for analysis and measurement.

Subsurface Evaluation Techniques

The most notable recent achievements in this category lie in the field of seismic data acquisition and processing. The advances are largely due to the digital computer together with developments in instrumentation. At the present stage, high resolution data usually can be obtained for interpreting the configuration of subsurface formations. Refinements, such as dense coverage to provide a three-dimensional effect, are well advanced and attention now has turned to devising techniques for identifying the actual character of the formations and even their fluid content. Petrophysical evaluations of rock types and fluids can be correlated with acoustic survey signals in the same borehole, which, under favourable conditions, can be matched with similar signals generated at the surface some distance from the borehole. The aim is to establish recognition patterns that could be applied for exploration ventures in undrilled territories. If successful, prediction of reservoir quality variations, even in individual reservoirs with limited well control, can be envisaged, and in the more distant future some quantification of these parameters may prove feasible.

Supplementary Recovery

Secondary recovery by the injection of water into reservoirs to supplement a natural water drive, or to create one, is an established technique; reservoir pressure can be maintained and oil swept towards the producing wells. If the water is not properly treated, it may prove incompatible with the reservoir rocks and cause impairment to their fabric. Continued experiments with appropriate mechanical and chemical treatments to improve their effectiveness and reduce costs are still worthwhile research objectives.

Similar considerations apply to the use of fluids other than water, and there is now a range of new techniques under the general heading of enhanced oil recovery. They are distinguished by the introduction of fluids that do not occur naturally in reservoirs. Thermal methods, using steam or hot water, have had a wide application to heavy oil reservoirs and depend upon reduction of oil viscosity to achieve their effect, but reservoir response has been difficult to predict until the advent of sophisticated numerical simulation models on high-speed computers. *In-situ* combustion, supplying air to a burning bank of oil in the reservoir, has yet to achieve any significant success and the method is still very much a continuing research problem.

Attention is currently focussed upon miscible and chemical flooding. In the former method an oil-soluble gas or liquid is injected to reduce the natural interfacial tension that traps oil in the rock pore space; some field applications are already operational. In the chemical method, polymers are used to thicken the water to a consistency similar to oil, thus reducing the tendency for oil to be by-passed. As yet, chemical methods are at the experimental and early field trial stage, with research aimed at cost-cutting and investigating any unforeseen undesirable side-effects.

Techniques for improving the productivity of individual wells, by treating the surrounding formations, are also of continuing importance. These include the artificial creation of large fractures around the well-bore in low permeability reservoirs and injecting materials to prevent the closure of such fractures. Wells in unconsolidated sandstone reservoirs are prone to plugging from the entry of loose sand. Instruments for the timely detection of sand entry have been developed, and chemical treatments to consolidate sands with the minimum loss of permeability are constantly under review.

Design of Offshore Equipment and Installations

The past decade or so has seen an immense effort by the oil industry to move into offshore environments made hostile by weather, distance from land and depth of water. Research is fundamental in developing new design concepts to cope with these conditions as well as in improving the design and reliability analysis of more conventional structures. Fixed platforms have been installed in ever greater water depths, with a consequent increase in the severity of the dynamic behaviour effects caused by the wave environment. In this respect the importance of fatigue in steel platform members is becoming increasingly apparent. For this reason, fixed platforms in such conditions may be limited to water depths around 450 metres.

For deeper water, various types of floating and compliant production systems are being studied. The latter are platforms with articulated columns, guyed towers and tension legs as connections to the seabed. The design and operation of underwater manifolds, remote control systems, deep diving research and manned submersibles are some of the associated areas of interest that require strong research support.

In research related to drilling, dynamic positioning of vessels has allowed the exploration of prospects in deep water, up to 1500 metres. Associated problems, such as the design of marine risers connecting the derrick floor to the seabed, have had to be solved. Since it is believed that a high proportion of reserves in deep water acreage may lie in numerous scattered small fields, there is a strong incentive for reducing drilling costs to allow economic development of such prospects. Typical of this effort is the development of improved drilling bits and mud-circulating pump design which increase footage rates (speed of penetration) by high-pressure drilling.

MANUFACTURING

The task of manufacturing research is to develop new refinery processes and to improve existing ones, as well as to reduce manufacturing costs and to enable the appropriate range of products to be made from a wide and increasing variety of crude oil and other hydrocarbon feedstocks. In all these operations, concern for safety and environmental acceptability play an important part.

Although many separation, conversion and treating processes have been applied for a considerable time, research is still needed to ensure their continual adaptation and improvement in order to meet the latest requirements. Changes in the performance requirements for automotive fuels and lubricants, for example, have important repercussions for the manufacturing processes involved. In a similar way, the industry has responded to (proposed) legislation in certain countries to restrict the lead content of motor gasoline; existing processes have been upgraded to increase the octane number of motor gasoline components without undue loss of product yield, and new processes such as C_5/C_6 paraffin isomerisation have been introduced with the same aim.

The increased cost of energy has stimulated much research into energy conservation and more cost-effective refining. The use of advanced computer-assisted supervision and control systems has contributed greatly to this development, and research continues into new and more sophisticated applications.

Two further factors are changing the whole face of manufacturing and are likely to command a substantial research and development effort in the coming years. One factor is the expectation that the long-term demand for fuel oil will decline relative to that for distillate fractions as the real price of crude oil increases. Whereas previously a substantial proportion of the refinery's product output was in the form of fuel oil, nowadays there is an increasing tendency to convert residues and heavy oils into distillate fractions (e.g. gasoline, jet fuel, gas oil components) rather than blending them into the fuel oil pool. These conversions present new challenges to the industry, with catalysts and process conditions having to be specially tailored, since the heavy feedstocks involved are far more prone to depositing coke during reaction (on catalyst and/or equipment) than is the case for lighter crude oil fractions. In addition, residues often have a relatively high percentage of sulphur and contain small quantities of metal compounds (nickel and vanadium), which block or poison the action of conventional catalysts. Thus research is in progress into methods of removing metals and sulphur from residues so that the resultant hydrocarbons can be processed more easily into products of higher value.

The second factor of major importance is the interest in raw materials other than crude oil, e.g. coal, tar sands and oil shales. The direct or indirect liquefaction of coal may in time compete at certain locations with crude oil refining as a source of transportation fuels. Research is being carried out on gasification of coal to synthesis gas (a mixture of hydrogen and carbon monoxide), which can then be converted either into methanol or into liquid hydrocarbons. Methanol in its own right may also gain in importance as a gasoline component or substitute. Routes for direct liquefaction of coal through hydrogenation also receive due attention.

Synthetic crude oil extraction from tar sands and shale, both of which occur abundantly in several countries, may offer prospects for the medium to long term, albeit that the technology required involves very large capital expenditure.

The potential environmental impact of waste water and process gas impurities, as well as of the solids remaining after processing or extracting the hydrocarbons from coal, tar sands or shale, is a major concern. A large part of the R&D effort

MANUFACTURING

Figure 12.2 **A pilot plant for process development, results from which can be scaled up to commercially sized plants.**

addresses these problems, which in many cases require advanced technologies for their solution.

Looking still further into the future, research staff are constantly searching out and examining the potential new technological developments that will determine the future of manufacturing operations in the oil industry.

Research and development of new processes usually starts off with small laboratory-scale experiments. This limits the financial risk at this early scouting stage. If preliminary results are encouraging, the activity has to be translated into a more costly pilot plant, from which the various chemical engineering parameters necessary to develop a full-scale operation have to be derived and operational experience gained. The pilot plant is important, not only for developing the process, but also to provide initial samples of the product for evaluation in the intended application. Before proceeding to commercial application, an intermediate-scale prototype or demonstration plant is often built, to scale up without undue risk. When the full-scale plant is commissioned, it is often necessary to continue research, both to improve the efficiency of the process and sometimes to overcome operational difficulties, which do not show up until full-scale operation has been achieved.

Figure 12.3 **Preparing a vehicle for test in a cold-weather chassis dynamometer, where fuels can be evaluated for performance relating to cold-start, warm-up and other critical low-temperature driving factors.**

OIL PRODUCTS

Oil Products R & D is largely determined by the needs of the market place and the identified prospects and attendant strategies being followed in the business. The

Figure 12.4 **Single-cylinder diesel engines, used to evaluate the performance of diesel engine lubricants against established specifications acknowledged throughout the world.**

Figure 12.5 **Testbeds used for the evaluation of engine lubricants, seen through the windows of a fully automated computer-linked control system.i**

main thrust of this R&D is in the following spheres:
- Adaptation and innovation in oil products to provide levels of performance of real value to consumers.
- Optimisation of manufacturing processes and market needs with cost-effectiveness for the consumer.
- Cost-effectiveness in all marketing distribution operations.
- Responsibility in all health, safety and environmental matters.

Product development directed at new or improved standards of performance is usually concerned with the isolation of key factors and the examination of these factors under carefully controlled conditions in laboratory rigs or test beds which simulate end-use. Ultimately, tests in the field are conducted to verify the various aspects of performance.

Calls on product R&D are various. For example, the demands for low-lead or unleaded motor gasoline have posed new and critical criteria for gasoline performance. Since the performance of a gasoline is a function not only of its octane value, but also of the volatility and distribution of octane value across the volatility range, it has been necessary to evaluate gasolines against the needs of both older and modern automotive engines to be sure of a high degree of vehicle and consumer satisfaction. Additionally, in a number of countries stringent exhaust emission standards have had to be observed. In the laboratories, very elaborate equipment is used which is capable of providing close control of operating conditions of the vehicle under dynamic conditions and, in some cases, under controlled climatic conditions. Such work is supplemented by road tests on fleets of cars under carefully monitored conditions.

Over the years, lubricants have been submitted to increasingly severe conditions calling for new levels of performance. Advances sought have been in terms of lubrication, or the ability to overcome more demanding conditions in machines, such as higher temperatures or contaminants from combustion. Longer oil life has often been a prerequisite. Additives technology, as well as improved manufacturing processes applied to the base oils, has done much to create the modern ranges of lubricants with their high oxidation stability, ability to suspend and prevent deposition of harmful deposits, and other necessary attributes.

R&D on fuels and lubricants has done much to facilitate the development of modern machines and equipment, and cooperation between manufacturers and the oil industry has been close.

In processing the crude oil or other feedstocks to meet the market demand, increasingly advanced chemical and chemical engineering technology has been employed, and R&D effort has been necessary to ensure that the needs of present and future machines, equipment, and other end-uses are fully met. Such work has embraced products across the range including LPG, motor gasolines, aviation jet fuels, diesel fuels, domestic fuels, industrial fuels and bitumen for roads and industrial uses.

In response to increased oil prices, alternative fuels are now beginning to be used in a number of countries, e.g. methanol from natural gas or coal, ethanol and vegetable oils from biomass. These fuels are presenting new challenges, both in combustion and in lubrication, which are being addressed by R&D.

Cost-effectiveness in storage, handling and distribution of oil products, where most of the costs of marketing lie, is an important area for research and development. Such work is not only concerned with such items as instrumentation, measurement and transfer of product, but has also become increasingly related to data-handling and retrieval and automation of operations employing advanced electronics technology combined with systems design.

Health and safety and environmental matters, both in own operations and in end-uses, has also been an increasingly important area for R&D. Toxicological screening of products and components, fire hazards and procedures, and emission standards, are all subjects of continuing R&D in support of a responsible business outlook.

Underlying all of the foregoing is basic work directed at better understanding of oil products' composition and behaviour, combustion characteristics of different fuels, and the fundamentals of lubrication, and it is from this platform that work of a development nature is tackled. Although oil products have a long history of development and the field is frequently characterised as being mature, innovation by crossing new technological thresholds is still possible and new ideas are pursued wherever they appear to have potential for step-wise improvements in product performance.

CHEMICAL PROCESSES AND PRODUCTS

The incentive for chemical process and product research based on petroleum fractions arose when the introduction of cracking to increase gasoline production resulted in the simultaneous production of large quantities of unsaturated hydrocarbons, which are potential raw materials for the manufacture of chemicals. However, the petrochemical side of the industry has grown so rapidly that its raw material requirements far exceed the availability of unsaturated hydrocarbons as by-products from oil refining, and major plants are now used for the production of unsaturated hydrocarbons and aromatics, gasoline being the by-product.

As a consequence of increases in oil and gas prices and national concerns about balances of payments, growing attention is at present being devoted to non-petroleum feedstocks, e.g. coal, which, after conversion to synthesis gas, can be used for the manufacture of valuable chemicals. Another source of chemical feedstock being studied is biomass. Fermentation of such material yields ethanol, which, apart from possible use as automotive fuel, is a versatile starting material for a number of chemical compounds.

Nevertheless, the chemical business of oil companies is still based to a very large extent on unsaturated hydrocarbons (olefins), and much research is devoted to the discovery of new products that can be produced from them and to the invention of new processes for the manufacture of such products.

The lower olefins such as ethylene, propylene and the butylenes, together with dienes such as butadiene, are produced by cracking petroleum fractions. These include ethane, LPG, naphtha and gas oil and even heavier products. The higher olefins (C_6–C_{18}) are produced by cracking wax. However, because of the shortage of suitable waxy feedstocks, the Shell Higher Olefins Process was developed. By oligomerisation of ethylene, higher olefins covering a wide range of fractions can be obtained. Much research is devoted to extensive studies of the chemistry of olefins, dienes and the many products that can be made from them. Processes have been developed for the production of alcohols from the corresponding olefins, and from these a wide range of solvents and other chemicals and intermediates can be produced. Research on olefin chlorination revealed that, instead of the classic addition reaction across the double bond, a substitutive chlorination of one of the hydrogens on the double bond could be achieved, and this has led to the development of a process for the production of allyl chloride and, from it, allyl alcohol and glycerol. From allyl chloride, epichlorohydrin, a key raw material for the production of epoxy resins, can also be produced.

Research on olefin oxidation led to a direct oxidation process for the production of ethylene oxide and to a process for the production of propylene oxide from propylene. The latter process also yields styrene as a coproduct. Propylene oxide is the main base material for the production of polyols, a component of polyurethane foams with important uses in insulation and the automotive industry.

Research pioneered the development of synthetic surfactants and continues to be active in developing new detergents as well as in investigating the problems resulting from their use. This has led to the development of biologically soft detergents, i.e. detergents that break down into simple compounds under the action of bacteria and therefore do not cause foam on rivers and waterways when effluent containing them is discharged. The Shell Hydroformylation process (SHF) enables the manufacturing of long-chain primary alcohols from olefins. Ethoxylation and subsequent sulphation or sulphonation of these products yields valuable components for detergent formulations.

Considerable research is devoted to the development of plastics and the processes by which they are made. A process using the chlorination of ethylene, followed by the dehydrochlorination of the dichloroethane thereby produced, has been developed for the production of vinyl chloride, from which polyvinyl chloride is manufactured. Research has also developed processes for the production of polyethylene from ethylene, polystyrene from ethylene and benzene, and polypropylene from propylene. Copolymers of, for example, ethylene and propylene showing improved impact resistance can also be produced. Specifically in the area of propylene polymerisation, the development of novel high-activity catalysts has led to improved process economics and better quality products.

The first commercial process for the large-scale production of styrene butadiene rubber was developed during World War II, but work is still going on in this field. Indeed, a new and unique process for an improved version of this key raw material has recently been developed. Passenger car tyres made from it have a low rolling resistance without sacrificing road-holding properties, particularly in wet weather (wet-grip). Other results of this research have been the development of polyisoprene, a synthetic rubber closely approaching natural rubber both in chemical structure and in properties, and polybutadiene, a new type of synthetic rubber with specialised properties and applications.

Thermoplastic rubbers, block copolymers consisting of, for example, styrene and isoprene, combine the properties of both plastics and rubbers. They do not need to be vulcanised and can be extruded and moulded. They find applications in shoes, electric wire coating and adhesives. Saturated thermoplastic rubbers that have increased weathering and resistance have been developed for other applications.

Research is also devoted to seeking new applications for existing products and technology, to improving the quality of those products, and to increasing the operating efficiency of the processes by which they are made. The latter consideration has led to considerable research being performed on improving catalysts used in processes in both the chemical and oil industry. As a result, high-quality catalyst carriers based on silica and alumina are currently being produced. Also, improved de-hydrogenation catalysts as well as a better catalyst for the production of ethylene oxide have been found.

Having entered the chemical business with the motive of deriving maximum value from petroleum fractions, it is natural that oil companies should extend their activities into other areas of the chemical business that are not dependent on oil feedstocks. Thus, oil companies have business in the agrochemical, oil additive, catalyst, and fine chemical areas, supported by their own research efforts.

One of the consequences of this move into other activities reflects back on the petroleum industry. Entry into agrochemicals required the establishment of a research effort in toxicology, since these products are designed to interact with chemical life mechanisms. Toxicology laboratories are also able to assess the properties of chemicals, oil or consumer products, and to ensure that they do not present unexpected hazards to health. At the time when the potentially harmful effects of chemical products on the health of workers, society at large, and the environment was first realised, the same laboratories were able to assess the toxicology of other chemical products. They were able, therefore, to respond rapidly to the new challenge by building up their toxicological research on the basis of established expertise. Today, toxicological investigations are carried out on all products, whether classified as chemical products or oil products, and research extends to the fundamental investigation of the *in vivo* mechanisms of toxicological effects: as stated earlier, industrial R&D is a highly interactive process.

NATURAL GAS

At the present time, natural gas is brought to the market via overland pipelines or by relatively short underwater pipelines, and where this is not possible the gas is liquefied and transported to more distant markets in insulated tankers. These methods are all capital-intensive in the production and transmission phases, and this has led to an R&D focus on the reduction of costs, without compromising engineering or safety standards.

The issue of safety standards is particularly important in relation to LNG schemes, which are subject to close scrutiny all around the world. Industry and government authorities are devoting considerable R&D resources to this subject. R&D is aimed at the further improvement of containment systems and to preventing any spillage of gas in the event of an accident. At the same time, research is carried out on the use of risk analysis to predict the frequency of possible mishaps and to predict the consequences of an accident. In this connection, Shell carried out extensive spillage tests in 1980 with LNG and LPG at Maplin Sands in the United Kingdom, and the Department of Energy in the United States also carried out trials at the Naval Weapons Center, China Lake. The objective of these trials was to test the validity of and to refine mathematical models, developed from smaller tests, for the prediction of the dispersion and combustion behaviour of flammable gas clouds.

Rising energy costs are leading to new prospects for the production of natural gas from smaller or more remote fields, or its use as a feedstock for chemical transformation into other products. Thus, natural gas can be converted to methanol, and this can be transformed into liquid hydrocarbon fuels or other chemicals. Such conversion would be of great benefit to many countries by enabling them to reduce expensive imports. The New Zealand Government has embarked on the first commercial scheme to use this type of technology.

There is also growing interest in the possibility of developing offshore gas finds, especially the smaller fields, by improving the economics of gas handling offshore, including LNG. Schemes are being devised, for instance, to mount liquefaction plant on platforms, and other efforts are directed to the increased use of process modules, and of barge-mounted plant. As shown by North Sea operations, offshore working can be both hazardous and expensive, and the above are only a few examples of the intense R&D being undertaken to find better means of producing both gas and oil from such localities and bringing them safely to land.

In the longer-term future, it seems likely that, in various parts of the world, future supplies of gas will once again be derived from coal as a feedstock, and first generation substitute natural gas (SNG) plants are now under development. Doubtless, further developments will be possible, but it remains uncertain which of the current ideas will ultimately prove successful.

In the even longer term, it is not impossible that heat from nuclear sources will be used in these processes and/or that the electrolysis of water at off-peak times could well provide hydrogen for use either as an industrial and domestic fuel, or as a component for synthesising hydrocarbons.

In relation to the more immediate future, much has been written about the possibility of producing relatively small quantities of methane from biomass or refuse. Undoubtedly, small schemes will be attempted and these will pose problems for the development engineer, particularly in regard to the amount of gas treatment required. While it is unlikely that these sources can make a major contribution to world gas supplies, they may well find important local uses. Whatever the source of gas in the future, it is likely to be relatively expensive, particularly if it is to be derived from remote or novel sources. Therefore there will be a continuing emphasis on fuel economy and on exploiting the cleanliness and controllability of gas in processes where its properties can lead to reduced overall costs.

TRANSPORT, STORAGE AND HANDLING

The movement of crude oil, petroleum products and gas to the final customer is a substantial element in the activities of companies and involves large amounts of capital. The research effort directed at this phase of operations is considerable and some aspects have been mentioned elsewhere in this chapter.

Research ranges from work on pipelines, sea and road tanker transport, to equipment for the dispensing of products to the customer. The FLAGS North Sea pipeline, which was designed to bring ashore natural gas and natural gas liquids co-produced from North Sea oilfields, provides an example of pipeline R&D. The design of this pipeline involved small-scale physical studies of the flow of mixed liquids and gases in pipelines and the development of mathematical models for the design of full-scale equipment. Research work on associated gas/liquid processing and handling equipment was also carried out to ensure a reliable and cost-effective design.

Marine research is primarily aimed at ensuring efficient, economic and safe transportation and is concerned with aspects ranging from the design of new ships, propulsion systems, and the reduction of internal and external corrosion, to the development of new antifouling paints that can reduce the energy required to transport oil at a given speed. The sizeable research establishments maintained by petroleum companies, and their actions through joint industry bodies, have enabled the industry to set high standards of operational efficiency and safety, which impact on other shipowners.

An example of research work on the dispensing of oil products is provided by the aviation fuels, which are dispensed at high rates to reduce aircraft turnround times. Shell laboratories developed the antistatic additive ASA3 to minimise the dangerous build-up of static during refuelling, and this is now used throughout the petroleum industry. Today, R&D is also addressing handling problems associated with the introduction of alcohols as engine fuels, e.g. in Brazil.

BASIC RESEARCH AND NEW TECHNOLOGIES

Research in the early days of the nascent petroleum industry originated from the need to devise specifications to control the quality of petroleum fractions, and it was soon found necessary to carry out work on the chemistry of petroleum, since the basic knowledge was not available from universities. This tradition of basic research and the development of new technologies has increased over the years. Nowadays, such work covers a wide range of specialisms in chemistry, biochemistry, microbiology, toxicology, physics, mathematics and engineering.

A strong basic research effort on the mechanisms of catalysis has been maintained for a number of years, since it is a key field for the petroleum and chemical industries and a field in which rapid advances are being made. Basic work on the mechanism of Ziegler–Natta catalysis played a large part in the successful development of new high-activity catalysts which lower the cost of polypropylene production.

The second example is to be found in the field of life sciences, which has developed rapidly in the last twenty years. Activities in natural product chemistry are pursued in relation to agrochemicals, whilst microbiology promises applications as widely apart as tertiary recovery of oil and the recovery of metals from ores. Genetic engineering will undoubtedly be influential.

This use of speculative basic research to generate new technologies involves an element of risk. Thus, although the combustion system of the early aviation gas turbines owed much to work by Shell engineers, work on the Fell locomotive transmission system, an alternative to diesel electric transmission, was unsuccessful. Similarly, while great advances were made in fuel cell technology by certain oil companies, a marketable product has yet to result.

Although industry devotes a substantial effort to fundamental research, it should perhaps be emphasised that this is in no way intended to be a substitute for the work done in university laboratories; in fact, it supplements it. Industry still needs universities, and it is to be hoped that they will continue to generate the fundamental breakthroughs in science that will generate the new industries of the future. Indeed, industry supports these institutions and encourages their research activities by means of grants for post-graduate research and donations to building funds. There is a considerable amount of collaboration, as can be seen from joint papers published by scientists from universities and industry, and the collaboration is likely to grow in the future.

PATENTS

Patents are granted by the state and provide monopoly rights of limited duration for inventions in return for a full disclosure enabling the invention to be used by anyone when the patent expires. The patent system is available to individual inventors or companies irrespective of size, and it has long provided the only real protection against those seeking to gain unfairly from the inventive ingenuity of others. It is general practice in industry to protect, by patenting, inventions arising from its R&D work. The publication effect of the disclosure requirement means that an industry can get to know about each new invention of relevance to it, and at an early stage. Duplication of research effort can be minimised, and the innovative nature of such disclosures itself stimulates technical progress. A fair return to patentees can be gained through licence arrangements, and the general willingness of oil industry companies to grant licences to others to enable them to use patented inventions, particularly in the area of oil processing, is a significant factor contributing to technical progress industrywide and often further innovation.

Many refinery processes are available under licence, and it is possible for a company to build a modern refinery by purchasing the required processes and the attendant operating information without it being necessary for that company to have its own large R&D organisation. Obviously, the royalties to be paid to patentees responsible for the original development of the processes will normally more than reflect the appropriate share of the research costs incurred, even though they may be less than would otherwise be borne "in-house". Moreover, the larger and more complex the operations concerned, the greater the economic advantage of developing "in-house" technical knowledge and invention, and the patenting of the latter. Apart from the main advantage of gaining time in achieving a desired modernisation and optimisation of operations, the research costs of doing so can be reduced by the licensing of resulting patented technology to others.

A similar attitude on licensing is generally taken by the oil and chemical industry in regard to the field of exploration and production, general equipment, and chemical processes and products. However, where the results of research contribute a unique advantage to the appeal that a company's product has for its customer or in the company's own operations, that company may not wish to license, since it will wish to keep the patented results to itself, at least initially, in order to protect its marketing or operational activities against competition. However, the limited life of a patent ensures that such inventions ultimately become available to industry in general. Also, competitors can, and often will, do their own research aimed at finding alternatives that will fall outside the patent.

Most large organisations in high-technology industries find it worthwhile to have their own professional patent staff, who keep in close contact with research and technological work conducted in laboratories and plants and are informed when novel and useful inventions are made so that appropriate action can be taken. Since each patent is effective only in the country that grants it, a series of patents for the same invention may have to be taken out in a number of countries. For example, in the agrochemicals business, in which R&D is lengthy and costly, patents can determine profitability or even the ability to compete effectively at all in a particular area, and wide foreign filing is the norm.

A major task of a company patent department is to see that, when plans are being made for the introduction or change of processes and products, no conflicts will arise with respect to patents held by others; and it has to advise on the scope and validity of outsiders' patents, on the possibility of avoiding or overcoming any potential patent obstacles, and on the desirability of seeking licences.

It is usual for such patent departments to contribute professional experience and expertise to those concerned with proposed changes in patent laws and with the various international arrangements and conventions, often co-operating with professional and business committees in that respect, with the object of maintaining a patent system that meets the contemporary needs of inventors and industry alike in a changing technological world.

Chapter 13

ENVIRONMENTAL CONSERVATION

INTRODUCTION

The petroleum and petrochemical industries keep society supplied with large amounts of energy and organic chemicals, and like any other major industrial organisations they have an impact on the environment. How great that impact is depends on how the environment is defined, and the acceptability of its effects is determined by the values of society. Neither the definition of the environment nor society's judgement of values is constant or uniform throughout the world. Both have changed considerably during recent decades. An increasingly broad area is coming to be regarded as the environment, and society's concern is shifting from the acute and direct effects of environmental impact to the potential long-term and indirect consequences.

When the environment is narrowly considered to comprise only one particular industrial installation and the people working there, its protection will be limited to the prevention of accidents and the assurance of safe working conditions for employees. These issues have traditionally been the concern of industrial managements and governments, and long-standing "safety at work" acts are in force in many countries.

By their very nature, most operations in the petroleum and petrochemical industries involve a fire hazard, and a number of them present toxic hazards as well. Means of containing such hazards have always been an integral part of the design and operation of installations and, as a result, the petroleum and petrochemical industries have a safety record that they can be generally proud of and that can be favourably compared with those of many other human undertakings.

A broader definition of the environment would include the surrounding area immediately affected by the operations of a particular installation—the atmosphere into which off-gases and vapours are emitted, water bodies into which liquid effluents are discharged, and the people in the neighbourhood who might be subjected to noise, odours and other pollutants emanating from the operations.

This definition applied to the situation in many countries during the late 1950s and the 1960s, when air and water pollution problems were recognised in rapidly growing centres of population and industry. The expanding petroleum and petrochemical industries gave considerable attention to the problem and made great efforts in developing effluent and emission control technology. This technology was usually applied from the very early stages in the construction of new installations, with the result that pollution from them was kept to a minimum.

Subsequently, emphasis shifted to include not only acute environmental effects in specific locations but also potential long-term effects on human health, the living environment and the world's climate. A particular response from the petroleum and petrochemical industries to this broader view of the environment has been an intensification of activities in the fields of industrial health and hygiene, and of toxicological studies of the substances they handle and market. The results are to be seen in improved measures for protecting both workers in the industries and users of their products against possible health hazards.

Around 1970, new anti-pollution laws began to come into force in many countries. Such laws generally made emissions and discharges from industrial installations subject to permission from regulatory authorities, who frequently required the application of new control technology for existing installations as well as for new ones.

Since that time environmental conservation laws and regulations have proliferated, and regulatory bodies with them. The current broad view of what is meant by the environment is clearly demonstrated by the content of the environmental impact statements required in many countries before major new developments can be undertaken. Issues that have to be studied include changes in the landscape, possible adverse effects on cultural values, consequences for local society of changed land-use plans, and increased risk of accidents affecting local inhabitants.

Another important development in many countries during the 1970s was increased public participation — through mandatory consultative procedures — in making decisions about new projects.

As a consequence of these developments, the petroleum and petrochemical industries now have to comply with a multitude of regulations aimed at protecting the health and safety of employees, of the local population, of customers and of other people handling their products, as well as with regulations aimed at protecting air, water and soil against pollution.

These regulations undoubtedly ensure that considerable attention is given to the issues involved. Obtaining permission for a new project, however, has become a complex, time- and effort-consuming procedure. Compliance with the control requirements places an increasing burden on investment and operating costs. Consequently, the regulations, and the manner in which they are executed, threaten to become an obstacle to further industrial progress.

In a few countries this threat is already recognised. Even so, it will remain a great challenge for governments and industry during the 1980s to streamline regulations and procedures so as to ensure proper protection of the environment in a cost-effective way whilst maintaining further industrial development.

The assessment of health, safety and environmental risks is of central concern to the petroleum and petrochemical industries, and a great deal of effort is devoted to it. Many aspects of this broad subject are discussed in detail in other chapters of this handbook. The aim of the following pages is to give a brief, general survey of the industries' environmental activities in the various sectors of their operations — ranging from exploration and production, through transportation and manufacturing to the distribution and marketing of oil and chemical products.

EXPLORATION AND PRODUCTION

In comparison with other major mining operations, exploring for oil and producing it have only a small direct impact on the environment. The area involved and the amount of rock displaced in gaining access to the oil are relatively small, and handling and moving crude oil by pipeline are basically clean operations. Accidental releases of oil, however, can have serious environmental consequences. The most important consideration in exploration and production operations, therefore, is accident prevention. In addition, specific measures, geared to local circumstances, have to be carefully considered. Some examples are given below.

Seismic Exploration

The traditional method of generating the vibrations needed for seismic investigation of underground rock formations is to explode small charges of dynamite, but this can create a nuisance, and even cause damage, in densely populated areas, and when the explosions are generated under water aquatic life may be endangered. Under such conditions, alternative techniques are used — employing for example, oscillators, airguns or gas exploders. Figure 13.1 shows a modern seismic generator.

Drilling

The problems associated with drilling exploration wells are basically the same as those encountered when appraising and developing oil fields, irrespective of the well location. There may, however, be considerable differences in potential environmental impact. Figure 13.2 shows a drilling operation in an urban environment.

Figure 13.1 **Vibroseis vehicles carrying out a seismic survey in the Landes department in south west France. Each vehicle lowers to the ground a hydraulically operated pad that generates sound waves capable of penetrating deep into the earth. Returning echoes are recorded by geophones and tapes for computer processing and analysis.**

Exploration drilling is by its very nature carried out in new locations, and in densely populated onshore areas is likely to conflict with existing land-use plans. If it is unsuccessful, it is a relatively short-term activity, but if a new field is discovered it becomes a long-term one. When this happens, it is essential that the chosen drilling location can be reached with minimum nuisance to the population or harm to the environment. Additional measures may be necessary, for example to reduce noise emissions and to minimise visual impact by camouflaging the drilling rigs.

Even in sparsely populated areas, or offshore, both technical and environmental factors are taken into account when choosing the most suitable drilling site. In nearshore areas, for example, the presence of vulnerable environmental assets such as coral reefs and mud flats may have to be considered when determining operating and disposal procedures.

Drill cuttings and used drilling mud have to be disposed of in an environmentally acceptable manner. Water-based muds rarely present any problems, since

EXPLORATION AND PRODUCTION 641

Figure 13.2 **An exploration well being drilled on the outskirts of Chester in the UK.**

the amount of chemicals in them is too small to have a significant environmental effect. Oil-based muds, however, usually have to be treated before they can be used for land-fill.

Production

The water drawn from the reservoir with the oil has to be separated and disposed of. Where water injection is used to maintain pressure, it may be appropriate to dispose of the produced water by treating it and then re-injecting it into the formation. Alternatively, the residual oil may be removed from the water by gravity separation, assisted by the application of demulsifying agents, and the treated water then discharged to suitable water bodies. Salts in the water may restrict discharge of the effluent to fresh water bodies with insufficient capacity to dilute it to harmless concentrations.

Figure 13.3　**The Shell Claus Off-gas Treating (SCOT) unit at the Waterton Gas Plant in Canada.**

Increasing use of chemicals, especially in enhanced recovery operations, may necessitate special attention to discharge procedures.

Another source of effluent water that may contain oil is rain water drained from surfaces subject to oil contamination. It is standard practice to collect this water and treat it before discharge.

Gas, either produced as such or separated from associated oil, often has to be treated before it can be used as pipeline gas or liquefied. Hydrogen sulphide removed from the gas is usually converted into elemental sulphur, which can either be used as raw material for producing sulphuric acid or stockpiled. Residual sulphur compounds in the off-gas from the sulphur plants are incinerated and discharged into the atmosphere or, under environmentally sensitive conditions, further recovered by, for example, the Shell Claus Off-gas Treating (SCOT) process (see Figure 13.3).

In isolated and sparsely populated areas there may be no direct use for associated gas. It may then be re-injected into the reservoir or, if this is undesirable, flared. Flaring is only practicable, however, where emissions of noise, light and possibly sulphur dioxide can be tolerated.

Accidental Oil Spills

Poor housekeeping in the operation of oil fields, or in transporting oil, may lead to accidental spillages. The quantities of oil involved are usually small, but even so such spills can cause a local nuisance.

When control over a well is lost, however, and it blows out, there may be a major oil spill. It is often possible to regain control of the well, and so quickly stop the oil flow, but sometimes relief wells have to be drilled to do this. Such an operation may take many months, in the course of which a large amount of oil may be spilled.

If a blow-out occurs onshore, it is generally possible to contain the spilled oil within a relatively small area and recover it. Offshore, however, the oil will quickly spread over a large area and, depending on distances and weather conditions, may contaminate beaches, shell-fish beds and fish-spawning grounds. On the other hand, if weather conditions are favourable, the oil may be dispersed naturally and cause little damage.

The possible outcome of an offshore spill can be assessed by a mathematical model, specially developed for the purpose by the oil industry. Other models can predict the movement of an oil slick under given weather conditions. Such models are valuable aids for contingency planning and for managing abatement action when an oil spill occurs.

Specific operations for combating an onshore spill are under the jurisdiction of a single country, but major spills at sea may affect a number of countries. The realisation of this fact has provided the incentive for international oil-spill abatement schemes, and oil companies operating in certain geographical areas have also joined forces in mutual assistance agreements.

In North-West European waters, financial responsibility for clean-up operations and for damage resulting from offshore spills is covered by the "Offshore Pollution Liability" agreement (OPOL). This is a voluntary agreement by the oil industry to ensure that, up to a certain limit, compensation is available for oil pollution damage, and to encourage prompt preventive and remedial action by the oil companies. It provides a mechanism for the speedy settlement of claims.

To enable the oil industry to respond internationally to the need for offshore environmental conservation the International Oil Companies Exploration and Production Forum has been established. Through a number of *ad hoc* working groups, this organisation, in which many oil companies participate, provides factual information, expertise and industry opinion. During the relatively brief period of its existence, it has already acquired a good reputation with international governmental agencies.

Supporting Services for Offshore Operations

Major offshore operations need the support of varied and extensive land-based activities, some of which must be provided from suitable locations on the coastline. Particularly at locations where there has been little or no previous industrial activity, the advent of an offshore support force can create environmental and social problems. Forward planning by governments is needed in selecting suitable development areas and in anticipating and alleviating possible problems. Environmental impact studies are frequently made to assess such issues so that the onshore activities can be properly planned.

TRANSPORTATION AND STORAGE OF CRUDE OIL AND GAS

Most crude oil is transported by pipeline on land and by tankers across the seas. Gas is preferentially piped to its destination but, when this is not feasible, it is liquefied and transported by refrigerated gas tankers. Many offshore fields too are linked to the shore by pipeline. Transport by pipeline is essentially a continuous operation, whereas shipment by tanker is carried out in batches. The transition between the continuous and batch operations takes place at the shipping and unloading terminals, where the required storage capacity is installed.

Pipelines

From an environmental point of view, transport by pipeline has the advantage over batch transportation in that it produces no vapour emissions. A pipeline, therefore, can affect the environment only by its presence or if it leaks.

To minimise the adverse effects of a pipeline's presence, its route has to be carefully planned so as to avoid, as far as possible, centres of population and other sensitive areas. Special attention is given to the design of shore and river crossings (see Figure 13.4) and stretches of line passing through areas presenting geological or other hazards.

Spillage from a pipeline can result from operational errors, mechanical failure, corrosion, or third-party activities. Whatever the cause, early discovery of the leak is essential to reduce the volume of oil, or gas, spilled.

Damage by accidental, or malicious, third-party activities, which has been the cause of more than half the volume of oil spilled from Western Europe's crude-oil pipeline system, is often beyond the control of the pipeline operators (see Table 13.1).

Figure 13.4 An area in North Wales where the land has been reinstated after laying of a section of the 36 inch crude oil pipeline from Anglesey to Stanlow Refinery in Cheshire.

Table 13.1 **Comparative data on pipeline spillages over the five-year period 1976-1980**

	1976	1977	1978	1979	1980	Total 1976-1980
Combined length (km × 1000)	18.1	18.4	18.5	19.0	19.0	–
Combined throughput ($m^3 \times 10^6$)	540	563	594	647	636	–
	Number of incidents					
Mechanical failure						
Construction	2	–	–	2	–	4
Material	3	4	3	–	2	12
Operational error						
System	–	–	–	–	–	–
Human	–	2	–	–	–	2
Corrosion						
External	2	2	7	4	3	18
Internal	–	1	–	–	–	1
Natural hazard						
Subsidence	2	1	1	–	1	5
Flooding	–	1	–	–	–	1
Other	–	1	–	–	–	1
Third party activity						
Accidental	4	6	4	3	3	20
Malicious	–	–	–	1	1	2
Incidental	1	1	–	–	–	2
	14	19	15	10	10	68
Gross spillage incidents						
Negligible	2	3	1	–	–	6
1–10 m^3	1	4	4	1	3	13
10–100 m^3	6	4	5	5	2	22
100–1000 m^3	4	7	4	4	4	23
In excess of 1000 m^3	1	1	1	–	1	4
Pollution resulting						
None	1	4	1	–	–	6
Soil						
Slight	9	13	13	10	9	54
Severe	–	1	1	–	1	3
Water courses						
Slight	4	2	4	1	–	11
Severe	1	3	–	1	2	7
Potable water	–	–	–	–	1	1

Source: CONCAWE report No. 2/82

Terminals

Environmental impacts from oil terminals may arise from vapour emissions, effluent-water discharges or accidental oil spills.

Vapour emissions from shore tanks have been greatly reduced by the use of floating, rather than fixed, roofs (see Figure 13.9 on page 654). A floating roof follows the movement of the oil-level in the tank, so that no vapour is emitted when the tank is being filled. Such roofs also prevent emissions resulting from the effect of ambient temperature fluctuations on cyclic expansion and shrinkage of the vapour in the tank's ullage.

Vapour can be emitted from vessels that are unloading if ballast water is loaded into tanks from which crude oil has been discharged but which still contain hydrocarbon vapours. On tankers equipped with an inert-gas blanketing system and designed for simultaneous cargo and ballast handling, such emissions can be avoided by using the gas line to connect the vapour space of the tank being ballasted with that of a tank from which oil is being unloaded. This method, however, cannot be used at a loading terminal. If the vessel has vented its tanks while still at sea, such emissions will be minimised, but some of the vapours generated whilst loading will be released to the atmosphere during topping-up.

Effluent water from oil terminals consists mainly of contaminated ballast water taken ashore from tankers and water drained from shore tanks in which it has settled out. Rain water drained from the site may also be contaminated with oil. All these effluent waters have to be treated before they can be discharged into surface waters, and acceptable quality standards can usually be achieved by gravity separation.

Operational errors or equipment failure can lead to oil spills at terminals during loading or discharge operations. The quantities of spilled oil, however, are usually small, and in calm waters can be contained by booms and collected by suitable skimmers (see Figure 13.5).

Single buoy moorings (Figure 13.6), where vessels are loaded or unloaded offshore rather than in a sheltered harbour, present special problems. A high degree of integrity is required of the flexible hoses connecting ships to the shore, and special measures have to be applied for the rapid detection of leaks. Effective ship/shore communications are essential. If there is a spill, sea conditions may be too rough to allow booms and skimmers to be used effectively. If the wind is blowing the spill away from areas vulnerable to pollution, then it may be left to disperse naturally; otherwise, chemicals may be applied to disperse it.

At large gas terminals, liquefied petroleum gas (LPG) and liquefied natural gas (LNG) are stored in refrigerated storage tanks. Smaller quantities of LPG may be stored in pressure tanks.

Since the boil-off from refrigerated storage is collected, neither refrigerated nor pressurised storage gives rise to vapour emissions. Accidental release of liquefied gas, therefore, with the consequent risk of fire or explosion, is the main environmental concern of liquefied gas storage and handling. Pressure tanks for LPG storage are protected against an external fire by pressure relief and automatic sprinkler systems.

Figure 13.5 The SOCK (Spilled Oil Containment Kit) Skimmer specially developed by Shell Development Company in the USA for use offshore.

Figure 13.6 The 550,000 deadweight tonne Shell tanker *Bellamya* loads crude oil through the single buoy mooring at Mina al Fahal in the Oman.

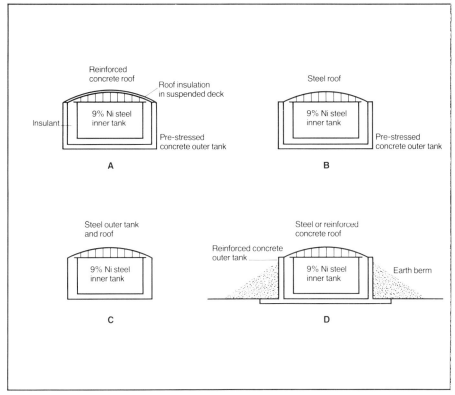

Figure 13.7 **Concepts for double containment refrigerated LPG storage tanks**

Special constructions of various kinds — all considered to be acceptably safe under the given conditions — are used for refrigerated storage tanks. Figure 13.7 shows a diagram of a double integrity tank.

The design of the loading and unloading facilities, and their operation, also need special care with a view to minimising the risk of fires and explosions. The greatest risk, however, is associated with the possible collision and stranding of tankers carrying the liquefied gas in the harbour approaches to the terminal. The siting of a new gas terminal, therefore, is usually the subject of extensive risk analyses, in which considerable attention is given to the tanker approach routes.

Oil Tankers

Tankers have, in the past, caused pollution by discharging into the sea ballast and tank wash water that contains oil — a common practice even after the 1954 International Convention for the Prevention of Pollution of the Sea by Oil came into force. This allowed such discharges to be made only at a distance of more than 50 nautical miles from land, where it was assumed that the oil would be dispersed in the general mass of water, biodegraded, and so have no harmful effects.

In 1958, the Intergovernmental Maritime Consultative Organisation (IMCO) *, a specialised agency of the United Nations concerned with maritime affairs, was established. Its objective is to facilitate cooperation among governments in order to achieve the highest practical standards of maritime safety and to protect the environment by preventing pollution of the seas by ships. In the same year, the 1954 Convention came into force and IMCO took over its administration.

It then became apparent that, with the rapid growth in oil transportation, the measures adopted by the Convention were insufficient to prevent serious pollution of the seas and shores, but it took many years for effective legislation to be adopted. During the next two decades, the voluntary application of measures developed by the oil and shipping industries preceded the formulation of international legislation by IMCO.

The International Chamber of Shipping (ICS) and the Oil Companies International Marine Forum (OCIMF), which have consultative status as observers at IMCO, have played an important part in providing technical recommendations and industry comment on the contents of the proposed legislation. In effect, after the inevitable delays caused by lengthy ratification procedures, IMCO in most instances adopted and made mandatory the measures that had been implemented for several years on vessels owned by the major oil companies.

In the early 1960s, in the absence of shore facilities capable of receiving and treating contaminated water from tankers, the oil industry developed a procedure for treating oily water aboard ship. Known as the "Load on Top" system (see Figure 13.8), oil companies introduced this procedure to their own fleets and also required its application on their time-chartered vessels. In 1969, the procedure was incorporated in amendments to the 1954 Convention and finally became mandatory when these amendments came into force in 1978. The same legislation prohibits the discharge of oil or oily effluents anywhere at sea, except under very stringent conditions and in such concentrations and quantities that no harm will be caused to the marine environment.

Another way of reducing considerably the need to wash tanks for ballast purposes is to use segregated ballast tanks, the water in which is never contaminated with oil. It is a legislative requirement (Marpol Convention 1973 and Protocol 1978) that all new large tankers be equipped with such tanks, and this

---

* In May, 1982, the name of this organisation was changed to the International Maritime Organisation (IMO).

TRANSPORTATION AND STORAGE OF CRUDE OIL AND GAS

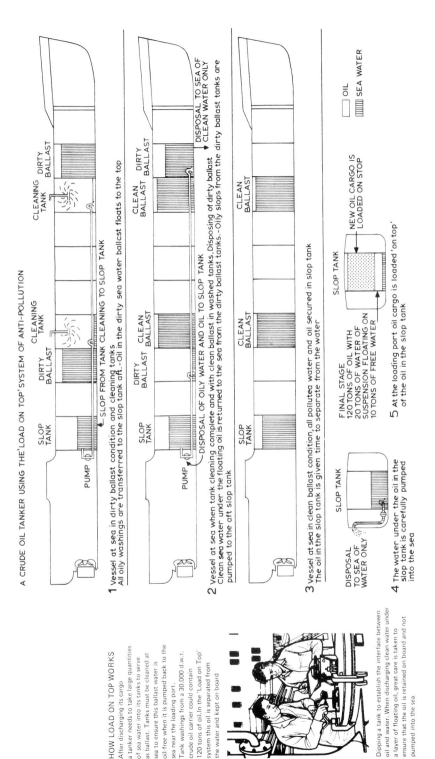

Figure 13.8 **The Load on Top system.**

will soon become mandatory for smaller vessels as well. As an alternative, applicable to existing ships, a technique has been developed for washing tanks with crude oil. It is more effective than water washing alone in preventing sludge build-up in tanks and leaves appreciably less residue, thus reducing the potential for oil pollution.

Operational pollution still contributes to more than two-thirds of the oil discharges to sea. But these discharges involve low concentrations over large areas of the world's oceans and are environmentally much less damaging than the accidental pollution which unfortunately still occurs, most seriously as a result of collisions and groundings. Such accidents characteristically occur close to the shore and in areas of high traffic density. Large amounts of oil may be released within a short time, causing severe pollution of the shores in the vicinity.

Accidents can be caused by failure of equipment, but most frequently they are due to human error. Adequate training of masters, officers and crews is the key issue in preventing them. Recommended sailing routes, traffic separation schemes and the use of sophisticated navigational equipment can help prevent collisions and strandings. Special traffic regulations frequently apply for oil and gas tankers in harbour approaches. Here again, legislation has been developed by IMCO — for example the "International Regulations at Sea" (1972), which are concerned with the "rules of the road" and traffic separation schemes, and the "Standards of Training, Certification and Watchkeeping for Seafarers" (1978), which cover the qualifications required of officers and crew.

To reduce the risk of explosion, large tankers have for some years been equipped with inert gas systems that blanket tanks with oxygen-deficient gas. In 1985, similar systems will be mandatory for tankers above 20,000 tons deadweight.

The amount of oil spilled as a result of a tanker accident can frequently be reduced by proper salvage action and/or transfer of the cargo to another vessel. Speed is the essence of successful salvage and no time must be lost in negotiating terms of a salvage contract before action is taken. The time-honoured Lloyds "no cure no pay" form did not address the problem of salvor remuneration for efforts to reduce pollution. The form was modified in 1980 and now, although the principle of "no cure no pay" still applies, in the event of an unsuccessful salvage operation a salvor whose efforts have prevented further pollution is recompensed on a cost-plus basis.

Despite every effort, tanker accidents and consequent pollution can never be entirely avoided. So as to provide fair and quick compensation for governments and people who suffer damage from such pollution, the tanker industry, in the late 1960s, entered into a voluntary agreement "The Tanker Owners Voluntary Agreement Concerning Liability for Oil Pollution" (TOVALOP). A complementary scheme introduced by the oil companies in 1971, takes the form of a "Contract Regarding an Interim Supplement to Tanker Liability for Oil Pollution" (CRISTAL).

Two intergovernmental schemes — the Civil Liability Convention (CLC), which is broadly similar in format to TOVALOP, and the Fund Convention, which mirrors CRISTAL, came into force respectively in 1975 and 1978 and are adopted by a growing number of nations. TOVALOP and CRISTAL will continue to apply to those countries that have not ratified CLC and the Fund Convention, but the conventions will progressively replace the voluntary schemes.

Oil-spill Clean-up

The best way to deal with an oil spill is to remove it physically, though this may not be practicable in open, unprotected waters, where equipment such as booms and skimmers is ineffective. In such circumstances, provided that there is minimal threat to the environment, the oil can be left to disperse and break down by natural forces. Should there be a potential threat to the environment in circumstances where the oil cannot be readily collected, dispersants sprayed from boats or aircraft can be used, subject to agreement from the relevant authorities.

Special emergency plans for dealing with large spills have been developed by individual oil companies and by governments. In addition, regional plans (usually joint efforts by groups of oil companies and/or governments — depending on the part of the world concerned) have been evolved to cover wider areas and incidents too large for one company alone to cope with effectively. They may also be applied by governments where a spill may cross national borders.

Gas Carriers

A high standard of safety is especially important during the marine transportation of liquefied gas, since the ships carrying it are usually required to discharge at ports near to industrial and urban areas. No major spillage of liquefied gas has yet occurred, but there has been much speculation about the possible consequences of such a spillage both to the ship and its surroundings. Long-term research programmes have been carried out by the oil industry to help clarify what might happen in such an unfortunate contingency. These have involved both small-scale tests in the laboratory and large-scale tests in which substantial quantities of liquefied gas were spilled on the sea. The results have provided a comprehensive set of dispersion data that enables safety distances to be calculated, as well as information that can be used to assess the vulnerability of gas carriers to fire or explosion.

OIL REFINERIES AND PETROCHEMICAL PLANTS

Large complexes of installations such as normally constitute an oil refinery or a petrochemical plant can affect the environment in various ways. They can, for example, cause pollution of the air, water and soil, and can create a local nuisance by the emission of noise and unpleasant odours. They may also present an accident risk to the neighbouring population.

Gaseous Emissions

Vapour emissions can originate from storage and handling of volatile raw materials, intermediates and final products. They can be expelled from fixed-roof tanks that are being loaded, and when changes in ambient conditions cause the tanks to "breathe". Tanks with floating roofs (see Figure 13.9), such as are commonly used at oil terminals, do not create such problems, though a small amount of vapour is emitted by evaporation of liquid sticking to the tank shell when the roof is lowered. Floating roof tanks are widely used in oil refineries for the storage of crude oil, naphthas, and gasoline. In petrochemical installations, where tanks are commonly smaller than in oil refineries, the application of floating roofs is usually restricted to large tanks.

Vapour emissions from fixed-roof tanks can be reduced by the installation of floating covers, which are frequently used in relatively small tanks.

Other sources of vapour are leaks from pumps, valves, flanges and sampling points, and evaporation from the handling and treatment of effluent water containing volatile material.

Figure 13.9 **Floating roof tanks at the Halul Island terminal offshore Qatar.**

Vapours may also be emitted from manufacturing processes. During start-up and shut-down of units it is usually necessary to release gases, and in normal operations this is sometimes necessary to prevent pressure build-up. Relief valves

Figure 13.10 **The 213-metre high stack on the east side of Pernis Refinery in the Netherlands. This is one of two tall stacks there designed to keep the air clean by dispersing sulphur dioxide into the upper atmosphere.**

are used for this purpose. In modern refineries the gases are, as far as possible, collected and flared. To avoid excessive flaring, however, with the associated emissions of light, noise and sometimes smoke, and at the same time to recover valuable fuel, systems may be installed that compress most of the flare gas for injection into the refinery's fuel-gas system.

Many refinery gas streams contain hydrogen sulphide. Such streams can be treated to remove the contaminant and the purified stream then used as refinery fuel. The concentrated hydrogen sulphide stream can be converted to elemental sulphur in recovery units that normally have a conversion rate of about 95 per cent. The residual gas is generally incinerated, but in environmentally sensitive situations further sulphur can be recovered. The processes for doing this are, however, costly.

By far the greatest quantity of emitted off-gas comes from the combustion gases produced in generating the energy, heat and steam used in a plant's operations. In oil refineries, and in many petrochemical plants, gaseous and liquid fuels are used. The fuel gas mainly originates from oil conversion processes and has usually been treated for the removal of hydrogen sulphide. The balance of the fuel requirement is liquid fuel, which is composed of the heaviest parts of the crude-oil barrel and, dependent on the original crude, may contain appreciable amounts of sulphur compounds and some nitrogen and metal compounds. This liquid fuel is therefore responsible for the greater part of the sulphur oxide, fly-ash and soot emissions — and contributes also to the nitrogen oxide emissions — from a refinery or petrochemical complex. Tall stacks (see Figure 13.10) are often used to ensure good dispersion of the combustion gases in the atmosphere, so as to avoid high pollutant concentrations at ground level.

Effluent Water

Crude oil delivered to refineries by tanker contains a few tenths of a per cent of water. This may be the residue of incompletely removed formation water and/or the remains of ballast water. Part of the water, containing inorganic salts and oil but little soluble material, is drained from the crude-oil tankage. The remaining, suspended, water is often removed by water washing in a desalting unit.

In distillation processes water is introduced in the form of stripping steam, and in some refining processes as a wash liquid, to prevent corrosion. The water separated from the hydrocarbon stream may contain hydrogen sulphide, ammonia, oil and soluble organic material such as phenols, mercaptans and organic acids. This "sour" water is usually stripped with steam to remove most of the hydrogen sulphide and some of the ammonia. A modern practice is to use the stripped water for washing crude oil in the desalting units.

Large process streams in an oil refinery have to be cooled. This can be done by means of water drawn from and returned to surface water. Such a "once-through" cooling system is generally used in older refineries located in a coastal area, and the amount of water needed may be as much as 20 m^3 per tonne of crude processed.

Figure 13.11 **The cooling tower that forms a central part of the circulating cooling water system at Clyde Refinery in Australia.**

Modern refineries have either air cooling, with trim cooling by "once-through" cooling water, or a combination of cooling by air and circulating water. Cooling towers are used to dissipate the heat from the circulating water. To prevent accumulation and deposition of salts, a slip stream (blow-down) has to be removed and make-up water supplied to compensate for blow-down and evaporation losses. In this manner the cooling water consumption of a refinery can be reduced to a few tenths of the crude oil throughput. Figure 13.11 shows a cooling water system.

As in production operations, another source of potential surface-water pollution is discharge of rain water drained from areas contaminated with oil.

The various categories of waste water require different kinds of treatment before discharge. To reduce treatment costs, therefore, they are often segregated.

The treatment processes commonly applied in oil refineries and petrochemical plants are:
- steam stripping, to remove volatile compounds;
- gravity separation, to eliminate floating oily material;
- flocculation/flotation to remove remaining traces of suspended oil;
- biotreatment, to reduce the amounts of dissolved biodegradable organic matter.

The waste-water problems of petrochemical plants are generally similar to those of oil refineries. Some chemical products, however, are more soluble in water than oil products. Consequently, the biological oxygen demand of effluents from chemical plants is usually much higher than for effluents from oil refining units. In some effluent streams, the dissolved organic load may even be too high for cost-effective biotreatment and require measures like extraction or wet oxidation.

Some chemical products are toxic and not easily broken down in the environment, and effluents from plants manufacturing them need special attention.

Waste Disposal

In comparison with many other industries, oil refining produces little waste material. Nevertheless, sludges from tanks and effluent-water treatment, and spent auxiliary material like catalysts and inorganic absorbents, have to be disposed of. The petrochemical industry also has to dispose of small amounts of highly toxic material.

In many countries, accidents at old dumping sites, where insufficient precautions had been taken, have led to strict regulations on disposal of toxic wastes. These can only be dumped, with the required precautions, on dedicated sites, and in some cases have to be detoxified. Combustible wastes are frequently incinerated. Oily sludges may, under favourable circumstances, be disposed of by "land-farming", a procedure in which the oil is converted in the soil by controlled biodegradation. With the consent of the competent authorities, however, non-toxic wastes can generally be used for land-fill.

Noise

High noise levels emitted by plant equipment can damage the hearing of people working at the plant and be a nuisance to the local population. The principal sources are rotating equipment, air-cooler fans, furnaces and flares, valves and vents. Much effort has been devoted in recent years to improving the design of rotating equipment to make it quieter and to methods for dampening the noise generated. Modern air coolers, for example, generate far less noise than earlier designs. Resonance in furnaces can now be largely avoided, and ground flares have been developed to lessen the light and noise emissions from flare stacks. High-pressure vents are fitted with mufflers, and vibrating valves are quickly replaced. Much improvement has already been achieved, and progress will undoubtedly continue to be made in noise reduction. Such progress will, however, become exponentially more expensive.

Distance is a major factor in reducing nuisance due to noise, and land-use plans should take account of this in allowing sufficient distance to be maintained between industrial complexes and inhabited areas.

Accident Hazards

Many of the products used in and produced at oil refineries and petrochemical complexes are flammable, and some are toxic. Serious attention, therefore, has always been given to safeguarding employees and installations. This is reflected by the generally good safety records of the petroleum and petrochemical industries. Even so, accidents do happen, and sometimes they have an effect outside the "factory fence". In particular, they arouse the concern of the public and of governments, who need to be reassured about the inherent safety of industrial projects. Major issues in accident-risk assessment are the choice of the plant's location relative to centres of population, the design of the installation, the safety organisation on site, the measures applied to reduce the effects of a potential accident and contingency plans for handling a major accident if it occurs. Installations from which flammable or toxic gases such as LNG, LPG, chlorine or ammonia can escape are the object of particular attention in this respect.

DISTRIBUTION AND MARKETING

Although oil refineries and petrochemical plants are important single sources of potential pollution, the users of their products emit overall far greater quantities of contaminants into the environment and often in a less controlled manner. Regulations aimed at reducing pollution by these emissions are in force in many countries. Other regulations aim at protecting the customer against possible hazards associated with the use of products. In the distribution and marketing of oil and chemical products all these regulations have to be observed.

Atmospheric Pollution

Petroleum fuels are used for energy generation in stationary, or mobile, installations that inevitably emit combustion gases. Stationary installations vary from domestic heaters burning gas or light and middle distillate fuels to industrial plants and electricity power stations burning residual fuel. Mobile sources comprise road transport with LPG, gasoline and gas oil as fuels; shipping, with middle distillates and residual fuels, and aircraft, with jet fuel (kerosine) or gasoline.

Combustion in stationary installations such as boilers and furnaces leads to virtually complete conversion of fuel to carbon dioxide and water. Sulphur compounds are converted into sulphur oxides, and some of the ash components can be emitted with the unburned particulate matter. Small quantities of nitrogen oxides are formed from oxidation of some of the nitrogen in the combustion air and the fuel. The formation of soot or carbon particulates depends largely on the design and maintenance of the plant, control of the burner and, to a lesser extent, on the crude oil used and the method of manufacturing the residual fuel.

In many countries, limits are set for the maximum allowable sulphur content in fuels. In others, the amount of sulphur dioxide that can be discharged to the atmosphere is limited. Further restrictions are often applied in zones of special protection.

The conditions of combustion in internal combustion engines are very different from those in boilers or furnaces. Exhaust gases from gasoline engines, for example, contain some carbon monoxide, hydrocarbons and nitrogen oxides. In properly maintained diesel engines, combustion is more efficient.

In most western countries, legislation exists (or is being developed) to control exhaust emissions from both gasoline and diesel engines.

In the USA, the Environmental Protection Agency has established Federal Standards for the emission of particulates from diesel-fuelled light-duty vehicles.

In Western Europe, heavy- and light-duty engines burning diesel fuels meeting current specifications also meet the present limits for particulate emissions and smoke. Any reduction of these limits could lead to demands from engine manufacturers for an improvement in fuel quality at a time when this would be critical from the manufacturing and supply viewpoint.

Under stagnant atmospheric conditions and in the presence of sunlight, the exhaust gases from motor vehicles have been found to be important contributors to photochemical smog. In the USA, this finding has led to regulations restricting the amount of carbon monoxide, hydrocarbons and nitrogen oxides allowed in vehicle exhaust gases. To meet the statutory limits, it is necessary to use catalytic converters to complete oxidation of carbon monoxide and unburned hydrocarbons to carbon dioxide and water. The introduction of these converters made it necessary to use unleaded gasoline, since lead compounds rapidly de-activate the catalyst.

In recent years, there has been a considerable move to reduce the amount of lead in gasoline (or, indeed to eliminate lead altogether) because of public concern about its possibly harmful effects on human health. The precise effect of lead on people, and the contribution to that effect made by gasoline engine exhausts, are still matters of serious debate. But legislation calling for a reduction in gasoline lead content to about 0.15 g/l is being introduced in many countries, and in some there are plans for the total exclusion of lead additives.

In the United States and Canada, a low-octane unleaded gasoline is available (in addition to leaded grades) and nearly all new cars in these countries are designed to run on this fuel.

In Japan, most cars now run on a low-octane unleaded gasoline. Before this was introduced, cars had for some time been designed and produced to run on unleaded gasoline. A leaded "premium grade" is retained for the few cars not designed to run on the unleaded fuel.

In Western Europe and Australasia, the reduction in lead levels has generally been more gradual, but with little associated reduction in octane number. Consequently, more expensive processing and more energy are required in maintaining octane levels.

In some parts of the USA and Japan, where there are severe problems with photochemical smog, measures have also been taken to reduce emissions of hydrocarbon vapour from the gasoline distribution system.

Spent Products

Unless they are disposed of by controlled dumping, incineration or recycling, products that have been used by the customer find their way into the environment. This is usually a social rather than an industrial problem, since — in most countries — the manufacturers lose control over a product after selling it. Present practices may change, however, as a result of increasing legislation — particularly in Europe — where disposability may be more highly controlled. This could encourage the industry to become more involved in the collection and disposal of used oil, and petroleum industry organisations are also continually looking into the environmental and health aspects of disposal of spent petroleum products.

Indiscriminately dumped lubricating oil, for example, can contaminate ground or surface water, or enter sewage systems. Such oil can, however, be re-used as fuel or processed to produce a regenerated base oil. The process most frequently used is acid/earth treatment, which gives rise to acid sludges, and it is a moot point whether the accumulation of such by-products from recycling might not present a greater problem in some countries than disposal of the oil in an environmentally acceptable manner.

Control over the disposal of used oil, either for environmental purposes or in order to conserve resources, will inevitably increase. Each case has to be looked at individually, taking into account the needs for environmental protection and resource preservation as well as the overall cost effectiveness.

INDUSTRY ASSOCIATIONS

In order to develop common industry views on environmental conservation issues and to collate relevant information on emissions, discharges, hazards and the effectiveness and cost of control measures in manufacturing, distribution and marketing, various national and international industry associations have been established.

The International Petroleum Industry Environmental Conservation Association (IPIECA) is the body through which the oil industry responds to the environmental initiatives of the United Nations Environment Programme (UNEP) and other agencies concerned with environmental protection.

In Western Europe, some 30 petroleum companies cooperate in the oil industry's international study group for Conservation of Clean Air and Water — Europe (CONCAWE). This organisation was established in 1963 and has produced many valuable reports on a wide variety of relevant subjects. The petrochemical and chemical industries cooperate in the European Council of Chemical Manufacturers Associations (CEFIC) and in the European Chemical Industry Ecology and Toxicology Centre (ECETOC), an organisation for the combined study of the toxicological aspects of commercial chemicals.

These organisations are providing an important industry input for legislative machinery aimed at achieving workable regulations that ensure protection of the public and the environment in a cost-effective manner.

NOTE ON UNITS OF MEASUREMENT IN THE WORLD ENERGY INDUSTRY

Over the course of time, as a result of the diverse physical nature of the raw materials and finished products handled by its component enterprises, the world petroleum industry (and, in a wider context, the energy business) has adopted a wide variety of units of measurement. While some rationalisation and harmonisation have been achieved by industry standards bodies, metrication legislation, etc., there remains a considerable number of units in common use. The object of this brief review is to categorise the principal expressions employed, to define their content where necessary and to specify the factors needed to convert from one major unit to another.

The discussion is couched in fairly broad terms, concentrating on the units generally encountered in production and consumption data. It does not attempt to list the numerous exceptions to general practice which can be found in individual countries or in particular sectors of the industry.

Throughout the energy supply industries, two distinct measurement systems are in use: an "Anglo-Saxon" approach used in the USA and until recently in the UK and most British Commonwealth countries, and a "metric" approach followed in Japan, Continental Europe and most other parts of the world.

Under the "Anglo-Saxon" system, amounts of oil are generally accounted for in volumetric terms with quantities expressed in multiples of the gallon, most usually in terms of *barrels* of 42 American gallons (equivalent to just under 35 Imperial gallons or about 160 litres). The components of oil supply and demand are often discussed in terms of average rates, the standard unit being *barrels per day* (b/d). One barrel per day is equivalent to approximately 50 tonnes per year, dependent on the specific gravity of the material. (The use of the barrel as a unit of measurement dates from the early days of the industry, when oil was collected, stored and transported in wooden barrels).

Certain oil products, in particular residual fuel oils, bitumen and petroleum coke, may be found expressed in terms of weight, generally in multiples of *tons* ("short tons" of 2000 pounds in US statistics; "long tons" of 2240 pounds or, in recent years, "metric tons" of 1000 kilograms, in British usage). Very small amounts of product may be expressed in pounds weight.

In the "metric" approach, oil statistics are shown volumetrically in *kilolitres* or *cubic metres* and by weights in terms of metric tons (*tonnes*) of 1000 kilograms, or occasionally kilograms as such. Japanese statistics usually employ kilolitres, while West European data often use cubic metres for light products, tonnes for black oils and tonnes also for overall amounts (total consumption, production, etc). Except in statistics of refinery operations and capacity, daily rates in metric units are seldom encountered, most data being expressed in absolute terms for the period concerned.

A similar diversity of treatment is found in the natural gas industry. North American usage favours multiples of the *cubic foot* on either an absolute basis or a daily rate. Alternatively, gas may be expressed in terms of its calorific content, using multiples of *British Thermal Units* (Btu).

Outside North America, the natural gas industry generally employs metric-based units. Volumes are normally expressed in *cubic metres*, usually in absolute terms, while calorific equivalents are quoted in multiples of the *calorie* (e.g. teracalories, equal to milliards (10^9) of kilocalories).

The conditions of measurement specified for gas volumes are not always the same. The *standard cubic foot* (Scf) is defined as the mass contained within one cubic foot, measured at 60°F and 30 in. of mercury (equivalent to a pressure of 14.73 psia). While the comparable standard cubic metre (measured at 15°C and 760 mm Hg) is sometimes encountered, the usual "metric" unit for statistics of gas volumes is the normal cubic metre (Nm^3), measured at 0°C and 760 mm Hg. Liquefied natural gas (LNG) is customarily discussed in terms of tonnes of liquid or as the volume (in cubic feet or cubic metres) resulting from regasification.

In the international gas industry, conventional abbreviations for multiples of units are: M for thousand, MM for million, mrd for milliard (10^9) and in US usage, b for billion (10^9) and t or T for trillion (10^{12}).

For comparison of gas volumes, the following rounded conversion factors may prove useful:

$$1 \text{ Nm}^3 = 37 \text{ Scf}$$

$$1 \text{ mrd m}^3/\text{yr} = 100 \text{ MM cf}/\text{d}$$

$$1 \text{ Tcf} = 27 \text{ mrd m}^3$$

$$\begin{aligned}&1 \text{ million tonnes LNG (as liquid) per year} \\ &= \text{(after regasification) } 1.4 \text{ mrd m}^3/\text{yr} \\ &\text{or } 140 \text{ MM cf}/\text{d}.\end{aligned}$$

The composition and calorific value of natural gases vary widely from one source to another (see Chapter 8). For ease of comparison, gas statistics are frequently quoted in terms of standard units, for example cubic feet of 1000 Btu heat content, cubic metres of "average" gas at 9500 kilocalories per cubic metre, or cubic metres of "Groningen gas equivalent" (8400 kcal/m^3).

The other branches of the energy industry also employ a variety of measurement units. US coal is accounted for in terms of short tons, while the rest of the world favours the metric approach. Variations in calorific content (for example, between hard coal grades and brown coal or lignite) are often tackled by expressing data in *tons of hard coal equivalent* (thce) (see below).

The electricity supply industry presents the most uniform and straightforward picture in the energy business. Throughout the world, output, consumption, etc. are expressed in *kilowatt-hours* or multiples thereof, with capacity data generally quoted in multiples of the *kilowatt* (e.g. megawatts or gigawatts, equivalent to thousands or millions of kilowatts, respectively).

The assessment and analysis of energy as a whole (e.g. the total consumption of fuels and electricity in a particular market sector, or of overall primary energy demand) require the use of common units to bring together the diversity of measurement systems employed by the supply industries. Here again, a number of approaches have evolved: these may be grouped under three headings:

(i) Units based on the British thermal unit;
(ii) Units based upon the kilocalorie;
(iii) Units based upon the joule.

Current US practice is to employ Btu-based units for energy statistics, the most frequently encountered ones being large multiples of the Btu — trillions (10^{12}) or quadrillions (10^{15}), the latter being sometimes referred to as "quads" — and the *barrel of oil equivalent* (boe), defined as containing 5.8 million Btu, gross calorific value. It is to be noted that the boe is only a convenient approximation to an average barrel of crude oil, actual values ranging from about 5.6 million Btu/barrel for a very light crude to around 6.2 million Btu/barrel at the other extreme.

As with oil measurements, energy data are often expressed as average rates, the customary unit being the *barrel per day oil equivalent* (b/doe). The other main Btu-based unit is the *therm* (100,000 Btu), which is still used extensively in the UK for gaseous fuels and energy aggregates.

National and international organisations wedded to the metric system for general measurements mostly still employ common energy units based upon the kilocalorie, although the joule and its derivatives are slowly gaining ground. Apart from straightforward multiples such as the gigacalorie (10^6 kcal) and teracalorie (10^9 kcal), the principal units linked with the kilocalorie are the *ton of oil equivalent* (toe) (normally taken as equal to 10 million kcal) and the *ton of coal equivalent* (tce) (7 million kcal), the latter being sometimes specifically related to hard coal. Both the toe and tce/thce are normally based upon net calorific values: the difference between net and gross heat contents is discussed below.

As with the barrel of oil equivalent, tons of oil or coal equivalent as conventionally defined are no more than convenient round numbers, "lumps of energy" which may be taken as *broadly* representative of typical oil or coal grades. While the net heat content of "average" crude oils corresponds fairly closely with that of the "ton of oil equivalent", current worldwide average coal quality is appreciably lower than 7 million kcal per tonne. Incidentally, although they are sometimes referred to as "metric tons" or "tonnes" of oil or coal equivalent, neither unit forms part of the metric system proper, so that this degree of precision in denoting the type of ton involved is not really appropriate.

The official SI (Systeme International) unit for heat and energy is the *joule* (J), and the use of joule-based units in national and international statistics is gradually increasing. However, the smallness of the joule as a unit means that for most practical purposes large multiples have to be employed, such as the terajoule (10^{12} J), petajoule (10^{15} J) or even exajoule (10^{18} J). An alternative approach is to use multiples of the kilowatt-hour (1 kWh equals precisely 3,600 kJ), which is arguably a more manageable and more easily understood arrangement, for the time being at least.

The *calorific value* of a fuel is measured "gross", that is including the latent heat of the water vapour produced in the course of combustion. However, in practice this latent heat is not generally recoverable during the combustion of a fuel; thus a "realistic" evaluation of its heat content should perhaps be based on an estimate of the "net" calorific value, excluding the latent heat of the water vapour. The difference between the net and gross measures of calorific value of a fuel is related to its chemical composition, and ranges from about 3 per cent in the case of coal to 10 per cent for natural gas with crude oils and refined oil products at between 6 and 8 per cent. As in other aspects of energy measurement, conventions differ in this respect between countries and between sectors of the supply industry. "Anglo-Saxon" units have been traditionally based upon gross calorific values, whereas the rest of the world tends to use net values. In the particular case of the gas industry, gross calorific values tend to be the norm, even in non-English-speaking countries.

Of the main common units encountered in international energy statistics the "barrel per day oil equivalent" used by the petroleum industry is normally based on gross calorific values, whereas the "ton of oil equivalent" adopted by the International Energy Agency and the "ton of coal equivalent" used by the United Nations Statistical Office are both defined in net terms. As the relative contributions of the various energy sources vary between countries, by market sector and over time, the relationship between energy data expressed in net calorific values and the corresponding figures in gross terms can be only approximately indicated by a single, unvarying conversion factor.

UNITS OF MEASUREMENT

The following ready reference tables of Energy Units and Conversion Equivalents provide multiplicative factors for transforming data expressed in one of the customary common units into a number of others. In the case of units with different calorific-value bases, approximate conversion factors are given, based on current worldwide weighted averages for total fossil fuels. For the sake of clarity, most factors are shown rounded to four significant figures. This degree of precision should suffice for most practical purposes.

Energy units and conversion equivalents

Basic energy units		Multiple	Power of 10	Prefix
1 British thermal unit	= 0.252 kcal = 1.055 kJ			
1 kilocalorie	= 3.968 Btu = 4.187 kJ	thousand	3	kilo
1 kilojoule	= 0.948 Btu = 0.239 kcal	million	6	mega
		billion or milliard	9	giga
1 barrel oil equivalent	= 5.8 million Btu	trillion	12	tera
1 ton oil equivalent	= 10 million kcal	quadrillion	15	peta
1 ton coal equivalent	= 7 million kcal	quintillion	18	exa
1 therm	= 100,000 Btu			
1 thermie	= 1000 kcal			
1 kilowatt-hour	= 3600 kJ = 3412 Btu			

Approximate conversion equivalents

equivalent values are arrayed vertically

	Unit (on per-annum basis unless stated otherwise)	Cal. value basis	Thous. b/doe	Mill. boe	Trill. Btu	Mrd. cu. m.	Tcal	Thous. toe	Thous. tce	PJ	TWh
Btu-based	Thousand barrels daily oil equivalent	Gross	1	2.740	0.472	17.81	0.002	0.020	0.014	0.478	1.612
	Million barrels oil equivalent	Gross	0.365	1	0.172	6.500	0.001	0.007	0.005	0.174	0.588
	Trillion British thermal units [1]	Gross	2.117	5.800	1	37.70	0.004	0.042	0.030	1.012	3.412
Kilo-calorie based	Milliard cubic metres natural gas [2]	Gross	0.056	0.154	0.027	1	0.0001	0.0011	0.0008	0.027	0.091
	Teracalories	Net [3]	500.0	1370	236.1	8901	1	10.00	7.000	238.8	805.7
	Thousand tons oil equivalent	Net [3]	50.00	137.0	23.61	890.1	0.100	1	0.700	23.88	80.57
	Thousand tons coal equivalent	Net [3]	71.41	195.6	33.73	1272	0.143	1.429	1	34.12	115.1
Joule-based	Petajoules	Net [3]	2.093	5.734	0.989	37.27	0.004	0.042	0.029	1	3.373
	Terawatt-hours electricity output [4]	—	0.620	1.700	0.293	11.05	0.001	0.012	0.009	0.296	1

(1) Equivalent to Billion cubic feet of natural gas at 1000 Btu/cu. ft
(2) At 9500 kcal/cu. m.
(3) Based on 1980 global average ratio of net/gross calorific value for total fossil fuels (0.937)
(4) Equivalent to Billion kilowatt-hours

GLOSSARY

A

Absorption process. A fractionation process, closely related to distillation, by which certain components of a gas are condensed in an absorption liquid (* lean oil) with which the gas is brought into contact. The absorption liquid with the absorbed components is called fat oil. The fat oil leaves the bottom of the absorber and is separated from the absorbed components in a following fractionator whence the fresh lean oil is returned to the absorber. For example, carried out to extract the heavier components from wet natural gas.

Acetonitrile, CH_3CN. By-product of the manufacture of acrylonitrile by oxidation of propylene in the presence of ammonia.

Acidity. The amount of free acid in any substance.

Acoustic velocity log, acoustic log (sonic log). A type of wireline *log which records the time of transit of a sonic impulse through a given length of rock along the borehole wall. The recorded transit time is used for the determination of lithology and *porosity.

Additive. A substance added to a product in order to improve its properties.

Adsorption process. A fractionation process based on the fact that certain highly porous materials preferentially adsorb certain types of molecules on their surface.

Aggregate. The mineral matter used together with bitumen in road construction.

Air gun. Chamber from which compressed air is released to produce shockwaves in the earth. Air gun arrays are the most common energy source used for seismic suveying at sea (see Seismic methods).

Alcohols. A class of organic compounds containing oxygen, of which ethyl alcohol (ethanol; the alcohol of potable spirit and wines) is the best known. They can react with acids to form *esters. They are largely used as solvents.

Alkanolamine process. A process for removal of hydrogen sulphide from (hydrocarbon) gases and LPG by a specific regenerable solvent. Carbon dioxide and, to a certain extent, carbonyl sulphide can be removed at the same time. The solvent used is an alkanolamine such as di-isopropanolamine (DIPA).

Aliphatic hydrocarbons. Hydrocarbons in which the carbon atoms are arranged in open chains, which may be branched. The term includes *paraffins and *olefins and provides a distinction from *aromatics and *naphthenes which have at least some of their carbon atoms arranged in closed rings.

Alkyd resins. A general term applied to synthetic resins formed from a polyhydric alcohol and a polybasic acid, of which there are at least three functional groups.

---

* An asterisk precedes those words that are defined in the Glossary.

Alkyl aryl sulphonates. Alkyl aryl sulphonates belong to the anionic types of detergents. Their production involves the manufacture of an *alkylate, sulphonation of this alkylate to a sulphonic acid and neutralisation of this acid with caustic soda.

Alkyl radical. Any *radical of the saturated paraffinic series, such as methyl, CH_3, having the general formula C_nH_{2n+1}.

Alkylate. Product obtained in the *alkylation process.

Alkylation. A reaction in which a straight-chain or branched-chain hydrocarbon group, which is called an *alkyl group or radical, is united with either an aromatic molecule or a branched-chain hydrocarbon.

Anhydrite. Anhydrous mineral form of the evaporite mineral gypsum. Composition: $CaSO_4$.

Aniline point. The lowest temperature at which an oil product is completely miscible with aniline in a 1:1 volumetric ratio.

Anticline. A fold in layered rocks in which the strata slope down and away from the axis, like the pitched roof of a house.

Anti-knock. An adjective signifying resistance to detonation (pinking) in spark-ignited internal combustion engines. Anti-knock value is measured in terms of *octane number for gasoline engines.

API. American Petroleum Institute. An association incorporated in the United States, having as its object the study of the arts and sciences connected with the petroleum industry in all its branches and the fostering of foreign and domestic trade in American petroleum products.

API gravity. In the USA an arbitrary scale known as the API degree is used for reporting the gravity of a petroleum product. The degree API is related to the specific gravity scale (15°C/15°C) by the formula:

$$\text{Degree API} = \frac{141.5}{\text{sp.gr. } 15°C/15°C} - 131.5$$

The degree Baumé is an antiquated form of the degree API and is based on a slightly different scale.

Appraisal wells. Wells drilled after a new oil or gas field has been discovered, in order to establish the limits of the hydrocarbon-bearing structure or the reservoir(s).

Aquifer. A zone of *reservoir rock in contact with the oil or gas accumulation which contains only water. In natural water drive it is expansion of the aquifer which displaces the oil or gas into the producing wells.

Aromatics. A group of hydrocarbons characterised by their having at least one ring structure of six carbon atoms, each of the latter having one valency outside the ring. If these valencies are occupied by hydrogen atoms, hydrocarbon radicals, or inorganic groups one speaks of mono-aromatics. If part or all of the valencies form other carbon atom rings, one speaks of condensed aromatics. These hydrocarbons are called aromatics because many of their derivatives have an aromatic odour. They are of relatively high specific gravity and possess good solvent properties. Certain aromatics have valuable *anti-knock characteristics. Typical aromatics are: *benzene, *toluene, *xylene, *phenol (all mono-aromatics) and naphthalene (a di-aromatic).

Asphalt. This term has two meanings: (1) it refers to a mixture of *bitumen and mineral *aggregate, as prepared for the construction of roads for other purposes; (2) in the United States it refers to the product which is known as bitumen elsewhere.

Asphaltenes. Constituents of (heavy) *residues characterized by being insoluble in aromatic-free low-boiling petroleum spirit, but soluble in carbon disulphide.

Asphaltic-base crude oils. Crude oils which contain little or no *paraffin wax but usually contain asphaltic matter. Now often referred to as naphthene-base crude oils.

Asphaltic bitumen. The full name for *bitumen adopted by the Permanent International Association of Road Congresses.

Asphaltic concrete. A carefully proportioned mix of coarse aggregate, fine aggregate and mineral filler, coated with bitumen.
Associated natural gas. *Natural gas associated with oil accumulations by being dissolved in the oil under the reservoir temperatures and pressures (solution gas) and often also by forming a gas cap of free gas above the oil (gas cap gas).
ASTM. American Society for Testing and Materials. An association incorporated in the United States for promoting knowledge of the properties of engineering materials and for standardising specifications and methods of testing.
AVCAT. Aviation Carrier Turbine Fuels. A high-flash-point kerosine-type jet fuel, used mainly in naval aircraft (US military JP-5).
AVGAS. Aviation gasoline, for use in piston-type aero-engines.
AVTAG. Aviation Turbine Gasoline. A wide-boiling-range (wide-cut) jet fuel, used mainly in millitary aircraft (US military JP-4) with only limited civil use (as Jet B).
AVTUR. Aviation Turbine Fuel. The standard kerosine-type jet fuel used worldwide (Civil Jet A-1/Jet A, JP-8).
Azeotrope. Two (or more) components are said to form an azeotrope if there is a mixture of those components which has no *boiling range but whose *boiling point and *dew point are the same.
Azeotropic distillation. A *distillation process characterized by the fact that the relative position of the components' boiling points is influenced by the addition of a compound which selectively forms an *azeotrope with one or a group of the components. The added compound is called the azeotrope former. For example, *furfural, used in the extraction of aromatics, forms an azeotrope with water.

B

Back-filling. This refers to the replacement of earth to the position from which it was originally excavated, e.g. back-filling a pipeline trench.
Backwash. Term used in the description of *extraction processes to designate the extract fraction which may be returned to the inlet end of the extraction system to increase the efficiency of the extraction process. The *extract fraction used must be freed, at least partially, from solvent.
Bailing. Removal of sand from a well by means of a bailer, an open-ended tube fitted with a valve at the bottom and lowered into the well on a wire line.
Barefoot completion. A *completion method in which the *casing is cemented down to a point immediately above the producing formation and the production zone is left unsupported.
Barrel. A standard measure of crude oil quantities; equivalent to 35 Imperial gallons, 42 US gallons or 159 litres.
Barytes. A mineral of high specific gravity consisting essentially of barium sulphate, which is mixed in powdered form in *drilling fluids to increase their density.
Bean. see Choke.
Bentonite. A naturally occurring colloidal clay earth used for treating petroleum products to improve their colour. Bentonite is also used as a component of *drilling fluid in order to improve the latter's properties. Together with cement and water or diesel oil it is used to plug off fissures or large rock pores which cause loss of drilling fluid. Bentonite is also used as a thickening agent for lubricating greases.
Benzene, C_6H_6. The parent compound of the aromatic hydrocarbon series. It is used in the manufacture of a large number of chemicals including *phenol, *styrene, detergent alkylate and insecticides.
Biotreater. A treatment unit in which waste water is contacted with a high concentration of active micro-organisms under intensive aeration, in order to make the water suitable for environmentally acceptable disposal.

Bit. The drilling-tool which cuts or grinds its way through the rock in drilling boreholes.

Bitumen. A non-crystalline solid or semi-solid cementitious material derived from petroleum, consisting essentially of compounds composed predominantly of hydrogen and carbon with some oxygen and sulphur; it gradually softens when heated. Bitumens are black or brown in colour. They may occur naturally or may be made as end products from the distillation of, or as extracts from, selected petroleum oils.

Bitumen blowing. See Blown bitumen.

Black products. (Marine) diesel oil and fuel oils.

Blending. Mixing of the various components in the preparation of a product of required properties.

Blowout preventers. An arrangement of rams and shear rams usually hydraulically operated, fitted to the top of the *casing series of a drilling well to close off the borehole if pressures are encountered which are not fully counterbalanced by the mud-column and could lead to a blowout.

Blown bitumen. A special grade of *bitumen prepared by the oxidation of *short residues, normally by blowing air at an elevated temperature.

Boiling point (at a given pressure). The temperature at which a liquid, contained in a closed vessel under a given pressure, will form a first bubble of vapour on the addition of heat. Further heating of the liquid at its boiling point results in evaporation of part or all of the liquid.

Boiling range. Petroleum products (which are mixtures of many compounds, each having a different boiling point) do not have a simple boiling point but have a boiling range instead, i.e. the temperature range from *boiling point to *dew point.

Bomb. A small pressure vessel, such as used for sample taking.

Bottoms. The residue from a distillation of petroleum; also the liquid layer left in a tank or similar container after draining to the level of the pump suction.

Breathing. When a storage tank containing volatile products is heated by solar radiation, some of the liquid contents evaporate. The excess vapour thus formed is blown out to the atmosphere. On cooling, the less volatile components of the vapour contents condense and a slight vacuum is created, causing air from outside to be sucked into the tank. This double action is referred to as "breathing" of the tank.

Bright stock. By *vacuum distillation of a paraffinic *long residue a waxy *short residue is obtained, which is deasphalted and *solvent-extracted and finally always dewaxed and sometimes earth-treated, giving a valuable *luboil component of high viscosity called bright stock.

British thermal unit (Btu). The quantity of heat required to raise the temperature of 1 lb of water through 1°F. 1,000 Btu = 252 kcal = 1055 kJ.

Bubble cap trays. *Fractionating trays consisting of a plate provided with holes and bubble caps. The latter cause the vapour to be distributed through the liquid. Now largely superseded by other designs (see Sieve tray, Valve tray).

Bunker fuel. Any fuel oil or diesel taken into the bunkers of ships.

Butadiene, $CH_2=CHCH=CH_2$. A colourless gas, obtained by the catalytic dehydrogenation of butane or n-butylene. It is principally used in *SBR (with *styrene) and, more recently, by steam-cracking hydrocarbon fractions in such a way as to produce large amounts of diolefins as well as olefins.

Butane, C_4H_{10}. Commercial butane is a mixture of two gaseous paraffins, normal butane and isobutane. When blended into *gasoline in small quantities it improves volatility and *octane number. Butane can be stored under pressure as a liquid at atmospheric temperatures ("bottled gas") and it is widely used for cooking and domestic heating.

Butene. The normalized name for *butylene.

Butylene, C_4H_8. Hydrocarbons of the *olefin series, used as raw materials for chemical solvents and butyl rubber, and also for the manufacture of *butadiene.

C

Cable tool drilling. Early method of oil well drilling consisting of making a hole by repeated blows with a bit attached to a "drill stem", which is a heavy length of steel suspended from a wire rope.

Calendering. A process of imparting the desired finish to, or ensuring the uniform thickness of sheet material by passing it under pressure through a machine generally consisting of a number of rollers.

Calibration. The determination of fixed reference points on the scale of any instrument by comparison with a known standard and the subsequent subdivision or graduation of the scale to enable measurements in definite units to be made with it. Also the process of measuring or calculating the volumetric contents or capacity of a receptacle.

Calming section trays. *Fractionating trays characterized by the presence of calming sections on a tray of the *sieve or *valve variety (hence the names: c.s. sieve tray and c.s. valve tray). Calming sections are actually downcomers, carefully designed and distributed over the tray area so as to ensure the best distribution of liquid.

Calorie. The amount of heat required to raise the temperature of 1 gram of water through 1°C (from 14.5° to 15.5°C). In calculations the kilocalorie, equal to 1,000 calories, is often used. 1,000 kilocalories = 3,968 Btu = 4,187 J.

Calorific value. The calorific value of a combustible material is the quantity of heat produced by complete combustion of unit weight of the material. The units in which the calorific value is usually given are (a) *calories per gram and (b) *British thermal units per pound. The systems may be converted by the relationship: 1 calorie per gram = 1.8 Btu per lb.

Cap rock. A formation which directly overlies a *reservoir rock and is impervious to the passage of fluids.

Casing. Heavy steel pipe used to line a borehole and secured in the formations by *cementing. It is used to seal off fluids from upper strata or to keep the hole from caving in. There may be two or more *strings of casing, one inside the other, in a single well.

Catalysis. The alteration of the rate of a chemical reaction by the presence of a "foreign" substance (catalyst) that remains unchanged at the end of the reaction.

Catalyst. In technology this word means a substance added to a system of reactants which will accelerate the desired reactions, while emerging virtually unaltered from the process. The catalyst allows the reaction to take place at a temperature at which the uncatalysed reaction would proceed too slowly for practical purposes.

Catalytic cracking. Process of breaking down the larger molecules of heavy oils into smaller ones by the action of heat, with the aid of a *catalyst. In this way heavy oils can be converted into lighter and more valuable products (in speech generally abbreviated to cat. cracking).

Catalytic reforming. Process of changing the molecular structure of the components of *straight-run gasoline or of a gasoline fraction by subjecting the gasoline to thermal treatment in the presence of a *catalyst (for example, platinum). By this process the anti-knock performance of the gasoline is improved.

Cathodic protection. Method of protecting tanks, ships, pipelines and jetties against corrosion. By reversing the electric current which flows away from a corroding metal, a corrosion process can be arrested.

Caustic soda. The name used in industry for sodium hydroxide (NaOH) on account of its property of corroding the skin. It is strongly alkaline.

Cementing of wells. Filling part of the space between *casing and borehole wall with cement slurry. The hardened cement keeps the casing in the hole stationary and prevents leakage from or to other strata that have been drilled through. *Conductor strings are cemented to the surface. Cement is also used to plug a well.

Centipoise, centistokes. A centipoise (cP) is 1/100th of a poise (P), which is the fundamental unit of dynamic viscosity in the centimeter-gram-second system of units. The viscosity of water at 20°C is approximately 1 cP. The centistokes (cS) is 1/100th of a stokes (S), which is the fundamental unit of kinematic viscosity in that system. The two viscosities are related by the density, i.e. number of centistokes = number of centipoise divided by liquid density (in g/cm^3).

Cetane. An alkane hydrocarbon ($C_{16}H_{34}$) found in petroleum, especially n-cetane.

Cetane number. The cetane number of a diesel fuel is a number equal to the percentage by volume of *cetane in a mixture with alpha-methylnaphthalene having the same *ignition quality as the fuel under test.

CFR engine. A standard single-cylinder variable compression engine developed by the Co-operative Fuel Research Council, to determine the *anti-knock value of motor gasolines or the *ignition quality of diesel fuels.

Char value. In the 24-hour kerosine burning test the amount of char formed on the wick under prescribed conditions is measured and reported as mg/kg.

Choke. A flow restricting device.

Christmas tree. A structure installed at the top of an oil well or gas-lift well which consists of a number of valves, by which the well can be opened or closed at the surface; named from the earlier complex arrangement of these devices, resembling a Christmas tree.

Circulation system. A system of circulating *drilling fluid, used in *rotary drilling.

Claus process. Process for the manufacture of sulphur from H_2S, comprising oxidation of part of the H_2S to SO_2 in a thermal reaction stage, followed by catalytic reaction of the remaining H_2S with the SO_2 formed to give sulphur.

Cloud point. The temperature at which a fuel, when cooled, begins to congeal and present a cloudy appearance owing to the formation of minute crystals of wax.

Cofferdams. The empty spaces fore and aft in a tanker, which traverse the whole breadth of the vessel and isolate the cargo tanks from the rest of the ship (fire protection).

Colorimeter. An instrument for determining the colour of oil products by measuring the percentage transmission of monochromatic colour through the liquid.

Completion methods. Methods of completing a well in such a manner as to permit the production of oil or gas. According to the nature of the producing formation, different methods are usually applied, depending on conditions. At the surface, a well may be completed by either a *Christmas tree or a pumping head and pumping unit.

Compression ignition. The combustion which takes place when fuels are injected in a fine spray into the hot compressed air (500°C) in the cylinder of a diesel (compression-ignition) engine. The heating of the air is due to its rapid compression by the piston.

Compression ratio. The ratio of the volume of air and fuel when the piston of an engine is at the outer end of the cylinder, to the volume when the mixture is compressed and the piston is in its deepest position.

Compressor. A pump which draws in air or other gases, compresses it and discharges it at a higher pressure.

Condensate. Liquid hydrocarbons which are sometimes produced together with *natural gas. In general: the liquid that is formed when a vapour cools.

Condensation (chemical). The coupling of organic molecules accompanied by the separation of water or some other simple substance, e.g. alcohol. A catalyst is usually required to promote the reaction.

Condensation (physical). The transfer of a material from the vapour phase into the liquid phase, for example by the withdrawal of heat.

Condensed aromatics. See Aromatics.

Condenser. A special type of *heat exchanger for the removal of heat from the top of a *fractionating column.

Conductor. The first *casing string of a borehole, also called the surface string. It is secured in the *formation by *cementing.

Connate water. Water which partially fills the pore spaces in an oil or gas reservoir. It is trapped there by capillary forces and does not move when the oil or gas are produced.

Conradson carbon test. Method of determining the amount of carbon residue left after evaporation and pyrolysis of an oil.

Continental crust. The type of crust that underlies most of the continents; lighter than oceanic crust the upper layer of the continental crust has a density of about 2.8 g/cm^3.

Continental drift. A hypothesis invoking the movement of continents over the earth's surface relative to each other. It is known that continents move passively away from each other because of the creation of new oceanic crust at median ridges by extrusion from deep within the crust.

Continental shelf. The shallow submerged platform, bordering the land and extending to the structural edge of the continent. The use of the term "shelf" is reserved for those features which terminate at a depth of around 200 metres.

Conventional products. Petroleum products which are manufactured from crude oil by physical *separation processes.

Conversion processes. Manufacturing processes which involve a change in the structure of the hydrocarbons.

Copolymer. Mixed polymers or heteropolymers, products of *polymerisation of two or more monomers at the same time to yield a product which is not a mixture of the separate polymers but contains both aforementioned substances in the same polymer molecule.

Coring. The taking of cylindrical samples of rock from a borehole by means of a hollow drill bit.

Corrosion. The gradual eating away of metallic surfaces as the result of chemical action such as *oxidation. It is caused by corrosive agents such as acids.

Cracked gas. Gas which is a by-product of *cracking processes. It is often used as raw material for the manufacture of chemicals.

Cracked gasoline. Gasoline produced by cracking heavy oils. It has a higher *anti-knock value than *straight-run gasoline.

Cracking. Process whereby the large molecules of the heavier oils are converted into smaller molecules. When this is brought about by heat alone, the process is known as *thermal cracking. If a *catalyst is also used the process is referred to as *catalytic cracking (in speech generally abbreviated to cat. cracking) or *hydrocracking if the process is conducted over special catalysts in a hydrogen atmosphere.

Cresylic acids. Chemical compounds of the same family as *phenol. They are derived mainly from coal tar, but can also be extracted from certain cracked distillates. They are used as gasoline *inhibitors.

Crown block. Assembly of sheaves at the top of a derrick over which the wire line from the *drawworks is reeved to connect with the travelling block.

Crude oil types. *Paraffin-base crude oils; *Asphaltic-base crude oils; *Mixed-base crude oils.

Crude wax. Crude wax, also called petroleum wax or slack wax, is an unrefined mixture of high-melting hydrocarbons, mainly of the normal straight-chain type, still containing a fairly high percentage of oil. It is obtained by filtration (as such or after addition of a solvent) from high boiling distillates or residual oils. Slack wax is primarily obtained as by-product in the manufacture of lubricating oils. The crude wax made from distillate oils is used for the manufacture of *scale wax and *paraffin wax or serves as a feedstock for the manufacture of petroleum chemicals such as synthetic detergents. The crude wax made from residual oils is refined to make a range of *microcrystalline waxes.

Crystallisation. A *fractionation process based on the difference in freezing point of the various constituents of the mixture to be fractionated. The process is, for example, used in the separation of *paraffins from *luboil (de-waxing).

Cut. Refinery term of a fraction obtained direct from a fractionating unit. Several cuts can be blended for the manufacture of a certain product.

Cutback bitumen. Bitumen which has been rendered fluid at atmospheric temperatures by the addition of a suitable diluent, such as *white spirit, *kerosine or creosote. The abbreviation "cutback" is often used.
Cyclic compounds. See Ring compounds.
Cycling. See Recycling.
Cyclisation. A reaction, for example, platinum-catalysed, by which a straight-chain paraffin hydrocarbon is converted into a naphthene and then into an aromatic.

D

Deadweight. The amount of cargo, stores and fuel which a vessel carries when loaded to the appropriate draught allowed by law. The difference between deadweight and displacement is the actual weight of the vessel.
Deasphalting. The removal of asphaltic constituents from a heavy residual oil, e.g. by mixing the oil with liquid propane; see also *propane deasphalting (= a process in which a short residue is split into bright stock components and asphaltic constituents, by means of liquid propane).
Deasphaltenising. The removal of *asphaltenes from a heavy residual oil, e.g. by mixing the oil with pentane or heptane.
Dehydration. The removal of water from crude oil, from gas produced in association with oil, or from gas from gas-condensate wells.
Dehydrogenation. A reaction process in which hydrogen atoms are eliminated from a molecule.
Depletion type reservoir. A reservoir from which the oil is displaced during production by the expansion of gas liberated from solution in the oil. The pressure in the reservoir goes down continuously as fluids are produced from it, and is not fully or partially maintained by influx of water.
Derrick. A steel structure, often about 140 feet in height, used to support the drill pipe and other equipment which has to be raised or lowered during well-drilling operations.
Detonation. Detonation or knocking is the sharp metallic sound emitting from the cylinders of spark-ignition engines under certain conditions. It occurs when conditions in a cylinder are such that self-ignition of an unburnt mixture of fuel and air takes place. It reduces power output.
Development wells. Wells which are drilled within the productive area defined by *appraisal wells after an oil or gas accumulation has proved sufficiently large for commercial production.
Deviated drilling. A method of drilling, also called directional drilling, whereby a well is drilled at an angle.
Dew point (at a given pressure). The temperature at which a vapour, contained in a closed vessel under the given pressure, will form a first drop of liquid on the subtraction of heat. Further cooling of the vapour at its dew point results in condensation of part or all of the vapour as liquid. The dew point of a normal gasoline is approximately the same as the temperature at which 70% by volume distils over in the *ASTM-distillation test. The dew point of a pure compound is the same as its *boiling point.
Dewaxing. The process of removing *paraffin wax from lubricating oils.
Diagenesis. Those processes affecting a sediment at or near the earth's surface that bring about changes such as cementation and solution of the original grains.
Diapir. A dome or anticlinical fold in which a plastic core material (usually salt or shale) has uplifted and even pierced the overlying strata by flowage. See salt dome, salt pillow, salt well.

Diesel fuel. A general term covering oils used as fuel in diesel and other *compression-ignition engines.

Diesel index. A measure of the *ignition quality of a diesel fuel; the index is calculated from a formula involving the gravity of the fuel and its *aniline point (API gravity times the aniline point [determined by ASTM D611-47T] divided by 100).

DIMERSOL process. A process developed by the Institut Français du Pétrole for the dimerisation of propylene and/or n-butylenes for production of high-octane gasoline or C_6 to C_8 olefins for the chemical industry.

Dip (in geology). The angle which a bed of rock makes with a horizontal plane as measured in a plane normal to the *strike.

Dipping. A process for measuring the height of a liquid in a storage tank. This is usually done by lowering a weighted graduated steel tape through the tank roof and noting the level at which the oil surface cuts the tape when the weight gently touches the tank bottom (see Ullage).

Directional drilling. See Deviated drilling.

Discovery well. An *exploration well which is successful in encountering an oil or gas accumulation which can be economically developed.

Distillate. The liquid obtained by condensing the vapour given off by a boiling liquid. Also the top product taken off a *fractionating column; and in its broadest sense: any *fraction other than the bottom product of the fractionator.

Distillate fuel oil. Fuel oil consisting mainly of *gas oil and (heavy) *distillates.

Distillation (fractional). A *fractionation process based on the difference in boiling point of the various constituents of the mixture to be fractionated. It is carried out by evaporation and condensation in contact with *reflux. When applied to the separation of gasoline, kerosine, etc., from a crude oil, to leave a residual fuel oil or asphaltic bitumen, the process is frequently called topping. Distillation is normally carried out in such a way as to avoid decomposition (*cracking); in the case of the higher boiling distillates, such as lubricating oils, this is accomplished by carrying out the distillation under vacuum.

Distillation range. See Boiling range.

Dope. A general name for a product which is added in small quantities to a petroleum *fraction to improve quality or performance (= *additive).

Drag bit. A drilling *bit used for soft *formations.

Drawworks. The power unit of a drilling rig which drives all systems including the hoisting winch and the *rotary table.

Drill collars. Lengths of extra-heavy pipe, several of which are placed directly above the drilling *bit. They serve to concentrate part of the weight of the drilling string near the bottom of the hole and to exert the necessary pressure on the bit, thereby preventing buckling of the upper part of the *string.

Drill pipe. Hollow pipe, normally made in 30-feet lengths, used in drilling.

Drill pipe elevators. A latched clamp, attached to the hoisting gear in a derrick and used to pull the *drilling string from the borehole.

Drilling fluid (mud). A fluid used in drilling wells. It is pumped down through the *drilling string to the bottom of the borehole, whence it rises to the surface through the space between drilling string and borehole wall. It serves the purpose of cooling the drill bit, removing drill cuttings supporting the borehole wall against collapse and stopping the entry of formation fluids into the borehole.

Drilling mast. A portable drilling structure. It may consist of two sections which are either telescoped or "jack-knifed", and lowered for transportation by truck and trailer.

Drilling platform. Any of several types of drilling units used in marine drilling operations. They fall into the following general categories:
 fixed platform with floating drilling tender;
 self-contained fixed platform (all equipment, pipe racks etc. on it);
 jack-up rig (supported on sea bottom by legs or spuds when in drilling position);
 floating platform (either "semi-submersible", or drill ship);
 submerged drilling barges.

Drilling string. The column of drill pipe and drill collars screwed together at the end of which the *bit is screwed, which is used as a drill in *rotary drilling.

Drop point. The temperature at which, when a grease is heated, a drop falls from the orifice in the bottom of the cup holding a sample of that grease.

Dry gas. Petroleum gas from which only insignificant quantities of liquid hydrocarbons condense during production.

Dry hole. A well which does not encounter hydrocarbons.

E

Electric logging – Electrical surveys. A family of wireline logging methods measuring electrical properties of rocks adjacent to the borehole wall. The electrical properties such as *spontaneous potential, formation resistivity/conductivity are used for correlation of rock strata, the determination of lithology and the calculation of formation water saturation in the pore space.

Electrical drilling. A drilling method, used to a certain extent in the USSR, whereby a *bit is rotated by a down-the-hole electric motor attached to the drill pipe or hanging from a cable in the borehole.

Electrical survey methods. Identification of subsurface rocks by measuring their resistance to electric currents.

Emulsifier. A substance used to promote or aid the emulsification of two liquids and to enhance the stability of the emulsion.

Emulsion. A dispersion of fine droplets of a liquid (the disperse phase) in the bulk of another liquid (the continuous phase) with which it is immiscible. A third substance, the *emulsifier, is sometimes necessary to keep the droplets dispersed as a stable emulsion.

Enhanced oil recovery (EOR). A process by which the yield of an oilfield is increased beyond that attainable by *primary recovery or by injection of water or natural gas. Fluids differing in their chemical or physical properties from those originally in the reservoir are injected. The commonly applied EOR methods fall into three categories:

Thermal, e.g. steam injection or *in-situ* combustion.

Chemical, e.g. injection of surfactant or polymer solutions

Miscible gas, e.g. injection of CO_2 or N_2 (at high pressures).

Ester. Compounds formed by the reaction between an organic or mineral acid with an *alcohol. For example, acetic acid and ethyl alcohol give ethyl acetate.

Esterification. The reaction of an *alcohol with an organic or mineral acid in the course of which water is eliminated and *esters are formed.

Ethane, C_2H_6. A colourless, odourless gas of the *methane series. Along with *methane one of the main constituents of *natural gas.

Ethene. The normalised name for *ethylene.

Ether. Organic compound in which two hydrocarbon *radicals are linked by an oxygen atom. The best known ether is $C_2H_5-O-C_2H_5$ (diethyl ether).

Ethyl chloride, C_2H_5Cl. An intermediate in the manufacture of tetraethyllead, an important *anti-knock additive for motor gasoline.

Ethyl fluid. Name for an *anti-knock compound, containing tetraethyllead and organic halides, such as 1,2-dibromoethane, serving as "lead scavengers".

Ethylene, C_2H_4. A hydrocarbon gas, the first member of the *olefin series. Important base material for the manufacture of resins and plastics (see Polyethylene).

Evaporite. Those sedimentary rocks deposited from a concentrated saline solution because of evaporation; includes *anhydrite, gypsum, rock salt.

Expectation curve. A graphical method of representing the uncertainty in an estimate of the value of any quantity, e.g. oil reserves. Any point of this curve gives the percentage chance that the individual value (e.g. reserves) will be obtained or exceeded.

Exploration. The search for undiscovered oil and gas, using geological and/or geophysical techniques, and by drilling exploration wells.
Exploration well. A well drilled to discover whether a previously untested trap contains oil or gas. Synonym: a wildcat.
Extract. The portion of an unrefined petroleum product (often a kerosine or a lubricating oil) resulting from a *solvent extraction process and consisting mainly of those components which are best soluble in the solvent. Generally the extract, after removal of the solvent, consists largely of *aromatic hydrocarbons. Those from lubricating oils are known in the paint trade as aromatic petroleum residues.
Extraction. A fractionation process based upon the difference in solubility, in a given solvent, of the various constituents of the mixture to be fractionated. The process is, for example, used in the separation of aromatics from gasoline or kerosine fractions.
Extractive distillation. A *distillation process characterized by the fact that the relative positions of the components' boiling points are influenced by the selection of an appropriate solvent. The process is, for example, used in the separation of butadiene from a mixture of *butanes, *butenes and *butadiene; solvent *acetonitrile.
Extreme pressure lubricants. A term applied to lubricating oils or greases which contain a substance or substances specifically introduced to prevent metal-to-metal contact in the operation of higher loaded gears and bearings. In some cases this is accomplished by the substances reacting with the metal to form a protective film.

F

F 1 octane number. See *Research octane number.
F 2 octane number. See *Motor octane number.
Fatty acids. *Aliphatic organic acids with straight alkyl chains; in combination with glycerol the fatty acids can constitute fats or *fatty oils.
Fatty oils. Oils which occur naturally in plants and animals. Typical vegetable oils are castor, rapeseed and olive oil; typical animal oils are lard, neat's-foot and whale oil.
Fault. A break in a body or layer of rock across which there has been vertical and/or horizontal displacement; e.g. normal, reverse, transcurrent, and thrust faults.
Fault trap. A *structural trap, favourable for the retention of oil, formed by a body of *reservoir rock bounded *up-dip by a fault. It is essential to the formation of the trap that the facing of the rock plane, up-dip, be sealed off by an impervious *formation.
FBP. Final Boiling Point. The maximum temperature observed on the distillation thermometer when a standard *ASTM or Engler distillation is carried out on gasoline, kerosine or gas oil.
Feed preparation unit. High-vacuum unit to split a *long residue into a *short residue and distillate fraction with a low metal content; the latter fraction is used as a cat. cracker feed.
Feedstock. Stock from which material is taken to be fed (charged) into a processing unit.
Fischer–Tropsch process. Term used for any one of several processes originating in Germany for producing hydrocarbons or their oxygenated derivatives from water gas or other mixtures of carbon monoxide and hydrogen.
Fishing. The retreiving of objects from the borehole, such as a broken *drilling string, parts of the *bit, or tools which may have fallen into the hole.
Flash distillation. The process of heating a liquid to a temperature within the *boiling range of the liquid which causes the evaporation of part of the liquid. The vapour may then be taken off and condensed.
Flash point. The lowest temperature under closely specified conditions at which a combustible material will give off sufficient vapour to form an inflammable mixture with air in a standardized vessel. Flash point tests are used to assess the volatilities of petroleum products.

Floating roof. A special tank roof which floats upon the oil. Applied to do away with the vapour space in storage tanks and reduce losses by *breathing and hazards of explosions.
Flotation. The process of treating a powder, such as a powdered metallic ore, with a liquid and blowing in air so as to cause a foam. Differences in the degree of wetting of the components by the liquid used (generally water containing some special chemical) cause the ore component of the powdered mineral to rise to the surface with the foam, while the impurities remain at the bottom.
Flow-bean. See Choke.
Flue gas. Gas from the combustion of fuel, the heating effect of which has been substantially spent and which is, therefore, discarded to the flue or stack. Its constituents are principally CO_2, CO, O_2, N_2 and H_2.
Fluidised catalytic cracking (FCC) process. A *cracking process whereby the finely divided *catalyst is continuously moved from *reactor to *regenerator and back to the reactor. The catalyst is kept in a fluid state by means of oil vapour, steam or air, in reactor, stripper and regenerator, respectively.
Fold. The bending of strata, usually as the result of compression.
Foraminifera. Simple animals (protozoans) living mostly in the sea, having a skeletal structure of calcium carbonate or other substance often preserved in sediment. They are often used for determination of geological age of rocks and their correlation.
Formaldehyde, HCHO. The first member of the class of organic substances known as aldehydes. It is made by oxidation of synthetic methanol.
Formation density logging. A wireline logging method using induced radioactivity (gamma rays) and measuring the decrease in strength of the induced gamma radiation. This decrease in strength is approximately proportional to the density of the material, which can be related to lithology and *porosity.
Fraction. A portion of petroleum separated from other portions in the fractionation of petroleum products. It is often characterized by a particular *boiling range.
Fractional distillation. See Distillation.
Fractionating column. An apparatus in which *fractionation is carried out. It consists of a vertical cylindrical metal vessel, containing equipment for the proper contacting of *flashed liquid and vapour. Heat is often supplied at the bottom of the column in a *reboiler, whereas heat is withdrawn at the top in a *condenser. Heat can also be supplied or withdrawn at intermediate heights of the column, if beneficial for the process (inter-heaters or intercoolers). The oil to be fractionated is fed into the column in one or more predetermined locations along the height of the column. The contacting equipment is formed by *fractionating trays in the oil and chemical industry in general, while for special applications packing material is used.
Fractionating trays. Equipment aimed at promoting contact between vapour and liquid in fractionation. For further information see *bubble cap trays, *calming section trays, *sieve trays and *valve trays.
Fractionation. The general name for a physical process of separating a mixture into its constituents, or into groups of these constituents, called fractions. Examples are: absorption, adsorption, azeotropic distillation, crystallisation, decanting, distillation, extraction, extractive distillation, flotation.
Freezing point. The temperature at which crystals first appear when a liquid is cooled under specified conditions. Freezing point is an important characteristic of aviation fuels.
Fuel cell. A device for generating electricity. In a fuel cell, chemical energy is directly converted to electrical energy by a process that is the reverse of electrolysis. A fuel gas is fed into one or two hollow porous electrodes in a liquid electrolyte whilst oxygen or air is supplied to the other electrode.
Furfural, $O(CH)_3CCHO$. A colourless organic compound — liquid at ordinary temperatures — having a *boiling point of about 160°C and a specific gravity of 1.16 g/cm^3. It is made by the action of dilute acids on bran, corn cobs and similar materials. It is used in an extraction process for upgrading of lubricating oils.
Furnace oil. A distillate fuel primarily intended for domestic heating.

G

Gamma ray logging. A wireline logging method measuring the variations in the natural gamma radiation resulting from the minute amounts of radioactive materials present in rocks. The log is used for correlation and the determination of lithology.

Gas cap. Free gas which is sometimes found in the highest part of a reservoir rock and overlies the oil.

Gas-cap-drive reservoir. A reservoir from which the oil is mainly displaced during production by expansion of the free gas present in a *gas cap.

Gas injection. Injection of gas into a reservoir to maintain the pressure in the producing formation to assist oil recovery. It is one of the methods of *secondary recovery.

Gas oil. A *distillate, intermediate in character between *kerosine and the light lubricating oils. It is used as a *heating oil, as a fuel for *compression-ignition engines and as feedstock for the manufacture of chemicals.

Gas/oil ratio. The volume of gas at atmospheric pressure produced per unit volume of oil produced.

Gas separator. Installation to separate *natural gas from the oil together with which it has been produced.

Gasoline. Light petroleum product, with a *boiling range between the approximate limits of 30 and 200°C. It is used as a fuel for spark ignition engines.

Gathering station. Oilfield installation which receives the production from several wells in its vicinity. It provides facilities to separate the gas and the water, to gauge the production of oil, gas and water, and to transport the oil to the main storage tanks.

Geochemical exploration. Method of exploration by analysing the gaseous content of near-surface soils or to detect the presence of bacteria which might indicate the presence of micro *seepages of hydrocarbon gases. Now rarely used.

Geophone. See Seismometer.

Geophysical exploration. Exploration by the use of geophysical methods, including the *gravimetric, magnetometric (*magnetic) and *seismic techniques, by which the internal geometry of the earth's upper crust is delineated.

Gilsonite. Substantially pure bitumen deposits found chiefly in Utah, USA.

Godevil. See Pig.

Graben. An elongate downthrown block between parallel faults. Antonym: horst.

Gravimeter. Instrument used to measure variations in gravity on or near the surface of the earth. Can be airborne or seaborne. A borehole gravimeter is also available.

Gravimetric method. Method of exploration based on measurements of variations in gravity.

Gravitometer. Instrument used for measuring changes in the specific gravity of oil flowing in a pipeline.

Gravity drainage. The mechanism by which oil within a reservoir segregates from the gas and flows downwards towards the producing wells under the action of gravity.

Grid. Common name for gas distribution network.

Gum formation. *Oxidation of gasolines may produce a sticky substance known as "gum". When unstable gasolines are stored for long periods the gum content may increase. Gum forming is retarded or prevented by using certain *inhibitors.

Gun perforation. Holes and perforations made by a "gun" or casing perforator through oil well *casing into the formation with a view to admitting the oil into the borehole. The holes are made by bullets or "shaped" charges which produce a high-powered jet of combustion gases and which are electrically fired from the gun suspended in the casing by a conductor cable from the surface.

Gypsum. See Anhydrite.

H

Heat exchanger. An apparatus for transferring heat from one fluid to another. Specifically, a piece of equipment having a tubular piping arrangement which effects the transfer of heat from a hot to a relatively cool material by conduction through the tube walls.
Heating oil. Any oil used for the production of heat.
Heaving. The habit of certain strata, notably clays and shales, to absorb water and so increase markedly in volume. Heaving clays will cause sloughing of the borehole wall and may close up the hole by plastic flow.
HF alkylation process. An alkylation process using hydrogen fluoride (HF) as a catalyst.
H_2SO_4 catalysed alkylation. An alkylation process using H_2SO_4 as a catalyst.
High vacuum unit. A unit for the production of vacuum distillates which can be further processed in a number of ways, e.g. as feed for cracking units, for the production of luboil, etc.
Homogeniser. A high-pressure mill in which the grease ingredients are broken up into minute size, thus increasing the ability of the grease to resist separation.
Horst. An elongate upthrown block between parallel faults. Antonym: graben.
Hot oil. Any oil used for the transfer of heat.
Hot-rolled asphalt. A very durable surfacing used for heavy duty roads. It consists of carefully graded sand, mineral filler and bitumen, to which a small percentage of stone chippings is added for greater stability.
Hydrate. A compound formed by the chemical union of water with a molecule of some other substance such as gypsum, from which water may be separated by a simple readjustment of the molecular structure. Gas hydrates, formed from water and, for example, *methane, may cause plugging of the tubing and flow lines of gas wells.
Hydration. The addition of water to a double bond, no breakdown of the molecular structure being involved.
Hydraulic fluids. Fluids used in the hydraulic systems of aircraft, industrial equipment, etc.
Hydrocarbons. Any organic compound, solid, liquid or gas, comprising carbon and hydrogen; e.g. coal, oil and natural gas.
Hydrodealkylation. A process to remove side-chains on aromatic molecules, either thermally or catalytically, under hydrogen pressure.
Hydrocracking. A process in which heavy distillate hydrocarbons are converted under hydrogen pressure into products of lower molecular weight, in the presence of an acidic catalyst.
Hydrodesulphurisation. The elimination of sulphur from sulphur-containing chain molecules in crudes or distillates by the action of hydrogen under pressure over a catalyst.
Hydrofinishing. A process in which hydrogen is employed over a catalyst to improve the properties of low-viscosity-index naphthenic (LVIN) and medium-viscosity-index (MVIN) oils to improve their properties; also applied to paraffin waxes and microcrystalline waxes for the removal of undesirable components.
Hydroformylation. The addition of a hydrogen atom and a formyl group to the molecule of a compound containing a double bond by reaction with hydrogen and carbon monoxide, the main product being one or more aldehydes.
Hydrogenation. The filling of the "free" places in *unsaturated structures by hydrogen atoms.
Hydrolysis. The decomposition of a molecular structure by the action of water.
Hydrometer. Instrument for measuring the specific gravity of oils.
Hydrophone. See Seismometer.
Hydrostatic head. The pressure exerted by a column of fluid, equalling the height of the column times the fluid density times the acceleration of gravity.
Hydrotreating. Usually refers to the hydrodesulphurisation process (*q.v.*) but may

sometimes be applied to other treating processes using hydrogen. (See also Trickle flow process.)

Hypoid gear. A combination of the spiral bevel and worm type which is very quiet in operation. The motion of the teeth is a comnation of rolling and sliding, causing high loading pressure on the tooth faces together with high rubbing speeds, and so demanding exceptional qualities of the lubricant (extreme pressure lubricants.)

Hysomer process. Shell's hydro-isomerisation process for pentane/hexane isomerisation (See TIP).

I

IBP. Initial boiling point, i.e. the temperature at which the first drop of distillate appears after commencement of distillation in the standard *ASTM laboratory apparatus.

Igneous rocks. Rocks which have been consolidated from hot liquid material (*magma).

Ignition quality. A measure of the ignition delay of a fuel in a diesel engine.

Inhibitor. A substance the presence of which in small amounts in a product prevents or retards undesirable changes in the quality of the product, or in the condition of the equipment in which the product is used. In general, the essential function of inhibitors is to prevent or retard oxidation. Examples of uses include the delaying of *gum formation in stored gasolines and of colour change in lubricating oils; also the prevention of corrosion, e.g. rust prevention by inhibitors in turbine oils and fuels.

Injection well. A well drilled for the specific purpose of injecting gas, water, steam or other chemicals into an oil (or occasionally gas) reservoir as part of a *secondary or *tertiary recovery scheme.

Injector. A mechanism which may be used in different forms for spraying fuel oil into the combustion chamber, or for feeding water into steam boilers.

In-line blending. A system in which all components are pumped simultaneously into a common discharge pipe (header) at rates of flow corresponding to the required proportions, the rates of flow being controlled. Blending takes place in the lines between the header and the storage tank into which the blend is discharged.

Intelligent pig. A device which can pass through pipelines and is fitted with appropriate sensors and telemetry or recorders which can survey for corrosion and other defects.

IP. Institute of Petroleum, the organisation in Great Britain primarily responsible for the advancement of the study of petroleum and its allied products in all their aspects. It is the recognised British standardisation authority for methods of testing petroleum products.

Isomer. Two substances composed of equal amounts of the same elements but differing in properties owing to variation in structure are called isomers.

Isomerisation. The conversion of a compound into its *isomer. For example, *butane may be converted into isobutane.

Isooctane, C_8H_{18} (2,2,4-trimethylpentane). A colourless liquid used with n-heptane to prepare standard mixtures to determine *anti-knock properties of gasoline.

J

JP-4, -5 and -8 type fuels. See AVTUR, AVTAG and AVCAT.

K

Kelly. Hollow, 40 feet long, square or hexagonal pipe attached to the top of the *drilling string and turned by the *rotary table during drilling. It is used to transmit the torque or twisting moment from the rotary machinery to the drilling string and thus to the bit.

Kerogen. The organic matter in *source rocks mainly derived from bacteria, algae or plants from which oil and gas are derived under appropriate conditions of temperature and pressure.

Kerosine. Any petroleum product with a *boiling range between the approximate limits of 140°C and 270°C which satisfies certain quality requirements (for lamp oil or jet fuel).

Ketones. A class of chemical compounds containing the group CO between two alkyl groups which are much used in industry as solvents and in the manufacture of certain artificial resins.

Killing a well. Overcoming the tendency of a well to flow by filling the wellbore with *drilling fluid of suitable density.

Knot. A unit of speed, equivalent to one nautical mile (6,080.20 feet) per hour.

L

Lake asphalt. Natural asphaltic material occurring in surface deposits.

Latex. A milk-like fluid in which small globules of particles of natural or synthetic rubber or plastic are suspended in water.

Lean oil. Absorption oil as it enters an absorber; see *Absorption process.

LHSV. Liquid hourly space velocity.

Light distillate. A term applied to distillates the final boiling point of which does not exceed 300°C.

Light ends. The lower-boiling components of a mixture of hydrocarbons.

Light tops. See Naphtha.

Liner. A *string of pipe suspended near the bottom, from the deepest *casing of a well, which may be perforated for production.

Liquefied natural gas (LNG). Natural gas can be liquefied, e.g. at atmospheric pressure by cooling to about $-160°C$ ($-256°F$).

Liquefied petroleum gas (LPG). Of the gaseous hydrocarbons, propane and the butanes can be liquefied under relatively low pressure and at ambient temperature and are then known as liquefied petroleum gas.

Lithology. The study of rocks (from Gr. *lithos* = stone).

LMO. Light machine oil.

LNG. *Liquefied natural gas.

Log. A detailed record of rock properties in a borehole with depth. The mud log is a record provided by a number of sensors installed on the surface supplemented by written information. The wireline log is a record obtained by lowering into the borehole a tool with a number of sensors on a steel cable containing several electrical conductors.

Long residue. The residue resulting from the atmospheric distillation of crude oil.

Load-on-top system. System of cleaning the tanks of a crude oil tanker by collecting washings from each tank in one tank, allowing the water to separate from the oil, then discharging the water overboard, leaving the oil residues in the tank. The next crude oil cargo is loaded on top of the residues.

LPG. *Liquefied petroleum gas.

Luboil. Lubricating oil

LVI. Low viscosity index; see Viscosity index.

M

Macadam. A collection of broken stone or similar material arranged in such a manner as to form a road surface in which the fragments of solid are interlocked and mechanically bound to the maximum possible extent.

Magma. Mobile, preponderantly liquid rock material generated within the earth and from which *igneous rocks are formed by solidification. It is capable of intrusion into the upper strata of the Earth's crust or of extrusion through volcanoes at the earth surface.

Magnetic method. Method of exploration based on measuring the intensity and direction of the earth's magnetic field and inferring the distribution of rocks possessing different magnetic properties from local variations in this field.

Magnetometer. An instrument used for measuring the intensity and direction of the earth's magnetic field. Commonly airborne or seaborne.

Mantle. That portion of the earth's interior which underlies the *continental and *oceanic crust and extends to a depth of about 3000 km.

Marine drilling. See Drilling platform.

Mastic asphalt. A mixture of bitumen, fine limestone aggregate and filler used in road construction, house building and hydraulic works.

MAV. Maleic Anhydride Value, a figure related to the content of conjugated dienes (dienes are most offensive for gum formation).

MCPA. 2-Methyl-4-chlorophenoxyacetic acid, used as a weedkiller.

Medicinal oils. Mineral oils drastically refined so that all unsaturated constituents, and those which impart colour and odour to the oil, have been removed.

Mercaptans. Mercaptans, thiols or alkyl-hydrosulphides are organic compounds of carbon, hydrogen and sulphur. They have a bad odour and frequently occur in unrefined gasoline. Mercaptans must be removed from gasoline or converted to the unobjectionable disulphides by suitable treating (*sweetening).

Merox process. A Universal Oil Products' fixed-bed process for sweetening mercaptans present in kerosine fractions, utilising an oxidation catalyst.

Metamorphic rocks. Rocks derived from pre-existing, igneous or sedimentary formations by mineralogical, chemical and structural changes, essentially in the solid state, in response to changes in temperature and pressure at depth within the earth's crust.

Methane CH_4. A light, odourless, inflammable gas. It is the chief constituent of *natural gas. It is also often produced by the partial decay of plants in swamps, so that its occurrence is not uncommonly misinterpreted by the layman as an indication of the presence of petroleum.

Methane series. A homologous series of open-chain saturated hydrocarbons of the general formula C_nH_{2n+2} of which methane (CH_4) is the first member and the type; generally called the paraffins.

Methanol. Methyl alcohol, CH_3OH. The first member of the class of organic compounds known as alcohols. It is a liquid boiling at 66°C. Methanol is inflammable and poisonous. It is used in the production of *formaldehyde, a chemical used in the manufacture of *thermosetting resins and of the thermoplastic polyacetal "Delrin".

Microcrystalline waxes. Waxes having a very fine crystal structure, and consisting mainly of iso- and cycloparaffins with some aromatics. They are produced mainly from heavy lubricating oil residues and have melting points from 60 to 90°C.

Migration (hydrocarbons). The movement of liquid and gaseous hydrocarbons from their *source rocks to *reservoir rocks.

Migration (seismic). The process by which dipping seismic events are moved to their true spatial position.

Mixed-base crude oils. Crude oils which contain both *paraffin wax and asphaltic matter in quantity.

MOHO (Mohorovicic discontinuity). The boundary surface between the earth's crust and the underlying mantle, which is marked by a sharp seismic-velocity discontinuity.

Mono-aromatics. See Aromatics.

MON. See Motor octane number.

Motor octane number (MON). The *octane number of a motor gasoline determined in

a special laboratory test engine under high "engine-severity" conditions, giving a rough measure of the high-speed knock properties of the gasoline.
Mud pump. Pump used in *rotary drilling for circulating the *drilling fluid.
Mud screen. Vibrating screen over which the *drilling fluid is conducted. The drill cuttings are retained on the screen, while the fluid passes through the meshes.
Multiple completion. The *completion of a single well producing from more than one oil/gas-bearing zone.

N

Naphtha. Naphthas are straight-run gasoline fractions boiling below kerosine. Being generally unsuitable as a blending component of premium gasolines, they are often used as a feedstock for catalytic reforming. Other important outlets for naphthas are their use as chemical feedstock (e.g. ethylene manufacture) and — mainly in the past — as feedstock for town gas manufacture.
Naphthenes. A class of saturated cyclic hydrocarbons of the general formula C_nH_{2n}.
Naphthenic acids. The accepted term for "petroleum acids", organic acids characterised by the presence of a naphthene ring and one or more carboxylic acid groups. Naphthenic acids are valuable by-products used in the manufacture of paintdriers, emulsifiers and cheap soaps.
Natural gas. The gas which occurs naturally with crude oils, but also in *reservoirs which contain only a few heavier constituents. It consists mainly of the lighter paraffin hydrocarbons. Natural gas is usually classified as wet or dry, depending on whether the proportions of gasoline constituents which it contains are large or small. Most gas reaches the surface through the *tubing, but in some pumping wells it is taken off at the top of the *casing (casinghead gas).
Natural gasoline. Gasoline extracted from wet *natural gas, consisting of *butane, pentane and heavier hydrocarbons. After *stabilisation — the removal of the lighter components — the gasoline is suitable for *blending into motor gasoline.
Neutron logging. Wireline logging methods using induced radioactivity created by a neutron source lowered in the borehole. The methods use either the gamma radiation or the neutron response of the rock on the induced radioactivity. It is used for the determination of lithology, porosity and formation fluid saturation.
Nitric acid, HNO_3. Nitric acid is made by the high temperature oxidation of ammonia over a platinum or platinum/rhodium catalyst. It has many uses in industry and some of its salts are used as nitrogenous fertilisers.
Non-associated natural gas. Gas accumulations which exist independently of any oil accumulation.

O

Obduction. The process by which *oceanic crust and ocean-floor sediments are emplaced over the margin of an adjacent plate (sometimes comprising continental crust) undergoing *subduction.
Oceanic crust. The type of earth's crust that underlies the oceans. Extruded at mid-ocean ridges from deep within the earth. With a density of about 3.0 g/cm³, it is heavier than continental crust.
Octane number. The octane number of a fuel is a number equal to the percentage by volume of *isooctane in a mixture of isooctane and normal heptane having the same resistance to *detonation as the fuel under consideration in a special test engine. It is a measure of the *anti-knock value of a gasoline and the higher the octane number the higher the anti-knock quality of the gasoline.

Offshore drilling. See Drilling platform.
Oil shale. A compact *sedimentary rock consisting mainly of consolidated calcareous muds and clays and containing organic matter (kerogen) which yields oil when heated.
Oil trap. The accumulation of oil in a reservoir rock under such conditions that its escape is prevented. Sealing is effected by an impervious caprock, or by the juxtaposition across a fault of an impermeable layer. Oil traps can be broadly divided into *structural and *stratigraphic traps.
Oiliness. The property of an oil to reduce the coefficient of friction under boundary conditions. *Fatty oils and *fatty acids are examples of substances possessing this property to a high degree.
Olefins. A class of *unsaturated, non-cyclic, *aliphatic hydrocarbons of the general formula C_nH_{2n} (mono-olefins) C_nH_{2n-2} (di-olefins), etc. *Ethene is the parent member of this group.
Outcrop. That part of a rock unit occurring at the surface of the earth.
Outstep well. A well drilled beyond the proved limits of a producing field in order to investigate a possible extension of the oil accumulation. See Appraisal well.
Oxidation. The reaction of oxygen with a molecule that may or may not already contain oxygen. Oxidation may be partial, resulting in the incorporation of oxygen into the molecule or in the elimination of hydrogen from it, or it may be complete, forming carbon dioxide and water (combustion).
Oxidizing flame. Term applied to a flame in which there is an excess of air or oxygen.

P

Palaeontology. The branch of science which deals with the fossil remains of animals and plants of the geologic past. The word is now mainly used for the study of fossil animals. Plants are dealt with under the heading of palaeobotany (from Gr. *palaios* = ancient).
Palynology. The study of the fossil pollen and spores of plants (from Gr. *palynein* = to disperse). It is a branch of palaeontology.
Paraffin-base crude oils. Crude oils which contain *paraffin wax but little or no asphaltic matter.
Paraffins. See Methane series.
Paraffin wax. Wax of solid consistency having a relatively pronounced crystalline structure, extracted from certain distillates from petroleum, *shale oil, etc. Refined paraffin wax has a very low oil content; it is white with some degree of translucency, almost tasteless and odourless and slightly greasy to the touch.
Partial pressure. Partial pressure of a component of a mixture in vapour–liquid equilibrium is that part of the pressure which is contributed by that component.
Pay zone. The *reservoir rock in which oil and gas are found in commercial quantities.
Peak shaving. Any of several methods of arranging production and storage of natural gas to deal with demand variations in the most economic way.
Penetration. A measure of the hardness and consistency of *asphaltic bitumen and lubricating greases in terms of the distance in tenths of a millimetre by which a weighted special cone or needle will penetrate the sample in five seconds, the temperature, unless otherwise stated, being $25°C$ ($77°F$).
Permeability (geological). The measure of a rock's faculty to allow a liquid or gas to flow through it. Normally expressed in millidarcies.
"Petroil" mixture. A lubricating system for small two-stroke gasoline engines, in which the lubricant is mixed in suitable proportions with the gasoline to make a "petroil" mixture. During its passage through the engine some of the heavier and unevaporated petroil fractions are deposited on bearing surfaces and so provide lubrication.
Petrolatum. A semi-solid material which is obtained from petroleum and consists essentially of *microcrystalline waxes in association with oil. Also called petroleum jelly.
Petroleum coke. A by-product of a special form of the thermal cracking process (coking).

If the feed to the unit is low in sulphur and metals the coke can be used for making carbon electrodes for the electro-chemical industry. Alternatively it can be sold as a cheap fuel.

Petroleum wax. See Crude wax.

Petrology. The study of the mineral and chemical composition of rocks (from Gr. *petra* = rock).

Phenol C_6H_5OH (carbolic acid). At one time wholly obtained as a by-product of coal tar manufacture, phenol is now made synthetically in various ways, starting from benzene. It is an important intermediate for the manufacture of synthetic resins, nylon, herbicides and disinfectants.

Phthalic acid, $C_6H_4(COOH)_2$. Aromatic carboxylic acid of which three isomers exist, viz. *ortho-*, *meta-* and *para-*phthalic acid. They are made by oxidation of *xylenes. The first two are important base materials for *alkyd resins, and the third, also called terephthalic acid, is used in the manufacture of the synthetic fibre Terylene.

Pig. (a) A device used for cleaning out a pipeline, consisting of a piston-type scraper usually pumped through the line. (b) A device used for the separation of two liquids which are pumped one after the other through the same pipeline (see intelligent pig).

Pilot plant. A small version of full-scale plant in which a laboratory pursues development of a new process, after bench-scale investigation has shown promise.

Plasticisers. Non-volatile liquids or low-melting solids which, when added to another material, change certain physical and chemical properties of that material, mainly imparting greater toughness, improved stability and increased flexibility.

Platforming. A *reforming process which makes use of a *catalyst containing platinum. *Catalytic reforming of *straight-run heavy gasoline produces a product which is richer in *aromatics and branched-chain paraffins and poorer in *naphthenes and straight-chain paraffins. The hydrogen produced in this process can be used for *hydrodesulphurisation or *hydrocracking.

Polyaddition. Polyaddition is the formation of a macromolecule either from monomers with reactive groups other than a carbon double bond in the molecule or by addition of active monomer molecules to such reactive groups.

Polyethylene, $(C_2H_4)_n$ (polymerised *ethylene). A group of thermoplastic resins used for making film and thin sheet, and extruded and moulded forms including squeeze bottles, domestic articles, piping and insulators for electrical equipment.

Polymer. A substance produced from another by *polymerisation.

Polymerisation and copolymerisation. The combination of a number of unsaturated molecules of the same or different compounds to form a simple large molecule, called a polymer when it is built up from a number of identical monomers, and a copolymer when it is a combination of two or more different types. In the oil industry the aim is making polymers from 2 to 4 monomers boiling in the gasoline range. In the chemical industry polymers made usually contain thousands of molecules and are used as plastics.

Polyvinyl chloride (PVC), $(CH_2=CHCl)_n$. A type of *thermoplastic resin which is made into sheet, film, foil, extruded form and moulded forms; it is also used as a coating compound applied to other materials.

Porosity (geological). That proportion of a rock volume that is occupied by the voids (pore spaces) between mineral grains. Expressed as a percentage of the total (bulk) rock volume. It is the pore spaces in a reservoir rock which contain the oil, gas and *connate water.

Pour point. The temperature below which an oil tends to solidify and will no longer flow freely.

Primary recovery. Production of oil under natural drive mechanisms, i.e. the oil is displaced from the reservoir by expansion of the fluids (gas cap gas, solution gas and *aquifer water) naturally present in the reservoir. There is no fluid injection of any kind.

Process integration. A term denoting the selection and arrangement of refinery processes making optimum use of the heat contained in the various plant streams.

Production string. A *string of pipe, set inside the last casing string of a completed well, and through which formation fluids are brought to the surface.
Promoter. A substance which may considerably increase the activity of a *catalyst. For example, the catalytic action of iron is greatly increased when the catalyst contains a small amount of oxides of aluminium or silicon, etc.
Propane, C_3H_8. A hydrocarbon of the paraffin series used for heating, welding and metal cutting. At ambient temperature it can be stored under pressure as a liquid.
Propane deasphalting. A process in which a short residue is split into an oil having a low asphaltic content and asphaltic constituents, by means of liquid propane.
Propene. The normalised name for *propylene.
Propylene, C_3H_6. A hydrocarbon of the *olefin series. Important base material for the chemical industry.
Prorationing (USA). Restriction of production by a state regulatory commission usually on the basis of market demand.
PVC. *Polyvinyl chloride.
Pyrolysis. A severe form of thermal cracking.
Pyrolysis gasoline. A byproduct of high-temperature (700–900°C) thermal-cracking processes aiming primarily at ethylene manufacture.

Q

Quenching oils. Specially refined high-flash mineral oils used for hardening alloy steels.

R

Radical. A group of atoms acting in chemical reactions as a unit which is replaced or introduced into a new compound without rearrangement of the atoms. It can only exist alone as a separate compound under very special conditions (for very short periods).
Radioactivity logging. A family of wireline logging methods measuring either natural or induced radioactivity of the rocks around the borehole. The forms of radioactive logging are: *gamma ray logging, *formation density logging and *neutron logging.
Raffinate. The product resulting from a *solvent extraction process and consisting mainly of those components that are least soluble in the solvent.
RDC. Rotating disc contactor. An apparatus consisting of a vertical cylinder divided by horizontal plates into a number of compartments in which discs rotate on a shaft. One of the uses of this device is to achieve better mixing of oil and solvent in extraction processes. For example: *furfural extraction of lubricating oil.
Reactor. Term applied to the part of a plant where a chemical reaction takes place.
Reboiler. A special type of *heat exchanger for the supply of heat to the bottom of *fractionating columns.
Recovery factor. That fraction of the original oil or gas-in-place that is expected to be ultimately withdrawn from a reservoir.
Rectifying absorption. An *absorption process combined with *stripping of the fat oil. Other names for this process are: rectified absorption, fractionating absorption.
Recycling. (a) Re-injection into a gas reservoir of the produced gas after extraction of the condensate. (b) Continuously feeding back part of a substance obtained or used in a process for further processing or use.
Reduction. A reaction whereby the number of oxygen atoms in a molecule is reduced or that of hydrogen is increased.
Redwood (seconds). See Viscometer.
Reflection shooting. A *seismic method of *exploration based on the principle that the

energy waves caused by a shock at or near the earth's surface are reflected at the boundaries between strata of differing density at different depths.

Reflux. A part (if the top product is in the liquid state) or all (if the top product is in the vapour phase) of the condensed top vapour of a *fractionating column, which is returned to the top of the column. The purpose is to create an extra downward flow of liquid; if properly applied this liquid acts as an absorbing agent for the relatively heavy components which are thus rejected from the top product.

Reforming. The operation of modifying the structure of the molecules of straight-run gasoline fractions in order to improve anti-knock quality. It can be achieved thermally (*thermal reforming) or with the aid of a catalyst (*catalytic reforming).

Refraction shooting. A *seismic method of exploration based on the principle that energy waves travel more rapidly in consolidated rocks than in less consolidated rocks. By measuring the speed and arrival times of those waves that have travelled nearly parallel to the bedding, the depths and inclinations of strata can be calculated.

Regenerator. Term applied to the part of a *catalytic cracking unit where the spent catalyst is regenerated by burning off the coke.

Reid vapour pressure (RVP). The pressure caused by the vaporized part of a liquid and the enclosed air and water vapour, as measured under standardised conditions in standardised apparatus: the result is given in pounds per square inch at $100°F$, although normally reported simply as "RVP in lb". RVP is not the same as the true *vapour pressure of the liquid, but gives some indication of the *volatility of a liquid, e.g. gasoline.

Relief well. Directional well, drilled to intersect a well that is blowing out, through which heavy drilling fluid is pumped down to "kill" the blow-out well.

Repressuring. The injection of fluids (usually *natural gas or water) into a reservoir to restore reservoir pressure towards its original value. (See secondary recovery).

Research octane number (RON). The *octane number of a motor gasoline determined in a special laboratory test engine, under mild "engine-severity" conditions, giving a rough measure of the low-speed knock properties of the gasoline.

Reserves. The volumes of oil or gas remaining in a reservoir at any given time which are expected to be producible, i.e. the total recoverable hydrocarbons originally in the reservoir minus the cumulative production to the time of the reserve estimate.

Reserves have traditionally been divided into three categories:

Proved reserves. The estimated quantities of hydrocarbons which geological and engineering data demonstrate into reasonable certainty to be recoverable in the future from known reservoirs under existing economic conditions.

Probable reserves. Quantities of hydrocarbons which are considered to have an overall 50% chance of being present and producible.

Possible reserves. Quantities of hydrocarbons which are considered to have on aggregate a 25% chance of being present and producible.

In the above system, the total "discounted reserves" are defined as "proven" + 0.5 × "probable" + 0.25 × "possible".

The recent trend has been to quote reserves in terms of an "*expectation curve" characterised by a "proven" value corresponding to an 85% or 90% cumulative probability and an "expectation" value corresponding to the area under the curve.

Reservoir rock. Porous rock units containing interconnected pores or fissures and voids which may contain oil or gas. Most commonly sandstone, limestone or dolomite.

Residual fuel oil. Fuel oil consisting mainly of long, short or cracked residue (in contrast to: *distillate fuel oil).

Residue (long, short, cracked). The bottom product from the atmospheric distillation of crude oil is called *long residue. If this residue is further distilled under vacuum a still heavier residual product results, which is called *short residue. The heavy residual product from thermal cracking operations is called cracked residue.

Residue hydroconversion. A further development to *residue hydrodesulphurisation, for the purpose of converting residual fractions into feedstocks which, after further processing, yield lighter products. Another application is the upgrading of tar sands, bitumens and heavy oils.

Residue hydrodesulphurisation. A catalytic hydroprocess for removing sulphur from the residual fractions of certain crudes. It operates at high hydrogen partial pressures and uses special catalysts.

Resistivity. Also called specific resistance. The electrical resistivity of a "formation" — which depends upon the type of rock and the fluid content of its pore space — is one of the properties recorded by *electric logging.

Rheology. The science dealing with the phenomena of flow and change of shape of matter under pressure.

Rig. The *derrick and surface equipment of a drilling unit.

Ring compounds. Organic compounds in which the atoms of a molecule are arranged so as to form at least one closed ring, for example, naphthenes and aromatics. Also called cyclic compounds.

Road octane number. The *octane number of a motor gasoline determined during actual road testing. Apart from the intrinsic quality of the gasoline tested, the Road Octane number depends also on the make of the engine of the vehicle but in general lies in between the *RON and *MON of the gasoline.

Rock bit. A drilling *bit used for hard formations.

RON. See Research octane number.

RON 100. The research octane number of the gasoline fraction distilled off at 100°C.

Rosin oil. Oil obtained by distillation of rosin and varying in colour from almost colourless to dark brown. The oil is used in grease making and in the manufacture of printer's ink.

Rotating disc contactor. See RDC.

Rotary drilling. Drilling procedure based on rotating a *bit and *drill pipe and use of a mud circulating system.

Rotary table. Chain or gear-driven circular table that rotates the *drill pipe and *bit.

Round trip. Pulling out and running in the *drilling string.

S

SAE. Society of Automotive Engineers (USA).

SAE classification. The *SAE devised a system for the classification of motor oils and transmission oils. It is based on the viscosity at 0 or 210°F.

Salt dome. A *diapiric or piercement structure with a central, vertical plug of salt. Related structures: salt pillow, salt wall.

Salt pillow. A pillow-like swell of diapiric salt uplifting the overlying strata; an embryonic salt dome.

Salt wall. A long linear *diapiric piercement structure with steep flanks; probably activated by movement of an underlying linear fault.

Sand mixes. Used in road-surfacing work in a variety of ways — hot sand mix, cutback sand mix, and wet sand mix — depending on climate and soil condition.

Saponification. The splitting of fats or *fatty oils by alkali to form soaps. The term is sometimes applied to the neutralisation of *fatty acids with alkali.

Saybolt. See Viscosimeter.

SBM. Single buoy mooring.

SBPs. Gasoline fractions, distilled to specially selected boiling ranges ("Special Boiling Point") and subsequently refined.

SBR. Styrene–butadiene rubber.

Scale wax. The paraffin obtained from slack wax or waxy distillate by deoiling. It contains up to 2 per cent of oil. Crude scale paraffins are yellow in colour.

Scavengers. Chemical additives which remove or inactivate impurities or undesired materials in a mixture or process.

SCOT process. A Shell process for treating the tail gas or off-gas from a Claus unit (see Claus process) to increase the overall sulphur recovery to as high as 99.9% and thus minimise emissions of sulphur compounds.

Sea-floor spreading. A hypothesis in which upwelling *magma is extruded at mid-ocean ridges to create new *oceanic crust. If, as a result, the ocean floor increases in area, attached continents drift passively away from the axis of spreading; if there is no increase in area, then older oceanic crust is consumed (subducted) in down-going subduction trenches. (See Continental drift.)

Secondary recovery. Displacement of oil from the reservoir by injection of water or gas through special *injection wells.

Sedimentary rocks. Rocks formed by the accumulation on land or in water of mineral or skeletal particles. They can be transported by air or by water as discrete particles (e.g. sand grains) or originate by chemical precipitation from water (e.g. rock salt). Sedimentary rocks generally have a layered structure, known as bedding or stratification.

Seepage. A naturally occurring escape of crude oil or gas to the surface.

Seismic methods. Methods of geophysical prospecting in which shock waves are generated at or near the surface by an explosion or other energy sources. The returning waves are picked up by detectors placed at increasing distances from the energy source. Two methods can be distinguished: *reflection shooting and *refraction shooting.

Seismogram. A graphic record of the vibrations recorded by *seismometers.

Seismometer. Instrument used in seismic surveys to detect energy waves. It is called a geophone when used on land and a hydrophone when used in water. Arrays (spreads) may be planted in the soil or towed behind a ship.

Self potential. See Spontaneous potential.

Separation processes. Manufacturing processes based on differences in the physical properties of the components of a mixture. See *Fractionation.

Setting point test. Laboratory test determining the temperature at which solidification of a molten wax begins.

Settling tank. A tank employed for separating two liquids which are not miscible. If the liquids do not form an *emulsion, they separate into layers according to their specific gravities, and these layers can be drawn off from different levels in the tank.

Shale oil. See Oil shale.

Shell Method Series. A private publication of the Group setting out special methods to be used in carrying out laboratory tests on petroleum products.

Side stripper. A fractionating column for *stripping undesired volatile components from a side stream which is drawn off as a liquid from a main fractionating column. Various fractions may be drawn off from one main column and be stripped in as many side strippers.

Sidetracking. Difficult *fishing operations sometimes necessitate deflection of the borehole to bypass the "fish". The operation, which is carried out by means of a *whipstock and special drilling tools, is known as sidetracking.

Sieve trays. *Fractionating trays in the form of a perforated plate, i.e. with holes for vapour passage.

Slack wax. See Crude wax.

Sludge. (a) Acid sludge or acid tar: material formed during refining of oils with sulphuric acid. (b) Engine sludge: insoluble product formed from fuel combustion products and from lubricating oils in internal combustion engines and deposited on parts outside the combustion space. (c) Tank sludge: material collected at the bottom of oil storage tanks.

Slush pump. See Mud pump.

Smoke point. The maximum height of flame measured in millimetres at which a *kerosine will burn without smoking when tested in a standard lamp for this purpose.

Smoke Point Improvement (SPI) process. Hydrotreating process for improving the smoke point of a kerosine fraction by converting aromatics into naphthenes.

Solvent dewaxing. The removal of *paraffin wax from waxy luboil feedstocks, usually by two solvents, toluene and methyl ethyl ketone. The process has three stages: mixing the oil with the solvents and chilling; filtration of the chilled oil to separate the wax; recovery and recycling of the solvents.

Solvent extraction. See Extraction.

Sonde. A downhole tool containing sensors, transducers and electronic equipment used in wireline logging to measure physical properties of the rocks around the borehole (see Log).

Sonic log. See Acoustic velocity log.

Sour crude. Crude oils containing sulphur and sulphur compounds.

Sour gas. Gas which contains contaminants e.g. hydrogen sulphide and other corrosive sulphur compounds.

Sour gasoline. Gasoline fractions which contain a certain amount of *mercaptans and therefore must be *sweetened.

Sour water. Water which contains objectionable amounts of dissolved contaminants, e.g. hydrogen sulphide, ammonia, phenols, etc.

Source rock. The sedimentary rocks containing organic material (*kerogen) from which hydrocarbons are formed.

Spill point. The point of maximum fill (retention) of oil or gas in a structural trap.

Spindle oils. Low-viscosity lubricating oils, produced as distillates, with a viscosity of 20/60 cS (*centistokes) at 20°C.

Spontaneous potential (SP). A record with depth of the spontaneous electrical potential resulting from the electrochemical effects due to the contact of drilling fluid with formation fluids of a different salinity present in permeable rocks.

Spread. Arrangement of *seismometers (geophones) in relation to the shot point in *seismic survey methods.

Spudding in. To commence drilling operations by making the first part of a hole.

SSU. See Viscosimeter.

Stability. Resistance of petroleum products to chemical change. Gum stability means the resistance of a gasoline to *gum forming while in storage. Oxidation stability means that the product is stable to oxidation, i.e. resists the action of oxidation which forms gums, sludges etc.

Stabilised gasoline. Gasoline after subjection to fractionation by which the vapour pressure has been reduced to a specified maximum.

Stabiliser. A *fractionating column designed to make a sharp separation between more volatile components and gasoline fractions, thus controlling the gasoline's *Reid vapour pressure.

Stands. Connected joints of *drill pipe racked in the *derrick when a *round trip is being made to change the *bit.

Static electricity. The electricity generated by the relative movement of unlike materials such as oil/pipeline, oil/water, plastic granules/vessel; or by the operation of equipment such as driving belts.

Stereospecific rubbers. A group of synthetic rubbers based on the polymerisation of diolefins such as butadiene by catalysts of the Ziegler/Natta type.

Stove oil. A *distillate fuel primarily intended for burning in kitchen stoves.

Straight-run. A term applied to a product of petroleum made by *distillation without chemical *conversion.

Strapping. The measurement of the external diameter of a cylindrical tank by stretching a steel tape around each course of the tank's plates and recording the measurement.

Stratigraphic trap. The type of trap where the hydrocarbons are enclosed as a result of a change in the rock from porous to non-porous (impervious), rather than by structural bending or faulting of rock layers; commonly occurs as result of an up-dip wedge-out of the reservoir rock. (Compare with Structural trap).

Stratigraphy. That branch of geology that deals with formation, composition sequence and correlation of stratified rocks.

Strike. The direction in which a horizontal line can be drawn on a dipping plane. It is at right angles to the maximum dip direction of a bed.

String. Name originally given to the suspended cable and tools of the *cable tool drilling method, but now applied equally to strings of drill pipe, casing, tubing, etc. in *rotary drilling. *Bit, *drill collars and *drill pipe are *drilling string" items.

Stripping. A *fractionation process, closely related to *distillation by which undesired volatile components are separated from a liquid mixture by fractional evaporation. The desired fraction is thus purified from lower-boiling components. Stripping is generally effected by the use of steam, by the reduction of pressure, by the vapour generated in a *reboiler or by a combination of these. In the laboratory nitrogen is often used as stripping agent.

Structural traps. Traps resulting from some local deformation, such as folding, *faulting, or both, of the *reservoir rock and *cap rock.

Styrene, $C_6H_5CH=CH_2$. A chemical made from benzene and ethylene by alkylation and dehydrogenation. It is used as base material for polystyrene plastics, synthetic rubber, etc.

Subduction. The process by which one crustal block descends beneath another. Antonym: obduction.

Success ratio. The ratio of the number of successful exploration wells drilled in a certain area or venture to the total number of exploration wells drilled there.

Sulfinol process. A process for removing contaminants such as hydrogen sulphide, carbon dioxide, mercaptans, carbonyl sulphide from gases by contacting with a regenerable solvent. The Sulfinol solvent is a three-component solvent combining the chemical properties of an alkanolamine solvent with physical properties of *Sulfolane and water.

Sulfolane, $(CH_2)_4SO_2$. A chemical which is increasingly being used as solvent in *extraction and *extractive distillation processes.

Sulphation. The reaction of an *unsaturated chain hydrocarbon (*olefin) with sulphuric acid. An *ester is produced in which the hydrocarbon group and the sulphur atom in the sulphuric acid are linked by an oxygen atom.

Sulphonation. The action of concentrated sulphuric acid on an *aromatic hydrocarbon, e.g. benzene, to form a sulphonic acid, in which a carbon atom is directly linked to the sulphur atom.

Supplementary recovery. Displacement of oil from the reservoir by injection of fluids (water, natural gas or *enhanced oil recovery chemicals). It is a term usually used to cover both *secondary and *tertiary recovery processes.

Surface string. See Conductor.

Surface tension. The force exerted by the particles of a liquid at its surface which maintains a continuous surface. The surface tension is determined by measuring the energy required to increase the surface by the unit of area.

Suspension. The state of a solid or liquid when its particles are mixed with and buoyed in another liquid but are not dissolved by it. A suspension of a liquid in a liquid is called an *emulsion.

Sweating. The operation of submitting crude paraffin wax to a very gradual increase of temperature with the twofold object of (1) removing the contained oil, which slowly oozes or sweats out and is drained away, and (2) preparing from the wax so obtained fractions with successively higher melting points.

Sweet gas. Hydrocarbon gas free from sulphur compounds. Antonym: sour gas.

Sweetening. The conversion of the *mercaptans present in *sour gasoline into non-smelling disulphides.

Swivel. A tool which is the connecting link between the hoisting gear in a *derrick and the rotating *kelly in a *drilling string. The weight of the string is carried by heavy roller bearings in the body of the swivel.

Syncline. A trough-like fold in layered rocks in which the strata dip down towards the axis.
Synthesis. The building-up of elements into compounds, or of compounds into more complex compounds.
Synthesis gas. A mixture of carbon monoxide and hydrogen produced, for example, from methane and steam. It is used in the manufacture of various alcohols and synthetic gasoline (Fischer–Tropsch process). The name "synthesis gas" is also used for a (3:1) mixture of hydrogen and nitrogen, used for the synthesis of ammonia.
Synthetic resins. Organic solid or semi-solid substances of high molecular weight produced by polymerisation or condensation from relatively simple compounds. They are basically either *thermoplastic or *thermosetting.
Synthetic rubber. Synthetic high polymers similar in structure to natural rubber. Their properties vary and can be tailored to fit special requirements.

T

Teepol. A synthetic detergent consisting of sodium secondary-alkyl sulphates. It is manufactured from straight-chain higher *olefins obtained by cracking slack wax. The process is developed by Shell.
TEL. Tetraethyllead, $Pb(C_2H_5)_4$, made commercially by treating a PbNa alloy with C_2H_5Cl. It is added to gasoline to improve the anti-knock quality of gasolines used in internal combustion engines.
Telemetering. Remote recording of meter readings.
Tertiary recovery. Production of the oil left behind by *primary and *secondary recovery processes. This normally involves injection of one of the *enhanced oil recovery (EOR) fluids through special *injection wells.
Thermal cracking. Process of breaking down the larger molecules of heavy oils into smaller ones by the action of heat. In this way heavy oils can be converted into lighter and more valuable products.
Thermal reforming. A thermal process for improving the anti-knock characteristics of gasoline fractions. Nowadays almost completely replaced by catalytic reforming.
Thermoplastic resins. Synthetic resins which, when heated, can be made to flow into a mould and, when cooled, regain their rigidity and retain their shape; they can be again softened and moulded by heating and cooling. Thermoplastics include *polyvinyl chloride (PVC), copolymers of vinyl chloride and vinylidene chloride, polyvinyl acetate *polyethylene, polypropylene, polystyrene and the polyacrylics.
Thermosetting resins. Synthetic resins which can be temporarily thermoplastic and moulded by heating and pressing, but then set permanently by virtue of chemical reactions which take place in the mould; they cannot be softened and remoulded by heating. They comprise *alkyd resins, phenol/formaldehyde resins (Bakelite), urea/formaldehyde resins and epoxide resins.
TIP. The total isomerisation process allows of virtually complete conversion of pentane/hexane mixtures into isoparaffin mixtures by integrating the *Hysomer process with the Union Carbide molecular sieve separation process.
TML. Tetramethyllead is added to motor gasoline to improve the anti-knock quality of gasolines used in internal combustion engines. It is more effective than *TEL in improving the *Road Octane Number of a gasoline at a certain *RON level, as a result of its higher volatility.
Toluene, $C_6H_5CH_3$. An *aromatic hydrocarbon, used in the manufacture of the explosive TNT (trinitrotoluene) and in the production of dyestuffs and pharmaceuticals.
Tonnage (marine). A marine measurement term. Gross tonnage is the total internal volume of the hull and all superstructures, such as deck houses, etc. being expressed in

tons of 100 cubic feet, or approximately 2.83 cubic metres. Deadweight tonnage (dwt.) is the weight of the cargo, stores, bunkers and water which the ship can lift, expressed in long tons (2,240 lb or 1,016 kg).

Tool joint. Coupling with a conical thread, used for making a tight, leakproof connection between two joints of drill pipe.

Toolpusher. Drilling supervisor of a drilling rig.

Topped crude. Also called *long residue, i.e. crude oil which by *topping has been freed of gases, gasoline fractions, kerosine and gas oil. It may be used as fuel oil; or it may be redistilled in a *vacuum distillation plant to produce, as distillates, gas oil, lubricating oils or catalytic cracking feedstock.

Topping. Nowadays called crude oil distillation, i.e. *distillation of crude oil at atmospheric pressure, yielding gases, gasoline fractions, kerosine, gas oil, middle distillates and *long residue.

Tops. The lightest gasoline *fractions obtained when distilling crude oils. Also generally: the top product of any *fractionating column.

Transformer oil. Oil used in transformers to remove the heat generated in the core and coils and to provide insulation between live parts. Transformer oil as a rule is a highly refined spindle type oil. A high degree of refining is required to give the oil good dielectrical properties.

Travelling block. Large tackle block to which a hook is attached and which is used in a derrick for running-in and pulling drill pipe, etc.; complementary to the *crown block.

Trays. See Fractionating trays.

Treating processes. Supplementary refining processes in which undesirable constituents (mainly sulphur-, nitrogen- and oxygen-containing compounds) are removed or converted into less harmful compounds so as to meet the product specifications for further processing or for marketing.

Trickle flow process. A hydrotreating process in which the oil to be desulphurised is contacted with hydrogen and made to trickle over a special *catalyst.

Tubing. See Production string.

Turbine oil. A specially refined, inhibited lubricating oil used to lubricate steam turbines.

Turbo-drilling. Drilling procedure whereby the bit is rotated by a turbine which is attached to the bottom of the *drilling string. This turbine is driven by the drilling fluid.

U

Ullage. The volume of space in a container unoccupied by contents. Hence ullaging, a method of gauging the contents of a tank by measuring the height of the liquid surface from the top of the tank. See Dipping.

Ultimate recovery (of oil or gas). The total cumulative production of oil or gas expected to be obtained from a reservoir. The "reserve" at any given time is equal to the ultimate recovery minus the cumulative production at that time.

Unconformity. A surface of erosion that separates younger strata from older rocks. In an angular unconformity the older strata generally dip more steeply than the younger strata.

Unsaturated. A term applied to organic compounds in which some carbon atoms are held together by double or triple bonds, so that these compounds are, under favourable conditions, capable of combining with other elements or compounds.

Up-dip. A direction that is upwards and parallel to the dip of the strata.

Underwater drilling. See Drilling platform.

V

Vacuum distillation. *Distillation of a liquid under reduced pressure, aimed at keeping the temperature level sufficiently low as to prevent cracking. For example, used to distill heavy gas oil, lubricating oil feedstocks or catalytic cracking feedstock from *long residue, leaving *short residue as remainder.

Vacuum forming. A process of producing objects from sheet material, in which a blank cut from the sheet is placed in position over a female mould, heated by radiation and then, after the mould has closed, is drawn into the lower part by the effect of vacuum induced below the blank.

Valve trays. *Fractionating trays consisting of a plate with holes for vapour passage, these holes being covered by metal discs (valves) which rise and fall with the gas flow rate.

Vapour lock. A condition which arises when a gas or vapour is present in the fuel line or fuel pump in sufficient volume to interfere with or prevent the flow of fuel to the carburettor of an engine.

Vapour pressure (at a given temperature). The lowest pressure at which a liquid, contained in a closed vessel at the given temperature, can remain in the liquid state without evaporation. Lowering the vessel pressure below the vapour pressure results in evaporation of part or all of the liquid. A compound or fraction with a high vapour pressure requires a high pressure to be kept as a liquid, thus it is *volatile.

Venturi meter. A specially designed tube for measuring the rates of flow of gases or liquids, having a constriction or throat with convergent upstream and divergent downstream walls, the angles of which are such that stream line or almost streamline flow through the tube is achieved. The rate of flow is measured by the pressure drop across the throat.

Visbreaking. A thermal cracking process aimed specifically at reducing the viscosity of *long or *short residues.

Vibroseis. A *seismic method of exploration in which the shock waves are created by a series of rapid vibrations.

Viscosimeter. Instrument for measuring *viscosities.

(a) Absolute viscosity is determined by a capillary type instrument. The time required for a sample to flow through a known length of glass capillary is registered. Results are often given in *centistokes or *centipoises.

(b) In the petroleum industry the viscosity is generally determined in standardised instruments consisting of a container with a hole or jet in the bottom. Various type are in use, viz. in the UK, the Redwood I and Redwood II, in the USA the Saybolt Universal (SSU) and Saybolt Furol and on the European continent the Engler viscosimeter. Results with the Redwood and Saybolt viscosimeters are expressed in seconds, those with the Engler in Engler degrees.

Viscosity. The dynamic viscosity of a liquid is a measure of its resistance to flow. It is defined as the force per unit surface required to shear a layer of unit thickness at a unit velocity. The kinematic viscosity is equal to the dynamic viscosity divided by the density of the liquid. If no distinction is made, the dynamic viscosity is usually meant.

Viscosity index. A method of indicating the *viscosity/temperature relation of an oil. Oils are generally classed as high, medium and low viscosity index oils (HVI, MVI, LVI).

Volatile. Term applied to materials which have a sufficiently high *vapour pressure at normal temperatures to evaporate readily at normal atmospheric pressure and temperature. It implies a high degree of volatility.

W

Wash oils. Petroleum fractions employed for the absorption of the relatively heavy and easily liquefiable components of a mixture of gases.

Water bottom. Water accumulated at (or sometimes added to) the bottom of the oil in a storage tank. In cases where the tank bottom is very uneven, the water level assists in the accurate measurement of the oil content of the tank.

Water-drive reservoir. A reservoir from which the produced oil or gas is displaced by expansion of the formation water in the *aquifer.

Water flooding. Injection of water under pressure into the formation via injection wells in order to produce artificial *water-drive. It is one of the methods of *secondary recovery.

Water string. A string of *casing set and cemented directly above the oil-bearing formation.

Weathering. The often undesired process of slow evaporation of *volatile fractions from a petroleum fraction during storage. It is promoted by *breathing.

Weight dropping. A *seismic method of exploration whereby vibrations are caused by dropping a weight.

Wet gas. Petroleum gas containing water and/or liquid hydrocarbons, i.e. the lower members of the paraffin hydrocarbon series (*propane, *butane, etc.). Where significant quantities of liquid hydrocarbons condense cut of it during production, these are normally extracted from the gas and sold separately.

Whipstock. Wedge-shaped device used in *deviated drilling to deflect and guide the bit away from vertical.

White oils. Oils produced by more drastic refining of MVI distillates to remove unsaturated compounds and constituents. They are usually *solvent extracted and then repeatedly treated with strong sulphuric acid or oleum and alkali. One of the best known is *medicinal oil or "liquid paraffin".

White products. Light petroleum products such as *gasoline, *white spirit and *kerosine.

White spirits. *Fractions intermediate between *gasoline and *kerosine with a *boiling range of approximately 150–200°C. They are used in paints and in dry cleaning.

Wide range distillate. A *distillate with a wide *boiling range. As a combination of *gasoline and *kerosine fractions it is used for jet aircraft.

Wildcat drilling. Drilling to discover oil or gas in an area (or location) where no previous discoveries were made.

X

Xylene, $C_6H_4(CH_3)_2$. An aromatic hydrocarbon of which there are three isomers (*ortho, meta,* and *para*). Oxidation of these isomers gives the corresponding *phthalic acids, base materials for important synthetic resins.

SUBJECT INDEX

Aberdeen, 191, 202, 204
Absorption, 237, 254
Accidents,
    collisions and groundings, 652
    gas carriers research, 653
    oil spill clean-up, 653
    prevention, in E & P, 639
    spillages, 643, 645, 646
Accounting, 195
Acetone, 594
Acetonitrile, 263
Acoustic impedance, 86, 89
Acrylics, 592
Additives,
    anti-rust, 450
    ASA-3, 405
    in engine fuels, 391
    for extreme pressure, 432
    functions of, in lubricants, 435
    in gasoline, 398
    TEL and TML, 397
    VI improver, 442
    in residual fuel, 422
Adsorption, 237, 266, 268
Aero-magnetic surveys, 64
Afghanistan, 30
Aggregates,
    in road construction, 466
Agriculture, 371
Airborne imagery, 61
"Albino" bitumens, 464
Alcohols, 399, 605
Algeria, 76, 517, 523
Aliphatic amines, 591
Alkanolamine process, 320
Alkylation, 230, 238, 276, 285, 300
    olefins used in, 302
    Phillips Petroleum Co. Unit, 300

Alternative fuels, 627
Aluminas used for drying gas, 268
American Petroleum Institute (API), 441
Ammonia, 610
Amoco Cadiz, 495
Anticlinal period, 37
Aquifers, 92
Argyll field, 169
Aromatics, 224
    chemistry of, 588
    for petrochemicals, 588
Asphalt, 464
    cold, 467
    concrete, 471
    hot-rolled, 472
    mastic, 465
Asphaltenes, 283, 311, 327
Athabasca tar sands 603
Automation in VLCCs 488
Automotive retail market, 353
Aviation,
    ASA-3, 359, 405, 632
    AVGAS and AVTUR, 356-359, 400-408
    civil aviation fuelling, 358
    fuel atomisation, 404
    gas turbine engines, 401
    marketing, 356
    piston engines, 406
    Shell water detector test, 404

Bailing, 150
Baku, 26
    "second" Baku, Volga/Urals, 26
    "third" Baku, Tyumen, 26
"Bare-foot" completion, 113
Barges, 376
Barrels,
    barrels of oil equivalent, 665
    barrels per day, 663

Barytes, 136
Batching oil, 456
Bentonite, 135
Benzene, 224
Bergius process, 609
Beryl field (North Sea), 164, 173
Biomass, 602, 628, 632
    low efficiciency of conversion, 614
    technologies, 610
Bit, drilling, 126
Bitumen, 373, 464
    blown grades, 325, 464
    high vacuum units, 251
Blending plants, 440
Blowouts, 136
Blowout preventers, 133, 145
Borneo and Sumatra oilfields, 480
Brent field, 164
Brightstock, 251
British Petroleum Company Ltd. (BP),
    Dunkirk refinery, 313
    isomerisation process, 278
British Thermal Units (Btu), 664
Brunei, 59, 76
Buchan field (North Sea), 169
Bunkers,
    cost affects tanker design, 488
    installations, 364
Buoyant towers, 163
Butadiene, 224
Butane, 261, 428
Butylene, 300

Cable tool drilling, 122
Cambrian rocks, 55
Canada,
    natural gas market, 529
    new crackers based on ethane, 588
    Rockies, 51
    tar sands, 313
Candles, 459
Capacitors, 452
Cap rock, 59
Car,
    accessories, 354
    carburation, 390, 395
    knocking, 390
Carbon dioxide, 106
Carbon removal, 610
Carbonate reservoirs, 90
Casing, 133
Caspian Sea, 161
Catalysis, 233

Catalysts,
    in manufacturing, 270
    research, 629
    silica–alumina, 291
    tungsten sulphide, 294
    "3-way catalyst", 389
    zeolite, 293
    Ziegler–Natta, 633
Catenary anchor leg mooring (CALM), 169
Caustic flooding, 109
Cellulose nitrate, 578
Centrally Planned Economies, 25
    exports to COMECON partners, 30, 33 53
Cetane number, 415 605
Chemicals,
    absorption treating, 319
    base chemicals, 585
    processes and products, 106, 624
Chilling, 254
China, Peoples Republic of, 34, 218
Chlorination, 231
"Christmas tree", 117, 171
Circulation, lost, 138,
Clastics sedimentology, 90
Claus process, 306, 319, 322, 335, 643
Clay treating, 324
Cloud point, 416
Coal,
    briquettes, 478
    costs, 21
    gasification, 610, 622
    hydrogen deficiency, 602
    reserves, 16, 53
    in South Africa, 508
    in USSR, 33
Coating materials, 504, 594
Cognac field (Gulf of Mexico), 160
Coking, 283, 284
Cold filter plugging point, 416
Column internals, 242
Combination carriers, 483
Combination tower, 252
Combustion, 387, 389, 390
    in diesel engines, 415
    in jet engines, 403
    in situ, 102
Completion, 112, 118
Compressed natural gas, 606
Compressors, 449
Computers,
    in E & P, 112, 209
    in manufacturing, 331, 333, 334
Computer graphics, 211

SUBJECT INDEX

CONCAWE, 507
Concorde,
    fuel system, 359
    supersonic speeds, 402
Condensate, 77
Condensation, 232
Conductor, 133
Congo, 163
Connate water, 91
Conservation,
    of energy, 20
    environmental, 637ff
Consuming countries, 9, 15, 566, 570
Continental crust, 39
Continental drift, 40
Contracting, 195
Coolants, 453
"Copying", 186
Coring, 80
Cormorant field (North Sea), 50, 173, 174
Corporation tax, 186
Corrosion,
    lubricants, 437
    in naval vessels turbines, 418
    research, 627
    of steel pipelines, 506
    in vehicles, 393
Costs,
    data acquisition, 188
    development, 190
    exploration, 188
    offshore seismic, 188
    total venture, 189
Cracking,
    catalytic, 256 284
    chemistry of, 229
    fluidised bed, 288
    hydrocracking, 294, 295
    Houdry process, 285
    steam crackers, 586
    thermal, 279
    Thermofor process, 285
CRISTAL, 652, 653
Critical path planning, 193
Crude oil,
    classification, 234, 566
    light and heavy, 568
    spot market, 573
Cryogenic ships, 425, 427
Crystallisation, 237, 263
Cumene, 594
Cushing field (Oklahoma), 37

Customers,
    bitumen, 374
    civil engineering, 370
    product specifications, 368
    requirements, 365
"Cutbacks", 468
Cutting oils, 453
Cyclisation, 230
Cycle oils, 296

Data handling, 211
Decompression sickness, 178
De Havilland Comet, 357
Dehydration, 152, 232
Dehydrogenation, 229
Delphic exercise, 215
Demetalising, 313
Deoiling, 264
Depletion, 192
Deposition, of wax, 151
Depots, 375, 378
Depreciation, 191
Desulphurisation, 548
Desuperheating, 256
Detergents, 588, 625
    "SHOP" process, 595
Developing countries,
    energy demand of, 19, 219, 566
Diagenesis, 50
Diesel, 413
    lubricants, 444
Discoveries, future, 219
Distillation, 237, 240
    vacuum, 248
Distribution,
    bulk filling, 383
    costs, 354
    drums and packages 383, 384
    planning, 375
    storage, 374, 384
Diving, 178
Diving bell, 178
Dolomite, 45
Domestic space heating, 359, 362, 546, 547
Drake, 14, 36
Drilling, 67, 122
    cuttings disposal, 208
    deep-water, 144
    deviated, 135
    drive water 92, 99
    gas caps, 93
    logs, 78

marine, 139
sidetrack, 138
solution gas, 93
steam, 100
technique, 133
Dry hole, 67
Dubbs process, 280
Dynamic positioning, 148

Edeleanu process, 257, 260
Effluents, 337
Ekofisk field (North Sea), 162, 209
Electric car, 614
Energy,
   efficiency, 366, 429
   comparative costs, 22
   conservation, 20, 360, 430, 622
   financial prospects, 20
   management of, in refineries, 327
   natural gas, 514
   oil and gas, 2, 520
   reserves, 18
   resource base, 16, 17, 19
   USSR exports, 33
Enhanced recovery, 97, 619, 620
Environment, 204 (see also Pollution)
   catalysts reduce emissions, 294
   CONCAWE, 662
   effluents, 337
   EPA, 660
   impact statements, 638, 644
   IPIECA, 662
   noise, 341, 659
   refinery operations, 335
   residual fuel emissions, 420, 421
   smog, 388
   tall stacks, 656
   waste disposal, 208, 658
   UNEP, 662
Environment Protection Agency (EPA), 660
Epoxy resins, 424
Erie Lake 171
Erosion, 43
Esterification, 232
Ethane, 424, 557
Ethanol, 599
Ethylene, 224, 378, 578, 586
Evaporites, 44
Exploration and production, 72, 618, 639
   accumulation of oil and gas, 55
   aim of, 36
   basin development, 47
   drilling, 122

   exploration economics, 187
   geology, 87
   historical background, 36
   marine risers, 176
   methods, 61, 67
   predevelopment studies, 73
   process facilities, 151
   production and development, 72
   production operations, 149
   results, 67
   sedimentology, 90
   stages, 69
   technology, 112

Faults and fractures, 48ff, 619, 620
   listric faults, 58
Feed preparation units, 251
Feedstocks, 291, 599, 628
Finance,
   capital intensiveness, 10
   implications of oil prospects, 20, 21
   LNG projects, 551
   NG liquefaction/pipeline costs compared, 534
   project financing, 23
"Fines", 290
Fingering, 103
Fischer–Tropsch process, 610
Fish, 138
Flags of convenience, 485
FLAGS pipeline, 501, 632
Flaring,
   of associated gas, 511, 561, 656
   in refineries, 329
Flash point, 247, 412,
Flocculation unit, 339
Flooding,
   in refinery operations, 243
"Flotels", 200
Flow, natural, 117
Flowlines, trenching, 174
Fluid, drilling, 131
Fluidised bed process, 288
Folds and faults, 48
Formation tests, 81
Fractionating, 241
France,
   Fos terminal, 525
   Gaz de France, 525
   natural gas, Lacq area, 524
Franchise system, 353
Frigg field (North Sea), 164

SUBJECT INDEX

Fuel,
  atomisation and volatility, 404
  aviation fuels, 359
  cloud point, 416
  economy, 390
  fuelling systems, 357
  gas oil and diesel, 413
  hot fuel handling, 395
  leaded, 407
  turbine, 402
  vegetable-derived, 372
Fulmar field (North Sea), 494
Furfural extraction, 261

Gamma ray log, 83
Gas fields evaluation, 76
Gas (see Natural gas)
Gasohol, 399
Gasoline,
  anti-knock properties, 389
  fuel volatility, 395
  marketing, 354
  spark ignition engine, 388, 393
  in transport industry, 387
  treating, 314, 315
Gears, 448
Geophones, 65
Gilsonite, 464
s.s. Glückauf, 679
Glycol, 153
Government take, 184
Grabens, 50, 64
Gravimetric method, 63
Greases, 435, 448
Groningen, 76, 154
Groningen gas equivalent, 665
Grouting, 467
Gum, 315, 408
"Gushers", 123, 136
Guyed towers, 168

Halogen, 269, 270
Hazards,
  assessment, 205, 342, 659
  fire and toxic, 342, 438, 627, 637, 659
Health and safety 438, 627
Heat pumps, 361
Heat transmission, 455
Helicopter traffic, 200
"Helium unscramblers", 181
H-Oil process, 296
Hoisting equipment, 128
Hoists, 50

"Hot spots",
  in machines, 431
Houdry process, 285, 286
Hulton field (North Sea), 168
Huntington Beach field (California), 139
Hydrant system 359
Hydrates, 153
Hydration, 232
Hydraulic fluids, 455
Hydraulic pumps, 120
Hydrocarbons, 222, 509, 618
Hydrocarbon gas, 104
Hydrocarbon kitchen, 55
Hydrocarbon solvents, 104
Hydrodesulphurisation, 233, 252, 256, 306, 307
  of residual fractions, 311
Hydrogen,
  addition technologies, 607
  deficiency in feedstocks, 602
Hydrogenation, 229
Hydrogen sulphide, 78, 227
Hydrolysis, 232
Hydrotreating, 238, 306
  catalytic, 284
  of distillates, 307
  of feedstocks, 293
  of luboils, 313
  of pyrolysis gasoline, 309

ICI,
  polyethylene process, 589
  second stage catalyst, 294
IEA (International Energy Agency), 16, 574
Ignition, 392, 393, 415
IMCO (Intergovernmental Marine Consultation Organisation) 650 652
Independent tanker owners, 486, 573
India, 573
Industrial fuel gas, 607
Information systems coordination, 214
Injection,
  of gas and water, 95
Installations,
  for distribution of oil, 374, 375, 378
  E&P, 156, 163
  floating, 168
  management of, 379
Institut Français du Pétrole DIMERSOL process, 304
Instrumentation, 238, 332
Insulating oils, 369, 451
Insulation, 360, 361
Insurance, 196

Iran, 30, 484, 553, 570
Isobutane, 276, 300
Isomerisation, 230, 238, 275, 276
Isooctane, 406, 407
Isoparaffins, 268
Isopropyl alcohol (IPA) 578

Jacket, 160
Japan, 76
    air pollution regulation, 550
    direct deals, 572
    LNG imports 527
    LPG imports, 428
    product requirements, 566, 568
    tanker tonnage, 485
Jobbers, 352
Joule, 666
Jurassic, 40, 70

Kelly, 125, 126
Kerogen, 52, 603
Kerosine,
    AVTUR, 406
    cookers, 409
    domestic, 408
    long-drum burners, 410
    marketing, 400
    properties of, 412
    short-drum burners, 411
    treating, 316
Kilocalorie, 665
Knock, 277
    in car engines, 391

Lanolin, 457
Lead,
    in gasoline, 302, 392, 397, 622, 626, 661
Less developed countries, 219
Liberia, 485
Licensing, 634
Life sciences, 633
Lift,
    artificial, 115
    gas 118
Light distillates, 244
Lightening at sea 484
Lighting market, 546
Linear programming, 334
"Load on top" system, 650, 651
Logs,
    acoustic, 84
    dipmeter, 85
    neutron, 84
    resistivity 84

LNG,
    carriers, 492
    research on spillage, 653
LPG
    as auto fuel, 400
    carriers, 491
    delivery vehicles, 378
    discharge lines, 380
    manufacture, 255, 322
    producing areas, 426
    uses, 425
Lubbock–Whittle collaboration, 403
Luboil, 251
Lubrication,
    agricultural machinery, 371
    electrical equipment, 451
    gas turbine, 445
    gasoline engine, 443
    general characteristics, 431
    hypoid, 449
    industrial, 446
    marine, 364
Lurgi process, 610

Macadam, 466, 467
Magnetic method, 64
Management,
    difficulties, in petrochemicals, 596
    energy manager in refineries, 330
    of installations, 379
Maps,
    structural contour, 88
Marine,
    bunkering, 364
    gas turbine, 367
    history and development, 479
    history of fuel use, 362
Marine completion, 114
Marine risers, 148, 176
Marketing,
    agriculture, 371
    construction industry, 370
    manufacturing industry, 366
    of market sectors, 351
    organisation and management, 349
    road and rail, 366
Major oil companies,
    future prospects, 24
    integration of activities, 11
    sales of oil products, 14, 15
    share of market declines, 12, 15
    supply operations, 563, 574
Mastic asphalt, 469

SUBJECT INDEX

Materials, 503
Maureen field (North Sea), 163
Measurement of energy, 663
Mercaptans, 315, 316, 322
Metal cutting, 452
Metal forming,
 rolling and drawing, 454
Methane, 222, 550, 632
Methanol, 321, 399, 558, 599
 "Rectisol", 321
Methyl ethyl ketone (MEK), 264
Mexico,
 exports to USA, 553
 gas production potential, 529
 Gulf of, 139, 156, 160
Middle East, 218, 514, 528, 588
Mobil Oil Co., 610
Modules, accommodation, 200
Mohorovicic Discontinuity 38
Mooring, 169
Muds, drilling 134
s.s. Murex 480

Naphtha 244, 296
Naphthenic acids, 228
Natural gas, 77, 96, 152
 associated and non-associated, 510, 511, 561
 Canadian market, 529
 chemicals feedstock, 550
 "Closed-loop system", 540, 551
 composition and origin, 509
 development of international trade, 551, 553
 distribution and marketing, 541
 in energy spectrum, by areas, 514
 exploration, 510
 future of, 559–561
 in Japan, 527
 LNG plants, 535, 539
 LNG shipment, 534, 539
 load balancing, 543, 544
 markets, 546, 547
 monopoly and government regulations, 545
 NGLs 555ff
 Pacific basin markets, 528
 pipeline networks, 527
 pricing, 548
 SCOTT process, 306, 319, 322, 335, 643
 town gas, 542, 543
 transportation, 530, 531
 in USA, 515
 in USSR, 518
 in W. Europe, 519, 521–526

Netherlands, 76, 154
 gas exports, 553
 gas market, 521ff
 Groningen field, 519
 Schoonebeek, 222
Nickel, 228
Niger, 50, 58
Nitrogen, 104, 227, 335
Noise, 341, 659
North Sea, 50, 54, 59, 70, 76, 90, 155, 162–164,
 168, 169, 173, 174, 193, 196 197, 201–203,
 209, 494, 508, 519, 520, 523, 524
 East Shetland basin, 201, 203
 facilities, 155
 Frigg and Brent fields, 164, 201, 202
 Leman Bank, 523
 logistics, 196
 reserves of gas, 54, 519
 submarine pipelines, 508
Norway, 524
Nuclear power, 18, 22

OCIMF (Oil Companies International Marine
 Forum), 650
Octane number, 389, 391, 396
 aviation gasoline, 406
 catalysts, 294
 of gasoline, 284, 302
 LPG, 400
 Platformate, 272
 "requirement increase", 392
 research octane number, 397
Offshore oilfields, 494
Oil,
 applications, 20
 changes in ownership of, 13
 development of Middle East, 5
 different types of, 368
 low-, medium-, high-cost, 20, 21
 markets, 350
 production and consumption by region, 564
 production figures, 3
 products research, 625
 properties of crude, 10
 spills, 209, 340
Oil products,
 handling and distribution, 353
 research and product performance, 625
 specification and testing, 386
Olefins, 302, 628
OPEC, 186
 barrel, breakdown, 571, 572
 control of production, 563

influence of Saudi Arabia, 9
1979 production, 570
rise of, 5
role in supply chain, 572
OPOL (Offshore Pollution Liability), 644
Orenburg,
  gasfield, in Volga/Urals, 29, 30, 519
Orinoco,
  tar sands, 603
Otto engine, 613
Oxidation, 231
  of petroleum wax, 460
  of transformer oil, 452
  of turbine oils, 451
Oxygen compounds, 227

Pacific plate, 52
Packaging, 584
"Pall" rings, 243
Paraffin, 223
  liquid paraffin, 458
  paraffin wax, 458
Patents,
  in agrochemicals, 635
Pentane, 224, 261
Perforating gum, 113
Petrochemicals, 577
  future of, 596
  manufacture, 585
  noble use of oil, 598
  production in WOCA, 579
Petrolatum, 457
Petroleum, 1, 2, 3
  chemistry of, 221
Petroleum industry 1, 5, 10, 11
  future propects of, 16ff, 23
Petrophysics, 83
Phenol, 228, 263
Pilot plant, 624
Pipelines,
  Brotherhood (Bratsvo), 30
  compressor stations, 533
  construction of, 505, 506
  economics of, 502ff
  and Environment, 645, 646
  FLAGS, 501, 632
  Friendship, 34
  gas pipeline economics, 531
  natural gas, 516
  Northern Lights, 30
  in Pennsylvania, 496
  product, 375, 500
  RMR system, 501

SAPPRO system, 501
Soyuz, 30
SPMR system, 501
Trans-Mediterranean, 533
Trapil, 376, 500
trunk, 374
  in W. Europe, 497, 499
Pipelaying barges, 505
"Placer" deposits, 38
Planning, 193
Plastics, PVC, 578, 582–584, 629
Platformate, 271, 272
Platforms, 160
  organisation, 203
  tension leg, 168
Polishes,
  use of wax, 361
Pollution, 546, 547, 548, 554 (see also Environment)
  air and water, 325, 638
  anti-pollution laws, 638
  atmospheric, 660
  clean up of spills, 653
  cost effective control, 335
  design, 336
  International Convention, 649
  Marpol Convention, 1973, 650
  Noise, 341 659
  OPOL, 644
  pipeline spillage, 645, 646
  smog, 661
  spent products, 438, 661
  TOVALOP and CRISTAL, 651
Polyacrylamides, 107
Polyester fibres, 277, 592
Polyethylene, 580, 589
Polymerisation, 231, 238, 303
  flooding, 107
  importance of polymers, 583
  Shell Development Company's sulphuric acid process, 304
  UOP phosphoric acid process, 304
Polysaccharides, 107
Polystyrene, 582
Polyurethane foams, 629
Pour point, 416
PVC, 582–584
Pressure control equipment, 132
Prices, 5, 8, 9
  of Bitumen, 474
  of crude oil, 568
  interfuel competitiveness, 350, 360
  marine, 363
  of natural gas, 545, 546, 548, 552

SUBJECT INDEX

Process oils, 458
Procurement, 194
Producing countries, 6, 8, 13, 20
    growth of Middle East production, 565
    North Sea and Mexico, 566
    Production sharing, 185
Production,
    engineer, 75
    natural mechanism, 92
    profile, 75
    tests, 81
Products carrier, 489
Profit forecasts, 75
    Windfall profits, 186
Project management, 193
Propane, 261
Propylene, 300
Prudhoe Bay,
    to Valdez pipeline, 497
"Pump-down" (PD) well servicing, 150
Pumps,
    bunkering, 380
    in E & P, 119
    for movements of oil, 380
Pyrolysis 230

Qatar, Halul field, 169
Quality assurance, 195

Rankine cycles, 329
Recontacting system,
    in manufacturing, 253
Recovery, conventional and enhanced, 91, 93, 94, 96, 97, 103, 109, 110, 217
Refineries,
    cost and staffing, 235
    environmental aspects, 656, 658
    fuel consumption, 327
    process control, 331
    supervision systems, 332
Reflux, 252
Reforming, 268
Refrigerators, 450
Regenerative process, 271, 274
Reid vapour pressure test, 393, 396
Research,
    chemical processes, 628
    design of offshore equipment, 621
    E & P, 618
    industrial research, 617
    manufacturing, 621
    marine, 632
    military aviation, 358

subsurface evaluation, 619
    transport of LNG and LPG, 508
    transport, storage and handling, 632
Reserves,
    gas in Middle East, 560
    natural gas, 512, 513, 559
    oil and gas, 18, 214
Reservoir
    data, 74
    model, 74, 81, 110
    rock properties, 88, 90
    simulation, 111
Resins, 578
    epoxy, 590
    phenolic, 590
    thermosetting, 590
Resource base,
    of fossil fuel, 601
Rift basins, 47
Rigs, 139
Risks, 182
Road emulsions, 465
Robots, 114
Rocks, 74
    igneous, 38
    metamorphic, 38
    sedimentary, 38
"Roll-over", 56
Romashkino (Volga/Urals) field 26
Rotary drilling, 122, 125, 133
Rotary vacuum filter, 265
Rotating disc contactor, 259
"Round trip", 133
Royal Dutch/Shell Group of Companies,
    expenditure on research, 617
    Hysomer process, 278
    interests in Russia 26
    SCOT process, 322, 335
    SHF process, 629
    SHOP process, 595, 628
    sulphuric acid polymerisation process, 304
    Teepol plant at Stanlow, 585
Royalties, 186
Rubber (SBR), 580, 593, 629
Rusting, 455

Safety, 204
    aviation, 356, 358
    contamination with water, 346
    design safety, 205
    fire fighting, 385, 343
    Institute of Petroleum Code 385
    pipeline, 507

refinery safety, 342ff
   static electricity, 347
   Torrey Canyon and Amoco Cadiz, 495
Samotlov field in Tyumen area, 27
Sand mixes, 468
Sandstone reservoirs, 109
SASOL, 609
Satellite well, 171, 174
Saudi Arabia, 8, 9,
   ethane potential, 557
   imposes supply embargoes, 563
   Iran, 570
   natural gas market, 528
SCADA systems, 213
Schoonebeek oilfield, 222
Scouring, 456
Sedco, 144, 145, 148
Sedimentary basins, 42, 47, 69
Seismic, 64
   data acquisition, 619
   production seismology, 89
Semi-Spar, 170
Separation, low temperature, 153
Separators, oil and gas, 152
Service stations, 352–355
   LPG, 426
Shale, 59, 83
   deposits USA and Australia, 53, 603
   feedstock, 602
   in situ technique, 609
   James Young's process, 4
   research, 622
   rock disposal, 18
   surface retorting, 609
Shale shaker, 132
Shellgrip,
   anti-skid road surfacing, 474
Shell Development Company,
   sulphuric acid polymerisation process, 304
Shell Thermopave,
   high-sulphur bitumen mix, 475
Siberia, 29, 519
Single buoy moorings,
   affects tanker design, 487
   environmental problems, 647
Single point mooring, 494
SLIKTRAK, 209
Sludge disposal, 340
Smoke point,
   improvement (SPI), 311
   of kerosine, 311
Solar energy, 614
Soluble oils, 453

Solvent extraction, 237, 257
   Edeleanu process 257
   extract phase, 258
   of luboils, 260
   raffinate phase, 258
Solvents, 369, 593
   physical, 321
   Sulfinol, 321
Sonde, 79
Source rocks, 52
South Africa,
   coal and petrochemicals, 598
Special product businesses, 372
   storage and handling, 385
Spindletop field (Texas) 123
Stabilisers, 138
Stadive, 182
Standard Oil of New Jersey (EXXON), 294
   first FCC unit, Baton Rouge, 288
State oil companies and independents,
   emergence in 1960s, 11
   in USA, independents, 11
State participation, 184
Static electricity,
   precautions against, 345, 405
   remedies, 347
Stavanger, 191
Stimulation of formations, 114
Stirling engines, 361
Stocks,
   luboil, 357
   stock levels, 379
   storage stability, 417
   working capital, 440
Storage, 417, 424
Straddle test, 83
Statfjord field, 164
Stratigraphic traps, 59
"Streamer", 65
Stripping,
   in refinery operations, 245, 247, 307
Structural traps, 56
Structures,
   concrete-piled 161
   steel-piled, 157
Stuck drill pipe, 138
Styrene–butadiene rubber (SBR), 580
Submersibles, 139, 178
Subsea tree, 174
"Sucker rods", 119
Suez canal, 480, 485
Sulfolane, 261
   Shell extraction process, 262

SUBJECT INDEX

Sulphation, 233
Sulphonation, 233
Sulphur, 226
    content in fuels, 403, 412, 416, 418, 422, 660
    emission, 239
    recovery, 318, 322
    sulphur–bitumen asphalts, 474
Sulphuric acid refining, 323
Sumatra and Borneo oilfields, 480
Supply,
    destination restrictions, 572
    diversity of product demand, 566
    factors and constraints, 564
    operations, 563
    security, 9, 23, 523, 561, 574
    transportation costs, 569
Surfactant flooding, 107
Sweating process, 266
Switchgear, 452
Synthetic fibres (see textiles)
Synthetic fuels, 356, 599, 604, 613

Tankers,
    bitumen carriers, 490
    chemical carriers, 490
    complements, 490
    discharging by pipeline, 379
    inert gas blanketing, 647
    organisation of fleets, 485
    parcels tankers, 490
Tanks
    double integrity, 649
    floating roofs, 381, 647, 654
    mixer/settler tanks, 258
    steam coils, 382
    storage, 381
    vents and manholes, 382
Tar, 465
Tarsands,
    production problems, 18
    research, 622
T2s 481
Taxation, 185
    affecting oil prices, 350
    automotive LPG, 426
Tazerka field (Tunisia), 169, 495
Tender
    supply tenders, 366
    terminals, 156
Testing,
    of oil products, 386
Textiles,
    machinery, 456
    synthetic fibres, 584, 591

Therm, 665
Thermal gas oil unit, 252
    cracking, 279
    gas oil production, 282
    recovery, 97, 99
Thermofor process, 285, 286
Thermoplastics, 590
"Through flowline" (TFL) well servicing, 150, 176
Tons,
    long tons, 663
    metric tons, 664
    short tons, 663
Tool pusher, 133
Torrey Canyon, 495
TOVALOP, 625, 653
Town gas, 285
Trans-Alaska crude pipeline, 497
Transformers,
    lubrication, 451
Transport,
    bulk lorries, 377
    LPG, 427
    rail, 378
    research, 632
    water and road, 376
Trapil pipeline, 376
Traps, oil and gas, 56, 59
Tray,
    in refining, 243
Treating,
    base oil, 323
    end of pipe units, 338
    sulphuric acid refining, 323
    tail gas, 322
Tribology, 431
Trinidad Lake,
    bitumen, 464
Tubing, 115
Turbines,
    lubricants, 445
    oil stability test (TOST), 436
    steam, 450
Turbo drilling, 125
Turkmen Central Asian gasfield, 29
Two-stroke engines, 444

"Umbilicals", 178
Unconventional raw materials, 599
Underwater equipment, 171
    manifold centre, 175
Union Carbide,
    ethylene glycol, 578
    molecular sieve separation, 279

Universal Oil Products Company (UOP), 269, 273, 300
  Butamer process, 277
  Merox process, 317, 318
  Penex process, 277
  phosphoric acid process, 304
United Kingdom,
  British Gas Corporation, 523
  imports LNG from Algeria, 523
  natural gas market, 523
  St. Fergus plant, 556
USA,
  first LNG export, 535
  gas consumption, 516
  gas market, 515
  imported gas, 514, 554, 560
  net importer of crude, 566
  production of ethane, 557
  reserves of gas, 512
USSR, 4, 5, 218
  energy consumption, 31, 518
  future prospects, 27, 30
  gas export potential, 514, 519, 553
  largest producer of oil, 25
  nuclear power, 33
  oil development, 26
  pipelines, 33
  second largest producer of gas and coal, 25
Uzbek Central Asian gasfield, 29

Vacuum distillation, 248
Vanadium, 228
Vapour,
  emissions, 647, 654
  lock, 396
Venezuela,
  Lake Maracaibo, 139, 161
  location premium, 569
  and oil Majors, 13
  prompts formation of OPEC, 5
  tar sands, 313
Vibroseis technique, 64, 208
Visbreaker, 251, 280
Viscosity,
  index, 433
  lubricants, 432, 441
  measurement, 422
  residual fuel oils, 420
Visual display units,
  in refineries, 332

Volatility,
  in aviation gasolines, 407
  in diesel fuel, 415
VLCCs, 483, 487

Water,
  drive, 92
  effluent, 643, 647, 656
  waste water treatment, 658
Waterproofing,
  using wax, 459
Waxes,
  candle making, 459
  danger in aviation fuel, 404
  for foodstuffs, 461
  and gas oils, 416
  in luboils, 434
  manufacture, 263, 314
  paper, 460
  petroleum waxes, 458
  testing of, 463
Wax deposition, 151
Weather, North Sea, 196
Welding,
  of pipelines, 503
Wells, 68, 149
  servicing, 150
  in USA, 68
Western Europe,
  energy supply, 520
  natural gas markets, 519, 525
West Germany,
  natural gas market, 522, 523
"Wet" units,
  in refining, 249
Whipstock, 135
White oils,
  medicinal, 458
  in textile machinery, 368
Whittle, Sir Frank, 403
"Wildcats", 68
Winkler process, 610
Wireline logs, 79, 81, 83, 86
  tests, 81
World Energy Conference, 430

Xanthan gum, 107

Ziegler process, 589